HANDBOOK
ON THE
PRINCIPLES
OF
HYDROLOGY

WATER INFORMATION CENTER, INC.
A subsidiary of Geraghty & Miller, Inc.
Enviromental Services

PERIODICALS

Water Newsletter/Research and Development News

The Groundwater Newsletter

International Water Report

BOOKS

The Water Encyclopedia, Second Edition—van der Leeden, Troise, and Todd
Drainage of Agricultural Land—Soil Conservation Service
Handbook on the Principles of Hydrology—Gray
Geraghty & Miller's Groundwater Bibliography, Fifth Edition—van der Leeden

HANDBOOK ON THE PRINCIPLES OF HYDROLOGY

with special emphasis directed to Canadian conditions in the discussions, applications and presentation of data.

Editor-in-Chief

Donald M. Gray, B.S.A. (UBC), M.S.A. (Toronto), Ph.D. (Iowa State). Chairman, Division of Hydrology, and Professor, Department of Agricultural Engineering, College of Engineering, University of Saskatchewan, Saskatoon, Saskatchewan, Canada.

TO THE ADVANCEMENT OF CANADIAN HYDROLOGY

CONTRIBUTORS

Donald M. Gray, B.S.A. (UBC), M.S.A. (Toronto), Ph.D. (Iowa State). *Editor-in-Chief.*
Chairman, Division of Hydrology, and Professor, Department of Agricultural Engineering, College of Engineering, University of Saskatchewan, Saskatoon, Saskatchewan.

LIST OF SYMBOLS.

SECTION I.   INTRODUCTION TO HYDROLOGY.

SECTION III.   ENERGY, EVAPORATION AND EVAPOTRANSPIRATION.

SECTION V.   INFILTRATION AND THE PHYSICS OF FLOW OF WATER THROUGH POROUS MEDIA.

SECTION VII.   RUNOFF–RAINFALL–GENERAL.

SECTION VIII.   PEAK FLOW–RAINFALL EVENTS.

SECTION XII.   STATISTICAL METHODS–FITTING FREQUENCY CURVES, REGRESSION ANALYSES.

Hugh D. Ayers, B.E. (Sask.), M.S. (Wash. State).
Professor and Director, School of Agricultural Engineering, University of Guelph, Guelph, Ontario.

SECTION X.   BASIN YIELD.

Kersi S. Davar, B.E. (Poona, India), M.C.E. (Colo. State), Ph.D. (Colo. State).
Professor, Department of Civil Engineering, University of New Brunswick, Fredericton, New Brunswick.

SECTION IX.   PEAK FLOW–SNOWMELT EVENTS.

Walter W. Jeffrey[1], B.Sc. (Hons) (Edinburgh), M.F. (Ore. State), Ph.D. (Colo. State).
Associate Professor, Faculty of Forestry, University of British Columbia, Vancouver, British Columbia.

SECTION XIII.   HYDROLOGY OF LAND USE.

---

[1]The death of Professor Jeffrey occurred while this textbook was being printed.

Gordon A. McKay, B.Sc. (Man.), M.Sc. (McGill Univ.).
Superintendent of Climatological Research, Canada Department of Transport, Meteorological Services, 315 Bloor Street West, Toronto, Ontario.

SECTION II.   PRECIPITATION.

SECTION III.   ENERGY, EVAPORATION AND EVAPOTRANSPIRA-
TION.

James M. Murray, B.E. (Sask.), M.S. (Utah State).
Associate Professor, Department of Agricultural Engineering, College of Engineering, University of Saskatchewan, Saskatoon, Saskatchewan.

SECTION VI.   GROUNDWATER HYDROLOGY.

Donald I. Norum, B.E., M.Sc. (Sask.), Ph.D. (Univ. of Calif., Davis).
Assistant Professor, Department of Agricultural Engineering, College of Engineering, University of Saskatchewan, Saskatoon, Saskatchewan.

SECTION V.   INFILTRATION AND THE PHYSICS OF FLOW OF
WATER THROUGH POROUS MEDIA.

John M. Wigham, B.Sc., M.Sc., (Alberta), Dipl. (Delft).
Associate Professor, Department of Civil Engineering, College of Engineering, University of Saskatchewan, Saskatoon, Saskatchewan.

SECTION III.   ENERGY, EVAPORATION AND EVAPOTRANSPIRA-
TION.

SECTION IV.   INTERCEPTION.

SECTION V.   INFILTRATION AND THE PHYSICS OF FLOW OF
WATER THROUGH POROUS MEDIA.

SECTION VII.   RUNOFF—RAINFALL—GENERAL.

SECTION VIII. PEAK FLOW—RAINFALL EVENTS.

SECTION XI.   SEDIMENT TRANSPORTATION.

FOREWORD

Over 50 centuries ago man was observing the movements of water. We know this because he recorded the duration times of rainfall and the rise and recession of flood waters in the Gilgamesh and Biblical epics of Ut-Napishtim and Noah. By some 25 centuries later man had developed a cosmogony and theory of life based on water; taught by the Ionian philosopher, Thales of Miletus, water was thought to be the one basic substance from which all came and to which all returned.

Today, then, it is somewhat surprising to have to plead the cause of this science, for Hydrology is only an 'infant' in growth in the modern-day family of sciences. I am sure our ancient forebears, if they could but see us now, would be shocked to find how lax we have been in neglecting the study of water.

Yet, in the short period of six years since my appointment in 1964 as Chairman of the Canadian National Committee for the International Hydrological Decade, I have seen this interest re-awaken and branch out into the many complex subject areas now linked with the study of Hydrology. The "Table of Contents" of this *Handbook* will exemplify how broad is the knowledge required.

Not only has this interest awoken in Canada, but in other countries too, across the world, through the singular and devoted enterprise of a few people who were largely instrumental in inaugurating the vast program of the International Hydrological Decade. Scientific research on the many phases of water is now ongoing in more than 100 countries. On behalf of my Committee, therefore, I wish this publication well and commend to you the efforts of all the contributors to this *Handbook* in their achievement of combining individual knowledges to further the study of Hydrology in Canada.

H.A. Young.
Chairman,
Canadian National Committee for
the International Hydrological Decade.

PREFACE

The material in this *Handbook* represents the contributions of several Canadian hydrologists who have felt the need and importance of a general text, integrated within academic disciplines, that encompasses the basic "Principles of Hydrology" with special emphasis on Canadian conditions. Because of the vast areal extent of Canada, and correspondingly, the wide variation in physiographic conditions encountered within the country, you can appreciate the difficulties of completely satisfying this basic objective within the covers of a single text.

The data used with the text are primarily those for Prairie conditions, the reason for this approach being explained subsequently. Although empirical results or data may change regionally or geographically, the fundamental principles governing hydrologic processes, when defined in mathematical terms, do not vary, and therefore have general application. And in another respect, because the hydrologist is frequently faced with the problem of extrapolating data from one area to the problem area (in which no data or only very crude data exist), a rational interpretation of results can only be made by the hydrologist who has a basic understanding of the physics of the hydrologic regime.

Historically, the impetus and stimulus for preparation of these notes originated in the presentation of a four-day Technical Development Course entitled "Hydrology for Engineers" that was sponsored jointly by the Engineering Institute of Canada, Saskatoon Branch, and the University of Saskatchewan, Extension Division. This course was held at the University of Saskatchewan from August to September of 1965.

In September 1966, the Canadian National Committee for the International Hydrological Decade elected to sponsor the first of a series of Familiarization Seminars on the "Principles of Hydrology" at the University of Saskatchewan, Saskatoon. The purpose of these seminars was to provide an opportunity for workers trained in diverse disciplines to gain a common understanding of the science of hydrology. It was for the 1966 seminar that the content of the *Handbook* was originally prepared, though several other interested hydrologists across the country have since supplemented this material with voluntary contributions. And the emphasis has primarily been to fulfill two purposes:

1. To suffice, to a certain degree as a supplemental text, the requirements for the teaching and training of students in hydrology, both at the advanced undergraduate and first-year graduate levels.

2. To provide a *Handbook* from which individuals, active in the practical aspects of hydrology, may gain some appreciation of the basic hydrologic principles, yet also have access to Canadian data which will be useful to them in design.

The main text of the *Handbook* is preceded by a LIST OF SYMBOLS, which includes (a) the symbol, (b) its meaning, (c) the page number on which it first occurs, and (d) its dimensions. A major attempt has been made to standardize those symbols used most consistently throughout the text to make the subject matter more readable, and to provide a readily accessible source where the meaning of the nomenclature can be found.

The material contained in SECTIONS I-IX follows a natural sequence, beginning with the definition of Hydrology and an outline of the Hydrologic Cycle. Following this segment of the text, the individual components of the Hydrologic Cycle are discussed, starting with Precipitation (as either rain or snow) and then considering its disposition as Evaporation and Evapotranspiration, Interception, Infiltration, Groundwater and Surface Runoff. In certain parts, detailed derivations have been presented of mathematical equations that may be used to describe certain phenomena. The purpose of this approach is twofold: to point out the ramifications and simplifications used in these derivations as applied to very complex systems; and to further the knowledge of the reader to improve his competence and confidence in selecting numerical values of empirical constants, which are used in standard design formulae. In addition, the interdisciplinary nature of the science of hydrology is shown by the wide range in sources from which the reference literature has been taken.

SECTION X on Basin Yield serves to summarize the material given in the preceding Sections by a discussion of the estimation of total water yield from a watershed; and SECTIONS XI-XIII are concerned with Sediment Transportation, the Statistical Analysis of hydrologic events and the effects of Land Use Hydrology on the hydrologic regime.

Speaking on behalf of the contributors, I recognize certain deficiencies in the *Handbook* in sufficing these objectives, but feel that its publication represents the initial step in overcoming the inertia of publishing documents oriented directly toward 'Canadian Hydrology'. For those — specifically Canadians — who may have reservations about such a publication and express very natural and viable comments, I challenge you to assist us in realizing our objectives by presenting your comments to me concretely and succinctly, so they may be used to supplement the contents of the text. To conclude by paraphrasing the Latin expression *tuum est,* it is up to you, the reader, to participate in furthering our knowledge of Canadian-oriented hydrology.

D.M. Gray.

ACKNOWLEDGMENTS

During the preparation of this *Handbook,* many individuals and agencies have enthusiastically contributed basic information, and extended constructive comments. I wish to acknowledge the contribution of these persons and to express my sincere appreciation for their assistance. I am especially indebted to the following:

- Canada Department of Energy, Mines and Resources, Inland Waters Branch, Ottawa; special thanks to P. Meyboom, Head of the Groundwater Subdivision, and his staff for their review of and comments on Section VI, Groundwater Hydrology.

- Canada Department of Transport, Meteorological Services, 315 Bloor Street West, Toronto.

- Canadian National Committee for the International Hydrological Decade. Particular mention is made of the participation of the Secretariat, CNC/IHD, and the contributions by J.F. Fulton, formerly Assistant Secretary; I.C. Brown, Secretary; and R.B.L. Stoddart, Scientific Assistant.

- Hydrology Section, Engineering Division, Prairie Farm Rehabilitation Administration, Department of Regional Development, Regina, Saskatchewan.

- National Research Council of Canada for their financial assistance.

- Secretarial staff of the Department of Agricultural Engineering, University of Saskatchewan, Saskatoon, for typing revisions of the final draft of the main script.

MAIN TABLE OF CONTENTS

SECTION I.   INTRODUCTION TO HYDROLOGY

SECTION II.   PRECIPITATION

SECTION III.   ENERGY, EVAPORATION AND EVAPOTRANSPIRATION

LIST OF SYMBOLS

The following list has been prepared in an effort to standardize the symbols which are used frequently or consistently throughout the text. The list thus provides the reader with a readily-accessible source where the nomenclature can be found.

Duplication in the meaning of certain symbols, and in some cases, of nomenclature, has been unavoidable because of the scope of the material and the wide range of reference sources used. Despite this, we have tried to be as consistent as possible with those symbols most often encountered in the field of study. Where there is duplication in meaning, the page numbers are listed on which the symbols and their descriptions first occur.

The preceding discussion does not apply to the numerous cases where certain symbols have been used in general equations to represent empirical or theoretical constants, exponents, etc. In these cases the symbols have been appropriately defined following each equation. Nevertheless, a certain amount of care should still be followed by the reader in interpreting the meaning of the different letters and characters.

The symbols have been listed alphabetically — first the English, pages *xx–xxxv;* then the Greek, pages *xxxv–xxxvii;* followed by the WMO list of recommended units for hydrometeorological elements, with conversion factors, pages *xxxvii–xxxviii.*

SYMBOL	DESCRIPTION	PAGE[1]	DIMENSION[2]
a	albedo of a surface — ratio of the reflected to incident radiation (usually expressed as a percentage.)	9.4	
A	area, *e.g.*, surface drainage area of a watershed	—	L^2
A	ratio of reference depth to flow depth	11.14	
A_i	initial abstraction	5.5	L
A_c	area of circle of equal perimeter as the basin	8.38	L^2
A_e	effective drainage-basin size	8.22	L^2
AE	actual evaporation or evapotranspiration in a given period	3.53	L
b	saturated thickness of the flow area or confined aquifer	5.29	L
b	average width of a watershed	8.14	L
B	Bowen's ratio	3.20	
B	thermal quality of a snowpack	9.4	
BD	bulk density of a soil	—	$ML^{-2}T^{-2}$
c	coefficient of runoff—Rational Method	8.2	
c	specific heat of a given material (water, soil)	3.22	$L^2T^{-2}\theta^{-1}$
cal	calories	—	ML^2T^{-2}
cfs	cubic feet per second	—	L^3T^{-1}
cm	centimeters	—	L

1. Refers to page on which symbol first occurs.

2. Refers to dimensions of parameter, using
   M = mass, L = length, T = time, and θ = temperature.

csm	cubic feet per second per square mile	–	LT^{-1}
C	Centigrade	–	θ
C	cover factor or crop and management factor	11.4	
C	sediment concentration of the fraction with a given fall velocity	11.12	$ML^{-2}T^{-2}$
C_a	sediment concentration at arbitrary reference level	11.12	$ML^{-2}T^{-2}$
C_d	coefficient of discharge	8.91	
CC	canopy characteristics	13.3	
C_v	coefficient of variation of non-transformed variable	2.86	
CV	coefficient of variation of transformed variable	12.13	
CU	consumptive use in specified time interval	3.42	L
d	diameter of a pore, tube, well, drops, etc	–	L
d	depth of overland flow	7.11	L
d	depth of flow in a channel	8.68	L
d(%)	per cent of total water of overlying snow-pack which infiltrates into a frozen soil	5.14	
d_i	average grain diameter of bed sediment within a given size range	11.22	L
d_{35}	average diameter of soil particles such that 35% by weight of the sample are of smaller size	8.80	L
d_{65}	average diameter of soil grains such that 65% by weight of the sample are of smaller size	8.80	L

D	grain diameter	—	L
D	effective surface retention	7.21	L
$\bar{D}$	mean date of formation or loss of a snow-pack	2.101	T
D_A	depth of water in overland flow or on a soil surface	5.12	L
D_{tr}	date of formation or loss of a snowpack of specified return period	2.101	T
DBH	diameter at breast height	13.7	L
DD	drainage density of a watershed	—	L^{-1}
DDF	degree-day factor	9.14	$L\theta^{-1}T^{-1}$
$D(\theta)$	diffusivity	5.39	$L^2 T^{-1}$
e	base of the natural logarithms	—	
e	emissive power or coefficient of emission	13.33	
e_a	actual vapour pressure at air temperature	3.3	$ML^{-1}T^{-2}$
e_f	emissive power of forest	13.33	
e_o	saturated vapour pressure at the temperature of the surface	3.3	$ML^{-1}T^{-2}$
e_s	saturated vapour pressure at air temperature	3.47	$ML^{-1}T^{-2}$
e_s	emissive power of snow	13.33	
exp	exponential	—	
E	absolute difference in precipitation by using lesser gauge network	2.50	L
E	evaporation during a given period, or evapotranspiration during a given time period	3.4	L or LT^{-1}

E_L	daily shallow lake evaporation	3.31	L
E_L'	daily deep lake evaporation	3.31	L
E_m	momentum transfer coefficient	11.12	LT
E_p	daily pan evaporation	3.30	L
E_s	soil erosion loss	11.4	L^3 or $ML^{-1}T^{-2}$
E_s	sediment transfer coefficient	11.12	L^2T^{-1}
ET	evapotranspiration	3.45	L or ML^2T^{-2}
f	monthly consumptive use factor	3.54	L
f	infiltration rate	5.1	LT^{-1}
f_{ave}	infiltration method	7.22	LT^{-1}
f_c	final or constant infiltration rate (t → ∞), or transmission rate through the control layer	5.5	LT^{-1}
fps	feet per second	--	LT^{-1}
ft	feet	--	L
F	fahrenheit	--	θ
F	ice stored on a reservoir	3.24	L
F	total porosity of a soil	5.27	
F	frequency factor	8.14	
F	forest cover factor	9.11	
F	forest view factor	13.36	
FC	field capacity of a soil	3.53	L

g	gravitational constant or acceleration due to gravity	—	LT^{-2}
gal	gallons	—	L^3
gm	grams	—	MLT^{-2}
G	rain gauge network density for reduced network	2.50	
G	soil intercepted in splash sampler	11.3	MLT^{-2}
G	rate of transfer of mass of suspended particles per unit area	11.12	$ML^{-2}T^{-1}$
h	cloud height	3.19	L
h	hydraulic head (potential energy in a fluid unit weight)	5.17	L
hr	hours	—	T
H	heat budget	3.45	ML^2T^{-2}
H_g	conduction of heat from ground	9.2	ML^2T^{-2}
H_r	heat content of rainwater	9.2	ML^2T^{-2}
i	rainfall intensity	5.1	LT^{-1}
i	hydraulic gradient	5.16	
i_b	fraction of bed material in a particular size	11.16	
i_B	fraction of bed load in a particular size	11.16	
i_t	total fraction of material in bed sediment load	11.17	
in	inches	—	L
I	inflow volume	1.2	L^3

I	inflow rate at a given time	8.49	L^3T^{-1}
I	heat index	3.57	
Imp.Gal.	Imperial gallon	–	L^3
I_s	inherent erodibility of the soil	11.4	L
I_{si}	snow interception loss	13.4	L
I_{ri}	rain interception loss	4.3	L
I_v	total interception loss	13.6	L
k	monthly crop coefficient (Blaney & Criddle)	3.54	
k	hydraulic conductivity, permeability	5.16	LT^{-1}
k	condensation-convection coefficient	9.11	
k	channel roughness factor	8.72	L
k'	intrinsic hydraulic conductivity, permeability	5.21	L^2
k'	basin shortwave melt coefficient	9.11	
k_o	von Kármán's constant	3.4	
$k(\theta)$	unsaturated hydraulic conductivity, capillary conductivity	5.38	LT^{-1}
K	frequency factor of a given frequency distribution which corresponds to a given level of probability	2.65	
K	sensible heat	3.45	ML^2T^{-2}
K	seasonal consumptive use coefficient	3.54	
K	slope of recession limb of a hydrograph	7.7	
K	storage constant	8.51	T
KE	kinetic energy	–	ML^2T^{-2}

J	length of flow path	5.16	L
l	mixing length	8.67	L
lb	pounds (weight)	—	MLT^{-2}
ln	natural logarithm	—	
log	logarithm to the base of 10	—	
ly	langleys	3.5	MT^{-2}
L	latent heat of vapourization	3.6	L^2T^{-2}
L	length of a watershed or the main stream	8.5	L
L_{ca}	length along the main stream to the center of area of a watershed	8.38	L
L_s	slope length and steepness factor	11.4	L
m	metres	—	L
m	rank of an observation when placed in series of decreasing order of magnitude	2.65	
m	mass of a water droplet	11.2	M
mb	millibars	—	$ML^{-1}T^{-2}$
mc	milli-curies	—	
mi	miles	—	L
min	minutes	—	T
mm	millimeters	—	L
mo	months	—	T
mpd	miles per day	—	LT^{-1}

mph	miles per hour	—	LT^{-1}
M	water equivalent of snowmelt	9.2	L
M	daily snowmelt, or basin snowmelt	9.11 9.13	L
M_c	daily snowmelt produced by convection exchange	9.7	L
M_{ce}	daily snowmelt produced by condensation-convection exchange	9.7	L
M_e	daily snowmelt produced by condensa- tion	9.7	L
M_f	mass infiltration in given time	5.6	L
M_p	daily snowmelt produced by rain	9.8	L
M_{rl}	snowmelt due to net longwave exchange	9.5	L
M_{rs}	daily snowmelt produced by net short- wave radiation exchange	9.4	L
M_s	farm management or conservation factor applied to soil erosion	11.4	
MAF	mean annual flood	8.19	L^3T^{-1}
MC	soil moisture content (% by weight)	5.14	
n	duration of bright sunshine	3.7	T
n	Manning's roughness coefficient	8.73	
N	size or number of observations taken or sampled	2.53	
N	maximum possible number of hours of sunshine at given location at a particular time of the year	3.7	T
N	energy used in photosynthesis	3.45	ML^2T^{-2}

N	number of days after peak of hydrograph when direct runoff ends	7.15	T
N_i	longterm normal precipitation recorded at a particular station	2.57	L
N_F	Froude number	8.72	
N_R	Reynolds number	5.24	
O	outflow volume	1.2	L^3
O	outflow rate at a given time	8.49	L^3T^{-1}
O_g	groundwater seepage rate from lakes, reservoirs, sloughs etc.	3.24	L^3T^{-1}
p	probability of occurrence	2.64	
p	pressure	–	$ML^{-1}T^{-2}$
p	monthly percentage of annual daytime hours	3.54	
psi	pounds per square inch	–	$ML^{-1}T^{-2}$
P_i	percentage by weight of bed sediment of average size, d_i	11.21	
$\bar{P}$	mean precipitation depth rain obtained by statistical analysis of a sample	2.53	L
P	total rainfall recorded by a true network, or average precipitation over a basin	2.50	L
P	wetted perimeter (hydraulics)	8.65	L
P_e	excess precipitation (rain)	5.5	L
P_i	depth of precipitation (rain) recorded by an individual gauge in given time	2.53	L

P_m	maximum rainfall recorded at a storm centre	2.90	L
P_n	net storm rain	8.27	L
P_r	depth of storm rainfall	13.3	L
P_{rn}	net precipitation (rain)	13.3	L
P_s	depth of storm precipitation (snow)	13.6	L
P_{sn}	net precipitation (snow)	13.6	L
P_{t_r}	point precipitation amount (rain) of specific duration of given return period	2.65	L
P_R	period of rise of a hydrograph	8.6	T
P_{24}	extreme clock-hr day precipitation from climatological records	2.58	L
$\bar{P}_{24}$	mean extreme precipitation amount during a 24-hr period	2.86	L
PE	potential evaporation or evapotranspiration in given time	3.44	L
PMP	probable maximum precipitation	2.99	L
PWP	permanent wilting point of a soil	3.53	L
q	specific humidity	2.96	
q	discharge per unit width (overland, channel, etc)	7.11	LT^{-1}
q	parameter of the incomplete gamma distribution	8.44	
q_B	bed load rate per unit width of channel	11.16	MT^{-3}

q_m	flood flow unit discharge rate (envelope curve)	8.13	LT^{-1}
q_p	peak discharge rate of a hydrograph	8.3	LT^{-1}
q_s	rate of suspended sediment load per unit width of a channel	11.13	MT^{-3}
q_t	total bed sediment load rate per unit width	11.17	MT^{-3}
q_{t_r}	unit discharge equalled or exceeded in a given return period	–	LT^{-1}
Q	discharge rate (hydrograph, flow to wells, etc.) at a given time	–	L^3T^{-1}
Q_m	flood flow (envelope curve)	8.6	L^3T^{-1}
Q_p	peak discharge rate of hydrograph	8.2	L^3T^{-1}
Q_{t_r}	discharge equalled or exceeded in a given return period	8.7	L^3T^{-1}
r	coefficient of correlation (statistical)	2.50	
r	reflection coefficient –ratio of the reflected to incident radiation	3.10	
R	total black body radiation	2.27	ML^2T^{-2}
R	ratio of area of vegetal surface to the projected area	4.3	
R	hydraulic radius	7.10	L
R_A	extra-terrestrial radiation received on a horizontal surface	3.7	ML^2T^{-2}
R_b	net longwave radiation exchange	3.5	ML^2T^{-2}
R_b'	net longwave radiation by the earth's surface under clear-sky conditions	3.17	ML^2T^{-2}

R_b'	hydraulic radius for grain roughness, or surface drag	11.17	L
R_e	radiant energy used for evaporation, or condensation	3.5	ML^2T^{-2}
R_h	sensible heat loss or gain by a body	3.5	ML^2T^{-2}
R_L	emitted longwave radiation	13.33	ML^2T^{-2}
R_{LD}	longwave radiation directed downward	3.17	ML^2T^{-2}
R_{LU}	longwave radiation directed upward	3.17	ML^2T^{-2}
R_N	net radiation	3.45	ML^2T^{-2}
R_o	insolation or radiation received on a horizontal surface of the earth in the presence of the average atmospheric conditions but cloudless conditions	3.8	ML^2T^{-2}
R_r	reflected shortwave radiation	3.5	ML^2T^{-2}
R_s	absorbed shortwave radiation	–	ML^2T^{-2}
R_{si}	shortwave radiation incoming to a surface	3.5	ML^2T^{-2}
R_θ	change in stored energy in a body	3.5	ML^2T^{-2}
R_v	net energy advected into a body	3.5	ML^2T^{-2}
R_w	energy advected out of body of water by mass of water evaporated	3.23	ML^2T^{-2}
RH	relative humidity	–	
s	standard deviation (statistical)	2.53	
s	drawdown in piezometric head as measured from the static water level	6.30	L
s	ratio of specific gravity of particle to that of the fluid	11.15	

s_D	standard deviation of dates of snowpack formation or loss	2.101	T
$s_{\bar{p}}$	standard error of estimate of precipitation amounts	2.55	L
s_s	specific gravity of the sediment	11.22	
s_p	standard deviation of precipitation amounts	2.53	L
sec	seconds	—	T
S	storage volume	1.2	L^3
S	soil heat	3.45	ML^2T^{-2}
S	potential soil moisture storage volume	5.6	L
S	'storativity' or storage coefficient	5.35	
S_A	sun's altitude	3.9	
S_c	slope of a channel	7.11	
S_d	depth of depressional storage	5.12	L
S_e	slope of total energy line	—	
S_f	friction slope	8.89	
S_f	stemflow	4.4	L
S_L	average land slope of a watershed	7.11	
S_v	storage capacity of vegetation on a given projected area of a canopy of vegetation	4.3	L
SRO	surface runoff in given time	5.12	L or L^3
t	time	—	T
t	student's 't' ... statistical	2.53	

t_b	time base of a hydrograph	8.33	T
t_c	time of concentration of a watershed	7.3	T
t_L	lag time—difference in time between the center of mass of effective precipitation to the peak of the hydrograph	8.6	T
t_o	unit-storm duration	8.32	T
t_r	return period (statistical)	2.64	T
t_t	travel time in flow of water between two points on a watershed	7.2	T
t_R	storm duration (rain)	4.3	T
T	temperature	—	θ
T	transmissibility	—	$L^2 T^{-1}$
T_b	base temperature	3.22	θ
T_c	temperature of the cloud base	—	θ
T_d	dewpoint temperature	3.51	θ
T_f	temperature of forest canopy	13.33	θ
T_h	throughfall precipitation	4.4	L
T_m	mean monthly temperature	3.54	θ
T_o	surface temperature of a body	3.19	θ
u	wind speed (velocity)	—	LT^{-1}
u'	average of velocity fluctuations normal to flow	11.11	LT^{-1}
u_t	ordinate of the instantaneous unit graph	8.34	$L^3 T^{-1}$

v	instantaneous velocity along a given streamline at a given point	8.68	LT^{-1}
V	average velocity of flow through a given cross-section (open-channel flow, porous media, etc)	–	LT^{-1}
V_{max}	maximum velocity along a given stream-line	8.68	LT^{-1}
V*	shear velocity or friction velocity	8.68	LT^{-1}
V_d	fall velocity of a raindrop	11.2	LT^{-1}
Vol	volume of flow	5.16	L^3T^{-1}, L^3
w	fall velocity of a particle	11.15	LT^{-1}
W	infiltration index	7.21	L
W_p	moisture charge in a vertical column of air	2.96	L^3
W(u)	well function	6.30	
W_d	dry weight of a soil sample	5.50	MLT^{-2}
W_w	wet weight of a soil moisture sample	5.50	MLT^{-2}
x	weight factor to account for reservoir storage	8.50	
x	correction factor for hydraulically-smooth conditions	11.14	
x,y,z	rectangular coordinate system	–	
X	independent variable (regression analysis)	–	
X	reference grain size for a given bed	11.17	L

Y	dependent variable (regression analysis)	—	
Y	correction for lift force (hydraulics)	—	
Y_s	specific yield of water by a soil	3.42	
z_0	roughness parameter	3.4	L
α	contact angle between a fluid and a surface	5.36	
α	compressibility index of the soil skeleton	5.35	$M^{-1}LT^2$
α_L	proportion of advected energy to a lake used for evaporation	3.31	
α_p	proportion of advected energy to an evaporation pan used for evaporation	3.30	
β	compressibility index of water	5.35	$M^{-1}LT^2$
β	momentum correlation coefficient in fluid flow	11.11	
∂	Lebnitz 'dee' (partial differential)	—	
δ'	thickness of the laminar sublayer	—	L
Δ	change	—	
Δ	slope of the saturated vapour pressure-temperature curve for water	3.46	$ML^{-1}T^{-2}\theta^{-1}$
Δt	routing period or interval	8.48	T
ϵ	emissivity	3.18	
ξ	correction factor of effective flow for grains	11.17	

γ	specific weight of a fluid	—	$ML^{-2}T^{-2}$
γ	constant in Penman's equation	3.46	$ML^{-1}T^{-2}\theta^{-1}$
γ	parameter of a two-parameter, incomplete gamma distribution	8.45	
Γ	gamma function	8.45	
μ	micron	—	L
μ	dynamic viscosity of a fluid	5.19	$ML^{-1}T^{-1}$
ν	kinematic viscosity of a fluid	8.65	L^2T^{-1}
ϕ	gravitational potential	5.37	L
ϕ	infiltration index	7.20	L
Φ	total soil moisture potential	5.37	L
Φ_*	bed load transport function	11.16	
π	pi = constant = 3.1416	—	
ψ	soil suction potential (unsaturated)	5.37	L
ψ_*	hydraulic parameter	11.16	
ρ	mass density of a fluid	—	ML^{-3}
ρ_e	mass density of water evaporated	3.23	ML^{-3}
ρ_s	mass density of sediment	11.16	ML^{-3}
σ	Stefan-Boltzmann constant	2.27	$MT^{-3}\theta^{-4}$
σ	surface tension coefficient	5.36	MT^{-2}
Σ	summation	—	
τ	isochronal line	7.3	T
τ	shear stress in a fluid	5.19	$ML^{-1}T^{-2}$
τ_c	critical shear stress	11.22	$ML^{-1}T^{-2}$

τ_O	shear stress on the bed of a channel	11.13	$ML^{-1}T^{-2}$
θ	angle (cylindrical coordinates)	—	
θ	soil moisture content (volumetric basis)	5.39	

RECOMMENDED UNITS AND CONVERSION FACTORS
(World Meteorological Organization, 1965).

Recommended units for hydrometeorological elements. Commonly used alternative units and corresponding factors for conversion to recommended units are also shown.

Element	Recommended unit	Alternative units	Factor for conversion from alternative unit (3) to recommended unit (2)
(1)	(2)	(3)	(4)
Water-level (stage)	cm	ft	30.5
Stream discharge	m^3/sec	cfs	0.0283
Unit discharge	$m^3/sec/km^2$	cfs/mi^2	0.0103
Volume (storage)	m^3	ft^3	0.0283
		ac-ft	1230
		cfs-days	2450
Runoff depth	mm	in.	25.4
Precipitation	mm	in.	25.4
Precipitation intensity	mm/hr	in./hr	25.4
Snow depth	cm	in.	2.54
Snowcover, area	%		
Water equivalent of snowpack	mm	in.	25.4
Ice thickness	cm	in.	2.54
Evaporation	mm	in.	25.4
Evapotranspiration	mm	in.	25.4
Soil moisture	%, volume	%, weight (conversion depends on density)	
Soil-moisture deficiency	mm	in.	25.4
Sediment discharge	MT/day	tons/day	0.907
Sediment concentration	kg/m^3	ppm (conversion depends on density)	
Chemical quality	ppm		
Energy (heat)	cal (gramme)	Btu	252
Radiation	cal/cm^2	ly	
Radiation intensity	$cal/cm^2/min$	ly/min	
Sunshine	% possible	hrs (conversion depends on possible sunshine)	
Temperature	°C	°F	$5/9(°F - 32)$
Wind speed	knots	mi/hr	0.869
	m/sec	mi/hr	0.447

Relative humidity	%		
Vapour pressure	mb	mm Hg	1.333
		in. Hg	33.86
Atmospheric pressure	mb	mm Hg	1.333
		in. Hg	33.86
Area	km²	mi²	2.59
		ac	0.00405
		ha	0.01

NOTE: Abbreviations used in the table are as follows:

ac	— acre	Hg	— mercury	mi	— mile		
Btu	— British thermal unit	hr	— hour	min	— minute		
°C	— degrees Celsius	in.	— inch	mm	— millimetre		
cfs	— cubic feet per second	kg	— kilogram	MT	— metric ton		
cm	— centimetre	km	— kilometre	ppm	— parts per million by weight		
°F	— degrees Fahrenheit	ly	— langley	sec	— second		
ft	— foot	m	— metre				
ha	— hectare	mb	— millibar				

Section I

INTRODUCTION TO HYDROLOGY

by

Donald M. Gray

TABLE OF CONTENTS

LIST OF FIGURES

Section I

INTRODUCTION TO HYDROLOGY

I.1   HYDROLOGY DEFINED

Hydrology in a broad and literal sense is the science of water, its properties, phenomena and distribution. For practical reasons, however, the word as used by scientists and engineers adopts a somewhat narrower connotation to exclude certain aspects. For example, it may not cover studies of oceans (oceanography). Nor do we think of a mechanical engineer when working with steam; nor a medical doctor when studying the medical uses of water; nor a botanist when he studies the movement of water in the leaves—as making use of standard, accepted hydrologic principles.

Chow (1964) indicates the definition of hydrology adopted in the United States:

> Hydrology is the science that treats of the waters of the Earth, their occurrence, circulation and distribution, their chemical and physical properties, and their reaction with their environment, including their relation to living things.

That is, the domain of hydrology embraces the full history of water on the earth.

Generally, five separate subdivisions of the science are recognized: (a) Hydrometeorology – the study of problems intermediate between the fields of hydrology and meteorology; (b) Limnology – the study of lakes; (c) Cryology—studies dealing with snow and ice; (d) Geohydrology—studies related to subsurface water; and (e) Potamology –the study of surface streams. However, very few hydrologic problems can be limited to just one of these branches. More often, because the phenomena are so interrelated, solutions to these problems can only be attained by a completely interdisciplinary approach by scientists from one or more of these branches. Implicit in these discussions is the fact that hydrology is an extremely broad science and therefore borrows heavily from other branches of science and integrates them for its own interpretation and use. Supporting sciences such as physics, mathematics, chemistry, geology, geography, agriculture, fluid mechanics, statistics, operations research, forestry, plant ecology, economics, wildlife management, sociology, law and computer science are but a few of these which may be used in hydrologic investigations.

Linsley *et al.* (1949) suggest that there are three broad problems in hydrology: (a) the measurement, recording and publication of basic data; (b) the analysis of these data to develop and expand the fundamental theories; and (c) the application of these theories and data to a multiple of practical problems. In essence, it can be envisioned that hydrology is not entirely a pure science, for the object of study is usually directed to a practical application. In these regards, the term 'Applied Hydrology' is often used to emphasize its practical importance.

I.2   HYDROLOGIC CYCLE

The hydrologic cycle is a concept which considers the processes of motion, loss, and recharge of the earth's waters. This continuum of the water cycle can be visualized as shown in Fig. I.1. As indicated in this figure, the cycle may be divided into three principal phases; (a) precipitation; (b) evaporation; and (c) runoff—surface and groundwater. Further, it is interesting to note that at some point in each phase there usually occurs: (a) transportation of water; (b) temporary storage; and (c) change of state. For example, in the precipitation (atmospheric) phase there occurs vapour flow, vapour storage in the atmosphere, and condensation or formation of precipitation created by a change from vapour to either the liquid or solid state. It follows that quantities of water going through individual sequences of the hydrologic cycle can be evaluated by the so-called 'hydrologic' equation, which is a simple continuity or water-budget equation defining the process. That is

$$I - O = \Delta S \quad \dots\dots\dots\dots\dots\dots\dots\dots\dots\dots\dots\dots\dots\dots\dots\dots \quad I.1$$

in which I = inflow of water to a given area
during any given time period;
O = outflow of water from the area during
the selected time period; and
ΔS = change in storage of the volume of
water in or on the given area during
the time period.

It should be recognized that the hydrologic cycle has neither beginning nor end, as water evaporates from the land, oceans and other water surfaces to become part of the atmosphere. The moisture evaporated is lifted, carried and temporarily stored in the atmosphere until it finally precipitates and returns to the earth – either on land or oceans. The precipitated water may be intercepted or transpired by plants, may run off over the land surface to streams (surface runoff) or may infiltrate the ground. Much of the intercepted water and surface runoff is returned to the atmosphere by evaporation. The infiltrated water may be temporarily stored as soil moisture and evapotranspired; or percolate to deeper zones to be stored as groundwater which may be used by plants; or flow out as springs; or seep into streams as runoff; and finally evaporate into the atmosphere to complete the cycle. A schematic sketch of the runoff phase is given in Fig. I.2.

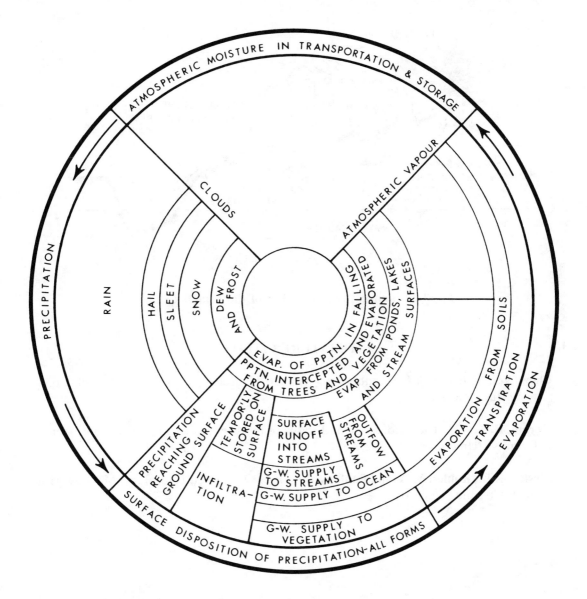

Fig. I.1 The Hydrologic Cycle — A Qualitative Representation (Horton, 1931)

From the preceding discussions, it is obvious that the hydrologic cycle is subject to the various complicated processes of *Precipitation, Evaporation, Transpiration, Interception, Infiltration, Percolation, Storage* and *Runoff*. The material presented in the following sections attempts to detail the basic theories or concepts underlying each of these processes and to outline standard procedures used to estimate the magnitude of each. In these notes, a particular effort has been made to present data and illustrative material applicable to Canadian and/or local Prairie conditions.

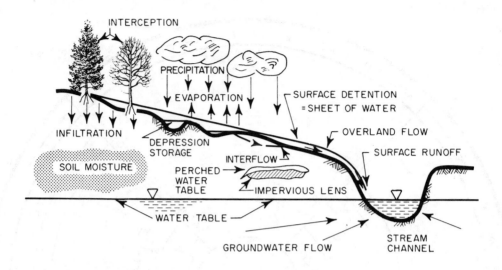

Fig. I.2 Simple Representation of the Runoff Cycle

I.3 **LITERATURE CITED**

Chow, Ven T. (ed.). 1964. Handbook of Applied Hydrology. McGraw-Hill Book Company, Inc., Toronto.

Horton, R.E. 1931. The field, scope and status of the science of hydrology. Trans. Amer. Geophys. Union, pp. 189-202.

Linsley, R.K., Kohler, M.A. and Paulhus, J.L.H. 1949. Applied Hydrology. McGraw-Hill Book Co., Inc. Toronto.

Section II

PRECIPITATION

by

Gordon A. McKay

TABLE OF CONTENTS

LIST OF TABLES

LIST OF FIGURES

Section II

PRECIPITATION

II.1   **INTRODUCTION**

Precipitation is the primary source of our water supplies. Its records are therefore the basis of many investigations and decisions relating to supplies, floods, drought, irrigation and regulating structures. The steadily increasing demand for water in the face of a relatively constant supply necessitates an efficient use of this resource. This can only be accomplished when we have a thorough knowledge of the character of precipitation.

II.2   **PRECIPITATION PROCESSES**

The hydrologic cycle, discussed in the preceding Section, is a concept which considers the processes of motion, loss and recharge of the earth's waters. The four major phases of this cycle are precipitation, the combination of evaporation and transpiration, streamflow, and groundwater. The first two phases are atmospheric and thus their study is usually considered to lie in the realm of hydrometeorology or that portion of the meteorological science which is concerned with hydrology. The latter two phases are referred to as runoff. In this Section the most important atmospheric phase, precipitation, is considered.

Precipitation includes all forms of moisture falling from the atmosphere to the earth's surface. It is produced primarily from water vapour present in the air. Atmospheric moisture occurs in vapour, liquid and solid form. It is replenished by evaporation and transpiration, and depleted by precipitation. Cooling of the air sufficient to cause condensation and growth of water droplets or ice crystals is necessary for the precipitation process to occur.

II.2.1   **Moisture Sources**

Atmospheric moisture is obtained mainly through evaporation from wet surfaces and through transpiration. Warm large bodies of water and zones of lush vegetation therefore provide excellent sources while cool and arid land areas are the poorest providers of atmospheric moisture. The Gulf of Mexico is an excellent example of good moisture source. Most major rainstorms follow sustained air transport from the humid source areas to the south. During the long summer days shallow northern lakes also become very warm and provide good local sources of moisture. Showers often follow rainy days because of evaporation from the wet terrain. The atmosphere's capacity of holding water is greatest in summer when it is warmest (see Table II.l) and it is in this period that heaviest showers can occur. It also varies with elevation such that about one half of the moisture in a saturated atmosphere is in the lower 7,000 feet of an air column.

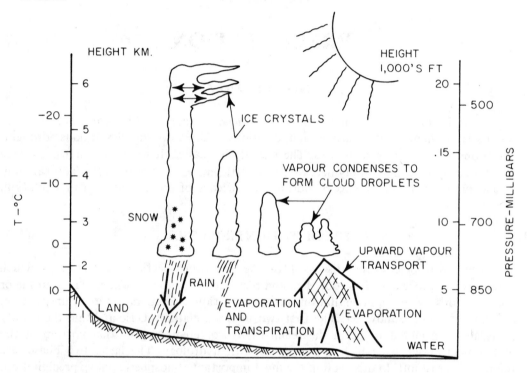

Fig. II.1 Atmospheric Phases of the Hydrological Cycle

II.2.2   Cloud and Precipitation Formation

Condensation of water vapour into cloud droplets precedes the precipitation process. Condensation occurs as the result of cooling of the air to a temperature which is below the saturation point for water vapour. Cooling rates sufficient for significant precipitation are achieved through the adiabatic (without any loss of heat to its environment) expansion of air through lifting to levels where the pressure is lower. The decrease in pressure with height is indicated on Fig. II.l. Although the atmosphere extends through great depths, because of its compressibility it is much denser near the earth's surface than at higher elevations. One half of the mass of the atmosphere lies within 3 1/2 miles of the earth's surface; in other words, the average atmospheric pressure at about 18,000 feet is only one half that found at sea level. Most atmospheric moisture and clouds are found in this lower half of the atmosphere.

Table II.1 The Capacity of the Atmosphere to Retain Moisture
as a Function of Temperature

Air Temperature °F	Relative Humidity %					
86	16	24	31	45	57	100
68	28	42	54	79	100	
61	36	53	69	100		
50	52	77	100			
43	67	100				
32	100					
Grams of water vapour per cubic meter	4.85	7.27	9.41	13.65	17.31	30.4

Also necessary for the formation of droplets are the presence of condensation nuclei. Such nuclei are abundant in the atmosphere and come from many sources such as the ocean, dust from clay soils, industrial pollution, volcanoes, etc. Cloud droplets average about 1/1000th of an inch in diameter. Such small droplets fall very slowly and it is only when the drop size exceeds 1/100th of an inch in diameter that significant precipitation occurs. Several million cloud droplets are required to make one raindrop.

Dr. T. Bergeron (1933) suggested that larger drops may be formed by the condensation of water vapour on or by the collision of droplets with ice crystals. Ice has a lower vapour pressure than water at the same temperature so that a crystal will grow at the expense of the water droplet when the two are adjacent. Rain-bearing clouds, to satisfy this theory, must extend into regions where ice crystals may form.

Ice crystals are formed at temperatures below freezing in the presence of sublimation or ice nuclei. Although condensation nuclei are common, this is not the case for nuclei which initiate the growth of ice crystals. None have been found in nature at temperatures above -4°C, and few are found at temperatures exceeding -12°C. Below -12°C, however, the number of such nuclei increases rapidly as the temperature decreases. Also, once ice crystals form, they may splinter and thereby create large numbers of nuclei which may aid the precipitation process.

Falling crystals continue to grow both through condensation and the capture of liquid droplets. They change into rain after entering air in which the temperature is above freezing.

Most contemporary rain-making techniques are based on increasing the number of nuclei which are active at temperatures above -15°C. Seeding of clouds is attempted when weather conditions favor precipitation occurrence, but natural nuclei are unlikely to be sufficiently abundant for optimum yield. Dry ice is used to create crystals by reducing temperatures to levels where natural nucleating agents are effective. Vonnegut (1949) observed the great similarity in crystal lattice between ice and silver iodide. This chemical was found to be effective as both a sublimation and ice forming nucleus at temperatures of less than -5°C; that is ten degrees warmer than the previously noted average level of -15°C for natural substances. Since this discovery, silver iodide has been used extensively to induce rain to fall from clouds. Concurrently, there has been a continued search for other agents which may be used to increase our water supplies from this source.

The effectiveness of ice crystals in producing raindrops has been established; however, rain sometimes falls from clouds in which the air temperature is not below 0°C, showing that ice crystals are not the only catalysts which may produce rainfall. Droplets may increase in size through coalescence, and by 'capturing', as well as by the action of hygroscopic particles such as minute particles of sea salt. Larger drops fall more quickly than smaller drops and in so doing may overtake and capture smaller drops in their paths. Drops larger than 5.5 millimeters in diameter are unstable and shatter, thereby providing many more raindrops.

Coalescence is considered to be the dominant process in the summer shower-type precipitation. Cloud droplets are very small, and tend to avoid collision or coalescence with each other. Only when the droplets are of appreciable size do the processes of coalescence and collision become significant. The initial growth required for this is considered to be achieved through condensation; this process is favored in very turbulent clouds, such as the swelling summer cumulus. When the coalescence process starts, it may proceed rapidly to produce rain within minutes over a wide range of heights in the cloud. If the updrafts, which maintain the cumulus clouds, equal or exceed the fall velocity of the raindrops, rain will not fall. A large mass of rain may then accumulate in the cloud if this condition persists.

The shower cloud is considered to be mature when rain starts to fall from it. At this stage the concentration of raindrops and ice particles is so great it cannot be supported by the updraft. Rain often falls in torrential bursts as the updraft collapses and 'pockets' of precipitation descend. The frictional drag of the falling precipitation creates a downdraft and the air in the downdraft is cooled by evaporation. Thus, often there is a sharp contrast in the temperature of this air to the warm air which it replaces.

II.2.3   Lifting Mechanisms

Cooling of the air that is sufficient to cause significant precipitation requires lifting, or rising currents of large scale. Four major processes are commonly distinguished.

II.2.3.1 Horizontal Convergence

Wind fields may act to concentrate the inflow of air into a particular area, such as a low pressure area at the earth's surface, thereby forcing the air to rise. In so doing the air is cooled by expansion.

II.2.3.2 Orographic Lifting

In orographic lifting the air is made to ascend a topographic barrier such as a mountain range. Precipitation resulting from this process is common along the west and east slopes of the Rocky Mountains whenever strong upslope winds bring moist air to the area.

II.2.3.3 Convection

Convection occurs when differential heating or advection results in air becoming more buoyant than its environment. The air may then rise to levels where it becomes saturated, forming cloud and precipitation. Thunderstorms are convective storms.

II.2.3.4 Weather Fronts

Most low pressure areas in temperate and polar regions have frontal systems associated with them. Fronts are the zones which separate large masses of air which have significantly different physical properties. The difference in density and motion of the two types of air associated with the front result frequently in large-scale lifting of an air mass, thereby causing extensive cloud systems and precipitation. Frontal storms tend to give very widespread precipitation. This is often heaviest to the northeast quadrant of surface low pressure areas, where the combined lifting due to convergence and frontal action is greatest. The passage of fronts in summer is often instrumental in producing 'lines' of vigorous thundershowers which give locally intense precipitation. The principal precipitation mechanisms are indicated in Fig. II.2

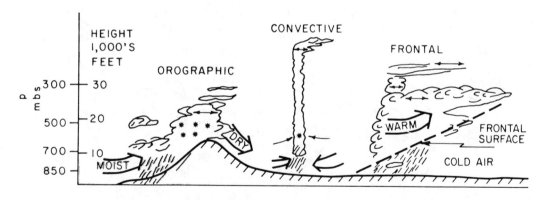

Fig. II.2 Precipitation Mechanisms

The above processes may act alone or in combination. In the latter case the precipitation release is generally more vigorous. Most major flood-producing storms in southern Alberta, for example, are due to the combined effects of intense orographic lifting of moist air and strong frontal lifting, and convergence which occurs to the northeastern quadrant of low pressure areas.

II.2.4 The Character of Precipitation

The precipitation character is dependent both on the character of the air which is being lifted, and the lifting mechanisms. Heaviest rainfalls occur when the moisture supply is high, the thermal structure of the air favours rising air currents and the lifting mechanism is vigorous. Since moisture supply and the nature of most mechanisms vary seasonally, precipitation character also undergoes a profound seasonal change in most areas.

Maps of mean annual precipitation, such as for the Prairie provinces in Fig. II.3, show integrated effects of the following:

1. Availability of atmospheric water vapour.
2. Topography.
3. The frequency and efficiency of non-orographic lifting process.

The orographic influence of the Rockies, and major hills or slopes, are quite apparent, as is the closer proximity of Manitoba to the major moisture source regions. An apparent belt of heavier precipitation to the north of the prairies would appear to be related to the relatively high frequency of occurrence of storms and frontal rainfall in that area.

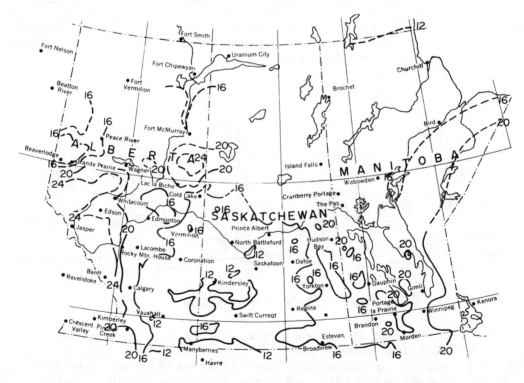

Fig. II.3   Mean Annual Precipitation in Inches, 1921-1950

While annual rainfall may appear relatively uniform, this is often due to the number and random occurrence of storms and showers which are highly variable in character. During a thunderstorm the occurrence of rain on one side of the street while the sun is shining on the other is not an uncommon event. Whenever rainfall is uniform, or long-term averages are being sought, measurement problems are relatively simple. However, where the detail of an individual thunderstorm must be known, the measurement problems are substantially increased.

II.2.4.1 Orographic Precipitation

Topography affects both large- and small-scale meteorological processes. The influence of mountain ranges is quite apparent. However, the small-scale topographic effects, which are not as obvious, may be quite significant.

In rugged terrain, precipitation is related to (a) elevation, (b) local and general slope, (c) orientation of the sloping surface, (d) the degree of local exposure, (e) topographic barriers to incoming moisture, and (f) distance from the moisture source. The relative, spatial distribution of orographic rainfall tends to be similar from storm to storm; and where orography is dominant, the individual storm patterns are similar to the annual-precipitation, isohyetal pattern.

Elevation and precipitation are highly related along a specific slope. The amount of moisture available for precipitation decreases with height, therefore, logically the amount of precipitation should decrease with height. This appears to be the case in short-duration convective storms. However, over a long period, higher elevations tend to receive more precipitation than is obtained at the lower elevations. This results from 'spillover' where precipitation formed in air rising along the windward slope is released in time to fall on the leeward slope of a mountain range, and from the relatively higher frequency of precipitation at high elevations. Spillover-effects for rain have been estimated to reach up to 10 miles.

Orographic barriers upwind of a mountain range may deplete the atmospheric vapour supply. Only that moisture which can pass over the upwind range is available for further precipitation release on the downwind range. As a consequence, sheltered-mountain valleys receive sparse precipitation. The influence of elevation on long-term precipitation as indicated by Curry and Mann (1965) is shown in Fig. II.4.

A major consideration in the occurrence of orographic precipitation is that while rain is falling at low elevations, snow may be falling at high elevations. Snow need not immediately contribute to streamflow. It can remain in storage and it is therefore important in the analysis of these storms to determine the freezing level or snow zone.

II.2.4.2 Convective Storms

The greatest short-duration rates of rainfall occur during convective storms. In fact, short-duration rates of the order of 5 minutes to 1 hour, which have been experienced in Saskatchewan, compare favourably with world-record values as noted in Fig. II.5.

Convective storms are usually of small dimension, and made of a cluster of cells which pass through stages of growth, release and decay. The occurrence of strong air-currents within these storms favours the growth and suspension of large raindrops which may subsequently be released in a torrential downpour. If the convective cells were stationary, some very remarkable rainfall amounts would probably be recorded. Fortunately the cells are generally moving with the upper level winds, so that the volume of water they release is spread over a substantial area.

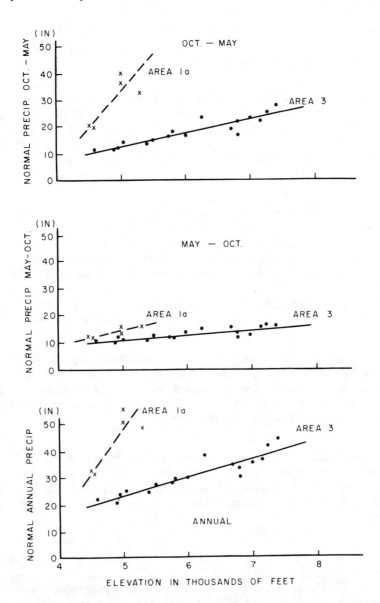

Fig. II.4   Precipitation-Elevation Relationships

Area 1a is south of Crownest Pass, Alberta.

Area 3 is north of Crownest Pass, Alberta.

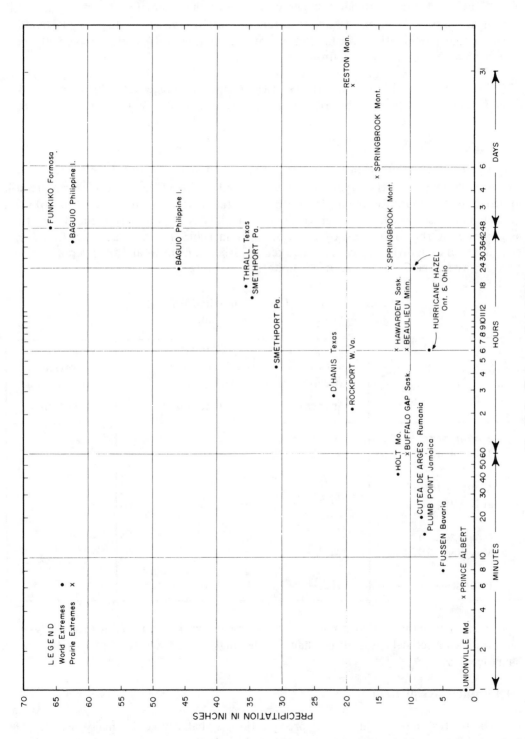

Fig. II.5   Extreme Rainfalls

Convective storms provide a large percentage of the summer-season precipitation in continental areas. They tend to occur in random fashion; and they shape the isohyets of mean annual precipitation. As random events, from time-to-time they repeatedly miss certain areas, giving rise to local droughts.

However, by and large, the effects of these storms average out. This is indicated in Figs. II.6 and II.7, and in Table II.2, which describe rainfall at Wilson Creek, Manitoba, where a large percentage of the rain is convective. The character of rainfall from a relatively simple thunderstorm cell is shown in Fig. II.6.

In Fig. II.6 the one-inch isohyet encompasses an area of about 1/2 x 3 miles or 1.5 sq. miles compared to the basin area of 8.8 sq. miles. The effects of this storm are still apparent in the monthly and four-month isohyetal charts (Fig. II.7), and the isohyetal gradients are also similar. However, the areal variability is substantially reduced with time as is shown by the coefficients of variation given in Table II.2. Averaged over four Julys, the effects of an individual storm are masked by other rainstorms and the overall effect of topography and gauge exposure on precipitation are more apparent:

Table II.2   Wilson Creek Watershed Rainfall
(Measurements from 25 gauges)

Period	Basin Mean (Inches)	Standard Deviation Of Point Values (Inches)	Coefficient Of Variation (Per Cent)
3-hour storm July 1961	0.42	0.36	86
Monthly total July 1961	2.34	0.40	17
Single season May-Sept. 1961	6.6	0.5	7
Four seasons	12.1	0.8	7

The above data indicate that a much more intense network is required for the analysis of individual convective-type storms, than for the analysis of the effects of topography or climatic stress.

II.2.4.3 Frontal Precipitation

Of greatest interest in frontal precipitation is that caused by major storms. The amount of cloud and precipitation associated with the front is, among other things, related to the rate of lifting of one air mass over another. This is usually greatest in the vicinity of 'waves' which may form on a front, and in thundershowers in the vicinity of rapidly moving cold fronts. Major rain storms in temperate latitudes are usually related to these frontal waves which, like ocean waves, increase in amplitude and then break as they mature. This is indicated very roughly in Fig. II.8.

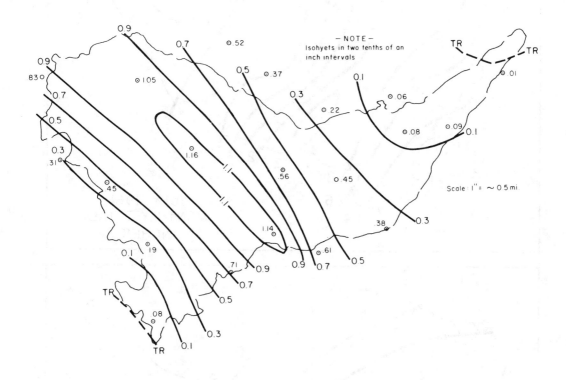

Fig. II.6   3-Hour Rainstorm, July 1961, Wilson Creek Watershed

Frontal waves move across a region at varied speeds, usually about 15 mph in summer and 25 mph in winter. Frequently with major storms the low pressure centre stagnates, resulting in sustained heavy rains over an area.

The storm precipitation pattern usually displays an orderly character. In the absence of instability within the air, rainfall tends to be steady, extending to within about 150 miles ahead of the warm front and within 25 to 50 miles of the cold front. Precipitation rates are generally heaviest within 50 miles of the wave in northeast quadrant of the low pressure area, and they tend to diminish with distance away from the front and the centre of low pressure. Conditions may differ markedly from this in any storm since the amount of cloud and weather is dependent on moisture supply and other factors as well as the vigor of the frontal wave. When the air is unstable, thunderstorms may be present within the general cloud system. Vigorous summer storms usually do have associated thunderstorms.

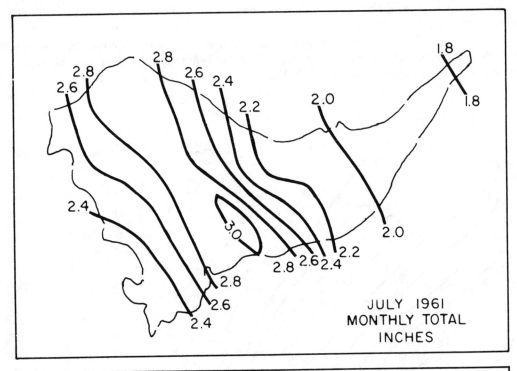

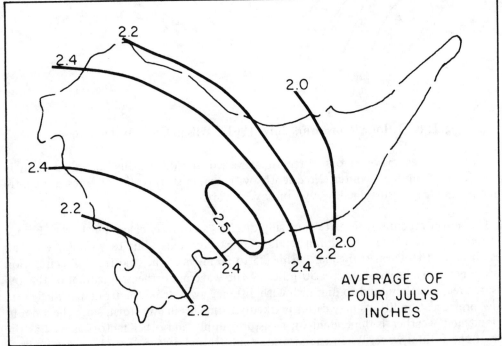

Fig. II.7   Precipitation,   Wilson Creek Watershed

The integrated widespread, and somewhat orderly character of precipitation from a vigorous frontal storm is indicated in Fig. II.9. Many centres of heavier precipitation are generally found in such isohyetal patterns due to variations in moisture supply, topography, storm vigor and airmass stability. (An airmass is a large body of air which has a relatively homogeneous character in the horizontal plane. An unstable airmass is one in which convective-type precipitation may occur and a stable airmass is one in which convective-type precipitation will not occur whenever the airmass is subjected to lifting, whether frontal, orographic or due to surface heating).

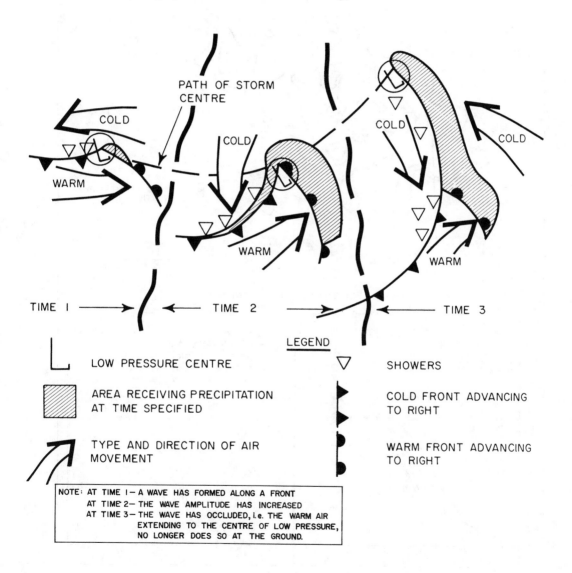

Fig. II.8   Areas of Rainfall at Various Stages During the Development of a Frontal Storm

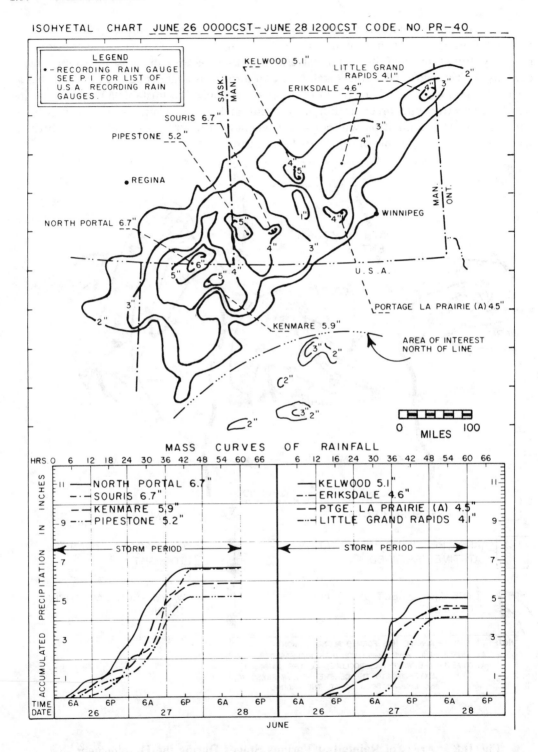

ISOHYETAL CHART JUNE 26 0000CST – JUNE 28 1200CST CODE. NO. PR–40

Fig. II.9   Rainstorm, June 26-28th, 1944

II.2.5   Snowfall and Snowcover

Snowfall is a vital source of water supply. It is particularly valuable in the semi-arid prairies where about 80 per cent of the streamflow in the major river systems and of the water stored in prairie dugouts and sloughs comes from the shallow snowpack.

II.2.5.1 General Character

Snowfall over an area tends to be more uniform than rainfall; but, because of the buoyancy of snow, its accumulation and retention on the ground is highly heterogeneous.

The broadscale character of Canadian snowcover is indicated in Figs. II.10, II.11 and II.12 (Potter 1965a). The number of days of snowcover increases northward, and with elevation. In some areas, such as the Fraser River Delta, snowcover is ephemeral, because of the mildness of the climate. Also, during winter in southern Alberta, warm chinook winds frequently remove the shallow winter snowpacks. However, over most of Canada the winter cold tends to persist, maintaining fairly continuous winter snowcover. The frequency of snowcover of specified depths is shown for selected locations in Fig. II.13 (Potter 1965a).

The specific case may differ significantly from the character indicated by generalized maps. For example, within mountainous areas, maps of snowfall and other parameters are generally based on weather statistics obtained at mountain pass and valley sites. The maps must therefore be interpreted accordingly, accepting that there may be major accumulations of snow at high elevations when the valleys are free of snow.

Table II.3   Seasonal Precipitation at Various Elevations
in the Eastern Rockies

Place	Elevation	Oct-May	May-Oct	Annual	No. of Years
	ft	in	in	in	
W.B. Castle River	5000	26.2	12.4	38.3	8
Bovin Lake	6800	30.4	13.4	43.0	8
Allison Pass	6500	24.6	15.1	40.3	8
Highwood Summit	7200	21.1	12.6	34.9	9
Kootenay Plains	4500	3.1	6.3	9.4	5

Within the Rocky Mountains snow may remain the year round at high elevations, particularly on north-facing major accumulation areas. Also at high elevations the precipitation character differs considerably from that found on the plains to the east. Whereas the Alberta foothills have most precipitation in summer, the most precipitation at high elevations occurs during the winter months, as noted in Table II.3.

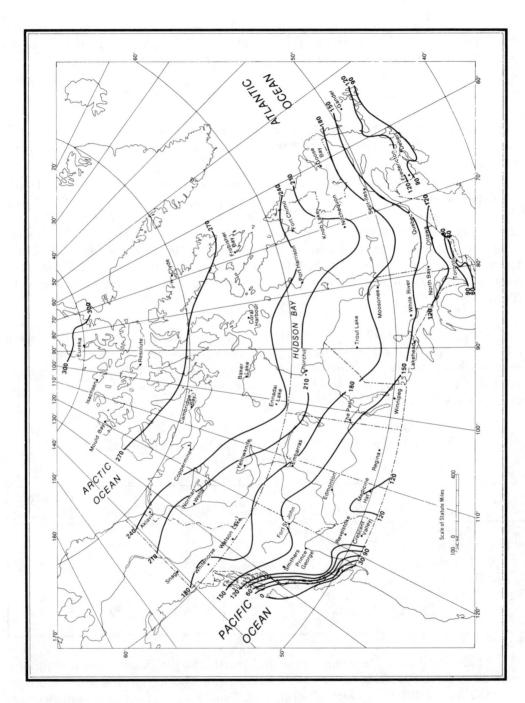

Fig. II.10   Median Number of Days with Snowcover (1 inch or more) for 20 Winters

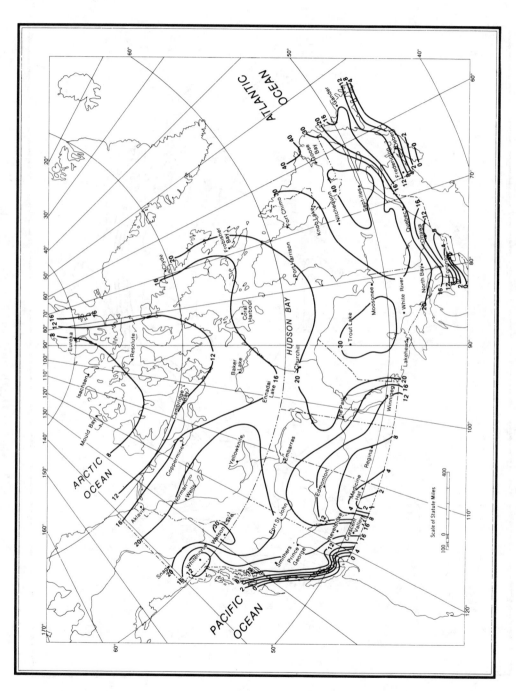

Fig. II.11 Median Depth (inches) of Snowcover, February 28

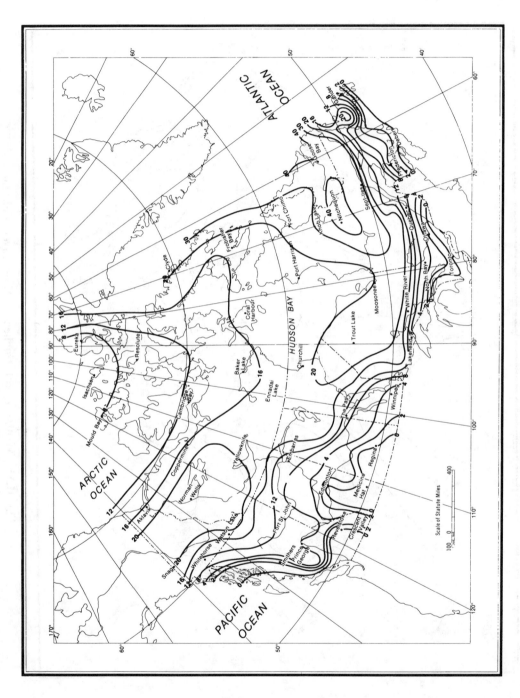

Fig. II.12   Median Depth (inches) of Snowcover, March 31

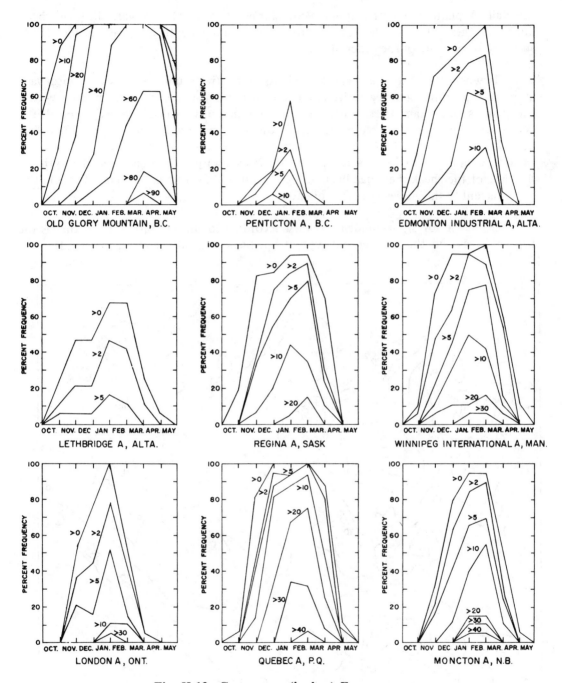

Fig. II.13 Snowcover (inches) Frequency

Snowcover within the mountains tends to be highly variable because of the combined effects of high winds and rugged topography. Slides and unusual disposition due to
local turbulence may result in remarkable accumulations which far exceed the annual

snowfall. A photogrammetric appraisal of Marmot Creek, Alberta watershed snowpack illustrates these effects. The results expressed as snow depth are given as Fig. II.14; accuracy is generally of the order of ± 5 ft.

Elsewhere the general snowpack tends to be relatively uniform; however, local variations in topography and vegetative cover cause major departures from the average. Within the Prairies, for example, shelterbelts and buildings may cause massive, deep drifts. Well-eroded gullies act as 'sediment traps' collecting snow which blows or drifts off of nearby, open fields. A check of snow accumulation in gullies near Regina in 1966 showed them to contain about 3-1/2 acre-ft. of water (as snow) per thousand feet of length. At the time of the check there had been no significant runoff although the grain fields in the area adjacent to the gullies were free of snow.

Within the forest, accumulations vary according to the vegetative cover, that is the type of trees and the number and type of openings. A solid evergreen forest, offers no

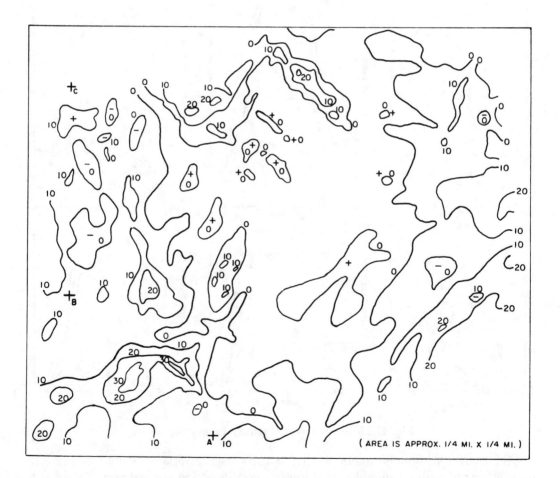

Fig. II.14   Snowpack Depth (feet), Marmot Creek, Aerial Survey, April 1964

'sedimentation trap' openings and is effective in intercepting snowfall. Significant amounts of the intercepted snow may evaporate and not contribute directly to spring runoff. A deciduous forest, lacking leaves in winter, permits the snow to accumulate on the forest floor. Clearings within a forest act as traps and are found to accumulate more snow than a surrounding evergreen forest. Such clearings are usually favored as snow--survey sites.

Although regional generalizations are possible, as previously noted, the accumulation within a region and indeed over a very small area, may be highly varied because of variations in catch and retention capabilities of the surface. Kuz'min (1960) suggests the following 'snow-retention' coefficients for different surfaces (see Table II.4). The retention coefficient is the ratio of the snow catch of a given surface to the catch by a virgin soil.

Table II.4   Snow-Retention Coefficients

Open ice surface of lakes	0.4 to 0.5
Arable land	0.9
Virgin soil	1.0
Hilly districts	1.2
Large forest tracts	1.3 to 1.4
River beds	3.0
Rush growth near lakes	3.0
Forest cuttings of a radius of about 100 to 200 m and edges of forests	3.2 to 3.3

He notes that strong winds may completely remove the snow off ice-covered lakes so that their coefficient then drops to zero.

A snow depth map for the Prairie provinces is shown in Fig. II.15. The character is relatively uniform showing the effects mainly of large-scale storms and topographic variation. It must be remembered in interpretation of these charts that snow-depth measurements are subject to systematic errors. Most Prairie measurements are made near farmyard shelterbelts, while forest measurements are made in forest clearings. Snowpack water-equivalent measurements are much more meaningful than depth measurements, but less accessible to the hydrologist. A water-equivalent map based mainly on the Province of Manitoba surveys March 15, 1965, is presented in Fig. II.16. Its character, too, is dependent on the occurrence of storms and topography.

II.2.5.2 Density

Snow depth information alone does not provide a good measure of the snowpack water equivalent (the depth of water in inches which would result from melting the snow). If depth measurements are used, it is advantageous to have some knowledge of density. In snow hydrology, snow density is taken to be the same as its specific gravity. It

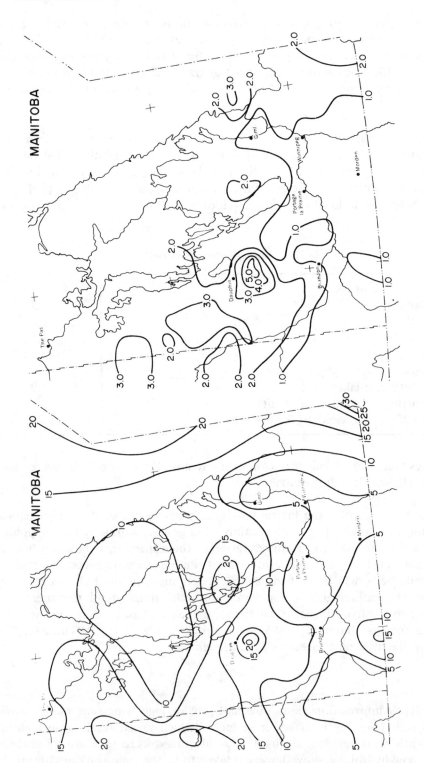

Fig. II.16 Snowpack Water Equivalent (inches),
March 16-18th, 1964 Based on Snow Surveys

Fig. II.15 Snow Depth (inches), February 29, 1964
Based on Climatological Measurements

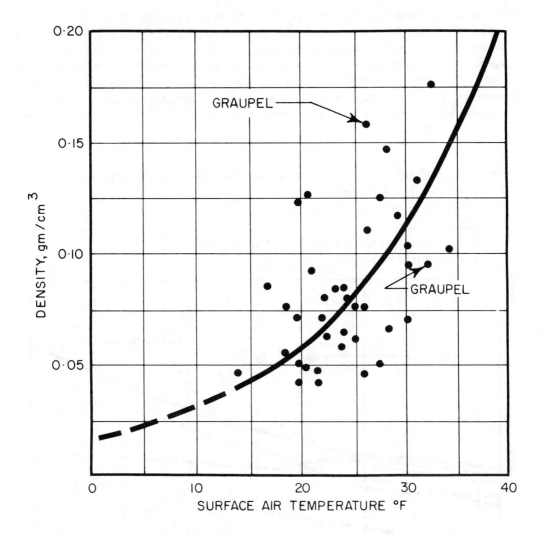

Fig. II.17 Density of New-Fallen Snow

NOTES:
1. Density measurements were made by SIPRE personnel of CSSL headquarters (E1.6900). Air temperatures were taken at about the time of density measurements at 4 feet above snow surface.
2. Times of accumulation of snow before observation were variable and were always less than 24 hours. The average time was probably in the range between 6 and 12 hours. The results, therefore, cannot be applied directly to the usually observed 24-hour snowfalls.
3. Variability is also introduced into the above relationship as the result of varying rates of snow accumulation which have not been considered.

is computed as the depth of the measured water equivalent of the snowpack, divided by the depth of the snowcover.

The snowpack is formed by falling snow crystals of various shapes and densities. As fallen snow ages and is subjected to climatic elements (primarily wind), it undergoes structural changes which alter its density. There is no definite time at which the change takes place from one form to another. The change from the 'loose-dry sub-freezing' pack of low density to a coarse, granular, moist pack of high density is called 'ripening'. A ripe snowpack is considered primed when it is ready to produce runoff. A snowpack is not necessarily homogeneous and may contain ice planes or lenses, and snow layers of different densities. The average density obtained in surveys is therefore a vertically-integrated value which averages out the effects of past weather events.

The factors which cause the density of the snowpack to change are as follows:

1. Heat exchange resulting from convection, condensation, radiation, and heat flow from the ground.
2. The pressure of the overlying snow.
3. Wind.
4. Temperature and water variations within the pack.
5. Percolation of melt water.

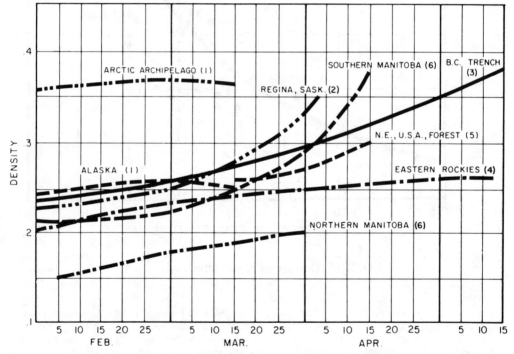

Fig. II.18 Seasonal Variation in Average Snow Density

DATA SOURCES:
1. Billelo, 1957.
2. P.F.R.A. Courses.
3. B.C. Water Resources Service.
4. Eastern Rockies and Forest Conservation Board.
5. Lull and Rushmore, 1960.
6. Manitoba Water Control and Conservation Branch.

Densities of freshly fallen snow have been observed to range from 0.004 to 0.34. Settling and crystal-change begin immediately after it has fallen. For example, Church (1943) observed that snow with a density of 0.036 to 0.056 increased in density to 0.176 after 24 hrs. of drifting. The relationship between density of new snow and air temperature as found at the Central Sierra Snow Laboratory is shown in Fig. II.17 (U.S. Corps of Engineers 1956). Since air temperature varies seasonally and geographically, it is to be expected that seasonal and geographical variations in the mean density also occur.

Because of changing climatic conditions and settlement, the average density of snow varies seasonally as noted in Fig. II.18. The greatest variation is found in sub-arctic snow where densities increase from about .25 in late winter to .35 or greater as the snow ripens. In midwinter large variations in density occur which appear to be related to wind.

During the melting period the snowpack density may be highly variable because of the presence and loss of melt water. Well-drained, ripe snow tends to have a density of about 0.35. Church (1941) noted that the density of a pack rose to as high as 0.49 when runoff began, and then dropped to 0.37. Under alternating freeze-thaw cycles, the pack becomes alternately drained and then primed; and its density can be expected to vary from near 0.40 to 0.48 accordingly (U.S. Corps of Engineers 1956). Surveyors will tend to find densities of about .35 in the morning and .45 in the late afternoon when sampling shallow packs in spring when diurnal temperatures fall below freezing at night and rise into the forties during the day. Some average densities for snowpacks suggested by Seligman (1962) are given in Table II.5.

Table II.5   Snowpack Densities

Snow Type	Density
Wild Snow	0.01 to 0.03
Ordinary new snow, immediately after falling in still air	0.05 to 0.065
Settling snow	0.07 to 0.19
Settled snow	0.2 to 0.3
Very-slightly-wind-toughened, immediately after falling	0.063 to 0.08
Average wind toughened snow	0.28
Hard wind slab	0.35
New firn snow*	0.4 to 0.55
Advanced firn snow	0.55 to 0.65
Thawing firn snow	0.6 to 0.7

*snow partly consolidated into ice.

II.2.5.3 Reflection and Radiation

The reflectivity and radiative properties of snow are of major concern to hydrologists when evaporation and the melting process are being considered.

Reflectivity (Albedo)

The reflective properties of snow have been observed to vary depending on the condition of its surface and the height of the sun (see Table II.6).

**Table II.6   Reflective Properties of Snow
(from K. Kondrat'ev 1954, quoting P.P. Kuz'min).**

Condition of Snow Surface	Height of Sun°	Reflectivity %
Compact, dry, clean	30.3	86
	29.7	88
	25.1	95
Clean, wet, fine grain	33.3	64
	34.5	63
	35.3	63
Wet clean granular	33.7	61
	32.0	62
Porous, very wet, greyish color	35.3	47
	36.3	46
	37.3	45
Very porous, grey, full of water, sea ice visible	32.8	43
	31.7	43
Very porous, light brown, saturated with water	29.7	31
Very porous, dirty, saturated with water, sea ice visible	37.3	29

Kondrat'ev (1954) concluded that reflectivity decreases with increasing solar elevation, and with increasing wave length. The diurnal change in the spectral composition of light, due to changes in solar elevation, is the most important factor in determining the change in reflectivity of an unchanging snow surface. When melting occurs, the change in snow quality must also be considered.

The change in reflectivity of melting snow with wave length of light is shown in Fig. II.19 (U.S. Corps of Engineers 1956).

The U.S. Corps of Engineers (1956) also obtained the relationship between reflectivity and temperature and time as given in Figs. II.20 and II.21.

Caution should be used in using such relationships in different climatic areas, or where other factors may influence the reflectivity, e.g. shallow-pack reflectivity is probably affected by the nature of the underlying surface and the pack depth.

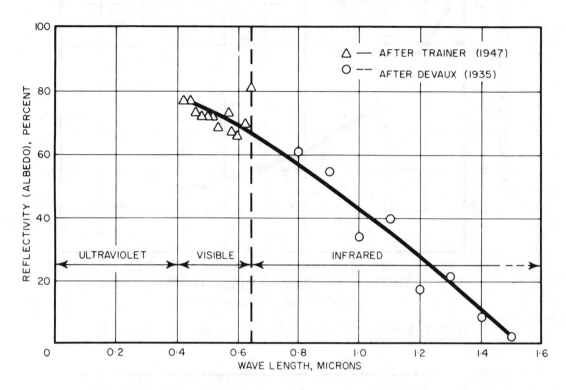

Fig. II.19 Approximate Spectral Reflectivity of Melting Snow
(From data given in Sipre Report)

Radiation

The snowpack acts very nearly in the same manner as a black body with respect to long-wave radiation. It absorbs most incident long-wave radiation, and radiates in accordance with Stefan's law:

$$R = \sigma T^4 \dots\dots\dots\dots\dots\dots\dots\dots\dots\dots\dots\dots\dots\dots\dots\dots \text{II.1}$$

where R = the total radiation in all wave lengths,
    T = temperature in °Kelvin, and
    σ = 0.826 x 10^{-10} langleys per minute /°K^4.

Since the upper temperature of the snow surface is limited to 32°F, the maximum intensity of radiation may be shown to be 0.459 ly/min or 27.5 ly/hr. (U.S. Corps of Engineers 1956).

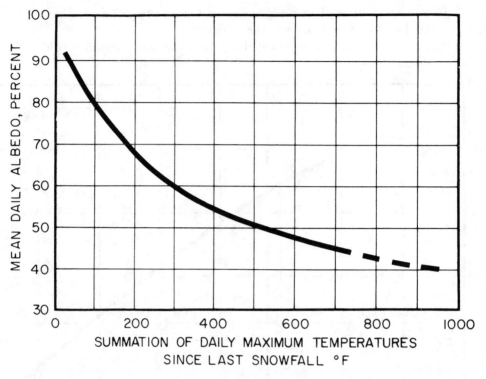

Fig. II.20 Variation of Albedo with Accumulated Temperature Index

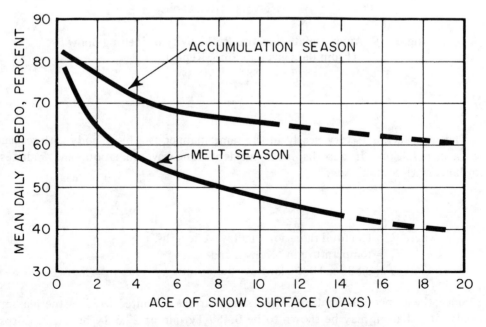

Fig. II.21 Variation in Albedo with Time

II.3.   PRECIPITATION MEASUREMENT

A recent study of five rain-gauge installations in the United States (Allis, *et al.*, 1963) concluded with the remarks that "true areal precipitation is unknown". This uncertainty arises because of errors associated with the gauges and because of the natural heterogeneity of precipitation. A knowledge of the limitations in precipitation measurement is clearly necessary in order to solve many hydrological problems.

The general purpose of meteorological measurement is to obtain values which are representative of the area. These are values which would have occurred at a select location in the absence of the gauge. Unfortunately this is not easy to accomplish. Meteorological parameters are influenced by the immediate environment, and also the instruments themselves may influence the measurement. When the values are used in index form, this may not be a problem. However, when absolute estimates for a watershed are required every effort must be made to ensure representativeness of the measurement.

II.3.1   Gauge Exposure and Design

It is unlikely that any one gauge will serve all requirements. Automated gauges are necessary to determine short-duration values as required for storm-rainfall analysis. Storage gauges usually give only seasonal totals, but require little servicing. Some gauges are useful only for the measurement of rain, others only for snow while still others may be used for both.

Precipitation gauges in their simplest form are hollow cylinders which are open at one end. Funnels have been added to aid in collecting and recording the catch. They also serve the purpose of minimizing evaporation and are shaped to control splash. The funnel is an obstacle to snow accumulation, and therefore funneled gauges are used mainly for rainfall. Heaters which melt the snow have been used, but these cause evaporation losses which are estimated to be as high as 35 per cent of the gauge catch in Maritime areas.

Large orifice gauges, without funnels, are necessary for measuring snow, hail and combined rain and snow. To minimize the storage requirement, snow is liquified by the use of antifreeze.Evaporation is suppressed by using an oil film. The tipping-bucket principle cannot be readily adapted for the measurement of snow and the weighing-type gauges are commonly used where its recording is required. Both the recording and storage gauges must generally be sheltered from the wind because of the height of their orifice level; this is of particular importance when they are used to catch snow.

II.3.1.1   Effect of Wind on Catch

Most precipitation gauges undercatch because of the turbulence they cause in the airflow. Turbulence produced by the immediate environment may also result in a highly varied precipitation fallout pattern in the vicinity of the gauge. Extreme care is therefore necessary in placement and operation of gauges for accurate and representative measurements.

The intensity of airflow turbulence varies with the wind strength and the ruggedness of the area. Buoyant particles such as drizzle and snow are more readily affected than heavy raindrops. Because of turbulence, very exposed rain gauges may have catch errors of up to 50 per cent, or more, generally deficiencies. Court (1960) found the catch of two identical recording gauges, with eight-inch diameter orifices which were placed ten feet apart on a ridge, differed consistently in catch by 50 per cent of the smaller catch. Black (1954) estimated that the measured precipitation at Point Barrow, Alaska was under-measured by 200 to 400 per cent. His claims were based on the accumulation of snow on the ground, and the increased moisture in the soil mantle. This difference was accredited to wind strength which averaged 12 and 15 mph at two sites.

Wilson (1954) has plotted graphically the effects of wind on catch found by many authors. Average values taken from his material are given in Table II.7

Table II.7 Gauge Catch-Deficiency as Percentage of True Catch

Wind Speed (mph)	Type of Precipitation	
	Predominantly Rain	Predominantly Snow
0	0	0
5	6	20
10	15	37
15	26	47
25	41	60
50	50	73

Since wind increases with height above ground it is advantageous to keep the gauge orifice as low as possible. This is illustrated by Symon's observations which are given in Table II.8 as quoted by Kurtyka (1963).

Table II.8 Variation in Catch with Height

	Inches				Feet				
Height above ground ...	2	4	6	8	1	1.5	2.5	5.0	20.0
Catch in % of that at 1 foot	105	103	102	101	100	99.2	97.7	95.0	90.0

Clearly, the lower the orifice is to the ground surface the better. Pit gauges essentially reduce the orifice level to that of the ground; however, they are seldom practical for watershed use.

II.3.1.2 Shields

An alternative way to overcome the wind is through the use of shields. Two basic types of shields are in general use, rigid (Nipher) and flexible (Alter, Shasta). Shields

Table II.9 Average Daily Catches of Five Rain Gauges
by Storm Type, 1946 - 1961

Type of Gauge	Type and Number of Events		
	Rain, 1059	Snow, 267	Mixed, 32
	Average Precipitation Caught, inches		
Recorder	0.322	0.090	0.274
Standard	.321	.099	.331
Alter	.326	.133	.370
Nipher	.332	.146	.382
On post	.321	.076	.302
Grand averages	.324	.109	.332
	Differences between Gauges Required for Significance, inches		
5% level	0.003	0.006	0.017
1% level	0.004	0.008	0.023

Table II.10 Average Rain Catch of Five Different Rain Gauges
by Five Size Classes of Daily Rainfall

Type of Gauge	Size Class of Daily Rain, inches				
	0-0.49	0.5-0.99	1.0-1.49	1.5-1.99	$\geqslant 2.0$
	Number of Events				
	849	124	44	24	18
	Average Rain Catch by Gauges, inches				
Recorder	0.129(5)*	0.704(3)	1.207(2)	1.687(5)	2.798(1)
Standard	.132(3)	.696(4)	1.178(4)	1.695(3)	2.752(5)
Alter	.136(2)	.704(2)	1.189(3)	1.702(2)	2.762(3)
Nipher	.138(1)	.724(1)	1.209(1)	1.724(1)	2.767(2)
On post	.131(4)	.696(5)	1.174(5)	1.690(4)	2.754(4)
Grand averages	.133	.705	1.191	1.699	2.766
	Differences between Gauges Required for Significance, inches				
5% level	0.006	0.009	0.015	0.020	0.023
1% level	0.008	0.012	0.019	0.026	0.030

*Rank of average size of rain catch by gauge in each size class shown
by numerals in parentheses—(1) highest to (5) lowest.

improve catches, reducing the error in measuring snowfall by about 40 per cent. But even shielded gauges are not suited for catching snow at locations exposed to strong winds. Allis *et al*.(1963) found very little difference in the performance of the Nipher and Alter Shields—as noted in Tables II.9 and II.10.

Warnick (1953) has shown that a modified Alter shield is very effective in winds in which the velocity is less than 15 mph. More recently Weiss (1961) compared unshielded and shielded (Alter) snow catches using winds at orifice level. His results indicated the following:

Table II.11   Ratio of Unshielded to Shielded Gauge Catch

Wind Speed (mph)	Ratio
5	.86
10	.65
20	.42
30	.30
40	.23
50	.19

The Nipher funnel may fill with snow, thereby permitting capping, or chunks of snow may be blown into the gauge; it therefore requires attendance under winter conditions. The flexible Alter-type shield is used on unattended gauges to obviate this problem.

The Tret'iakov shield used in the Soviet Union has the shape of the Nipher shield, yet is flexible like the Alter shield. Weiss and Wilson (1957) reported it to be quite efficient for use with snow in winds having velocities up to 25 mph.

II.3.1.3 Orifice Effects

The diameter and attitude of the gauge orifice affect catch. Precipitation acquires a slanted trajectory because of the wind. If the trajectory of precipitating particles tends to become parallel to the orifice due to strong winds, the gauge may undercatch because of a rim effect. The catch deficiency decreases as diameter increases.

The variation of catch with gauge diameter found by Huff (1955) is shown in Table II.12. Variations in tube diameter, or slight errors in orifice diameter may cause large per cent errors in small-diameter gauge measurements. If precision is required it is wise to calibrate such gauges. Adherence of droplets to the inner wall of plastic tubes makes it difficult to obtain fine precision with some small-diameter tubes.

Gauges are usually mounted so that their orifices are level, since the depth of precipitation falling on a horizontal surface is conventionally used in climatological analyses. This value is inadequate for quantitative studies of precipitation on steep slopes, where it may be necessary to adjust for the trajectory of the raindrops. This may be done by simple geometry if the trajectory, i.e. slant of the falling rain, is known. In some areas where major storms occur with upslope winds, the raindrop trajectories tend to be perpendicular to the slope of the ground, a factor which has resulted in the use of stereo- or sloped-orifice gauges, i.e. gauges placed with their orifices parallel to the general slope of the ground.

Table II.12 Median Catch Ratios for Noted Storm Sizes and Wind Speeds

Gauges Compared With 8″ Diameter Gauge	Wind Speed (mph)	0.01-0.19	0.20-0.49	0.50-0.99	1.00-2.00	All Storms
Wedge 2.5″ x 3.8″ orifice	0-10	1.00	0.98	0.99	1.02	1.00
	11-20	1.00	0.99	0.98	0.99	0.99
Cylinder 3″ diameter orifice	0-10	1.00	1.00	1.00	1.00	1.00
	11-20	1.09	1.03	1.00	0.97	1.01
Cylinder 1.15″ diameter orifice	0-10	1.15	1.03	1.00	1.00	1.04
	11-20	1.25	1.02	0.96	0.97	1.02
Cylinder 0.8″ diameter orifice	0-10	1.30	1.09	1.05	1.08	1.13
	11-20	1.25	1.11	1.05	1.05	1.10

Stereo-gauges were used by Hamilton who found that for a watershed where major rainstorms were upslope in character, the vertically installed gauges caught 15 per cent less than stereo-gauges. The effects of both the stereo orifice and shield are clearly demonstrated by Helmers' (1950) observations presented in Table II.13.

The catch of stereo-gauges approaches the true value only when the previously described conditions are met, and therefore these instruments are of limited value. For example, Leonard (1962) found stereo-gauges caught less than vertically mounted gauges in an area where precipitation was more likely to be associated with downslope winds. Highly variable results were obtained at Wilson Creek watershed, Manitoba since upslope winds do not accompany all of the significant storms. Catch ratios for paired stereo and vertical gauges at Wilson Creek in 1963 and 1964 were 1.14, 1.11 and 1.09.

II.3.1.4 Evaporation From Gauges

Losses due to evaporation may introduce substantial errors when relatively large surfaces are exposed to the free air over long periods. Funnels restrict vapor flow and therefore reduce evaporation. Vented-plastic-bottle receptacles and oil may be used to minimize losses. Other devices used to reduce evaporation from precipitation gauges are highly-reflective gauges and those containing multiple inner containers to insulate the catch.

Many non-conventional gauges experience significant losses to evaporation because of the direct exposure of catch to the air. Gill (1960) found 0.05 to 0.30 in/day loss from wedge-shaped gauges with 0.30 in/day lost consistently when the gauge was read at the four-to-six-inch level. The 1.15-inch diameter gauge which has a relatively small exposed-catch-area lost 0.04 in/day when moderately full, and 3.57-inch diameter gauges have lost 0.60 in/week in midsummer when the funnel was filled with water.

Table II.13   Precipitation Measurements on Windswept Slopes

Type of Installation	Measured Precipitation in Inches	Comparison in %
Recording gauge with stereo-orifice and tilted Alter shield	62.53	100
Unshielded 8 in. x 24 in. stereo gauge	54.38	83
Recording gauge with horizontal orifice and Alter shield	40.09	61
Unshielded U.S.W.B. 8 in. x 24 in. gauge	27.36	42

II.3.1.5   Placement

The chief purpose of measurement is to obtain a sample that is representative of the precipitation falling over an area. Thus, gauges must be located to meet this requirement. The generally-accepted measurement is the depth to which the precipitation would cover a horizontal surface. To obtain this, the gauge orifice should be in a horizontal plane. The site should be selected so that the wind at the orifice level is as low as possible without resorting to using objects which block the catch or disturb the air flow near the gauge. The site should also be flat since slight rises in the ground, bushes, etc., may cause selective drop-out patterns for raindrops of different sizes. The height of the orifice should be low to minimize the effects of splashing or overtopping by snow. Splashing, which may reach four feet from a hard surface (Kurtyka 1963) may be reduced by the use of grass or a matted surface.

Gauges should be separated from nearby objects by a distance of at least four times their height. They should be sheltered from the full force of the wind by a permeable windbreak (World Meteorological Organization 1961). Brown and Peck (1962) defined a well-protected gauge as sheltered in all directions by objects subtending angles of 20 to 30 degrees from the gauge orifice, none being greater than 45 degrees, with the objects being of sufficient breadth to minimize eddy effects, e.g., an open area in a coniferous forest. Caution should be used in forests to avoid drip, shelter and snow shedding. The centre of small forest openings, where drifting does not occur are very suitable.

Care should be taken to ensure that observations are not lost by neglect. All instruments should be regularly serviced and calibrated. Leaky containers and distorted orifices should be replaced. Gauges must also be protected from human and animal interference.

II.3.2.   Types of Gauges

II.3.2.1   The Standard Rain Gauge

The standard rain gauge is cylindrical and has three components: a collector, a funnel and a receiver. The rim of the collector has a sharp edge which is bevelled on the outside, and falls away vertically on the inside. The collector is deep and the funnel has at least a 45-degree slope to prevent splash-loss. The receiver has a narrow neck, and is protected from radiation so as to minimize evaporation loss.

The Canadian gauge (Fig. II.22) has a 3.57-in. diameter orifice, and when mounted has its orifice at one foot above the ground. Its capacity is 4.5 in., a value which has been found to be too small for some hydrologic purposes. Consequently, redesign is being considered so that it will be able to hold most of the 24-hour-extreme rainfalls. Measurements are made by pouring the catch from the receiver into an etched glass graduate.

II.3.2.2   Storage or Total Precipitation Gauges

Storage gauges are used to measure seasonal precipitation in remote sites. The gauges have a large capacity, usually 60 to 100 ins., the amount depending on the climate of the area and the interval between readings. They have large, 8 to 12 in. orifices to minimize capping. The collector is placed on a stand at a height several feet above the maximum expected snow depth. Because of its elevation and exposure to the wind, the gauge should be shielded. Unbridled Alter shields are used. These are mounted with the ring level, and the leaves extending one-half inch above the rim of the orifice. Previously the slats of the shield were joined to each other by a yoke (bridled); this practice has been discontinued since it was conducive to capping.

An antifreeze charge, preferably ethylene glycol, is placed in the gauge to convert snow falling into the gauge to the liquid state; a small charge of oil, preferably transformer oil, is also added to suppress evaporation loss. Ice sometimes forms in the diluted surface of the antifreeze mixture. Inexpensive nitrogen-gas bubblers have proven successful in preventing ice formation.

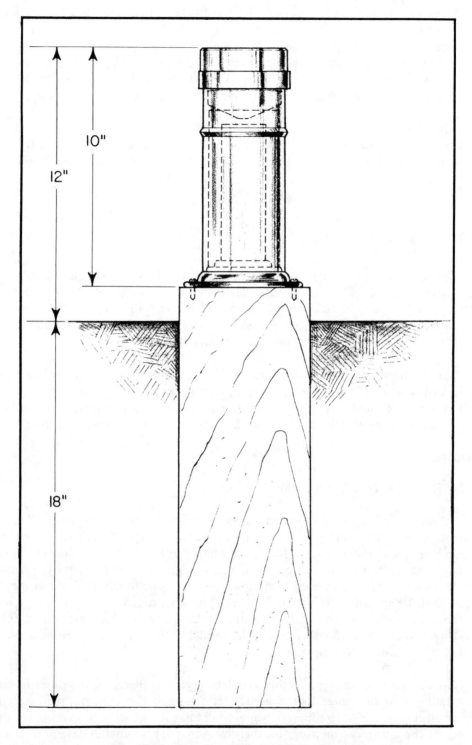

Fig. II.22 Canadian Standard Rain Gauge

Gauge readings are obtained with a ruler, or 'T' square type measure, or by weighing; also some storage gauges are equipped to give a continuous tracing, or for telemetering (see Fig. II.23). Readings must be adjusted to allow for the antifreeze and oil charges.

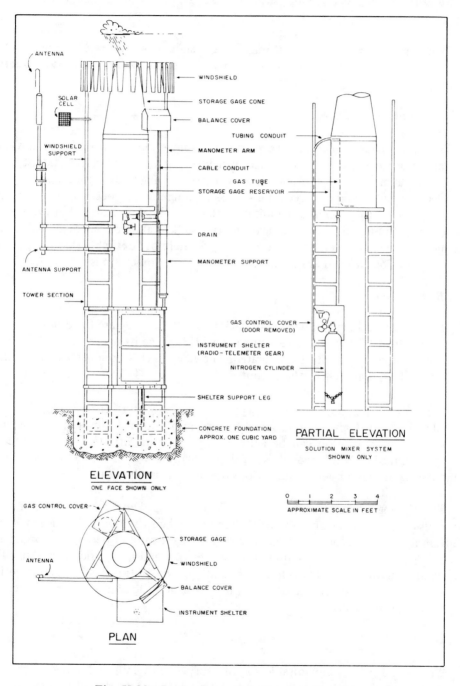

Fig. II.23   Radio-Reporting Storage Gauge

Two types of storage gauges are in common use. These are the "Sacramento" (which is a truncated cone) and the "Standpipe". Fibreglass standpipes of various heights are now being used extensively by the United States Weather Bureau. The conical section of the truncated cone gauges should subtend an angle of not more than 25 degrees with the vertical.

Where precision is required, storage gauges should be calibrated. Concurrent gauge-level wind measurements are advocated where the degree of exposure is of interest and comparability between gauges is desired.

II.3.2.3   Recording Gauges

Recording gauges in common use are the tipping-bucket gauge, the weigh-type gauge and gauges using a float type recorder. Weigh-type gauges and certain float-type gauges may be operated in all types of weather, whereas tipping-bucket gauges are used only in the measurement of liquid precipitation. Recording gauges, because of their bulk, exposure, etc., tend to catch less precipitation than the standard gauge. For example, Sharp et al., (1961) found that on the average the U.S. standard caught 5 per cent more precipitation than the weigh-type recorder. It is therefore desirable that the performance of these gauges be checked against a standard whenever precision is required.

Tipping Bucket Gauges

Two balanced buckets, which tip back and forth as they are filled in turn by rainfall directed to them by the collecting funnel, give this gauge its name. As the balance swings about its pivot, a relay contact is operated. The gauge is therefore particularly suitable for remoting and for digitizing of the output.

Commonly used models of these types of gauges require chart changes on a one or seven day basis, and give 5 min. and 20 min. resolutions. Extension of this time and improvement of resolution can be achieved by slight modification. The Meteorological Service of Canada (MSC) model is equipped with a heater which prevents ice from accumulating on the funnel during periods of freezing precipitation.

The tipping-bucket requires about 0.2 sec. to tip. During very intense rainfall this may cause significant errors (about 10 per cent when the rate is 10 in. per hr.). Other disadvantages of the gauge are related mainly to light rainfall because (a) one-hundredth of an inch of precipitation is required to tip a bucket and it is difficult to define the time of start and ending of light precipitation; and (b) evaporation losses in hot weather may be a relatively large percentage of light rainfalls. The gauge must be serviced regularly to ensure good performance.

Weigh Type

Weigh-type gauges are suitable for measuring both rain and snow. Because of the large exposed water surface area, an evaporation suppressant should be used. Most commercial gauges have a 20-in. capacity; they operate over periods from 6 hrs. to 17 mos., and have resolution varying from about 15 mins. to 2 days. Wind action may cause the receptacle to oscillate and the trace to be thick. This can be controlled somewhat by dampening devices.

In weigh-type gauges a spring is compressed as precipitation accumulates in a container mounted on top of the spring. This compression is converted to inches of a precipitation on a chart; it may also be converted to a change in electrical resistance, voltage, inductance, etc., so that the gauges may be readily adapted to telemetry. Digital output which gives recording intervals ranging from 5 min. to 12 hr. is available. Spring and power operated units are in use, but power is necessary for digital and telemetering equipment. Weighing accuracy of these units is to within ±0.1 in.

Float Type

Two types of float gauges are in use. In one type the rain from the collector is led to a chamber in which a float is used to sense the water level. A syphon is used to remove the rain from the chamber. This gauge is suitable only for the measurement of rainfall. A much more versatile type is the storage gauge modified as a long-duration recorder. This will take all types of precipitation, and can be maintained in freezing weather using antifreeze and bubbler attachment (Castle 1965)—see Fig. II.23. A stilling well is attached to the storage gauge, and a float is operated in the stilling well to obtain stage, or alternatively a manometer may be used. The float elevation can be recorded directly or telemetered as with river stage data. Gauges must be operated with the same precautions as specified for storage gauges.

II.3.2.4   Radar

Radar is being used with increasing frequency in the solution of hydrological problems. Water droplets and ice crystals cause a backscatter of radio waves, and may therefore be observed through radar. Hydrologists relying on reports from a sparse gauging network may be misled particularly in thundery weather. Radar can be used to delineate the area and relative intensity of storm activity over large areas (usually within 125 mi. depending on radar characteristics) and is therefore a very useful hydrometeorological instrument.

Attempts have been made to equate radar echoes into quantitative precipitation measurements. One method is to count the number of echoes observed over an area over a period of time and relate the count to measured precipitation. Photographic film has been used as an integrating device in another approach. Electronic area-echo integration techniques, where echoes from different drop sizes are received, are also being used. Attenuation resulting in under estimation is the principal problem to be overcome before radar can give satisfactory quantitative-precipitation values.

II.3.2.5   Throughfall

Standard installations show how much precipitation may reach the ground in the absence of obstructions such as trees. Trees intercept precipitation, some of which reaches the ground as stemflow or drip, and some of which evaporates. The evaporated portion may be effective in reducing normal evapotranspiration, nevertheless it does not reach the ground. The amount of throughfall varies with the precipitation type and duration; and the tree type, density, etc. The reported amount of stemflow is equally varied ranging from zero to about 75 per cent of the interception. This diversion of rainfall must be considered in interpreting rainfall data. Troughs fitted to tree trunks are used to obtain stemflow. Randomized gauge networks operated below the forest canopy and gauges operated at the top of the canopy and in adjacent clearings, are used to estimate the magnitude of interception.

II.3.3   Snow

II.3.3.1   Climatological Measurements

The general practice at climatological stations in Canada is to measure 24-hr. snow-fall with a ruler, and to obtain an estimate of precipitation amount by assuming the snow has a density of 0.1. By and large this assumption is fairly good, but in individual instances it may be appreciably in error. The density of freshly fallen snow has been found to range from 0.004 to 0.34. Investigations by Currie (1947) and Potter (1965b) indicate that an average density of 0.08 to 0.09 may be superior for the prairies. Average densities as found by Potter show a marked variation with climate as noted in Fig. II.24.

In the ideal case, ruler measurements are made at several points in an uniform, representative snowcover. To ensure separation of fresh snow from old snow, snow boards which are cleaned after a measurement are recommended. Drifting and blowing of snow make it very difficult to maintain a suitable measurement surface, and to get depth measurements which are truly representative of freshly-fallen snow.

II.3.3.2   The MSC Snow Gauge

The Meteorological Branch began to install shielded snow gauges at principal stations in 1953. These gauges consist of a hollow metal cylinder, 5-in. diameter, open at one end, and a shield called the "Nipher Shield". A drawing of the gauge and its mounting is given in Fig. II.25. The adjustment mechanism on the stand permits the gauge orifice to be maintained at 72 in. above the snow surface. Wind-tunnel tests have shown the shield to be effective in maintaining horizontal air flow over the orifice. Snow caught in these gauges, is melted and measured to obtain the water equivalent. Measurements are made at least daily, and usually every six hours.

II.3.3.3   Snowcover Water Equivalent

Snow surveying with sampling tubes is the conventional way of obtaining measure-ments of the water equivalent of the snowpack. Because of certain physical difficulties— such as the time consumed by and the cost of operating many of these surveys—there have been many attempts to develop alternative methods, particularly those which lend themselves to telemetering. These other methods generally can be used to obtain measurements of the rate of accumulation and ablation of the snowcover as well as the water equivalent. They may, therefore, provide much other information which is useful in design operations and research.

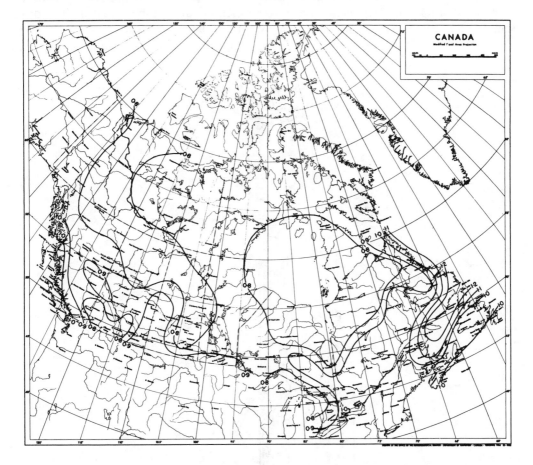

Fig. II.24   Ratio of the Measured Water Content of Snowfall to an Estimated Water Content Using 10 Inches of Snowfall as Equivalent to 1 Inch of Water

Snow Surveys

Snow survey measurements are generally used as an index of basin precipitation. Snow, because of its buoyancy, accumulates in a heterogeneous manner, so that it is difficult to obtain a truly representative measure. When data are used as indices, snow courses need not be representative. They may give superior results if located in large accumulation areas which contribute substantially to runoff, and which retain snow until survey time in most years. Unrepresentative sites are frequently the major or only areas contributing to runoff. If a relationship between snowpack and runoff is sought, such locations should be sampled. For example, the runoff from Prairie watersheds is often entirely from gully accumulation, and the snow disappears from the representative sampling points well before there is any measurable runoff through the stream-gauging station.

A snow course should be selected with the same care as a precipitation-gauging site. It should be sheltered, flat, well-drained, free of stumps and debris, on clean litter or soil. Sample points should be clearly marked, away from the radiative and other influences of trees, in areas of uniform accumulation. The course should be free of marked secular

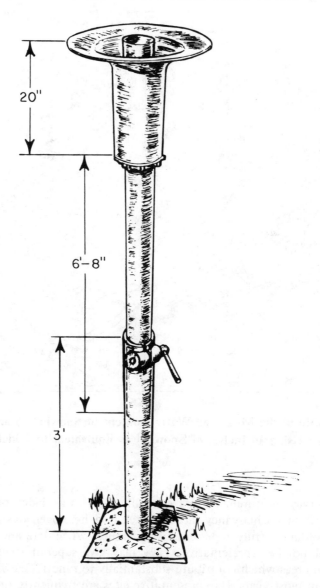

**Fig. II.25 Meteorological Service of Canada Standard Snow Gauge
with Shield and Adjustable Stand**

change such as tree growth or changing land use; and free of pollutants. Ready access is very important. Snow photography for the early melt period may prove very useful in selecting suitable locations.

Six good sampling points generally provide as good an index as a much larger number, however, it is advisable to start with a larger number, and then reduce to the best six. The points may, but need not be in a straight line, but should be 100 to 200 ft. apart. Stakes should be used to mark the sampling points. Natural grass provides a good surface cover. Cultivated fields are less satisfactory since the catch and retention characteristics of these fields vary with different cropping practices.

A large-diameter (3-in.) sampler (a cylindrical tube with a cutting edge) is preferred for use on the shallow snowpacks. Small diameter samplers, such as the Mount Rose (1.49 in.), are preferable in deep snowpacks (3 ft. or greater). Samples from very shallow packs cannot be accurately weighed directly, because they are usually combined with a high percentage of earth or litter. More precise values are obtained by subtracting the weight of the dried residue from that of the total sample.

There are systematic errors in most snow courses because of site selection. The forest-meadow areas used in mountain surveys, act as snow traps and tend to overcatch. Court (1963) found in the study of snow courses on the King's River Basin that the results were representative of average unbiased snowpack conditions at elevations about 2600 ft. higher than the sampled area; that is, the water equivalent was about 2 in. too high. Prairie courses are overexposed relative to climatological stations which collect about 50 per cent more snow.

The snow samplers in common use do not give true absolute values. A comparison made of the Mount Rose and a 3-in. diameter sampler gave the following results (McKay and Blackwell 1961) under Prairie conditions (see Table II.14).

Table II.14 Performance of Snow Samplers

Test No.	Sampler	No. of Samples	Av. Water Equivalent(in.)	% Difference from actual	Standard Deviation (in.)
1	Actual	1	4.98	0	—
	Mount Rose	18	5.40	+ 8	.28
	3-inch				
	Sampler	17	5.18	+ 4	.14
2	Actual	1	5.19	0	—
	Mount Rose	22	5.84	+ 12	.32
	3-inch				
	Sampler	22	5.57	+ 7	.27

Radioisotope Snow Gauge

Radioisotope snow gauges are now in use in France, Japan, Russia and the United States. Gamma radiation from radioactive cobalt and Geiger-Muller counter are usually employed. Attenuation of the emission from the source is related to the water equivalent of the accumulating snowpack. The units are costly and therefore used mainly in remote locations.

A 'twin-probe' Gamma-transmission gauge with a scintillation counter (Smith, Willen and Owen, 1965) provides detailed information not only on the mass of the snowpack, but also of its density and soil moisture detail. The 5mc Cesium 137 source and a Sodium Iodide detector are raised or lowered through the snowpack in twin probes. Gamma protons, emitted constantly by the source, are partly transmitted and partly back-scattered by collision with the electron field of the surrounding snow. The backscattered protons have less energy and a discriminatory amplifier circuit amplifies only high-energy impulses which are received directly. The amplified signal is therefore proportional to the density of the snow between the two probes.

A major advantage of these probes is the way in which backscatter is handled. Commonly used neutron and gamma gauges, which contain the source and counter as a single unit, are sensitive to backscatter. They measure the density of a sphere of 12 to 16-in. radius, and are not reliable when the snow-air interface lies within this sphere. The twin probe which is based on attenuation can be collimated to allow measurements of density of one-half-inch layers of snow. Profiling is possible right up to the snow surface and sampling times are 5 sec. per inch when one-inch increments are used.

Aerial readout from radiation sources is of particular value in remote isolated areas. Radiation is measured from an aircraft before the snowpack forms and periodically during the winter. A digital counter carried in the aircraft records emission intensity. The cobalt gamma radiation source lasts from 15 to 30 years.

Pressure Pillow

The pressure pillow gauge is essentially an air mattress, filled with an antifreeze solution, and fitted with a manometer. The weight of the overlying snow is indicated by the manometer. Pillows of various shapes and sizes have been used; the larger pillows—those 12 ft. in diameter—have given best performance.

Pillows are of interest for several reasons:
1. They are inexpensive and easy to operate.
2. Observations may be readily recorded or telemetered.
3. They are fairly sensitive; a 12-ft. pillow being able to sense snow rates as low as 0.03 in. per hour.
4. The response is almost instantaneous and the measurement probably is much more meaningful since it is the average value for a large area.

The pillow therefore offers considerable promise in recording the depth of snow, the rate of snowfall and the rate of snowmelt.

On the other hand, the pillow is unsatisfactory for areas with shallow packs; and pillows of different diameters have different response times. Thus calibration of individual units is necessary. The manometer readings may show a diurnal fluctuation due to thermal contraction and expansion of the antifreeze mixture and instrument. For example, Tarble (personal communication) in Sacramento, Calif., reported this fluctuation may be about 0.40 in.

II.3.3.4 Snow Depth Measurements

Measurements of snowcover mass or water equivalent are usually time consuming and costly. If assumptions can be made concerning the density of the cover, then it is possible to use less demanding depth measurements. Aerial snow surveys are now used extensively to do this. Depth measurements are made at many locations and density relationships are used to estimate the water equivalent. Usually measurements for accessible survey locations are used to provide a good aerial index of snow density.

Locally, it is frequently found that pack densities are quite uniform. The use of a few density samples combined with a large number of stakes, in such areas, provides a way of obtaining very useful volumetric estimates of the snowpack water equivalent.

Snow Stakes

Ground stakes are about 1.75-in. square and are graduated in inches. They are painted white and have black markings. Stakes are placed in accumulation areas where the catch is fairly uniform and representative. North sides of hills are preferred in hilly terrain. The area should be free of trees, buildings and obstacles which may unduly influence wind and melting. The ground should be clear of bushes and tall grass. Stakes should be mounted vertically with the zero mark at the ground or compressed-grass level.

Aerial Snow Markers

Aerial snow markers are usually established in locations which would serve as sites for ground snow surveys. The markers are made of durable metal; and a compromise is reached in cross-sectional area which can be read from the air or by telescope, and one which will not significantly influence the drifting of snow. One type of marker is made of 2.5-in. galvanized pipe with crossbars of 6 in. x 24 in. x 1/8 in. steel plate. The unit is painted black and set in concrete. Readings are made visually and verified photographically to a measurement accuracy of ± 6 in. Usually, the best accuracy in measurement can be obtained by this method when the marker casts a shadow. Aircraft access (and departure) must be considered in planning such a network. Aerial photography and infield inspection are of great value in determining suitable locations of snow markers.

II.3.3.5 Photographic Methods

Photography has provided valuable assistance in assessing the extent and volume of snowcover. Areal cover is readily discernible from standard photography, and volumes may be estimated using photogrammetry. Color infrared photography and photography using different portions of the visible light spectrum offer further promise in distinguishing types of snow and the presence of melt water.

The areal extent of snow cover has been used in several stream-flow forecasting routines. Sequential photography in mountainous areas shows the steady shrinkage of the covered area as the melt season continues. Since the shrinkage is related to temperature, the snowline during this period tends to be at a single elevation. Both ground observations and photographs and aerial photography may be used to document the recession or advance of cover. Skill has been developed using pictures from very high-level aircraft; the value of satellite photography to hydrologic investigations is now being appraised.

Photogrammetry

Photogrammetry provides a means of depth sampling at an infinite number of points. In practice, sampling is generally performed on a grid basis. Using photogrammetric methods, the snow depth is obtained by subtracting the elevation of the ground, determined by photography obtained without snow cover, from that obtained for the snow surface. Recent well controlled experiments have indicated measurements of snow accumulation could be made within accuracies of ±0.5 ft., on relatively flat ground and ±2 ft. on steeply sloping terrain. The accuracy of the method is dependent on photographic procedures, the photogrammetric plotter, the photographic scale, topography and other factors. Errors as noted above may be decreased as improved equipment and technology become available.

The potential of the method may be appreciated by considering a model based on 1:6,000 scale photography, covering an area of 90 acres. Depths may be measured at 2,000 grid points on this model, and the data treated by computer. The cost of physically sampling this number of points by ground survey would be prohibitive. By contouring with a plotter, and by comparing differences in elevation, a much larger number of points are considered in the analysis; but this analysis is somewhat cumbersome.

Photogrammetric measurements are not possible where there is dense, uniform forest. Weather and rugged terrain may also restrict its use. In addition, density measurements must be obtained when the total water equivalent is desired.

II.3.4   Networks

The character of precipitation varies not only with each storm and with topography, but also with the general climate. Convective showers are common to the interior of a continent, but the dominant rainfall along west coasts of continents is orographic, and therefore of different character to that found in the interior.

The torrential rains from aging hurricanes are characteristic of the eastern provinces and are not experienced on the Prairies. To be informed on the varied character of precipitation, networks of stations are required. The nature of each required network depends on the phenomenon being observed, the purpose of the observations, economics, and other factors.

II.3.4.1   National Networks

National networks are intended for broad-scale, general interpretation to service a highly varied need. Their development is influenced by economics, observer availability and access. Many of the areas of interest to hydrologists are sparsely settled, have few qualified observers, and limited access; consequently the national network is not likely to satisfy the intensive gauging required for many hydrologic problems. National networks do, however, provide general information of climatic statistics and long-term records which are essential for design and planning. These records are valuable for the planning of a more intensive network. If there are no data directly available or applicable for the study area, records obtained from adjacent, climatically-similar locations may provide the answers.

At the beginning of 1965 there were about 2,000 precipitation stations in the Canadian network, of which approximately 750 were in the Prairie provinces. Fifty-four of these latter stations had recording rain gauges. Precipitation records from these stations are published by the Meteorological Branch in the "Monthly Record of Meteorological Observations".

Private companies operating in the Prairies maintain about 1,300 additional precipitation stations.

Network densities, where sampling locations are representative, and the elements randomly distributed, may be determined using sampling-error theory. The standard error of the mean is, under these circumstances, inversely proportional to the square root of the sample size or $1/\sqrt{N}$. Over a period of time, the mean error is inversely proportional to the square root of the length of record or $1/\sqrt{t}$ (t = years). However, network problems are seldom easily solved. Sampling sites are not always representative, nor are the elements always randomly distributed. Also, records are used for purposes other than the determination of the mean; for example, extremes, trends, etc. Landsberg and Jacobs (1951) have recommended the following lengths of periods-of-record to obtain a stable frequency distribution:

Mountains — 50 years
Plains — 40 years
Shore — 30 years

In Canada, the attempt has generally been to achieve a density with about a 15-mi. gauge-separation for standard precipitation gauges.

The automatic-rain-gauge network provides information for sewer and drainage design, as well as for the analysis of major rainstorms. A network density of less than

50-mi. separation is desirable for storm analysis. Special networks are required for sewer design in cities of over 50,000 population. It is current practice to place a gauge in cities of 10,000 or greater population.

II.3.4.2   Mountain Networks

The heterogeneous topography in mountainous areas results in an equally hetero-geneous rainfall pattern. Gauge separations considered adequate for the plains are not adequate in the mountains. The only way to truly identify the precipitation character in rugged terrain is to cover the whole area with gauges; this is of course neither feasible nor practical. The alternative is to determine the physical relationships between precipitation and topography and to interpret these throughout regions of relatively similar climate. This is achieved by defining climatic zones, and by operating chains of weather stations in each zone. Sampling stations should be located in valleys, along slopes and on ridges. The chains contain both 'permanent' type stations and satellite stations which may be moved when they are found to correlate highly with the permanent stations. By judicious use of these stations the major features of climate may be defined with a minimum of expend-iture.

II.3.4.3   Watershed Networks

Watershed networks differ from national networks in that they are intensive whereas the national network is extensive. They should be carefully planned to ensure that all required measurements are taken. The initial stages of planning should consist of a liter-ature survey and problem analysis. Procedures may be devised and tried to see what measurements are needed, and what accuracy and definition are required in order to obtain the desired results.

An analysis of past weather and runoff records for the area, or similar areas, may provide guidance information concerning the type of instruments to be used, and how they should be located. This may obviate such errors as placing the precipitation gauge orifice below the average maximum height of the snowpack. It may aid in operational problems such as determining the capacity required for the gauge and the antifreeze charge needed for a storage precipitation gauge.

Aerial photography, topographic charts, vegetative cover and other data should be reviewed along with the climatic and hydrometric records. These may provide some indication of measurement requirements, possible gauging sites and access. Access merits special consideration since a gauge which cannot be serviced provides little information.

The number, type and placement of gauges depend on the analytic approach, climatic stress, economics, and access. When the data are used in index form the network requirements are minimized, but it may be necessary to run the experiment for long periods to establish significant results. When data are used as 'absolute estimates' the analysis period is usually shortened, but much more precision is required to ensure that the measurements are truly representative. The density of a network is determined mainly by the variability of precipitation and the research objectives. Long-duration and large-area precipitation is much less variable than short-duration and small-area precipitation; therefore, the density requirements for the latter cases are much more stringent. For example, it was found for the 1,130 sq. mi. Chickasha watershed that at the 5 per cent significance level, 10 gauges were as good as 158 in assessing mean-daily areal rainfall. However, over subnetworks of 61 and 208 sq. mi., the greater density was barely adequate at the 1 per cent level. Not only was the reduced network of questionable value for the smaller basins, but it was inadequate for other objectives such as sampling peak storm-rainfall, and consequently in developing fine-scale depth-area-duration curves (Nicks 1965).

At the Reynolds Creek experimental watershed, the actual network density is one recording gauge for each of the 93 sq. mi. of area of the basin. As the gauge density is decreased, the accuracy of estimating the areal mean is likewise decreased. For example, Neff (1965) found the estimates to have the following 95 per cent confidence limits:

No. of gauges	95% Confidence Limits
90	± 0.8 in.
30	± 1.4 in.
10	± 2.7 in.
4	± 5.2 in.

These results and those given in the preceding paragraph indicate the need of stating the research objectives when designing a network.

II.3.4.4   Approaches to Watershed Network Design

Generally, data from watershed networks will be used either as indices, or as absolute values for some water balance study. As indicated previously when only indices are needed, network requirements are minimized, but it may be necessary to run the experiment for a long period to establish significant results. When data are used as absolute estimates the analysis period is usually shortened, but much more precision is required to ensure that the measurements are truly representative.

The ideal network cannot be properly planned, however, until after an initial period of measurement and inventory. A preliminary network must first be operated to determine the areal variation within the watershed. Also, where soils, vegetation and physiography are being considered, the final design cannot be completed until other pertinent surveys are adequately advanced.

The Index Approach

The index approach is based on correlation. Sensors should be exposed so that their measurements have the highest possible correlation with the effects which are being studied.

The number and location of gauges used in an index-type network is dependent on the complexity of the cause-and-effect relationship and the degree of association which is required. Generally, at least one gauge should be placed in each homogeneous contributing-area. Also the network should provide indices for sub-watersheds which may be used for experimentation. It should be remembered that there is no good purpose served in attempting to develop a correlation procedure beyond the accuracy of the streamflow measurements, or other measurements which the meteorological records are being compared with.

Saturation

Saturation of a watershed with instruments may permit a measurement which approaches the true value and which is frequently called the 'absolute estimate'. Attempts have been made to effectively saturate a basin with gauges; for example, on a watershed at San Dimas, California, (area 17,000 acres) 322 gauges were used. Fortunately, it is usually possible with time, to scale down the saturation network to a small number of gauges which predict the 'absolute estimate' within the desired confidence limits. At San Dimas it was found that a minimum network of 21 gauges satisfactorily would serve the project requirements.

Transposition

If a 'true value' is given by a dense network, then it is possible to estimate the error resulting from the use of a less dense network. A dense network is unlikely to exist at the start of a project; however, if it may be assumed that the rainfall characteristics for a proposed basin are similar to those of other densely gauged basins, then it is justifiable to design the network on the experience of the densely gauged watersheds. Essentially, this type of design is based on transposition of a determined model to the proposed site.

McGuinness (1963) has computed the error resulting from using a lesser network for Coshocton, Ohio as follows:

$$E = 0.03P^{0.54} G^{0.24} \dots\dots\dots\dots\dots\dots\dots\dots\dots\dots\dots\dots\dots\dots\dots\text{II.2}$$

where

E = the absolute difference in inches,
P = the rainfall in inches for the "true" network, and
G = the gauging network density in square miles per gauge
for the reduced network.

This equation provides a means of estimating the accuracy of a proposed network. It was developed using data from watersheds of less than 25 sq. mi. in area but was shown to be consistent with relationships for larger areas. As it is based on average Ohio rainfall conditions, it may not be suitable for unusual thunderstorm-type rainfall; also, its use in Canada beyond the Great Lakes would be highly speculative.

The average spacing between gauges required to obtain a correlation of r = 0.9

between gauges for storm rainfall events may be estimated from the 24-hr. and the 1-hr.-2-year-return-period rainfall. These latter values may be computed from local precipitation records. Estimation procedures developed by Hershfield (1965) using data collected from 15 widely separated American watersheds are given in Fig. II.26. This method uses local rainfall characteristics, and should therefore provide reasonable estimates for most Canadian watersheds.

Holtan *et al.* (1962) have recommended the following densities as minimum standards for agricultural watersheds:

Table II.15   Number of Rainfall Stations Required

Size of Drainage Area (acres)	Gauging Ratio (mi² /gauge)	Minimum Number of Stations
0-30	.05	1
30-100	.08	2
100-200	.10	3
200-500	.16	1 per 100 acres
500-2500	.4	1 per 250 acres
2500-5000	1	1 per square mile
over 5000	3	1 per each 3 square miles

These minima are based on common practice. Special conditions, such as localization of storms, would impose higher minima on the concentration of gauges.

Unit Source Areas

When watersheds are very large, complete network coverage may be impractical; and 'unit source area' procedures are sometimes used to obtain 'absolute estimates'. Unit source areas are essentially small watersheds which are representative of larger, hydrometeorologically homogeneous areas. These areas are densely gauged and their performance is used to estimate the performance of the whole basin. It is, therefore, necessary to maintain a reduced general watershed network as well. Detailed inventories of soils, vegetation, topography, etc., are necessary for the selection of 'unit source areas' and subsequent prediction.

Stratified Random Sampling

Areal estimates of variable parameters such as snowfall can be obtained by stratified random sampling procedures. Such methods are particularly useful when 'unit source area' concepts are applied. According to Ezekiel and Fox (1963):

Where the items are so selected as to represent different portions of the universe, it may be called a "stratified sample"; where they are all selected from one portion of the universe, it may be called a "spot" sample. Where the universe is not completely uniform, a stratified sample tends to be more reliable than a random sample, and a spot sample tends to be less reliable than a random sample.

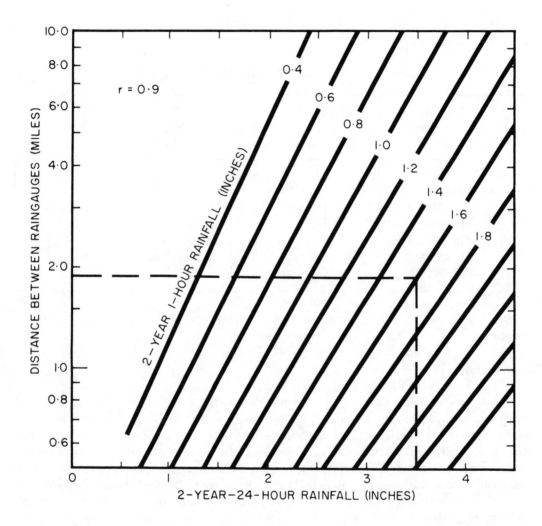

Fig. II.26 Diagram for Estimating the Distance Between Gauges as a Function of the 2-Year-24-Hour and 2-Year-1-Hour Rainfall (After Hershfield)

The values of a random sample must be so selected that (a) there is no relation between the size of successive observations; (b) successive items are not definitely selected from different portions of the universe in regular order, but are selected in random fashion; and (c) the samples are not picked all from one portion of the universe, but are scattered throughout the universe by chance selection.

Within an area which is hydrologically homogeneous, (i.e. the snow depth and density are relatively uniform in the horizontal plane and systematic sampling errors are unlikely), the distribution of values is normal and therefore the true mean lies between $\overline{P} \pm (s_p/\sqrt{N})$ t where N is the number of samples, $\overline{P}$ is the observed sample mean, s_p is the standard deviation $= \sqrt{\sum_{i=1}^{N}(P_i - \overline{P})^2/(N-1)}$

P_i is precipitation recorded at an individual station and is a distribution which is dependent on N, (Students 't'), the values of which are available in tabular form. For the true mean to lie within $\pm\Delta\overline{P}$ inches of the sample mean

$$(s_p/\sqrt{N})\ t \lesssim \Delta\overline{P} \quad \ldots\ldots\ldots\ldots\ldots\ldots\ldots\ldots\ldots\ldots \text{II.3}$$

Using tables for t, the number of observations required for any value of $\overline{P}$ may be determined. Munn (1953) provides an example of this calculation. If 20 snow depth measurements have a standard deviation of 2.0 in., then the mean snow depth has a 95 per cent probability of being within 1 in. of the true value. In this case N = 20, s_p = 2.0, $\Delta\overline{P} \lesssim \pm 1$. The relationship of these values for the 95 per cent fiducial limits are given in Fig. II.27.

Unit source areas are hydrologically homogeneous and therefore suitable for this type of measurement. Within these areas unbiased sampling can usually be achieved by using a predetermined network grid. The number of sampling areas may become substantial if stratification is made according to all physiographic, and vegetative features (eg. elevation, slope, aspect, etc.).

Gauge Spacing

When absolute estimates are required a uniform grid network has generally been found superior. However, the grid spacing should be shortened across zones of unusually intense climatic stress, such as may result from uniformity of storm tracks, changes in elevation, etc. For unit-source-area studies, and for rugged terrain the general network grid should be integrated with special networks. This will facilitate the application of source area research results.

An alternate to the grid approach for networks, used in rugged terrain is to sample each area of hydrologic homogeneity; that is, areas which are uniform with respect to vegetation, climate and topography. Gauging density should be sufficient to adequately sample the areal variability of precipitation.

II.3.4.5   Reappraisal

Reappraisal is essential to determine:
(1) the adequacy and possible refinement of the network and observation program,
(2) the time required to achieve experimental significance,
(3) the possible need for restating or redefining objectives, and
(4) the presence of faulty gauges, poor instrument locations and exposures.

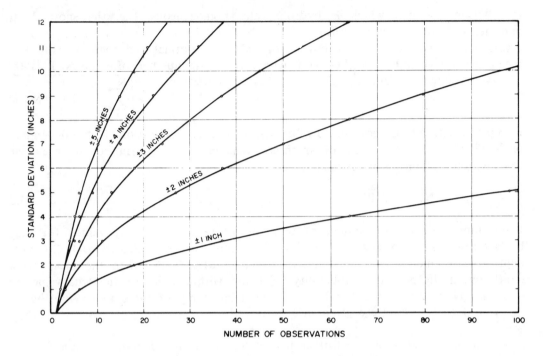

Fig. II.27 95th Percentile Chart for Deviation from the Mean

The network density should be adequate for problem solution. For example, in the study of erosion at Wilson Creek, Manitoba, it was concluded that the erosion was the result of summer thunderstorms of one inch or more rainfall. The network was designed to have an average error of eight per cent for one-in.-24-hr. storms, and 5 per cent for five-in.-24-hr. storms. These levels were considered adequate for problem analysis giving consideration to the runoff coefficient, streamflow measurement accuracy, and the frequency of storms of this magnitude.

The analysis of individual storm data may show sufficiently high correlations within a few seasons to permit reduction of a large network to just a few gauges. Day-to-day comparisons of other measurements may disclose redundancy or the need for further instrumentation.

Hamilton (1958), referring to the San Dimas Watershed, indicates the value of reappraisal:

> This more operational arrangement of 77 gauges instead of 322 was still costly in time and money. To effect further economies, available data were examined and we tested the idea that we might be able to select from many gauges, one gauge in each watershed which would integrate watershed rainfall . . . Accordingly, a so-called 'minimum' raingauge network was selected . . . Preliminary comparisons of measurements

from the 21 gauge network and the accredited 77 gauge network . . . agreed within 3 per cent.

Where detail on storm rainfall is essential, the adequacy of the network may be determined from inter-gauge correlation. The representativeness of a particular gauge is indicated by the size of the area over which its catch is found to correlate highly. The effectiveness of each gauge in this regard can be determined by correlation, and the network may then be adjusted to suit project requirements.

In the reappraisal of the Wilson Creek Watershed network, an analysis of storm records disclosed that the correlation was lower between catches of rain gauges on the slope than on level terrain at the top and bottom of the watershed. It was also found that, approximately, the standard error of estimate, between gauges was in direct proportion to the distance between them (See Fig. II.28). This relationship was used to define the area about a gauge with which the precipitation could be predicted with a known average standard error of estimate, as shown in Fig II.29. This analysis shows both areas requiring gauges, and gauges which are redundant or not contributing significantly to the study. An analysis of storms of design intensity would have been preferable for the reappraisal; however, this was not feasible because of the paucity of data available at that time.

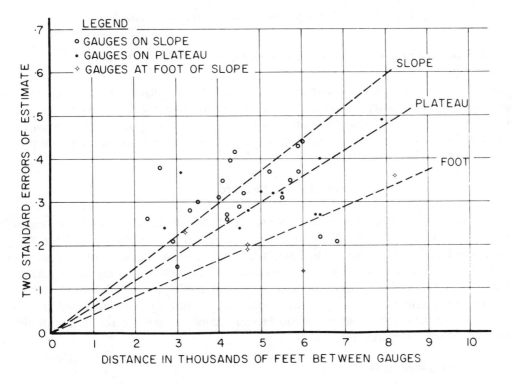

Fig. II.28 **Standard Error of Estimate $s_{\bar{p}}$ vs Gauge Separation**

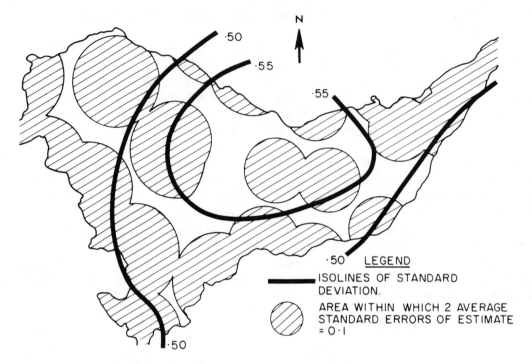

Fig. II.29 Network Reappraisal, Wilson Creek Watershed

II.4   THE ANALYSIS AND PRESENTATION OF PRECIPITATION DATA

Hydrologists' interest in precipitation data relates to engineering design, water supply and river forecasting, watershed management and research. The data are usually subjected to analysis of frequency characteristics, extremes, regression and physical relationships. Some of the methods and problems relating to data interpretation and processing are discussed in this Section.

II.4.1   Climatological Records

The quality and completeness of records should be considered in analysis. Homogeneous, complete series are required for many studies, but are not always available. The observer, the location and exposure of instruments, the type and shielding of instruments may change from time-to-time, thereby necessitating checks on homogeneity. Observational errors are probably less common in current publications, because of present quality control methods; however, historic data may require close scrutiny and adjustment. Voids continue to appear in climatological records because of the nature of the co-operative observation program.

II.4.1.1   Estimating Missing Data

Missing precipitation data may be estimated from values measured at about three other locations, near and evenly spaced around the station with the missing record. If the normal period precipitation at the stations used for reconstruction is within 10 per cent of that with the missing record, a simple arithmetic average provides an estimate. Otherwise the 'normal ratio' method is used:

$$P_4 = 1/3 \left[\frac{N_4 P_1}{N_1} + \frac{N_4 P_2}{N_2} + \frac{N_4 P_3}{N_3} \right] \quad \ldots \ldots \ldots \ldots \ldots \ldots \ldots \ldots \ldots \ldots \text{.II.4}$$

Where P_4 is the station for which the missing precipitation depth is required; $P_1, P_2, P_3,$ is the precipitation recorded during the period of study at different stations; and $N_1, N_2,$ N_3 and N_4 is the long-term normal precipitation of the different stations.

The above procedure is satisfactory when there is a very high correlation between precipitation for the considered sites. It is unsatisfactory when this is not the case, such as may occur for single, convective-type storms. Also a single nearby station may give a superior estimate to the average of three, two of which are several hundred miles distant. Under conditions of highly variable rainfall and incomplete network spacing, it is preferable to sketch a simple isohyetal chart from which the missing values may be estimated. For estimates of rainfall during short durations, interpolation between mass curves of rainfall will give a reasonable estimate.

Many floods have occurred on small streams when the network stations have not revealed significant precipitation. Considering the spacing of gauges in the general network, it is quite likely that some storms and the peak-storm rainfall will not have been recorded. This fact should be kept in mind in interpreting reconstructed rainfall values.

Monthly precipitation totals are sometimes estimated from seasonal values obtained from storage-gauge records. This is usually done by prorating the monthly precipitation record obtained from nearby, daily-reporting stations. Caution should be used in this procedure to ensure that the considered sites do have similar precipitation regimes. For example, the precipitation regime varies rapidly when proceeding eastward from the Continental Divide to the plains, as noted in Fig. II.30; consequently prorating records from foothill stations may not give valid estimates of monthly precipitation for mountain sites.

II.4.1.2   Time Adjustments

The times at which observations are taken are very varied. If this fact is ignored it may lead to reconstruction and analysis errors. When daily or storm values are being considered, the times of beginning, ending and measurement should be carefully noted. Climatological records list the precipitation which occurred over the 24-hr. period preceding the time of observation. A one-hr. rainstorm bridging the observation time would, if the observer were precise, be accredited to two adjacent days, and in the absence of

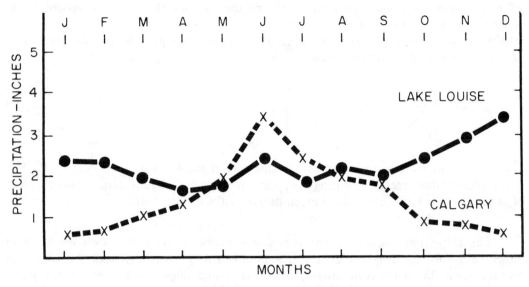

Fig. II.30 Monthly Precipitation

other data the analyst would have to consider the possibility of it being a 48-hr. rain. The observer may have a good weather eye and wait the storm out, in which case the rainfall would be accredited only to the first day, but the analyst would make a slight error in interpreting the time.

It has been shown that on the average, the peak 24-hr. precipitation in a storm should be increased by a factor of 1.13 to obtain a superior estimate of 24-'clock-hour' day. This adjustment cannot be applied indiscriminately to an individual event. The following equation has been shown to be useful in estimating extremes of 24-hr. (1,440-min.) rainfall from climatological records:

$$P_{24} = \text{maximum climatological-day P} + 1/2 \text{ the}$$
$$\text{maximum adjacent-climatological-day P} \quad \dots\dots\dots\dots\dots\dots \text{II.5}$$

where P_{24} is the 24-hr. precipitation amount.

II.4.1.3 Publication Hazards

The exact location of the observation station should be determined. This may not always agree with the user's conception of the gauge location since in some instances stations are named according to a distant post-office address. Lundbreck, Alta., for example, is a gauging station about 25 miles north of the town; similarly, Manyberries refers to a range-experiment station about 20 miles southeast of the town of Manyberries, Alta.

Some published figures are in error, and these errors are difficult to identify and correct. Most original records are retained in archives, and if necessary, reference can be made to them. A failing in some older records is in the publication of precipitation amounts accumulated over several days, as having occurred on a single day. Modern data processing, with quality control, is minimizing the possibility of these types of errors and other hazards.

II.4.1.4   Double Mass Curves

Double-mass curves are used as a check on the consistency of precipitation records. A substantial change in the relative catch of precipitation may result from change in observer, location, observation procedure and exposure such as caused by tree growth or construction. This is particularly true in rugged terrain, but perhaps not quite as serious within the Prairies if exposure standards are maintained.

Consistency checks should begin with a review of the station's history, then by a double-mass comparison with adjacent stations whose history has also been checked—preferably stations with perfectly homogeneous values for the station under review against the mean of the accumulated precipitation for the surrounding check stations.

An example of a double-mass-curve analysis is given in Fig. II.31 (Curry and Mann 1965). Here a storage gauge was moved in 1959 to a location with a superior exposure. The improved catch obtained by locating the gauge in a more sheltered location is apparent. Earlier records, considered deficient because the gauge was previously overexposed to the wind, were adjusted upward by the ratio 1.33 (see adjustment ratio in Fig. II.31).

Caution should be exercised in using this technique. Changes in slope should be accepted only when very marked, or substantiated by historical evidence of a change and continuing over a period of 5 yrs. The natural variability of precipitation may cause changes of the same order of magnitude as those arising from the previously mentioned factors.

II.4.1.5   Mass Curves of Rainfall

Certain periods in a rainstorm's life are of much more interest to the hydrologist than others; it is desirable to segment the storm, and analyze only the critical portions. To define what periods of a storm should be analyzed and to obtain the maximum of useful information from storm-rainfall records, it is customary to prepare mass curves, examples of which are given in Fig. II.32.

Mass curves are equivalent to recording-rain-gauge traces. One of those shown in Fig. II.32 is a direct transfer of data from a recording gauge, the others are based on a wide variety of fragmentary data. Newspaper reports, bucket surveys, streamflow data, weather charts and data from other sources are used to prepare these curves for all stations which recorded the total storm rainfall. The mass curves which are 'reconstructed' are drawn to pass through plotted values of accumulated precipitation, and to fit characteristics

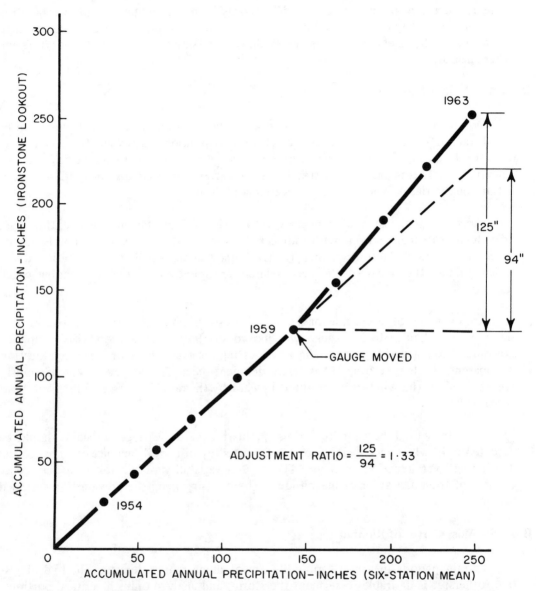

Fig. II.31 Double-Mass Curve for Ironstone Lookout Storage Precipitation Gauge

determined by isochronal and other analysis. Isochrones of the beginning, ending, and other major changes in rainfall intensity may be constructed from data contained in original climatological records, and from recording-rain-gauge traces.

Mass curves are the basis of storm-rainfall analysis. Incremental values of rainfall for specified time intervals (e.g. 1, 3, 6 hr., etc.) extracted from mass curves, are used to compute maximum average depth-area-duration curves and are also useful in preparing hyetographs. In these analyses the extreme rates of rainfall are given full weight; thus, the

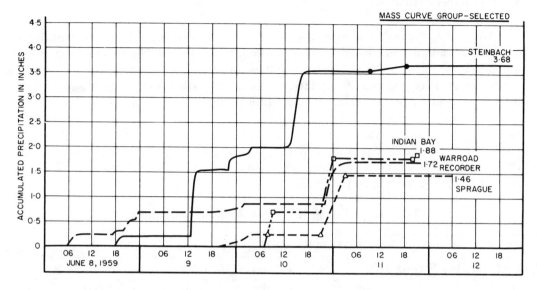

Fig. II.32   Mass Curves for Steinbach (Manitoba) Area, June 8-12, 1959

incremental time interval should encompass the period of most intense rainfall. Since this period varies throughout the area of the storm, and also since there may be several periods of intense rainfall, interval selection is sometimes difficult. Usually the most suitable interval is obtained by scanning those curves experiencing the heaviest total storm rainfall with a transparent overlay.

II.4.2   Mean Precipitation Over an Area

When precipitation is relatively uniform and measurements are representative, the arithmetical average of several values may provide a satisfactory estimate of the areal mean rainfall amount. However, the effects of rugged terrain, and the natural variability of rainfall make it necessary to resort to more sophisticated solutions for many problems.

II.4.2.1   The Thiessen Polygon Method

This method is superior to the simple average calculation since it adjusts for the non-uniform distribution of gauges, by weighting each observed value. The procedure is as follows. The stations are located on a map, and connecting lines drawn. Perpendicular bisectors of these connecting lines are drawn to form a polygon about each station. To determine the mean, the value for each station is multiplied by the area of its polygon, and the sum of these products is divided by the total area. Where the network is fixed, this procedure is fairly convenient and can readily be handled by machine methods. However, when the network is variable, the method may be laborious. Linear variation between stations is assumed in this method. It therefore does not permit the analyst to use his skill in interpreting topographic effects on precipitation, runoff or other useful records.

II.4.2.2   Isohyetal Analysis

Generally, the isohyetal approach is the most preferable method of computing the areal mean rainfall, since it meets the objections noted for the Thiessen method. Prior to drawing isohyetal maps, all records should be checked for consistency, and a specific period selected for analysis. Precipitation accumulations for this period are then plotted on a map, preferably one which contains topographic detail. Isohyets, or lines of equal rainfall amount are then placed on the map to fit the plotted data along with topographic relationships, storm character, observed runoff and other information available to the analyst. The average precipitation is determined by measuring the area between isohyets, by multiplying this by the average precipitation between isohyets, and then by dividing the sum of these products by the total area.

Table II.16 Average Depth of Rainfall Computation

STORM PR-68 JUN. 8-11 1959										
Plan No. 15192 Setting 3178					Coeff. 170/625 mi²				681	
Line No.	Rainfall Centre or Zone	Isohyet Inches	Area Enclosed		Net Area in Sq. Mi.	Average Depth of Rainfall in In.	Volume of Rainfall in Inch/Sq. Mi.		Average Depth of Rainfall in In. (Col. 9 - Col. 5)	Check
			Planimeter Reading	Area in Sq. Mi.			Increment (Col. 6 x Col. 7)	Accumulative		
1	2	3	4	5	6	7	8	9	10	11
	A	7	27	99	99	7.1	702.9	702.9	7.1	√
		6	65	239	140	6.5	910.0	1612.9	6.7	√
		5	149	548	309	5.5	1699.5	3312.4	6.0	√
		4	270	993	445	4.5	2002.5	5314.9	5.4	√
		3	444	1632	639	3.5	2236.5	7551.4	4.6	√
		2	1123	4129	2479	2.5	6242.5	13793.9	3.3	√
	B	3	35	129	129	3.1	399.9	399.9	3.1	√
		2	422	1551	1422	2.4	3412.8	3812.7	2.5	√
	A + B	3		1791				7951.3	4.4	√
		2		5680				17606.6	3.1	√

A computation of average rainfall is given in Table II.16. Examples of the Thiessen Polygon and Isohyetal Analysis are given in Fig. II.33.

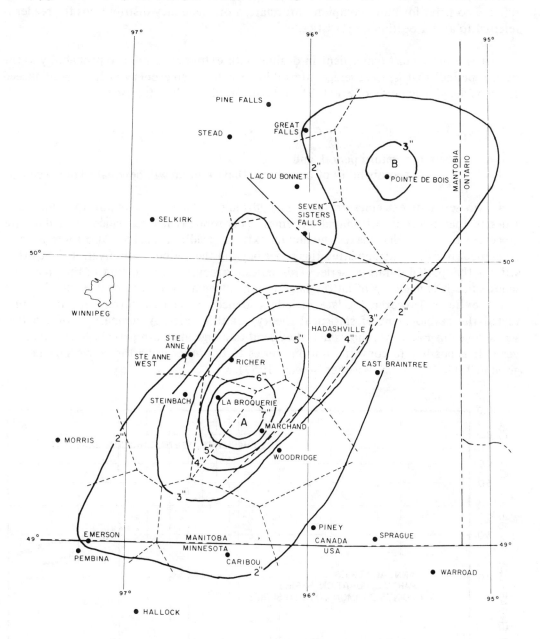

Fig. II.33 Isohyetal Chart Showing Zones and Polygons

II.4.3   Statistical Analysis

Meteorological series commonly used in hydrologic studies deal with both routine observations and extremes. The extreme values such as the greatest 24-hr. rainfall in a specific year are extensively used in design problems. The analysis of extremes is briefly referred to here; for more complete information on frequency distributions the reader is referred to a later Section on statistics.

It is usually most convenient in dealing with extremes to refer to probability as the 'return period' that is, the average interval in years between events which equal or exceed the considered magnitude of event. The return period may be expressed:

$$t_r = \frac{1}{p} \quad \cdots \cdots \cdots \cdots \cdots \cdots \cdots \cdots \cdots \cdots \cdots \quad \text{II.6}$$

where t_r = return period, and
p = probability of occurrence that the event will be equalled or exceeded.

Two series of extremes are used in hydrology. The most common extreme value series is that made up of annual extremes. This series is favored since most data are processed in a way that makes the annual extreme readily available. Also there is good theoretical basis for extrapolating this series beyond the period of observation. The other series is the 'partial duration' series. This series overcomes an objection to the preceding series since it is made up of all the large events above a given base; that is, not just the annual extreme. However, the lack of independence between concurrent events has prevented the development of statistical theory for this series. A comparison of the two series, given in Fig. II.34, shows that they tend to merge at return periods greater than 10 years. It is possible to convert from one series to the other using the conversion factors given in Table II.17 (U.S. Weather Bureau, 1957).

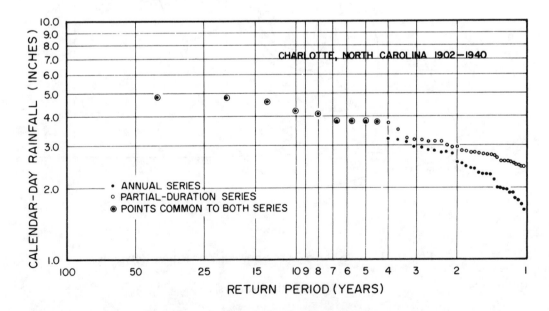

Fig. II.34 Rainfall Frequency Data

Table II.17 Factors For Converting Partial-Duration Series to Annual Series

2-year return period	0.88
5-year return period	0.96
10-year return period	0.99

Various methods have been used in the analyses of extreme precipitation amounts. Their distributions are skewed and therefore not fitted to the normal probability curve. The skewed data may be normalized by introducing a logarithmic transformation, and log-normal graph paper has been frequently employed for this purpose. Extreme-value procedures as developed by Gumbel (1954) provide a simple two-parameter solution, such that any return period value may be predicted from the mean and standard deviation using the straight line relationship:

$$P_{t_r} = \overline{P} + K s_p \dots \dots \dots \dots \dots \dots \dots \dots \dots \dots \dots \dots \dots II.7$$

where P_{t_r} = precipitation with return period, t_r

K = the frequency factor,

s_p = the standard deviation of P, and

$\overline{P}$ = the sample mean.

The frequency factor, K, is the number of standard deviations by which the considered extreme exceeds the sample mean. It is related to the sample size and the return period (Kendall 1959).

A set of extremes may also be fitted to the Gumbel distribution using extreme-value probability paper (Fig. II.35). Plotting positions are usually determined by the formula

$$p = \frac{m}{N+1} \dots \dots \dots \dots \dots \dots \dots \dots \dots \dots \dots \dots \dots \dots \dots II.8$$

where N is the number of years of record, m is the rank of the observation in the series, and p is the probability of being equalled or exceeded. If the points plot as a straight line, then the data may be accepted as fitting the Gumbel-type distribution.

II.4.4.   Intensity-Duration-Frequency

Statistical studies of point-rainfall extremes are used extensively in small-area studies. Storm sewers, drains and highway culverts may all be designed more economically when the frequency of high-intensity rainfall is known.

Annual rainfall extremes tend to plot as a straight line on extreme-value-probability (Gumbel) paper. The Gumbel distribution, however, does not fit partial duration series which approximate a straight line on logarithmic graph paper (Linsley *et al*., 1958). Fig. II.35 shows such an analysis of 6-hr. annual rainfall extremes for Regina. Frequency analysis of this type may be used to get rainfall intensities for different durations and return periods. When the return-period values are plotted against duration on a logarithmic paper they also tend to form a straight line. This latter form of presentation is commonly used in hydrologic studies. As an example, the intensity-duration-frequency curves for Regina, developed in this manner, are shown in Fig. II.36.

Generalized or regional rainfall-intensity-duration analyses incorporate a form of 'quality control' through intercomparison of data. They may therefore provide more reliable information as well as data for locations for which there are no measurements. Canadian regional studies have depended on correlations, since there are but few long-term records for recording rain-gauge stations in Canada. Bruce (1959) has developed "Rainfall-Intensity-Duration-Frequency Maps for Canada" using available recording gauge data, and augmenting this by estimates made from regression relationships between more plentiful 6-hourly rainfall and other duration values. The intensity-duration-frequency characteristics for different locations in Canada as derived by Bruce (1968) are given in Figs. II.37-II.54.

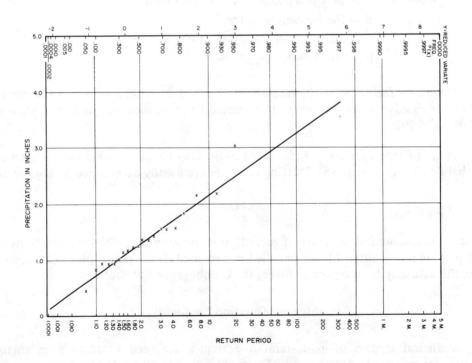

Fig. II.35   6-Hour Rainfall Extremes, Regina, Annual Series, N = 20

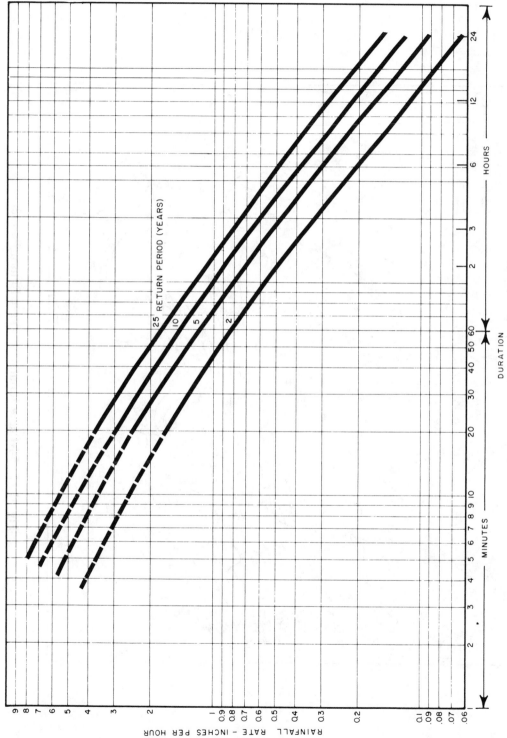

Fig. II.36 Rainfall-Intensity-Duration-Frequency, Regina, Sask.

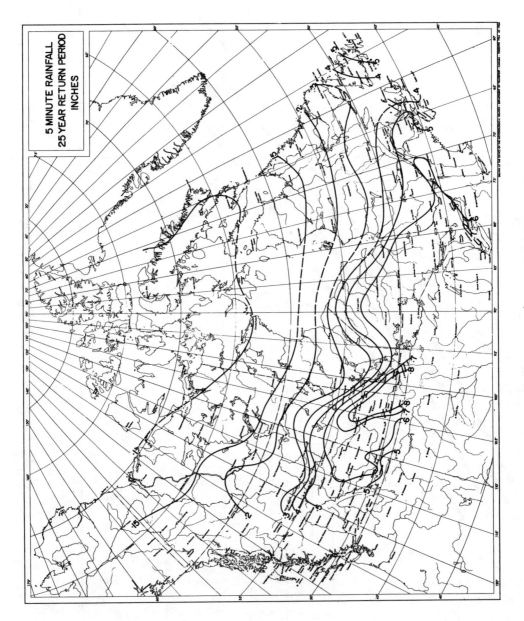

Fig. II.37   5-Minute Rainfall, 25-Year Return Period

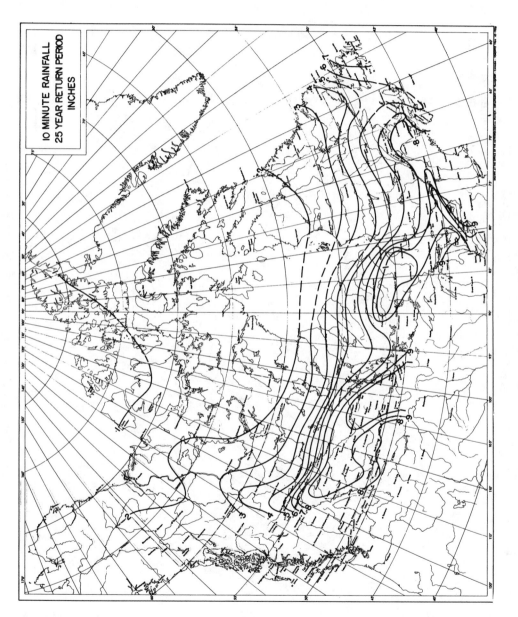

Fig. II.38 10-Minute Rainfall, 25-Year Return Period

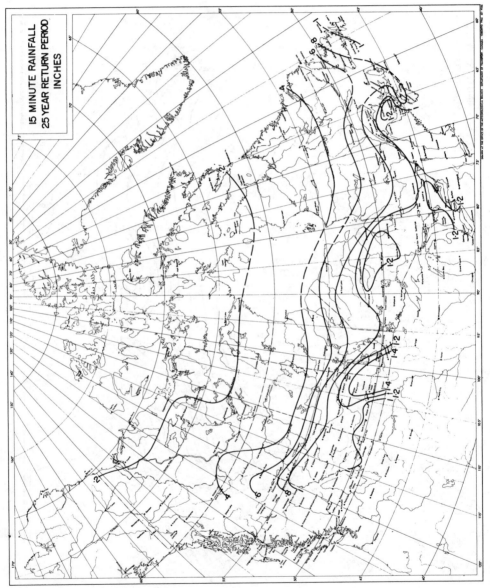

Fig. II.39   15-Minute Rainfall, 25-Year Return Period

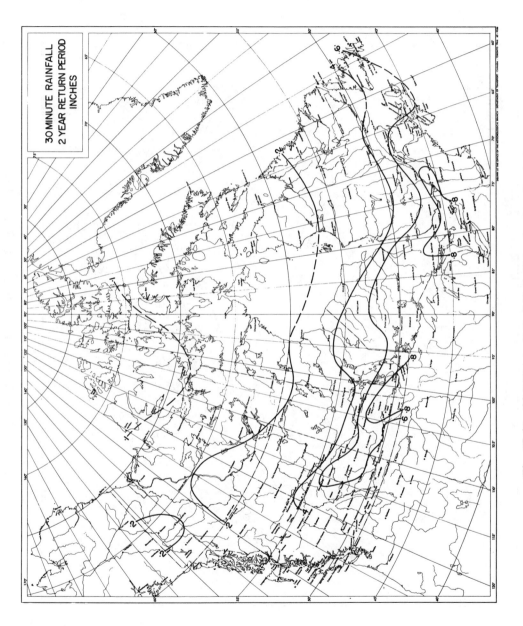

Fig. II.40 30-Minute Rainfall, 2-Year Return Period

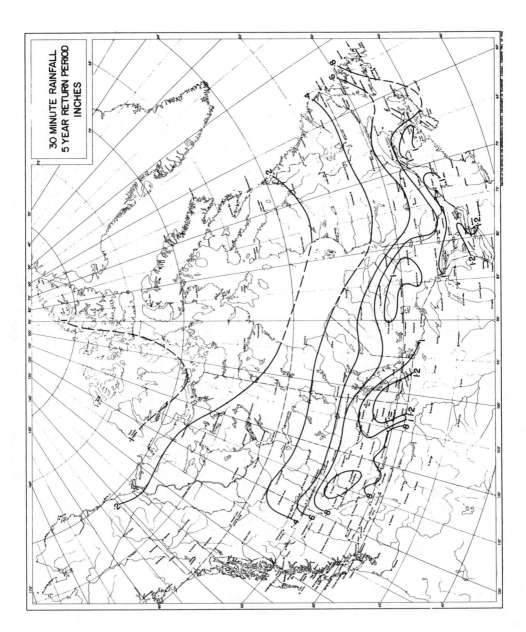

Fig. II.41   30-Minute Rainfall, 5-Year Return Period

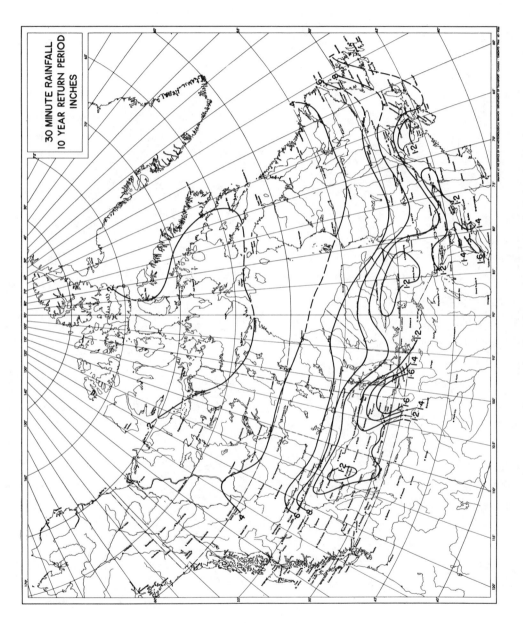

Fig. II.42   30-Minute Rainfall, 10-Year Return Period

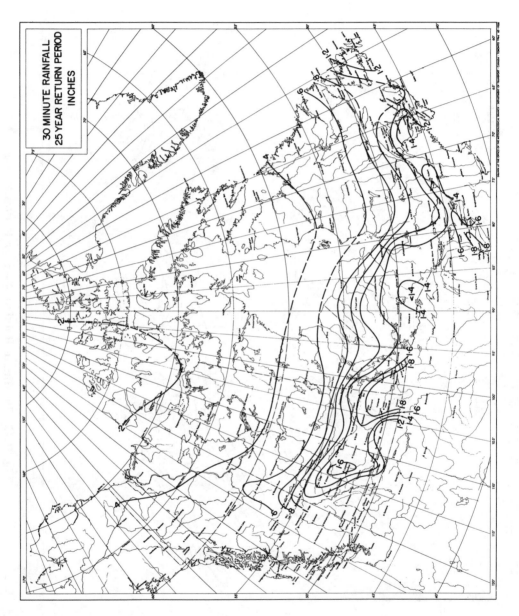

Fig. II.43   30-Minute Rainfall, 25-Year Return Period

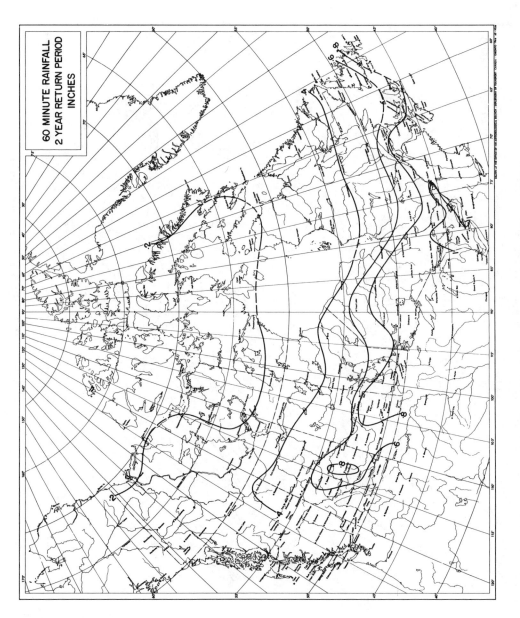

Fig. II.44   60-Minute Rainfall, 2-Year Return Period

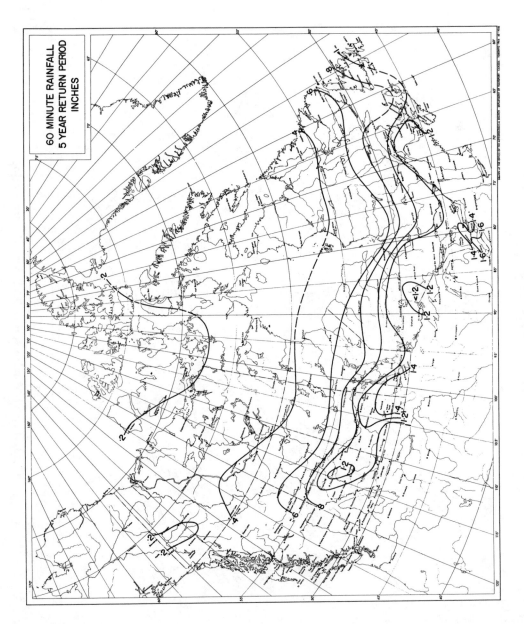

60 MINUTE RAINFALL
5 YEAR RETURN PERIOD
INCHES

Fig. II.45   60-Minute Rainfall, 5-Year Return Period

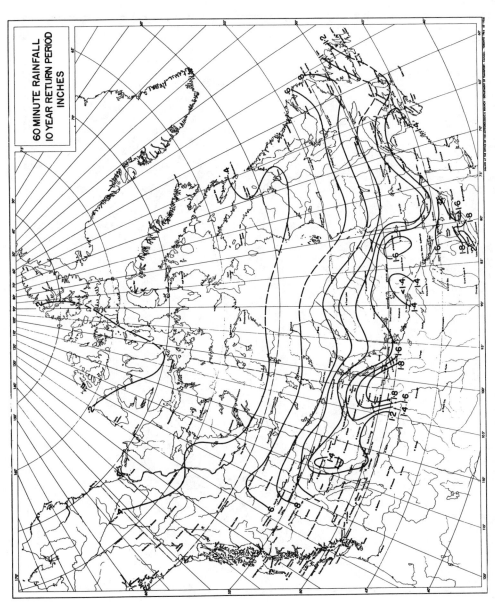

Fig. II.46   60-Minute Rainfall, 10-Year Return Period

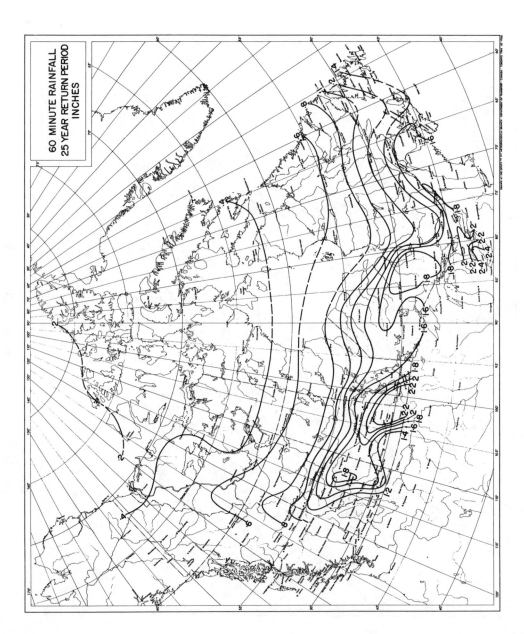

Fig. II.47   60-Minute Rainfall, 25-Year Return Period

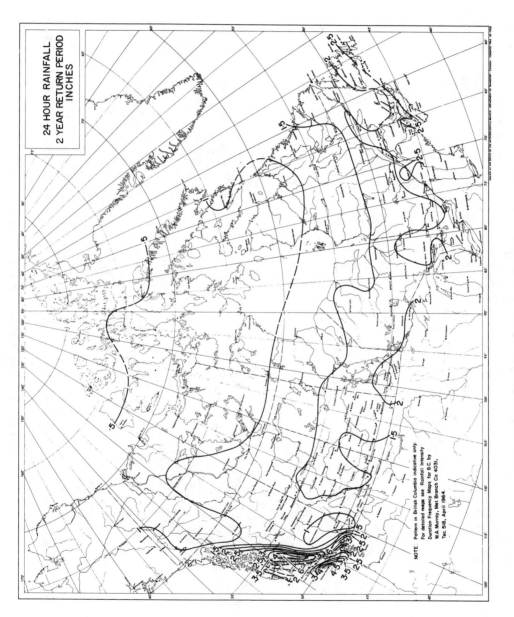

Fig. II.48   24-Hour Rainfall, 2-Year Return Period

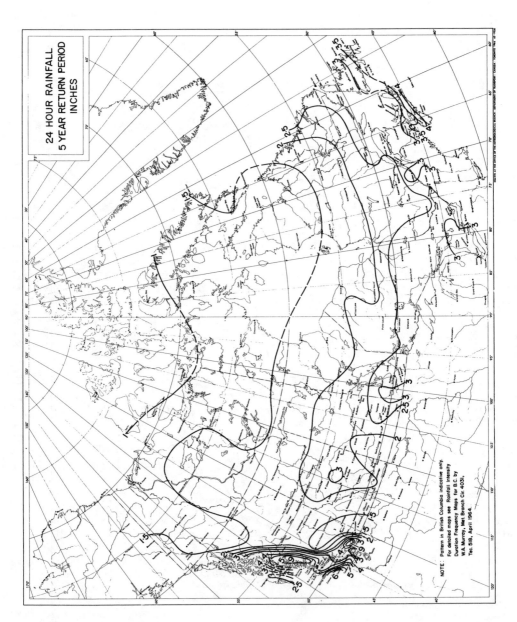

Fig. II.49   24-Hour Rainfall, 5-Year Return Period

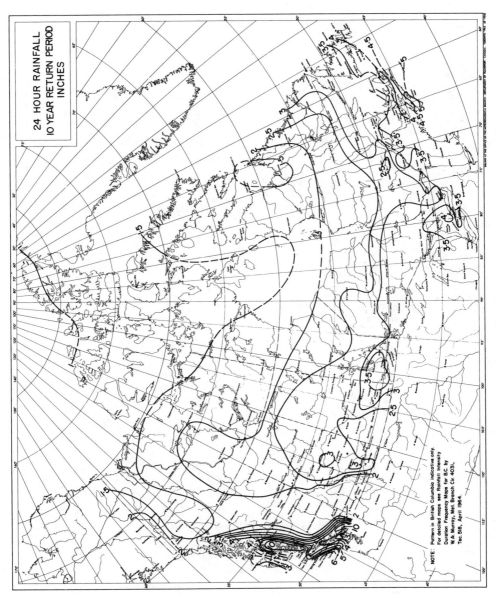

24 HOUR RAINFALL
10 YEAR RETURN PERIOD
INCHES

NOTE: Pattern in British Columbia indicative only.
For detailed maps see Rainfall Intensity
Duration Frequency Maps for B.C. by
W.A. Murray, Met. Branch Cir. 4031,
Tec. 518, April 1964.

Fig. II.50   24-Hour Rainfall, 10-Year Return Period

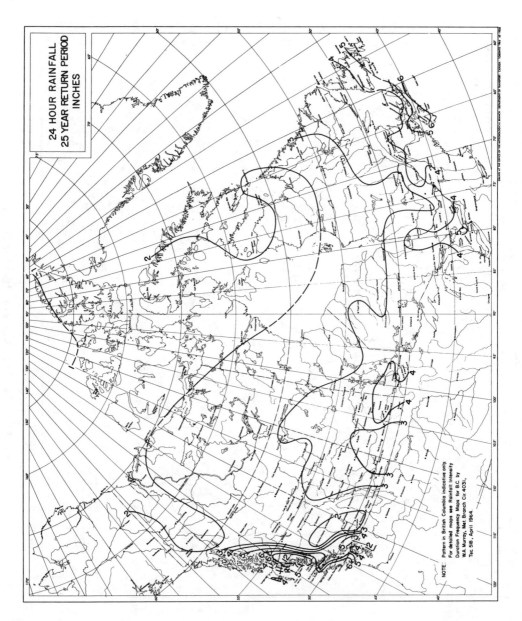

Fig. II.51 24-Hour Rainfall, 25-Year Return Period

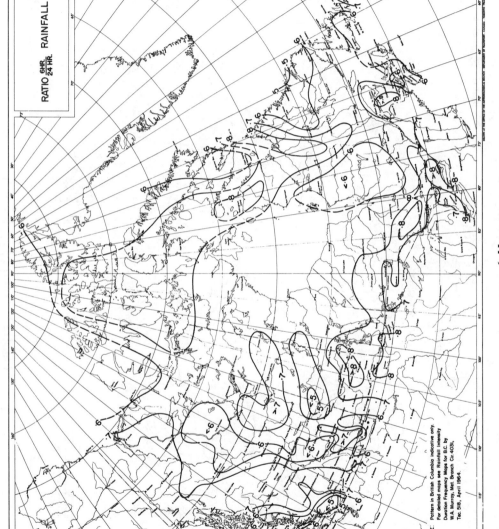

Fig. II.52   Ratio $\dfrac{\text{6-Hr.}}{\text{24-Hr.}}$ Rainfall

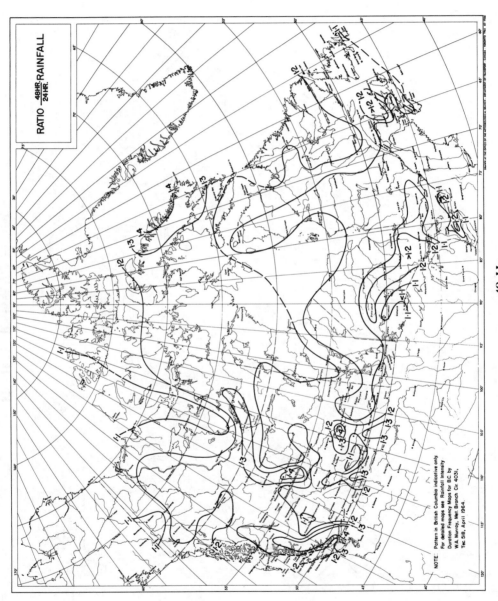

RATIO $\dfrac{\text{48HR. RAINFALL}}{\text{24 HR.}}$

NOTE: Pattern in British Columbia indicative only. For detailed maps see Rainfall Intensity Duration Frequency Maps for B.C by W.A. Murray, Met Branch Cir 4031, Tec. 518, April 1964.

Fig. II.53   Ratio $\dfrac{\text{48-Hr. Rainfall}}{\text{24-Hr.}}$

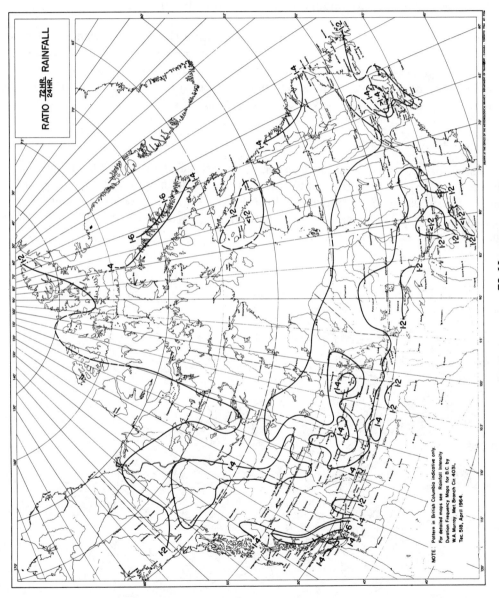

Fig. II.54   Ratio $\dfrac{\text{72-Hr. Rainfall}}{\text{24-Hr.}}$

Depth-duration estimates may also be obtained using Equation II.7. Twenty-four hr. rainfall is the most commonly observed rainfall statistic, and its analysis offers an opportunity for the application of this equation. Using 24-hr. extreme statistics, the equation yields estimates for the climatological day, and for a point. The estimates should be increased by 13 per cent to obtain a value for the 'clock-hour' day. Local time relationships may be used to obtain estimates for other durations.

The statistics required for the solution of Equation II.7 are the mean and standard deviation. The coefficient of variation, C_v, which is the ratio of the standard deviation to the mean, is more conservative within an area which is climatologically homogeneous, and therefore is more amenable to regional analysis. Consequently it is substituted for the standard deviation in the extreme value equation to get:

$$P_{t_r} = \overline{P}_{24}(1 + KC_v) \dots\dots\dots\dots\dots\dots\dots\text{II.9}$$

Values of $\overline{P}_{24}$ and C_v are given in Fig. II.55. Estimates abstracted from these charts may be solved by entering them into Equation II.9 or its nomographic solution in Fig. II.56 to obtain estimates of the climatological day 24-hr. precipitation extreme of given frequency. These estimates must be increased by a factor of 1.13 to adjust to the clock-hour (1,440 min. series) day.

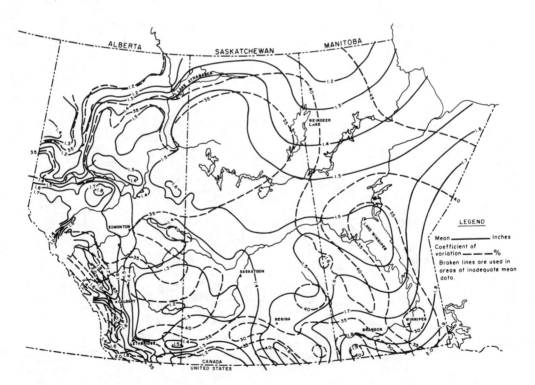

Fig. II.55 Means and Coefficients of Variation of Annual 24-Hour Precipitation Extremes

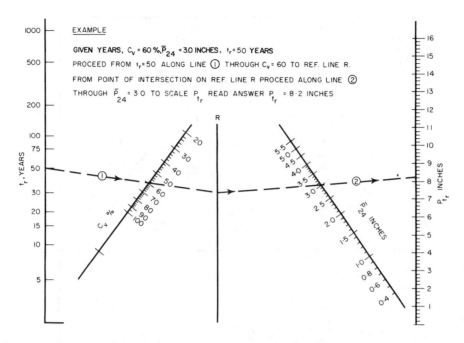

Fig. II.56 Nomographic Solution of $P_{t_r} = \overline{P}_{24}(1 + KC_v)$

Fig. II.57 provides conversion factors which may be used to convert from 24-hr. values to other durations. These factors do not appear to be related to the return period.

II.4.5   Depth-Area

Depth-area relationships may be either storm-centered or geographically-centered. Storm-centered values are useful in studies of probable-maximum-precipitation, whereas geographically-centered values are useful in frequency studies. Frequency-derived, geographically-fixed curves are based on observations taken in different parts of the storm and consequently tend to be flatter than storm-centered curves. Using storm-centered type data, Hershfield and Wilson (1960) found the following ratios of average rainfall to that falling over 10 sq. mi. (tropical and non-tropical storms).

Table II.18   Ratio of Average Rainfall to that Over 10 Square Miles

Area mi²	100	200	500	1000	5000
Time (hr.)					
6	0.85	0.80	0.73	0.65	0.46
12	0.89	0.85	0.79	0.72	0.51
24	0.92	0.88	0.82	0.76	0.58
48	0.93	0.90	0.84	0.80	0.63

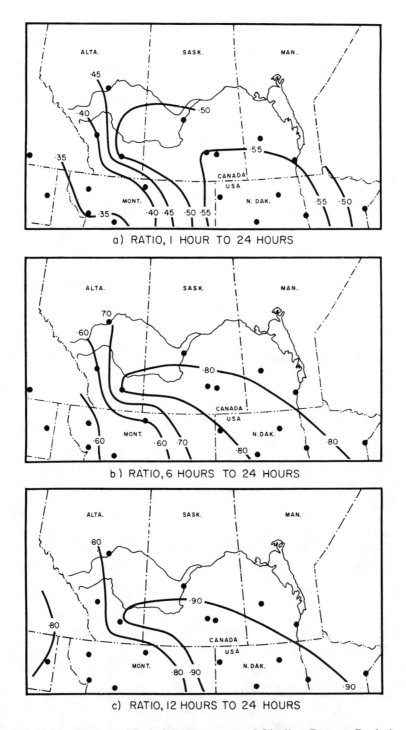

a) RATIO, 1 HOUR TO 24 HOURS

b) RATIO, 6 HOURS TO 24 HOURS

c) RATIO, 12 HOURS TO 24 HOURS

**Fig. II.57   Ratios of Rainfall Extremes of Similar Return Period
(based on point rainfall frequency distribution)**

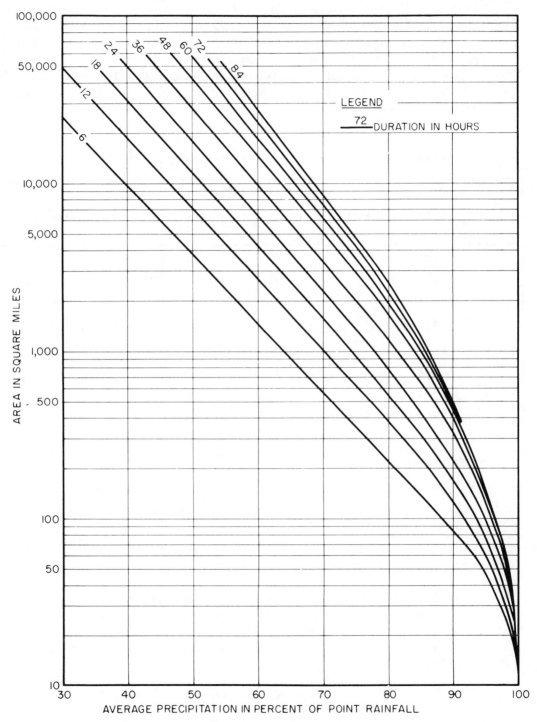

**Fig. II.58 Conversion of Rainfall Extremes According to Area
(based on storms of 24 hours or longer duration)**

Formulas for smaller areas have been developed, but they give variable results depending on the nature of precipitation that the area (from which the data have been collected) has experienced. Court (1961) found a Gaussian curve to be best in fitting the average cross-section through the centre of heaviest rainfall, but notes that short-duration storms have steeper isohyetal gradients than longer-duration, larger-area storms. He tabulated, as shown in Table II.19, the following areal relationships found by various authors.

Table II.19 Formulas for Average Rainfall Over Area A(mi^2) Centred at Point of Maximum Rainfall, P_m, or Within Circular Isohyet of Radius, x

Average Precipitation, P	Area, A	Author
$P_m (1 - 0.14A^{1/4})$	1.6	Fruhling
$P_m - 0.14A^{3/5}$	18.5	Woolhiser-Schwalen
$P_m - bA^{1/2}$ ($b = .07; .025; .03P_m; .06P_m$)	5-280	Huff-Stout
$P_m + a_1 A^{1/2} + a_2 A + \ldots$	<300	Chow
$P_m \exp(-kA^n) = P_m \exp(-0.01A^{1/2})$	20-20,000	Horton
$2P_m b^{-2} x^{-2} [1 - (1 + bx) \exp(-bx)]$ ($b = 0.0235$)	>100	Boyer
$P_{500} (2.11 - 0.41 \log_{10} A)$ $P_{500} (2.35 - \log_{10} A^{1/2})$	50-500	Light
$(\pi P_m / Aab)[1 - \exp(-Aab/\pi)]$	any	Court

NOTE: a and b define the scale and ellipticity: when $a = b$, the isohyets are circular.

Hershfield and Wilson (1960) found no systematic variation in areal concentration between tropical and non-tropical storms.

Over 70 large Prairie rainstorms have been analyzed, and these analyses provide data from which average spatial characteristics have been determined. These are shown as Fig. II.58. There is justification in using depth-area relations of this nature in converting point rainfall of a given frequency to areal rainfall of a similar frequency at least for watersheds less than 400 sq. mi. in area.

II.4.6   Depth-Area-Duration

Depth-area-duration analyses provide time as well as spatial information on rainfall. A typical set of these curves for a given storm are shown in Fig. II.59.

These curves are obtained as follows:

1. Analysis of the mass curves for a rainstorm provides point rainfall extremes for durations up to that of the length of the storm. It is generally assumed for storms which encompass large areas that the maximum point value is representative of a 10 sq. mi. area.

2. The mass curves are then analyzed to obtain increments of precipitation for a specified time interval which is usually six hours (see Table II.20). These increments are used to determine the maximum average precipitation for the areas contained within each isohyet, and for time intervals varying from the incremental interval to the storm duration (for an example see Table II.21).

3. The maximum average values are plotted on graph paper according to depth and area. Envelopes for each time interval are then fitted to the plotted values as in Fig. II.59.

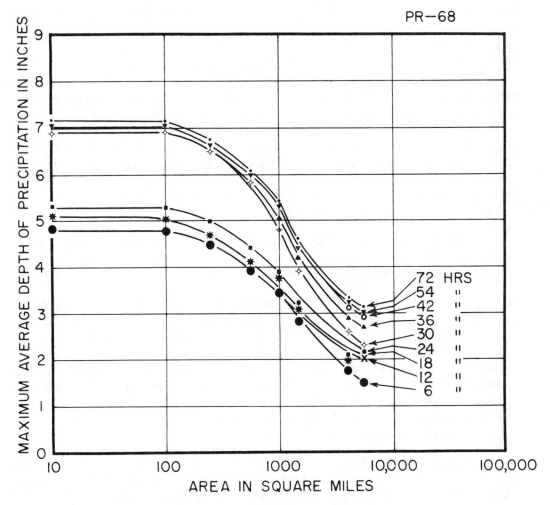

Fig. II.59 Maximum Depth-Area Curves
(storm of June 8-10, 1959)

The above analysis is carried out by zones for complex storms, and within a single major isohyet for simpler storms. The average storm-precipitation is computed following the isohyetal method. Thiessen-polygon procedures are usually followed in determining average rates of fall; for instances, in basins in which there are less than six stations available from which the average depth can be calculated (U.S. Weather Bureau, 1946).

II.4.6.1   Statistical Depth-Area Duration

Point rainfall intensities may be expressed in terms of area and varied duration using the statistical average ratios noted in preceding sections. There are obvious hazards in using such relationships. The use of time and area ratios also introduces probability, and the resulting model has therefore a compound probability which may be difficult to assess. Also the ratios are generally developed for specific situations and may not have general application. For example, the ratios indicated by Fig. II.58 were developed for large-area, long-duration storms. They are therefore not suitable for isolated thunderstorms, orographic and winter-season storms.

II.4.7   Sequence

Having selected a design rainstorm, the hydrologist is often confronted with the problem of what time-sequence of rainfall should be used during the storm interval. In major project design, several historical storms are frequently used to provide the answer for a single storm duration. The recurrence interval of storms may also be determined from historic data. A study of Mississippi Valley storms show that heavy precipitation in that area can readily begin again 72 hours after the preceding event (Myers 1959). The average time distributions of rainfall for shorter duration storms have been determined for climatic regions within the United States (U.S. Soil Conservation Service 1957); but their transposability is subject to question.

Average time distributions may readily be computed from precipitation intensity records. Table II.22 provides averages based on 12 one-hr. summer storms at Regina, and 15 long-duration (24 hrs. or more) storms which occurred on the Alberta plains.

Both samples are too small to provide conclusive results. Individual storms varied markedly from the average in both sets of data. In reviewing the analysis of many long storms it is apparent that the critical period often occurs over a very small fraction of the total storm period. Also it was evident for the Alberta storms that the peak could occur at almost any time within the first five-sixths of the storm interval. Averaging of these storms smooths out what may be very important design characteristics of the rain-storm. An improvement in describing character may be obtained by referencing values to the time of occurrence of the peak. Treated in this manner, the data for the 15 Alberta storms yield the average characteristics shown in Table II.23.

It is interesting to note that, on the average, over one-third of the long-duration storm rainfall occurred within a one-hr. period. The yield from rainfall of this intensity would probably far surpass that obtained from the time distributions noted in Table II.22. If a storm is to be reconstructed from statistical relationships, this fine-scale structure should be considered.

Table II.20 Mass Curve Abstracts

Sheet No. 10

Line No.	Zone or Centre	PR – 68 June 8-11/59 Station	Abs. Max. 6	12	18	24	6 (0800)	12 (1400)	18 (2000)	24 (0200)	30 (0800)	36 (1400)	42 (2000)	48 (0200)	54 (0800)	60 (1400)	66 (2000)	72 (0200)	78	84	90	96	102	108	114	120
1	A	Marchand	4.8	4.8	5.1	5.3	0.1	0.1	0.1	0.1	0.1	0.1	1.8	2.1	2.3	2.3	7.1	7.1								
2		La Broquerie	4.8	4.8	5.0	5.4	0.0	0.0	0.2	0.3	0.3	0.3	1.7	2.0	2.2	2.2	7.0	7.0								
3		Richer	3.1	3.1	3.1	3.3	0.2	0.3	0.3	0.4	0.4	0.4	1.8	2.1	2.1	2.1	5.1	5.1								
4		Hadashville					0.0	0.0	0.0	0.4	0.4	0.4	0.4	0.4	0.7	0.7	3.4	4.5								
5		Steinbach					0.0	0.0	0.0	0.2	0.2	1.5	1.5	1.9	2.0	2.3	3.6	3.6								
6		Woodridge					0.0	0.0	0.0	0.0	0.0	0.0	0.5	0.6	0.8	0.8	1.9	2.8								
7		Caribou					0.2	0.2	0.2	0.2	0.2	0.2	0.4	0.4	0.6	0.6	2.2	2.3								
8		Emerson					0.0	0.1	0.2	0.6	0.6	0.6	0.7	0.8	0.9	1.8	2.3	2.3								
9		Ste. Annes'(p)					0.0	0.0	0.2	0.2	0.2	1.4	1.5	1.8	1.9	1.9	2.2	2.2								
10		E. Braintree					0.0	0.0	0.0	0.0	0.0	0.0	0.6	0.8	0.8	0.8	1.3	2.1								
11		Ste. Annes'(w)					0.0	0.0	0.1	0.1	0.1	1.4	1.5	1.8	1.9	1.9	2.1	2.1								
12																										
13																										
14	B	Pointe Du Bois					0.0	0.0	0.2	0.2	0.2	0.2	0.2	0.2	0.5	0.5	1.4	3.1								
15		Great Falls					0.0	0.0	0.0	0.2	0.2	0.2	0.2	0.2	0.4	0.4	1.6	2.2								
16		Lac Du Bonnet					0.0	0.0	0.0	0.0	0.0	0.0	0.0	0.0	0.0	0.0	1.4	1.5								
17		Seven Sisters					0.0	0.0	0.0	0.0	0.0	0.0	0.0	0.0	0.0	0.0	0.0	1.2								
18																										

Table II.21 Maximum Depth-Duration Data

Centre or Zone	Storm Period — Station	Encompassing Isohyet Av. Pepn. In.	Av. Pepn.	Area Sq. Mi.	Effect. Area Stn. Plan RDG K=	Effect. Area Stn. Wt. %	6	12	18	24	30	36	42	48	54	60	66	72	78	84	90	96	102
							Accum. Stn. Rainfall x Stn. Weight Unless Noted											(Day and/or Hour at end of Period unless noted)					
	Marchand (Absolute Maximum Precip.)						4.8	4.8	5.1	5.3	6.9	6.9	6.9	6.9	6.9	6.9	7.0	7.1					
	La Broquerie						4.8	4.8	5.0	5.4													
A	Marchand	7	7.1	99	.015	55	.06	.06	.06	.06	.06	.06	.99	1.16	1.27	1.27	3.91	3.91					
	La Broquerie				.012	45	.00	.00	.09	.14	.14	.14	.77	.90	.99	.99	3.15	3.15					
	Total				.027	100	.06	.06	.15	.20	.20	.20	1.76	2.06	2.26	2.26	7.06	7.06					
	Adjusted to Average						0.1	0.1	0.2	0.2	0.2	0.2	1.8	2.1	2.3	2.3	7.1	7.1					
	6-Hour Increment						0.1	0.0	0.1	0.0	0.0	0.0	1.6	0.3	0.2	0.0	4.8	0.0					
	Maximum Precipitation						4.8	4.8	5.0	5.3	5.9	6.9	6.9	6.9	7.0	7.0	7.1	7.1					
A	Marchand	6	6.7		.033	51	.05	.05	.05	.05	.05	.05	.92	1.07	1.17	1.17	3.62	3.62					
	La Broquerie				.027	42	.00	.00	.08	.13	.13	.13	.71	.84	.92	.92	2.94	2.94					
	Richer				.005	7	.01	.02	.02	.03	.03	.03	.13	.15	.15	.15	.36	.36					
	Total				.065	100	.06	.07	.15	.21	.21	.21	1.76	2.06	2.24	2.24	6.92	6.92					
	Adjusted to Average						0.1	0.1	0.1	0.2	0.2	0.2	1.7	2.0	2.2	2.2	6.7	6.7					
	6-Hour Increment						0.1	0.0	0.0	0.1	0.0	0.0	1.5	0.3	0.2	0.0	4.5	0.0					
	Maximum Precipitation						4.5	4.5	4.7	5.0	6.5	6.5	6.5	6.6	6.6	6.6	6.7	6.7					

Table II.22 Time Distributions of Rainfall

Interval as Sixth Total Period	Per Cent Falling in Interval		Accumulated Percentage	
	Regina one-hour storms	Alberta long-duration storms	Regina one-hour storms	Alberta long-duration storms
1	15	4	15	4
2	17	19	32	23
3	21	55	53	78
4	21	13	74	91
5	15	6	89	97
6	11	3	100	100

The time sequence of rainfall is highly variable and consequently a storm interval with a rainfall amount of a given return period may contain shorter-duration amounts with higher return periods. It is possible, however, in constructing storm models to have all values consistent. There is a high probability that, for a specific location, the 5-min. rain event occurred within the 10-min. event with the same return period; and also that the 10- and 20-min. events bear the same relationship. On the average, then the one-hr. storm of given return period should have a time distribution which contains shorter duration values of the same return period. By computing these values independently and allowing for the characteristic time of peaking of rain for the area and storm type, the short-duration values may be distributed in time to construct a design storm. This storm will have a sequence of shorter duration rainfall amounts which is consistent with design requirements.

Table II.23 Distribution of Rainfall Referenced to the Time of Peaking

Period (hours)	-7	-6	-5	-4	-3	-2	-1	0	1	2	3	4	5	6	7	8	9	10	11	12	13 etc.
% of total	.3	.5	3.0	3.0	3.4	6.2	8.7	37.2	10	6	6.3	5.2	4.6	2 7	3.0	2.8	1.2	0.8	0.5	0.5	0.2	0.2 etc.

II.5   TRANSPOSITION AND MAXIMIZATION

Flood frequency analyses are commonly used in the design of small structures. However, streamflow statistics are generally inadequate for the design of major structures when the design recurrence interval may be several hundreds of years, or when the probable-maximum flood must be passed. Major rainstorms of history, and the probable-maximum-precipitation are used to determine the design flood in these situations. Transposition and maximization of storms are techniques which are used in this process.

II.5.1   Transposition

Transposition is the transfer of rainfall values from the place where they occurred to a river basin for which the design is proposed. It assumes that the locations of storms has been fortuitous within certain geographical limits. By transposing storms, the meteorological experience of the considered basin is effectively increased.

The transposition limits of a storm are determined both by topography and climate. In the absence of significant topographic barriers, a storm may be transposed within that area in which a synoptically similar storm could have occurred. Transposition of storms is normally not used in mountainous areas because of the difficulty in adjusting for orographic effects; however, it is used at times, along a windward slope where the orographic depletion of moisture and increased lift tend to balance out. Extremes of surface dewpoints, the characteristics of point rainfall and wind are useful aids in determining transposition limits.

II.5.1.1   Atmospheric Moisture Supply

The transposition and maximization of storms requires an understanding of the processes which affect precipitation. Precipitation results from the lifting of moist air. The rate of precipitation depends on the amount of moisture present and the rate of lift. Atmospheric moisture may be measured, and assigned a return period or limiting value, but the lifting mechanism is not usually sufficiently understood, and it is therefore not generally possible to estimate its limiting or a return-period value.

A vertical column of air has a moisture charge, W_p, equal to:

$$W_p = \int_0^p \frac{q}{\rho g} \, dp \dots \dots \dots \dots \dots \dots \dots \text{II.10}$$

where q = specific humidity,   g = acceleration due to gravity, and

ρ = density of water,   p = pressure at the bottom of the column.

This mass is usually referred to as effective precipitable water, even though it is recognized that no natural process will remove all the vapour from the air. A column extending upward from sea level, and saturated with respect to water vapour contains the equivalent of 2 in. of water when the sea level temperature is $67°F$. The sizes of many rainstorms therefore show that winds are continuously supplying moisture to the storm during its lifetime. The amount of this moisture varies geographically, seasonally and with elevation and these variations must be considered in transposition and maximization. Surface measurements and atmospheric soundings of water vapour provide the information which is necessary for these adjustments.

Although atmospheric soundings provide measurements of atmospheric moisture, they are not commonly used in storm maximization because

(1) the network of observation stations is sparse and may miss important detail on moisture inflow;

(2) the network did not exist prior to 1940;

(3) until recently the computation of precipitable moisture from the soundings was not routine, consequently statistical data have not been readily available; and

(4) there is a high correlation between the values of precipitable water measured in the atmosphere and that estimated from dewpoints which are readily available.

Since surface dewpoints are highly correlated with the amount of water vapour contained in the atmosphere above the measuring point, particularly under circumstances which favour atmospheric lifting, they may be used in transposition and maximization. For this purpose, the value usually used is the 12-hr. maximum-persisting dewpoint reduced to 1000 mbs. This value was selected because a major portion of storm rainfall tends to occur in this interval. Dewpoints are used which are representative of the air moving into the storm. The time is selected so that it suitably precedes the period of most intense rainfall. The maximum-persisting dewpoint is the highest value which is equalled or exceeded throughout a specified period. McKay (1963c) conducted a statistical study of this parameter for the Prairie provinces and recommended that the 100-yr. return-period value be used in estimating probable-maximum precipitation.

Moisture adjustments are made by multiplying the observed rainfall by the ratio of effective precipitable water for the basin, to that observed for the storm. Tables and nomograms have been prepared to facilitate the computation of precipitable water from atmospheric measurements, Ferguson (1962), Peterson (1961) and U.S. Weather Bureau (1951). Values for each watershed should be adjusted for latitude and proximity to a moisture source by reference to the regional isotherms of 12-hr. persisting dewpoints. Elevation adjustments are made by deducting the amount of moisture contained in the column bounded by the observed elevation where the storm occurred, and the basin elevation.

Orographic barriers increase precipitation on the windward side and decrease it on the leeward side. As previously noted, however, storm precipitation may be distributed some distance beyond the ridge. Air which is forced over barriers may lose most of the atmospheric moisture which was contained below the crest level upstream from the barrier. If a barrier is crossed by the transposition process, an adjustment must be made for the loss of atmospheric water vapour. No major barriers to moisture flow are found within the Great Central Plain, and transposition within this area is fairly straight-forward.

II.5.1.2   Mechanical Efficiency

As previously noted, atmospheric lifting associated with storms is generally not adequately understood to permit adjustment for changes in storm efficiency due to tranposition. In some instances, adjustment can be made for orographic lifting. Also where there are no significant orographic effects, the ratio of precipitation to the available moisture charge (P/Wp) provides a means of comparison of the mechanical efficiency. If orographic effects are significant, their impact too must be evaluated in order to obtain an estimate of the efficiency of the non-orographic (convergence) component of the storm. This may be done in a generalized sense for large areas which are relatively homogeneous with respect to slope, aspect and elevation. However, where the terrain is very rugged the problem is complex.

II.5.1.3   Other Adjustments

It is not acceptable to arbitrarily fit an isohyetal pattern borrowed from an adjacent area to the basin under study. Storm patterns for the basin should be studied to determine their characteristics, thereby disclosing what patterns are acceptable for transposition. Rotation of about 20 degrees is usually permitted in transposition.

Several storms are usually transposed in a study. The selection of these storms depends, among other things, on the size of the watershed. A storm which produces a flood on a small basin may not be critical on a large basin. Also, different storms will be critical for different durations within the same basin.

The storm moisture supply and factors affecting the mechanical efficiency of the storm vary seasonally, and therefore there are limitations to transposition in time. It is generally considered that any storm may have occurred 15 days earlier or later than the recorded date without modification. However, it is not likely to have the same probability of occurrence at the earlier and later date. A good guide on time-transposition limits and/or adjustment is obtained from an analysis of the seasonal variation of 24-hr. point rainfall extremes.

II.5.2   Probable-Maximum-Precipitation

Horton once wrote: "A small stream cannot produce a major Mississippi flood for much the same reason a barnyard fowl cannot lay an egg a yard in diameter" (U.S. Weather Bureau 1960). That is, there is a rational upper-limit to precipitation rates, and this amount is called Probable-Maximum-Precipitation (PMP).

Two approaches are currently used in determining this value. The rational approach as suggested by Horton, and a statistical approach which accepts the transposed values of greatest recorded rainfall as an approach to the upper limit.

II.5.2.1   The Rational Method

The most commonly used procedures for estimating PMP assume

(1) storm rainfall can be expressed as a product of (a) available moisture supply, and (b) the combined effect of storm mechanical efficiency and inflow wind; and
(2) the most effective combination of efficiency and inflow wind have occurred or have been closely approached.

An index of the maximum possible moisture supply is obtained by entering tables of 'effective-precipitable water' with the 100-yr. 12-hr.-maximum-persisting dewpoint for the basin. Maximization is achieved by multiplying the observed rainfall values by the ratio of this figure to that for the air supplying the atmospheric moisture to the actual storm.

By transposing all major storms, the conditions for mechanical efficiency are met for a specific location. Consequently through transposition and maximizing moisture supply, probable-maximum-precipitation is approached.

II.5.2.2 Precipitation Moisture Ratios

The above method presupposes that there is a long record of storm history. For many areas there are no satisfactory records and other methods must be used. Also in many areas the frontal (convergence type) storm characteristics are confused by orographic effects. It is desirable therefore to separate the two components and obtain models which may be transposed and maximized.

One approach in splitting the two components has made use of the precipitation/moisture ratios. Enveloping values of the P/M ratio, determined from a seasonal analysis, are accepted as representative of the greatest observed efficiency of storm processes. It is then assumed that, (U.S. Weather Bureau 1961):

$$\left(\frac{\text{Precipitation}}{\text{Moisture}}\right)_{max} \ x \ \left(\text{Moisture}\right)_{max} = \text{convergence} \quad \text{PMP} \ \ldots \ldots \ldots \text{II.11}$$

The remaining precipitation is that due to orographic effects. Where there is no orographic effect, enveloping P/M ratios, when combined with maximum moisture charge, will give an estimate of convergence PMP.

In orographic models, precipitation is generally considered to be the difference between inflow and outflow of moisture within a specified volume, along a path ascending a slope. Models developed on storm experience may be maximized by maximizing inflow wind and moisture. The development of orographic models for California is discussed in detail in Hydrometeorological Report No. 36, U.S. Weather Bureau (1961).

II.5.2.3 Statistical Estimates

Probable-maximum-precipitation in statistical studies is the upper limit of rainfall which is justified climatologically. It is reasonable to expect that, within the Prairie provinces, rainstorms approaching the probable-maximum have occurred and have possibly been recorded—nevertheless these extremes still form the lower boundary condition in estimation of PMP. The estimates of probable-maximum obtained by rational methods have not always been satisfactory, there being at times wide divergence between the estimates and experience. Statistical methods do give results which are consistent with experience.

Precipitation extremes for a stated return period may be estimated from Equation II.9 $\left[P_{t_r}=\overline{P}(1+KC_v)\right]$ in which K is the frequency factor by which the considered rainfall exceeds the mean value. By computing K for all past recorded rainstorms, the extreme value of K is obtained. This, given a sufficient period of record should approach a value relating to probable maximum precipitation. The enveloping value of K is dependent on $\overline{P}$. Using envelopes of K and values of C_v and $\overline{P}$ it is then possible to compute statistical estimates of probable maximum precipitation for a point. Enveloping K for annual series of Prairie 24-hr. precipitation extremes is given in Fig. II.60. This figure may be used in conjunction with Fig. II.55 to estimate the probable maximum precipitation. For example, the values of C_v and $\overline{P}_{24}$ for Saskatoon are .38 and 1.35 in. respectively; the enveloping value of K for $\overline{P}_{24}$ =1.35 is 22.5. Entering these values into the equation, one obtains

$$PMP = 1.35 \ (1 + 22.5 \times .38) = 12.9 \text{ in.}$$

where PMP is the probable-maximum-precipitation for a point.

When dealing with large storms, the statistical relationships for time and area, which were discussed previously, may be used to obtain estimates for other durations and areas.

Probable-maximum-precipitation varies seasonally. Since both PMP and storm rainfall are often used in conjunction with snowmelt floods, a value consistent with the snowmelt season is required. This is achieved rationally by maximizing and transposing spring storms, and by adjusting the moisture supply to a level consistent with the snowmelt period. Statistical estimates may be adjusted to the proper date using envelopes of K which have been analyzed according to both $\overline{P}_{24}$ and to date of storm occurrence (Fig. II.61).

II.6   THE MAXIMUM SNOWPACK

Project-design and probable-maximum-flood studies for the Prairies must generally consider the case of a rainstorm occurring simultaneously with the spring snowmelt freshet. The size of the freshet is dependent on the mass or water equivalent of the snowpack, and the melt rate as well as other hydrological factors. Maximum snowpacks are likely to be the product of the longest and stormiest winters. The later the melting date, the more rapid the melt rate is likely to be, and also the greater the size of the probable-maximum rainstorm. The length of the accumulation period and the probable date of disappearance of the snowpack must therefore be considered in determining the maximum snowpack.

II.6.1   Duration

Daily measurements of snow depth are made at 'synoptic-type' weather stations, and this information may be used to determine the duration and the average or return-period values of the date of pack appearance and disappearance. Usually the last date of snow cover having at least two-week duration is accepted as the date of snowpack disappear-

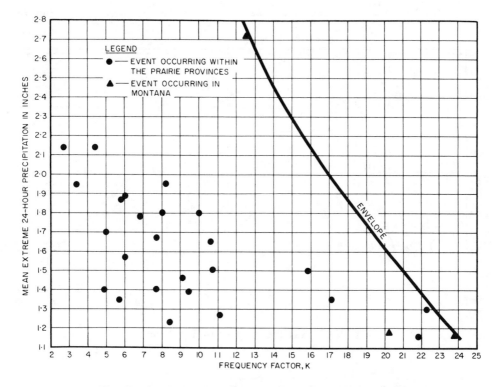

Fig. II.60 Frequency Factor Envelope, Annual Series

ance. The series for snowpack formation (earliest date) and disappearance are normally distributed within the Prairie provinces (Fig. II.62). The date of formation or loss of pack may therefore be computed for a specified frequency using the equation:

$$D_{t_r} = \overline{D} + Ks_D \quad \dots\dots\dots\dots\dots\dots\dots\dots\dots\dots \quad \text{II.12}$$

where D_{t_r} = the date of formation (or loss), s_D = the standard deviation, and $\overline{D}$ = the mean date of formation (or loss), K = the frequency factor.

The values of K for various return periods are listed below in Table II.24, and the values of $\overline{D}$ and s_D are given in Figs. II.63 and II.64.

Table II.24 'K' Factors for Various Return Periods

Return Period, years	Frequency Factor, K
5	0.84
10	1.28
20	1.65
50	2.05
100	2.33

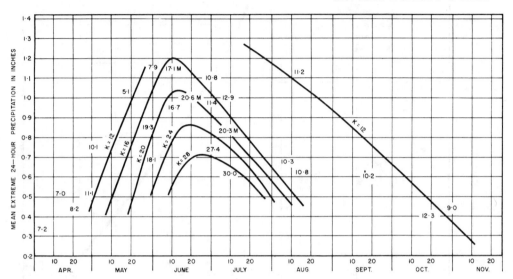

Fig. II.61   Frequency Factor Envelopes

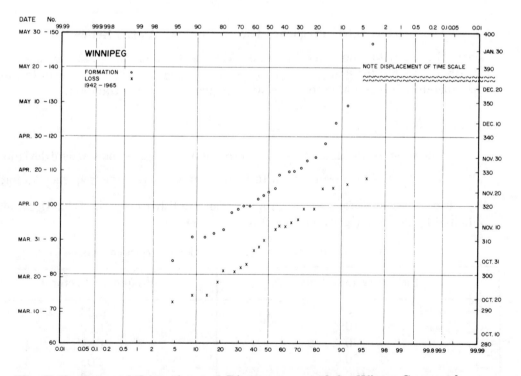

Fig. II.62   Dates of Formation and Disappearance of the Winter Snowpack

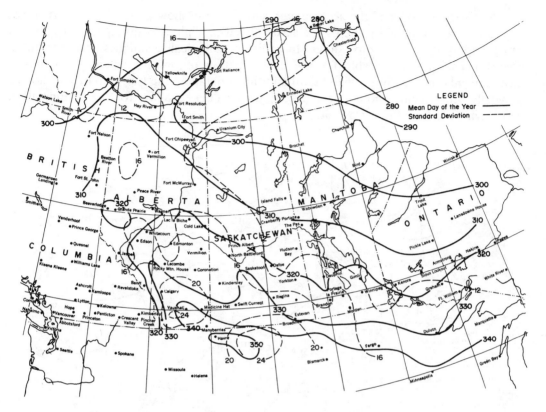

Fig. II.63 Mean Date and Standard Deviation of Snowpack Formation

II.6.2   Estimating the Maximum Water Equivalent

A simple 'lower-estimate' of the maximum pack is given by the observed maximum accumulation of snow. To do this the accumulation period must first be defined from meteorological records. This seasonal extreme is dependent on the length of record, and in time it will be exceeded. Very long records may be necessary to obtain acceptable values; for example, 124 in. of snow fell at Estevan in the season ending 1904—this value has not been approached since, not even at Regina where the highest value in 74 years is 77 in.

A 'synthetic' approach in which the maximum pack is built up of a series of unrelated maximum occurrences, is also commonly used. As an example of this approach, the maximum November precipitation of record is added to the maximum December precipitation of record and so on until a synthetic, seasonal value is obtained. Meteorological persistence tends to favour this method. It is apparent that the synthetic, seasonal value increases as the time-unit (day, week, etc.) decreases in size; that is, the sum of peak daily values exceeds the sum of peak monthly values. The time-unit should, therefore, be selected in a realistic manner to be consistent with the storm frequency of the area.

Statistical analysis of the accumulated snowfall, and snow-survey data are also used; the 100-yr. return period value of the accumulated precipitation during the snow season being employed in some procedures. In addition, statistical techniques similar to those used for PMP are acceptable for the determination of limiting values. Snow-survey measurements frequently provide useful additional detail on the topographic variation of snowfall. Adjustments should be made to survey data to allow for the accumulation period which follows the survey date.

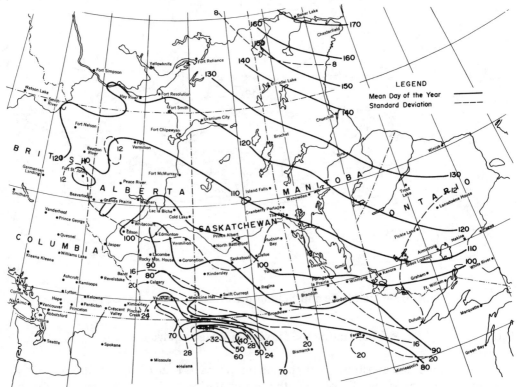

Fig. II.64 Mean Date and Standard Deviation of Snowpack Disappearance

The enveloping-ratio of the extreme to mean-seasonal precipitation may be used for transposition, thereby increasing the snowcover 'experience' of the area. The enveloping-ratio is obtained by fitting an envelope to the ratios plotted against the mean; and transposition is achieved by applying the appropriate enveloping value to the known seasonal mean values.

Where there are marked changes of elevation within a region, it is necessary to determine the variation of seasonal snow accumulation with height for each of the major, topographically homogeneous areas. It may usually be assumed that these relationships hold for the major accumulation season, thereby permitting the extrapolation of estimates from lower to higher elevations. The increase in the length of the accumulation season with elevation and the variation of accumulation with vegetative cover, wind, etc., should be kept in mind in transposition.

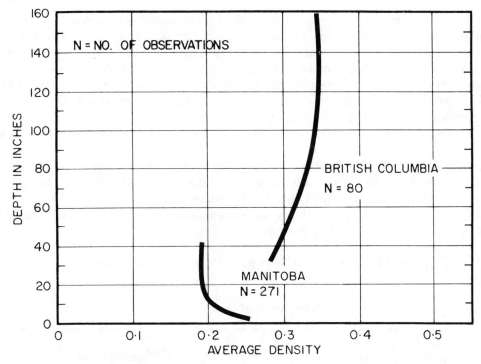

Fig. II.65 Depth vs Average Density, March

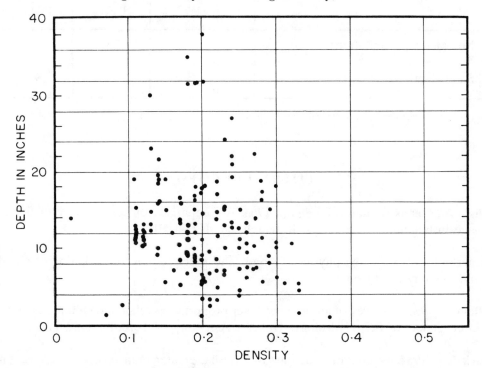

Fig. II.66 Depth vs Average Density, Manitoba, March 10-15

Summated precipitation values may tend to overestimate the snowpack because of ablation which may occur during the summation period (and possibly due to the density assumption in snow measurement). Snow survey measurements overcome these objections, but they must be adjusted for subsequent precipitation and ablation when used in design studies. Snow-depth measurements, available for many climatological stations, provide positive evidence of the existence or non-existence of the snowpack, and therefore provide a check on the reasonableness of summated values. They can be used to obtain very approximate estimates of snow cover water equivalent, but must be used with caution. As an example, a comparison of snowpack water equivalents estimated by summation of precipitation, and by assuming the March 31st pack has a density of 0.2 is given below in Table II.25. An assumed density of 0.3 would make the two sets of values more compatible; on the other hand, the summated values are undoubtedly an overestimate. The seasonal variability of snowpack density in Figs. II.65 and II.66 indicates the hazard in estimating snow density.

Table II.25 Estimates of Snowpack Water Equivalent (inches)

N=40 years of record			
Estimated Water Equivalent for Dates	Return Period		
	1:25	1:50	1:100
Nov-March precipitation*	6.3	7.0	7.7
Snow depth March 31st	3.6	4.0	4.8

*Assuming an average density of 0.09 for freshly-fallen snow.

II.7   LITERATURE CITED

Allis J.A., Harris B., and A.L. Sharp. 1963. A comparison of performance of five rain-gage installations. J. Geophys. Research, 68, 4723-4730.

Bergeron T. 1933. On the physics of clouds and precipitation. Inter. Union of Geodesy and Geophysics, Lisbon.

Billelo M. 1957. A survey of arctic snow cover properties as related to climatic conditions. Inter. Assoc. of Sci. Hydrology, General Assembly of Toronto, 1957 IV pp. 63-77.

Black R.F. 1954. Precipitation at Barrow, Alaska, greater than recorded. Trans. Amer. Geophys. Union, 35:2, 203-206.

Brown M.J., and E.L. Peck. 1962. Reliability of precipitation measurements as related to exposure. Jour. of Applied Meteorology, 1:2, 202-207.

Bruce J.P. 1959. Rainfall intensity-duration-frequency maps for Canada. Canada Dept. of Transport Meteorological Branch, Technical Circular 308, 9 p., 14 charts.

Bruce J.P. 1968. Atlas of rainfall intensity-duration frequency data for Canada. Climatological Studies No. 8, Toronto.

Canada Dept. of Transport Meteorological Branch. Storm rainfall in Canada. A continuing publication, Toronto.

Canada Dept. of Transport Meteorological Branch, 1963. Precipitation, Toronto, 12 p.

Castle G.H. 1965. Telemetering precipitation gages for remote areas. Proc. Western Snow Conference, Colorado Springs.

Church J.E. 1941. The melting of snow. Proc. Central Snow Conference. East Lansing, Mich., U.S.A. 21-32.

Church J.E. 1943. The human side of snow. The Scientific Monthly, 1943, Vol. LVI, 211-231.

Court A. Dec. 1960. Reliability of hourly precipitation data. J. Geophys. Res., Wash., D.C., 65(12), 4017-24.

Court A. 1961. Area-depth rainfall formulas. J. Geophys. Res., 66. 1823-31.

Court A. 1963. Snow cover relations in the King's River Basin. California, J. Geophys. Res., 68:16, 4751-4761.

Curry G.E., and Mann A.S. 1965. Estimating precipitation on a remote headwater area of Western Alberta. Proc. Western Snow Conference, Colorado Springs.

Currie B.W. 1947. Water content of snow in cold climates. Bull. Amer. Met. Soc., 28:3, 150-151.

Ezekial M. and Fox K.A. 1963. Methods of Correlation and Regression Analysis. John Wiley and Sons Inc. New York. p. 15.

Ferguson H.L. 1962. A tephigram overlay for computing precipitable water. Canada Dept. of Transport, Meteorological Branch Tech. Circular 409, 12 p.

Gumbel E.J. 1954. Statistical theory of extreme values and some practical applications. National Bureau of Standards Applied Mathematics Series, 33. Wash. D.C.

Gill H.E. 1960. Evaporation losses from small orifice rain gages. J. Geophys. Res., 65:9, 2877-2881.

Hamilton E.L. and Reimann L.F. 1958. Simplified method of sampling rainfall on the San Dimas Experimental Forest. Tech. Paper No. 26, California Forest and Range Experiment Station Forest Service, U.S. Dept. of Agriculture, 8 p.

Helmers A.E. 1950. Precipitation measurements on wind-swept slopes. Northern Rocky Mountain Forest & Range Experiment Station, Missoula, Mont., 4 p.

Hershfield D.M. 1961. Estimating the probable maximum precipitation. J. Hydraulics Division, Proc. A.S.C.E., Sept., pp. 99-116.

Hershfield D.M. 1965. On the spacing of rain gages. IASH & WMO Symposium on Design of Hydrometeorological Networks, Quebec City, 7 p.

Hershfield D.M. and Wilson W.T., 1960. A comparison of extreme rainfall depths from tropical and non-tropical storms. J. Geophys. Res., 65.969-982.

Holtan H.N., Minshall N.E. and Harrold L.L. 1962. Field manual for research in agricultural hydrology. SWCD, ARS. Washington, D.C., 214 p.

Huff F.A. 1955. Comparison between standard and small orifice rain gages. Trans. Amer. Geophys. Union, 36:4, 689.

Kendall G.R. 1959. Statistical analysis of extreme values. National Research Council Subcommittee on Hydrology First Canadian Hydrology Conference, Ottawa. pp. 54-78.

Kondrat'ev 1954. Radiant energy of the sun. Gidromet, Leningrad 1954.

Kurtyka John C. 1963. Precipitation measurements study. State of Illinois, State Water Survey Division, Urbana Illinois, 178 p.

Kuz'min P.P. 1960. Snow cover and snow reserves. Gidrometeorologicheskoe Izdatelsko Leningrad 1960. Trans. National Science Foundation Wash. D.C. 1963. pp. 99-105.

Landsberg H.E. and Jacobs W.C. 1951. Applied climatology. Compendum of Meteorology, Amer. Met. Soc., Boston, p. 979.

Leonard R.E. and Reinhart K.G. 1962. Some observations on precipitation measurement on forested experimental watersheds. 4 p.

Linsley R.K., Kohler M.A. and Paulhus J.L. 1958. Hydrology for Engineers. McGraw-Hill, New York, p.36.

Lull H.W. and Rushmore F.M. 1960. Snow accumulation and melt under certain forest conditions in the Adirondacks. Station paper no. 160. N.E. Forest Experiment Station, Darby, Pa. 16.

McGuinness J.L. 1963. Accuracy of estimating watershed mean rainfall. J. of Geophys. Research, 68:p 6. 4763-4767.

McKay G.A. 1963a. Relationships between snow survey and climatological measurements, International Assoc. Sci. Hydrol., IUGG Assembly, Publication No. 63, Berkeley Calif., 214-227.

McKay G.A. 1963b. Climatic records for the Saskatchewan River headwaters. Canada Dept. of Agriculture, P.F.R.A., Regina, for East Slopes (Alberta) Watershed Research Program, Calgary, 102 p.

McKay G.A. 1963c. Persisting dewpoints in the Prairie Provinces. Canada Dept. of Agriculture, P.F.R.A., Regina, 23 p.

McKay G.A. 1964a. Statistical estimates of probable maximum precipitation. Canada Dept. of Agriculture, P.F.R.A., Regina, 76 p.

McKay G.A. 1964b. Meteorological measurements for watershed research. N.R.C. Associate Committee on Hydrology, 4th Hydrology Symposim on Watershed Research, Guelph, Ont.

McKay G.A. 1965. Statistical estimates of precipitation extremes for the Prairie Provinces. Canada Dept. of Agriculture, P.F.R.A., Regina, 11 p., 18 maps and diagrams.

McKay G.A., and Blackwell S.R. 1961. Plains snowpack water equivalent from climatological records. Proc., 29th Annual Meeting, Western Snow Conference, Spokane Wash. 1961 pp. 27-42.

Munn R.E. 1953. The measurement of snow depth. Canada Dept. of Transport, Met. Division, Cir 2391, Tec 171, 5 pp.

Myers V.A. 1959. Meteorology of hypothetical flood sequences in the Mississippi River Basin. U.S. Dept. of Commerce Weather Bureau, Hydrometeorological Report No. 35, Wash., D.C.

Neff E.L. 1965. Principles of precipitation network design for intensive hydrologic investigations. WMO - IASH Symposium on Design of Hydrometeorological Networks, Quebec. 9 pp.

Nicks A.D. 1965. Field evaluation of rain gage network design principles. WMO - IASH Symposium on Design of Hydrometeorological Networks, Quebec. 18 p.

Paulhus J.L.H. and Kohler M.A. 1952. Interpolation of missing precipitation records. Monthly Weather Review. 80: pp. 129-133.

Peterson K.R. 1961. A precipitable water nomogram. Bull. Amer. Met. Soc., 42:2, pp. 119-121.

Potter J.G. 1965a. Snow cover. Canada Dept. of Transport, Met. Branch Climatological Studies No. 3, Toronto. 69 p.

Potter J.G. 1965b. Water content of freshly fallen snow. Technical Circular 569. Canada Dept. of Transport, Meteorological Branch, Toronto. 12 p.

Seligman G. 1962. Snow structure and ski fields. R.R. Clarke Ltd. Edinburgh 555 p.

Sharp A.L., Owen W.J. and Gibbs A.E. 1961. A comparison of methods for estimating precipitation on watersheds. Oral presentation, AGU Annual Meeting, Washington D.C., 1961.

Smith J.L., Willen D.W., and Owen M.S. 1965. Measurement of snowpack profiles with radioactive isotopes. Weatherwise Vol. 18, No. 6, 247-257.

U.S. Corps of Engineers. North Pacific Division. "Snow Hydrology," Summary Report of Snow Investigations, 1956. 433 p.

U.S. Dept. of Commerce Weather Bureau 1946. Manual for Depth-Area-Duration Analysis of Storm Precipitation. Washington D.C. 72 p.

U.S. Dept. of Commerce Weather Bureau 1951. Tables of Precipitable Water, Tech. Paper No. 14, Washington, D.C.

U.S. Dept. of Commerce Weather Bureau 1957. Rainfall Intensity Frequency Regime, Part 1 - Ohio Valley. Technical Report No. 29, Washington, D.C., 44 p.

U.S. Dept. of Commerce Weather Bureau 1960. Generalized Estimates of Probable Maximum Precipitation for the United States West of the 105th Meridian. Technical Paper No. 38, Washington, D.C., 66 p.

U.S. Dept. of Commerce Weather Bureau. 1961. Interim report probable maximum precipitation in California. Hydrometeorological Report No. 36. Washington, D.C. 202 p.

U .S. D.A. Soil Conservation Service 1957. "Hydrology" Engineering Handbook Supplement A, Section 4, U.S. Dept. of Agriculture, p. 3.21-3.

Vonnegut B. 1949. The nucleation of ice formation by silver iodide. J. Appl. Phys. 18:593-595.

Warnick C.C. 1953. Experiments with windshields for precipitation gauges. Trans. Amer. Geophys. Union 34, 379-388.

Weiss L.L. and Wilson W.T. 1957. Precipitation gage shields. Intern. Assoc. Sci. Hydrol., General Assembly of Toronto. 1:462-484.

Williams P. and Peck E.L. 1962. Terrain influences on precipitation in the intermountain west as related to the synoptic situation. Jour. of Applied Meteorology 1:3. pp. 343-347.

Wilson W.T. 1954. Discussion of precipitation at Barrow, Alaska greater than recorded by R.F. Black. Trans. Amer. Geophys. Union, Vol. 35 pp. 206-207.

Wolbeer H.J. 1964. Drainage basin study report No. 3. Sask. Research Council, Saskatoon, 10 p.

World Meteorological Organization, 1961. Guide to Meteorological Instrument and Observing Practices, Second Edition, Geneva, Switzerland, 20 p.

World Meteorological Organization, 1965. Guide to Hydrometeorological Practices, WMO-No. 168. TP. 82.

Section III

ENERGY, EVAPORATION
AND EVAPOTRANSPIRATION

by

Donald M. Gray, Gordon A. McKay
and John M. Wigham

TABLE OF CONTENTS

LIST OF TABLES

LIST OF FIGURES

Section III

ENERGY, EVAPORATION
AND EVAPOTRANSPIRATION

III.1   INTRODUCTION

Evaporation is an important process of the hydrologic cycle inasmuch as, on a continental basis, approximately 75 per cent of the total annual precipitation is returned to the atmosphere by evaporation and transpiration.

The amounts of water evaporated constitute a direct loss from both surface and subsurface reservoirs and therefore estimates of these losses are needed in practically all studies concerning water apportionment or balance and reservoir operation. Some appreciation of the average annual free water surface evaporation on the Prairies is given by Fig. III.1 (Prairie Prov. Water Board, 1952).

III.2   EVAPORATION PROCESS

Evaporation is the change of state from a liquid to a gas. The process occurs when water molecules, which are in constant motion, possess sufficient energy to break through the water surface and escape into the atmosphere. Similarly, some of the water molecules contained in the water vapour in the atmosphere that are also in motion may penetrate the water surface and remain in the liquid. The net exchange of water molecules per unit time at the liquid surface determines the rate of evaporation. Further, continued evaporation can take place only when there is a supply of energy to provide the latent heat of evaporation, (approximately 540 calories per gram of water evaporated at 100°C), and there is some mechanism to remove the vapour, so that the vapour pressure of the water vapour in the moist layer adjacent to the liquid surface is less than the saturated vapour pressure of the liquid, (that is, a vertical gradient of vapour pressure exists above the surface).

III.2.1   Methods of Estimating Evaporation from Free Water Surfaces

There are several methods which may be used to estimate the amount of evaporation from a free water surface. In general these may be grouped into the following categories:

1. Mass transfer methods.
2. Energy budget methods.
3. Water budget.
4. Empirical formulae.
5. Measurements from evaporation pans, etc.

III.2.1.1   Mass Transfer

The mass-transfer approach for calculating evaporation from free water surfaces originated from the aerodynamic law first presented in 1802 by Dalton. Dalton recognized that the relationship between evaporation and vapour pressure could be expressed by

$$E = b \, (e_o - e_a). \dots\dots\dots\dots\dots\dots\dots\dots\dots\dots\dots\dots \text{III.1}$$

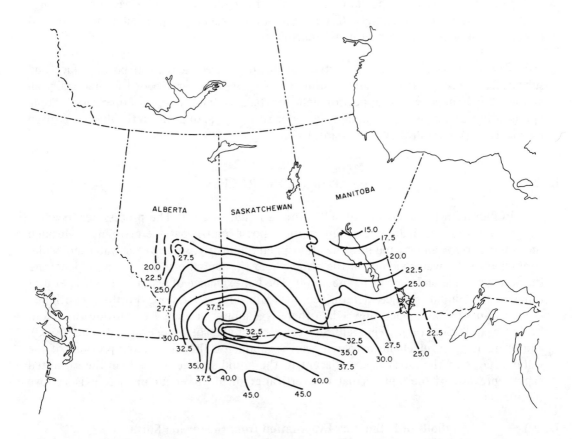

Fig. III.1   Mean Annual Evaporation from Lakes and Reservoirs in Inches of Depth

in which E = evaporation,

        b = empirical coefficient, and

    $e_o - e_a$ = vapour pressure difference in which e_o is
the saturated vapour pressure and e_a is the
actual vapour pressure of the overlying surface.

With slight modification of Equation III.1, one can obtain a simple mass-transfer formula for calculating evaporation of the form,

$$E = \frac{K f(u) (e_o - e_a)}{f(z_o)} \quad \ldots\ldots\ldots\ldots\ldots\ldots\ldots\ldots\ldots III.2$$

in which f (u) = wind velocity function,

    $f(z_o)$ = roughness parameter and,

      K = parameter which includes the effects of
air density and pressure.

Equation III.2 is not readily solved because of the difficulty of evaluating the components, K, f (u) and f(z_o).

In effect, the mass-transfer approach, as the name implies, is based on the determination of the mass of water vapour transferred from the water surface to the atmosphere. Recent advances in the understanding of boundary layer phenomena have provided means of calculating these amounts. However, because of the complexity of the physical process, a rigorous mathematical treatment of the subject is beyond the scope of these notes. For reference to the mathematics of the process the reader is referred to the works of Anderson et al. (1950).

Inasmuch as an understanding of the mass-transfer approach provides an insight to the physics of the evaporation process, at least a cursory treatment of the method is warranted. When air passes over land or water surfaces the lower atmosphere may be divided into three layers; the laminar layer (near the surface), the turbulent layer and the outer layer of frictional influence. The laminar layer, in which flow is laminar, is only in the order of a few millimeters in thickness. In this layer, the temperature, humidity and wind velocity vary almost linearly with height, and the transfer of heat, water vapour and momentum are essentially molecular processes. The overriding turbulent layer can be several meters in thickness depending on the level of turbulence. In this layer, temperature, humidity and wind velocity vary approximately linearly with the logarithm of height, and the transfer of heat, vapour and momentum through this layer are turbulent processes.

The mass-transfer approach can be used to evaluate evaporation if the eddy diffusion in the turbulent layer can be evaluated. That is, one is confronted with the problem of relating velocity distribution, vapour distribution and temperature distribution to momentum and/or to vapour transfer. The process of eddy diffusion is generally related to the wind velocity profile because it is usually assumed that eddy diffusion of water vapour is identical to the eddy diffusion of momentum.

Two general approaches have been taken to evaluate eddy diffusion; the 'mixing length' theory and the 'continuous mixing' theory.

1. The mixing length concept developed by Prandtl (1904, 1934) and Schmidt (1925) considers that in the eddy diffusion process an eddy carrying its properties moves from one position to another where it dies or loses its identity as a distinctly different mass from its surroundings. The length travelled which is analogous to a molecular mean free path is the 'mixing length'.
2. The continuous mixing theory of Sutton (1949) considers the eddy as a mass of a fluid which moves as a discrete mass and blends continuously with its surroundings. The properties of the mass are shared with neighbouring masses until mixing is complete.

Boundary layer mixing processes are still incompletely understood; in addition, all components of the processes cannot be measured accurately by present-day techniques. It would appear, however, that a complete understanding of the process and precise measurements are not required to estimate evaporation by these methods over periods longer than a day (Evans, 1962).

Two of the most widely used mass-transfer equations are those given by Sverdrup (1946) and Thornthwaite and Holzman (1939):

Sverdrup—Mixing Length Theory

$$E = \frac{0.623 \, \rho \, k_O^2 \, u_8 (e_O - e_8)}{p(\ln 800/z_O)^2} \quad \dots\dots\dots\dots\dots\dots\dots\dots\dots\dots\dots \text{III.3}$$

Thornthwaite and Holzman

$$E = \frac{0.623 \rho k_O^2 (u_8 - u_2)(e_2 - e_8)}{p(\ln 800/200)^2} \quad \dots\dots\dots\dots\dots\dots\dots\dots\dots \text{III.4}$$

in which E = evaporation (gm/cm^2/ sec or cm/sec),
ρ = density of the air, (gm/cc),
k_O = von Karman's constant,
u_2, u_8 = wind speeds at heights of 2m and 8m respectively, (cm/sec),
e_O = saturated vapour pressure at temperature of water surface, (mb),
e_2, e_8 = vapour pressures at heights of 2m and 8m respectively, (mb),
p = atmospheric pressure, (mb), and,
z_O = roughness parameter, (cm).

III.2.1.2   Energy Budget

The energy-budget approach has received wide application for estimating the amount of evaporation from a body of water. It is based on the principle of the conservation of energy whereupon the amount of evaporation is calculated from consideration of the thermal budget of the body of water; (see Fig. III.2). In equational form this budget may be expressed as

$$R_{si} - R_r - R_b - R_h - R_e + R_v = R_\theta \quad \dots\dots\dots\dots\dots\dots\dots\dots\dots\dots\dots\dots III.5$$

where R_{si} = total solar radiation incident to the water surface,
           R_r = reflected solar radiation,
           R_b = net longwave radiation exchange between atmosphere and the body of water,
           R_h = sensible heat loss from the body of water to the atmosphere,
           R_e = energy utilized for evaporation,
           R_v = net energy advected into the body of water, and
           R_θ = increase in energy storage in the reservoir

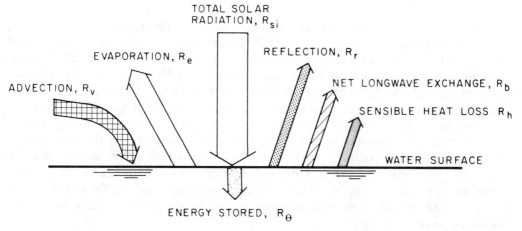

Fig. III.2   Schematic Diagram of an Energy Budget over an Open Water Surface

In Equation III.5 it is assumed that the amounts of energy used in the transformation of kinetic energy to heat, heating due to chemical or biological processes, and conduction of heat through the bottom of the lake or reservoir are negligible.

All the energy terms in the equation are usually expressed in units of cals/cm² or langleys. Thus, the amount of evaporation can be calculated as

$$E = \frac{R_e}{L} \quad \dots\dots\dots\dots\dots\dots\dots\dots\dots\dots\dots\dots\dots\dots\dots\dots III.6$$

in which E = amount of evaporation (cm) and

L = latent heat of evaporation (on the average
about $585-590$ cals/cm³) or $L = 596 - 0.52\,T_0$
where T_0 is temperature of water surface in °C.

Although the 'energy-balance' approach for determining evaporation is basically sound, the application of the budget has been seriously restricted because of the difficulty in evaluating (measuring) the various components. In addition, of course, the method (Equation III.5) should not be applied for estimation of evaporation of shallow bodies of water without consideration of the transfer of sensible heat to the underlying surface.

Evaluation of the Components of the Energy Budget—Radiation from the Sun and Sky, R_{si}

At the outer limit of its atmosphere, the earth receives normal to its surface, solar radiation (short-wave $<1\mu$) amounting to about 2 cals/cm²/min (the 'solar constant'). However, at most latitudes normal incidence does not occur and therefore a horizontal surface at the earth's atmosphere would only receive a portion of the solar constant. Considering the radius of the earth as r, then the earth as a whole receives a total flux of $2\pi r^2$ cal/min. However, since this amount is distributed over the entire earth's surface, then the average flux density at the limit of the atmosphere is $2\pi r^2/4\pi r^2 = 0.5$ cal/cm²/min.

Of the sun's energy available at the outer surface of the earth's atmosphere, only a small portion affects the heat economy of the air and ground. Geiger (1950) suggests that for the northern hemisphere approximately 33 per cent of the total annual incoming radiation is reflected back to universal space. Further, another portion of the radiation is scattered diffusely by air molecules and substances suspended in the atmosphere (dust, plankton). Part of this scattered radiation goes back to universal space and is lost to the terrestrial heat exchange. On the average, reflections from clouds and diffuse scattering into universal space (referred to as the reflecting power or albedo of the earth) account for approximately 42 per cent of the total incoming radiation. Yet another loss is the absorption of radiation by ozone, water vapour, nitrogen compounds and carbon dioxide in the air. This loss in energy is used to increase the temperature of the absorbing body. An illustration of the disposition of solar energy to the earth's surface for northern latitudes is given in Fig. III.3.

From the preceding discussions, it is obvious that the total amount of short-wave radiation reaching the earth's surface is partly direct radiation from the sun and partly reflected or scattered radiation from the sky. Further, the amounts of incoming radiation incident upon a horizontal surface depend mainly on the altitude of the sun, the amount of absorption by the atmosphere and the type and amount of cloud cover.

Two approaches are used to evaluate the solar radiation incident upon the earth's surface:

(1) direct measurement using a pyrheliometer, and
(2) indirect evaluation using easily observable or measurable quantities.

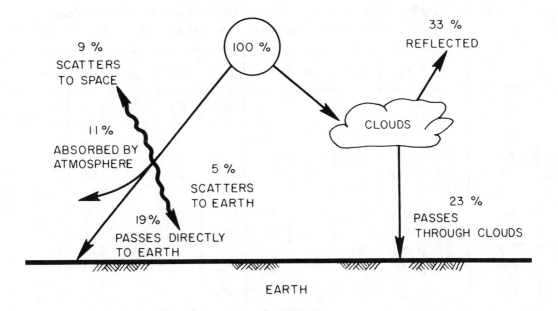

Fig. III.3   Short-Wave Radiation at the Earth's Surface

In the latter method, use is frequently made of tabulated values of the radiation received at the outer surface of the earth's atmosphere as listed in the Smithsonian Tables. For example, Table III.1 prepared by Criddle (1958) lists the mean monthly extra-terrestrial radiation on a horizontal plane in mm of water/day at different latitudes and times of the year.

As outlined in previous discussions, the values of R_A (given in Table III.1) must be adjusted to account for losses due to absorption, reflection and diffuse scattering to arrive at estimates of R_{si}. Generally, this is accomplished using a simple empirical expression for the area of the form:

$$R_{si} = R_A\left(a+b\frac{n}{N}\right) \quad\dots\dots\dots\dots\dots\dots\dots\dots\dots\dots\dots\dots\dots\text{III.7}$$

in which R_{si} = solar radiation received on a horizontal
surface of the ground,

R_A = extra-terrestrial solar radiation received
on a horizontal surface,

n = duration of bright sunshine (hr.), and

N = maximum possible hours of bright sunshine (hr.).

If the value of 'n' has been measured (Campbell - Stokes sunshine recorder), Equation III.7 can be readily solved using values for R_A, N, a and b listed in Tables III.1, III.2 and III.3 respectively.

Table III.1 Mean-Monthly Extra-Terrestrial Radiation on a Horizontal Plane (mm water/day). Northern Hemisphere (L ≅ 560 cal/gm).

North. Lat.	90°	80°	70°	60°	50°	40°	30°	20°	10°	0°
Jan.	-	-	-	1.3	3.6	6.0	8.5	10.8	12.8	14.5
Feb.	-	-	1.1	3.5	5.9	8.3	10.5	12.3	13.9	15.0
Mar.	-	1.8	4.3	6.8	9.1	11.0	12.7	13.9	14.8	15.2
Apr.	7.9	7.8	9.1	11.1	12.7	13.9	14.8	15.2	15.2	14.7
May	14.9	14.6	13.6	14.6	15.4	15.9	16.0	15.7	15.0	13.9
June	18.1	17.8	17.0	16.5	16.7	16.7	16.5	15.8	14.8	13.4
July	16.8	16.5	15.8	15.7	16.1	16.3	16.2	15.7	14.8	13.5
Aug.	11.2	10.6	11.4	12.7	13.9	14.8	15.3	15.3	15.0	14.2
Sept.	2.6	4.0	6.8	8.5	10.5	12.2	13.5	14.4	14.9	14.9
Oct.	-	0.2	2.4	4.7	7.1	9.3	11.3	12.9	14.1	15.0
Nov.	-	-	0.1	1.9	4.3	6.7	9.1	11.2	13.1	14.6
Dec.	-	-	-	0.9	3.0	5.5	7.9	10.3	12.4	14.3

Table III.2 Mean Possible Hours of Bright Sunshine Expressed in Units of 30 Days of 12 Hours Each.

North. Lat.	J	F	M	A	M	J	J	A	S	O	N	D
0°	1.04	0.94	1.04	1.01	1.04	1.01	1.04	1.04	1.01	1.04	1.01	1.04
10°	1.00	0.91	1.03	1.03	1.08	1.06	1.08	1.07	1.02	1.02	0.98	0.99
20°	0.95	0.90	1.03	1.05	1.13	1.11	1.14	1.11	1.02	1.00	0.93	0.94
30°	0.90	0.87	1.03	1.08	1.18	1.17	1.20	1.14	1.03	0.98	0.89	0.88
35°	0.87	0.85	1.03	1.09	1.21	1.21	1.23	1.16	1.03	0.97	0.86	0.85
40°	0.84	0.83	1.03	1.11	1.24	1.25	1.27	1.18	1.04	0.96	0.83	0.81
45°	0.80	0.81	1.02	1.13	1.28	1.29	1.31	1.21	1.04	0.94	0.79	0.75
50°	0.74	0.78	1.02	1.15	1.33	1.36	1.37	1.25	1.06	0.92	0.76	0.70

Considerable care and caution should be exercised before applying Equation III.7 to determine incident radiation over small time intervals. As indicated in Table III.3, the magnitude of the coefficients, a and b, vary with both geographical location and time of the year.

The calculation of R_{si} for different locations in Canada is expedited by using values of the average mid-monthly, daily, short-wave radiation received on a horizontal position of the earth's surface under cloudless skies, R_o (insolation) reported by Mateer (1955a). The values shown in Figs. III.4 to III.9, inclusive, include the average effects of such factors as water content, air mass, etc., of the atmosphere on R_A for different geographic locations.

R_{si} for summer months can be computed from these data by substituting the value of R_O obtained from the figures into the expression: $R_{si} = R_O (0.355 + 0.68 \ n/N)$ (Mateer, 1955b).

Table III.3 Typical Annual Values of Constants 'a' and 'b' of Equation III.7 for Different Geographical Locations

Reference	Geographical Location	a	b
Kimball (1914)	Virginia, U.S.A.	0.22	0.54
Prescott (1940)	Canberra, Aust.	0.25	0.54
McKay	Southern Sask.	0.25 May, Aug.	0.60
		0.34 Sept., Oct.	0.52
Penman (1948)	England	0.18	0.55

Reflected Solar Radiation R_r

The amount of the incoming radiation which is reflected by the surface may constitute a major loss of energy. This amount varies with such factors as the spectral wave length, the solar height and angle, atmospheric turbidity, and the turbidity of the water or the colour and roughness of the surface.

According to Schmidt, with the sun at 90, 60, 30 and 10 degrees above the horizon, the relative percentages of direct solar radiation reflected by a water surface were found to be 2.0, 2.1, 6.0, and 34.8, respectively. Similarly, studies undertaken at Lake Hefner (U.S. Dept. of Interior 1952) showed that the variation in reflectivity, R_r, of a natural flat water surface under a clear sky varied with the sun's altitude, S_A, according to the following expression:

$$R_r = aS_A^b \dots\dots\dots\dots\dots\dots\dots\dots\dots\dots\dots\dots\dots\dots\dots\dots \quad \text{III.8}$$

Values of 'a' and 'b' determined for Lake Hefner were 1.18 and -0.77 respectively. In effect, the studies indicated that the reflectivity of a flat water surface under clear skies is a function primarily of the sun altitude, and that the effects of wind and air-mass turbidity are negligible.

In addition, different bodies exhibit preferential absorptive characteristics for radiation of different spectral lengths. For example, Geiger (1950) cites the reflectivity of leaves and plants for different wave lengths, shown in Table III.4. Also, Schmidt calculated the theoretical reflectivity of diffuse radiation of sky and clouds as 17 per cent.

From the preceding discussions, it is evident that to evaluate completely the reflectivity of a surface to cover all conditions is an exceedingly difficult task. Generally, where direct measurements of this quantity have not been made, it is taken as a percentage of the total incoming radiation. That is

$$R_r = r\, R_{si} \dots\dots\dots\dots\dots\dots\dots\dots\dots\dots\dots\dots\dots\dots\dots\dots \text{III.9}$$

where r = reflection coefficient.

Values for the coefficient, r, of Equation III.9 for different surfaces taken from the works of Geiger (1950), Williams (1961) and Brunt (1944) are reported in Table III.5.

Table III.4 Reflectivity of Leaves and Plants for Radiation of Different Wave Lengths (after Geiger, 1950)

Spectral Range	Wave Lengths	Reflectivity %
Ultraviolet	$<0.36\,\mu$	<10
Visible Light	$0.36\,\mu - 0.76\,\mu$	8 - 20
Infra-red	$0.8\ \mu$	45
	$1.0\ \mu$	42
	$2.4\ \mu$	9
	$10.0\ \mu$	5

Table III.5 Values of the Reflection Coefficient of Different Materials to Short-Wave Radiation

Material	r [Reflection Coefficient]		
New snow ⎫	0.80	—	0.90
Old snow ⎬ refer also to	0.60	—	0.80
Melting snow ⎭ Table II.6	0.40	—	0.60
Ice	0.40	—	0.50
Water	0.05	—	0.15
Forests, Green Leaves	0.05	—	0.20
Tilled soil	0.15	—	0.30
Sand	0.10	—	0.20
Dry Grass	0.15	—	0.25

Net Long-Wave Radiation Exchange, R_b

Most bodies absorb long-wave radiation. For example, the reflection coefficient of sand for long-wave radiation is only 0.11, and for snow, 0.005. Whenever possible the long-wave radiation exchange should be considered in terms of its components: (a) incoming long-wave radiation from the atmosphere; (b) reflected long-wave radiation; and (c) long-wave radiation emitted by the surface. A typical budget for the latter component is shown in Fig. III.10.

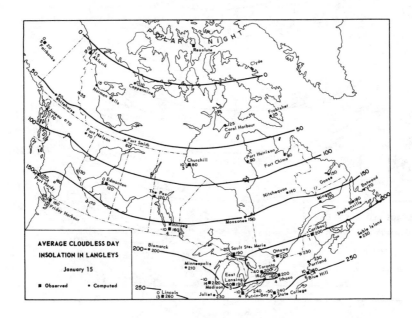

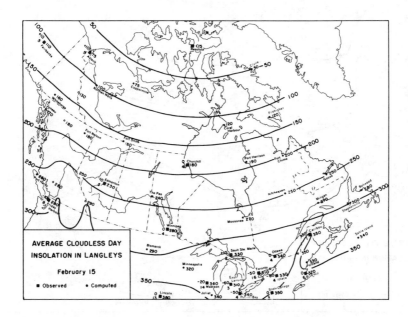

Fig. III.4   Average Cloudless Day Insolation in Langleys — January 15 and February 15

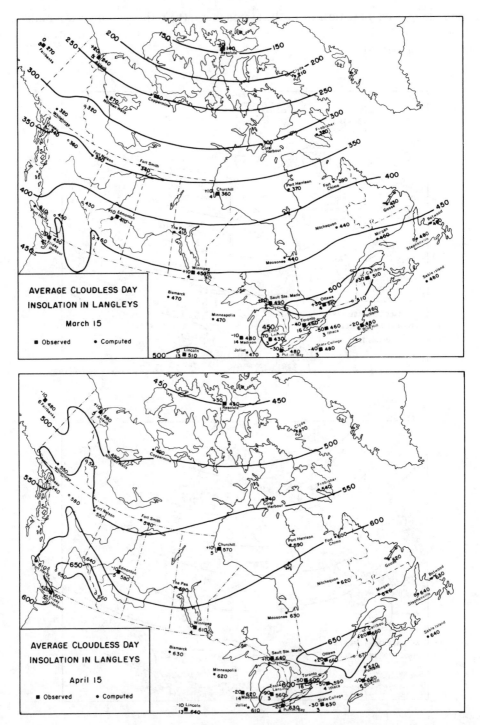

Fig. III.5   Average Cloudless Day Insolation in Langleys — March 15 and April 15

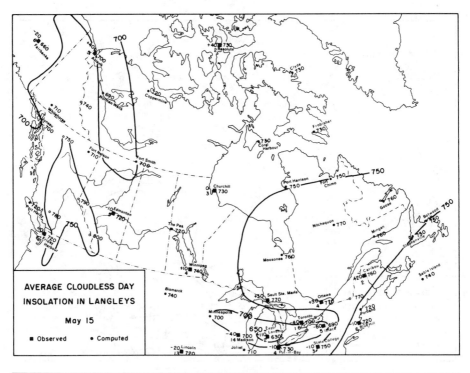

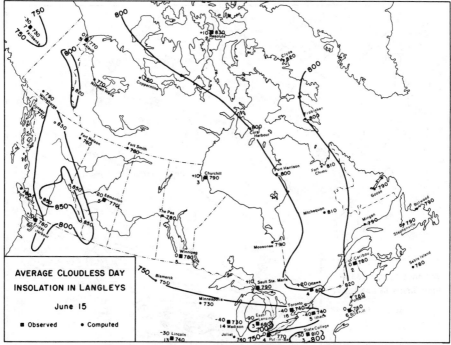

Fig. III.6   Average Cloudless Day Insolation in Langleys – May 15 and June 15

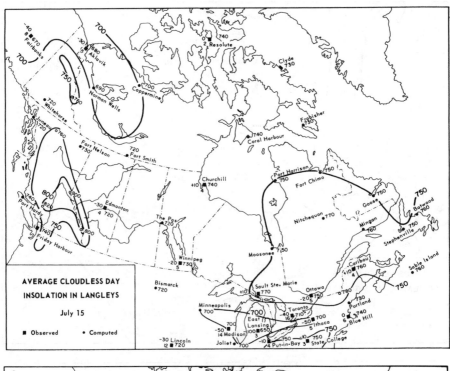

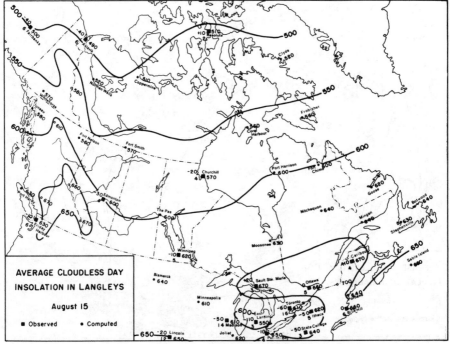

Fig. III.7 **Average Cloudless Day Insolation in Langleys — July 15 and August 15**

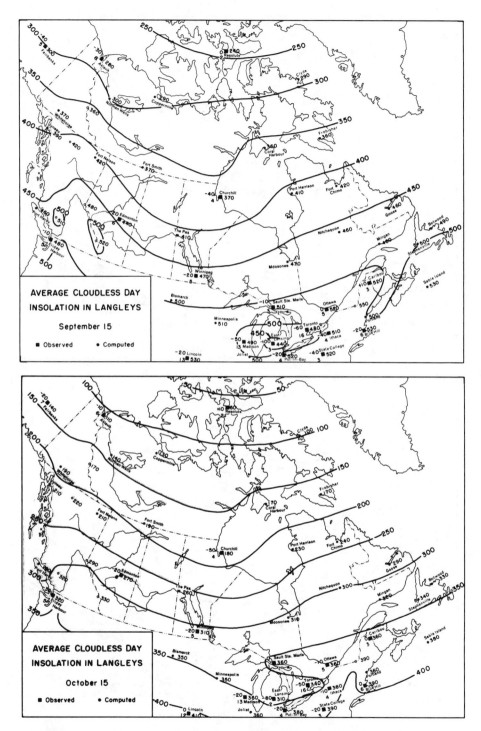

Fig. III.8   Average Cloudless Day Insolation in Langleys – September 15
and October 15

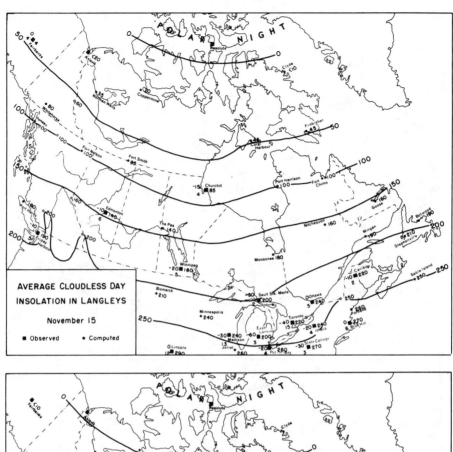

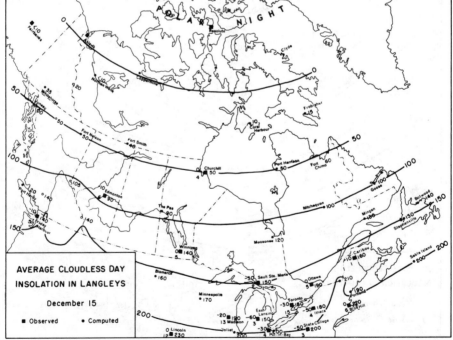

**Fig. III.9   Average Cloudless Day Insolation in Langleys — November 15
and December 15**

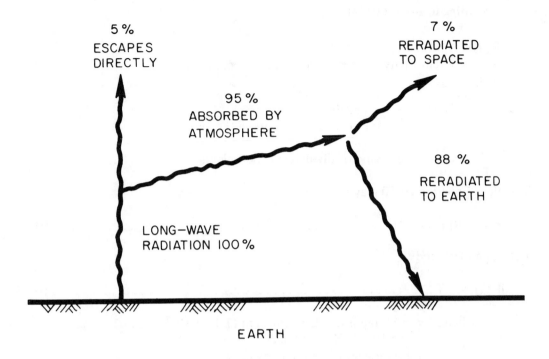

5 %
ESCAPES
DIRECTLY

7 %
RERADIATED
TO SPACE

95 %
ABSORBED BY
ATMOSPHERE

88 %
RERADIATED
TO EARTH

LONG–WAVE
RADIATION 100 %

EARTH

Fig. III.10   Long-Wave Radiation Exchange – Northern Latitude

In Fig. III.10 it can be observed that the net long-wave loss is much less than the total amount of long-wave radiation emitted by the earth. This exchange is greatly influenced by the amount of water vapour in the atmosphere and the cloudiness which effectively blocks the long-wave escape. In arid regions the net long-wave loss is greater at night, and the net short-wave gains greater during the day, than in humid regions (few clouds and little water vapour).

When adequate records are not available, empirical formulae may be used to estimate the net long-wave exchange. These equations may be developed in the following manner:

Under Clear Skies

$$R_b' = R_{LD} - R_{LU} \quad \dots\dots\dots\dots\dots\dots\dots\dots\dots\dots\dots\dots\dots\dots\dots\dots \quad \text{III.10}$$

where R_b' = net long-wave radiation to the surface
under clear skies,

R_{LD} = long-wave radiation directed downward,
and

R_{LU} = long-wave radiation directed upward.

According to Brunt (1944)

$$\frac{R_{LD}}{R_{LU}} = \epsilon \text{ (emissivity)} = a + b\sqrt{e_a} \quad \dots\dots\dots\dots\dots\dots\dots\dots\dots\text{III.11}$$

in which ϵ = emissivity,

e_a = mean vapour pressure in the air (mb)
usually taken at a height of 2 m, and

a,b = empirically-derived coefficients.

Therefore Equation III.10 may be written as

$$R_b = R_{LU} (\epsilon - 1) \quad \dots\dots\dots\dots\dots\dots\dots\dots\dots\dots\dots\dots\text{III.12}$$

But, for a black body

$$R_{LU} = \sigma T^4 \quad \dots\dots\dots\dots\dots\dots\dots\dots\dots\dots\dots\dots\dots\dots\text{III.13}$$

where σ = Stefan-Boltzmann constant $(1.17 \times 10^{-7}$
$cal/cm^2/°K^4/day)$, and

T = temperature of the body (°K)

Substituting Equation III.13 into Equation III.12 one obtains

$$R_b{}' = \sigma T_a^4 (\epsilon - 1) \quad \dots\dots\dots\dots\dots\dots\dots\dots\dots\dots\text{III.14}$$

Also, Penman (1948) has shown that the ratio of actual long-wave radiation to the amount of 'clear sky' long-wave radiation can be related to cloudiness as

$$\frac{R_b}{R_b{}'} = (0.1 + 0.9 \, n/N) \quad \dots\dots\dots\dots\dots\dots\dots\dots\dots\dots\text{III.15}$$

Substituting Equation III.14 into Equation III.15 one obtains an expression for R_b as

$$R_b = \sigma T_a^4 (\epsilon - 1) (0.1 + 0.9 \, n/N) \quad \dots\dots\dots\dots\dots\dots\dots\text{III.16}$$

Equation III.16 must be adjusted to correct for the emissive characteristics of the surface (as it relates to the Stefan-Boltzmann law of black body radiation). It has been found by various investigators that the emissivity factor for water varies from 0.906 to 0.985. Further, it should be pointed out that variations in the magnitude of this factor are extremely important in computing evaporation by the energy budget. As indicated in the Lake Hefner investigation (U.S. Dept. of Interior, 1952, p.90) a change of emissivity from 0.985 to 0.94 would increase the evaporation by 30 per cent when the evaporation rate was 0.20 cm/day and by 20 per cent when the evaporation rate was 0.40 cm/day.

Usually, however, for practical purposes an emissivity factor of 0.97 is used. Thus Equation III.16 becomes

$$R_b = 0.97\sigma T_a^4 \, (\epsilon\text{-}1) \, (0.1 + 0.9 \, n/N) \dots\dots\dots\dots\dots\dots\dots\dots\dots\dots III.17$$

The equation above is an empirical result inasmuch as the coefficients included in the relationship for emissivity Equation III.11 include the effects of several climatic factors. Values of these coefficients determined by several investigators reported in the Lake Hefner report (p.93) are listed in Table III.6. Further, to facilitate the use of Equation III.17, values of the product σT^4 expressed in terms of mm water/day are given in Table III.7.

Table III.6 Different Values of Constants Used in Brunt's Atmospheric-Radiation Equation (Equation III.11)

Investigator	Place	a	b	Correlation Coefficient
Dines	England	0.53	0.065	0.97
Asklof	Sweden	0.43	0.082	0.83
Angstrom	Algeria	0.48	0.058	0.73
Kimball	Washington, D.C.	0.44	0.061	0.29
Eckel	Austria	0.47	0.063	0.89
Raman	India	0.62	0.029	0.68
Anderson	Oklahoma	0.68	0.036	0.92

One other method, which may be used to calculate the net long-wave radiation loss, uses the empirical equations developed by Anderson (1954) that are summarized by Tabata (1958) as

$$R_b(cal/cm^2/day) = 1.141 \left[T_0^4 - T_2^4 \, (a + be_2) \right] \times 10^{-7} \dots\dots\dots\dots\dots\dots III.18$$

$$a = 0.740 + 0.025 \, Ce^{-0.0584 \, h}$$

$$b = 0.0049 - 0.0054 \, Ce^{-0.060 \, h}$$

where T_0, T_2 = temperatures of water surface and air at 2-meter height (°K),
e_2 = vapour pressure at 2 meters (mb),
C = cloud cover (tenths),
h = cloud height (thousands of ft), and
e = base of Naperian logarithms.

Sensible Heat Loss from the Body of Water to the Atmosphere R_h

The direct measurement of R_h is most difficult. To evaluate this component, use is made of the ratio of energy utilized by evaporative processes and that energy conducted to or from the body of water by the air as sensible heat. In 1926, Bowen attempted to relate this ratio to easily measurable quantities. He concluded this ratio could be expressed as

$$B = \frac{R_h}{R_e} = c \left[\frac{T_o - T_a}{e_o - e_a} \right] \frac{p}{1000} \quad \dots \dots \dots \dots \dots \dots \dots \dots \dots \dots III.19$$

where B = Bowen Ratio,

T_o, T_a = temperatures of the water surface and air ($^\circ$C) respectively,

e_o = saturation vapour pressure (mb) corresponding to temperature, (T_o),

e_a = vapour pressure of air, (mb),

p = atmospheric pressure, (mb), and

c = constant.

Table III.7 Values of σT^4 for Various Temperatures.

Temperature, T °F	σT^4 mm water/day
35	11.48
40	11.96
45	12.45
50	12.94
55	13.45
60	13.96
65	14.52
70	15.10
75	15.65
80	16.25
85	16.85
90	17.46
95	18.10
100	18.80

Note: Heat of vaporization was assumed to be constant at 590 cal/gm of H_2O.

The limiting values of 'c' were found to range from 0.58 – 0.66, with an average value under normal atmospheric conditions of 0.61.

Considerable controversy has evolved about the general use of the Bowen ratio for its intended purpose. Unquestionably, the validity of the relation depends fundamentally on the assumption that the eddy diffusivities of the two processes of convection and evaporation are identical. Nevertheless, as pointed out in the Lake Hefner Investigation (U.S. Dept. of Interior, 1952, p. 109):

> The Bowen ratio appears to be sufficiently accurate for computing energy utilized by evaporation for most conditions. In exceptional cases, for example, when evaporation rates are small and the difference in vapour pressure of the atmosphere and that of the air saturated at the surface-water temperature approaches instrumental accuracy the Bowen ratio is inadequate.

Further, as pointed out by Williams (1961) it should be stressed that, unless R_h and R_e can be measured independently, the errors in measuring or estimating all the components in the energy balance equation are accumulated in the remainder $R_h + R_e$. Thus if R_h and R_e are evaluated using Bowen's ratio, it becomes difficult to assess the estimate of R_e.

Net Energy Advected into the Body of Water, R_v

R_v is a measure of the amount of energy which is gained or lost by the inflow of volumes of water to the lake by surface inflow, rainfall, seepage, bank storage, controlled outflow, evaporation and condensation. To evaluate this term, it is necessary to measure the volumes of the various components and their respective temperatures during the time interval for which the budget is performed.

Obviously, in certain situations all the advective components need to be considered, whereas in other situations some may be negligible and thus neglected. In effect, the feasibility of applying the energy budget to determine evaporation from a given lake will be governed largely by the ability to evaluate the quantity of advective energy. Thus, on some lakes, it may be impossible to apply the energy budget approach to obtain reasonable estimates of evaporation.

Change in Energy Storage in the Reservoir, R_θ

In most applications of the energy-budget concept to large bodies of water the heat storage term is difficult to evaluate, because of the problems involved in determining the total volume of the water body and the temperature structure throughout this volume. Essentially, the basic equation in which we are interested is

$$R_\theta = \rho_1 c V_1 (T_1 - T_b) - \rho_2 c V_2 (T_2 - T_b) \dots\dots\dots\dots\dots\dots\dots\dots \text{III.20}$$

in which ρ_1, ρ_2 = density of water at beginning and end of the period,
$\qquad\qquad$ c = specific heat of water,
$\qquad$ V_1 and V_2 = volumes of the body of water at the beginning and end of the period,
$\qquad\qquad$ T_1, T_2 = temperatures of body of water at the beginning and end of the period, and
$\qquad\qquad$ T_b = base temperature.

Where vertical temperature profiles and an area capacity curve for the body of water are available then the amount of energy stored can be computed by the numerical integration method (Lake Hefner Report). That is, the energy storage above a base temperature for a column of water one square centimeter in area and extending to the bottom may be expressed as

$$\text{Energy in column} = \sum_{i=1}^{n} \rho c \, [\overline{T}_i - T_b] \, \Delta h_i + C \quad \ldots\ldots\ldots\ldots\ldots\ldots \text{III.21}$$

Where ρ = density of water,
$\underline{c}$ = specific heat of the water,
$\overline{T}_i$ = average temperature of a layer of water
    h in thickness,
T_b = base temperature, and
C = the amount of energy, above the base temp-
    erature, and in a layer of unit cross section
    whose thickness extends from the bottom
    of the "n" th layer to the bottom of the body
    of water.

Thus, from Equation III.21, the energy storage in a body of water above a base temperature is

$$\text{Energy in body} = \sum_{i=1}^{n} \rho c \overline{A}_{\Delta h_i} \left[\overline{T}_i - T_b\right] \Delta h_i + A_c \, C \ldots\ldots\ldots\ldots\ldots \text{III.22}$$

where $\overline{A}_{\Delta h_i}$ = mean horizontal area of each increment
    of thickness Δh_i, and
A_c = mean horizontal area of area below "n"th
    layer.

And, from Equation III.22,

$$R_\theta = \sum_{i=1}^{n} \rho c \overline{A}_{\Delta h_i} (\overline{T}_i - T_b) \Delta h_i - \sum_{i=1}^{n} \rho c \overline{A}_{\Delta h_i}' \left[\overline{T}_i' - T_b\right] \Delta h_i' + A_c C - A_c' C' \ldots \text{III.23}$$

in which the 'prime' notation refers to final conditions.

As pointed out in the Lake Hefner Report (p. 101), it is probable in any body of water that regions of similar energy storage are defined by approximately the same depth of water. That is, shallow depths of water would probably experience a change of energy storage different from that of deeper water. Therefore, it is suggested that the lake is divided into areas representing equal depths, and that the total energy stored is calculated by summing the energy stored in the individual areas. This technique assumes that representative vertical temperature profiles are available for the areas which demark locations of different depths.

Summary of the Applicability of the Energy
Budget Approach for Use in Estimating
Evaporation from Lakes

In summary, the following points should be recognized when applying the energy-budget approach for estimating the evaporation from lakes:

1. The approach does not consider the flow of heat through the bottom of the lake. This quantity may be an important item in the budget of shallow lakes.
2. The approach assumes that use of the Bowen ratio provides a sufficiently accurate estimate of the amount of energy conducted to or from the body of water to the air as sensible heat.
3. The approach neglects the effects due to radiative diffusivity, stability of the air and spray.
4. The approach is strongly affected by the ability to evaluate the advective energy components. It is essential that the basic equations should be modified to correct for the advected loss of the evaporated water. This can be done in the following manner:
    Rearranging Equation III.5 to consider the amount of advected loss one obtains

$$R_e + R_h = R_{si} - R_r - R_b - R_\theta + R_v - R_w \dots \dots \text{III.24}$$

in which R_w = energy advected out of the body of water by the mass of water evaporated.

But

$$R_e = EL\rho_e \dots \dots \text{III.25}$$

where   E = amount of evaporation,
L = latent heat of vaporization usually taken
as $L = 596 - 0.52\, T_o$ (cal), and
ρ_e = density of water evaporated.

And

$$R_h = B\, R_e \dots \dots \text{III.26}$$

where B = Bowen ratio.

And

$$R_w = c\rho_e\, (T_o - T_b) \dots \dots \text{III.27}$$

where T_o = temperature of evaporating surface, (°C),
T_b = base temperature (0°C), and
c = specific heat of water (cal/gm).

Substituting Equations, III.25, III.26, and III.27 into Equation III.24 one obtains an expression for evaporation:

$$E = \frac{R_{si} - R_r - R_b - R_\theta - R_v}{[\; L(1 + B) + c\,(T_o - T_b)]} \quad \dots\dots\dots\dots\dots\dots\dots\dots\dots\dots\dots\dots\dots\dots \quad III.28$$

5. The energy-budget equation must be used with caution for periods of less than seven days. When applied to periods greater than seven days it will yield an accuracy approaching $\mp$ 5 per cent of the mean evaporation, provided utmost care is exhibited in evaluating the various terms.

For more sophisticated and precise techniques of evaluating the radiative flux components used in the energy budget, the reader is referred to the works of Elsasser and Culbertson (1960).

III.2.1.3   Water Budget

From previous discussions concerning the advected terms of the energy budget, it becomes obvious that an indirect method may be used to compute the evaporation from a body of water. This method, referred to as the Water Budget, is based on the continuity equation, which in its simplified form may be written as

$$E = I - O - \Delta S \dots\dots\dots\dots\dots\dots\dots\dots\dots\dots\dots\dots\dots\dots\dots\dots\dots\dots\dots\dots\dots \quad III.29$$

where   E = evaporation,
        I = inflow,
        O = outflow, and
        ΔS = change in storage.

Generally, Equation III.29 is resolved into its different components and expressed as

$$E = I + P - O - O_g + \Delta S + F \dots\dots\dots\dots\dots\dots\dots\dots\dots\dots\dots\dots\dots\dots\dots\dots \quad III.30$$

in which   P = precipitation,
          O_g = seepage, and
          F = ice stored on the reservoir.

The water-budget method, although having the obvious advantage of being simple in theory has a decided disadvantage; that is, the errors in measurement of the terms entering the equation are reflected directly in the computed amounts of evaporation. Some of the difficulties of applying this method are:

1. If the surface inflow, I, and the outflow, O, are large, relative to the amount of evaporation, then large errors in the calculation of evaporation can occur. Stream flow measurements which provide records that are accurate within 5

per cent are considered excellent. Remember, in the budget, consideration must be given to all sources of inflow and outflow, including municipal uses, etc.

2. Although precipitation in the form of rain can usually be measured reasonably well, precipitation in the form of snow can create problems. Firstly, the actual water equivalent of the snowfall added to the lake is difficult to assess. Further, a lake generally acts as a type of trap, so that considerable discrepancy usually exists between the amount of snow recorded on a gauge located on land and the amount which actually falls on the lake.

3. Probably the most difficult term to evaluate in the water budget is the seepage, O_g. This component can be estimated knowing the hydraulic conductivity of the bed of the lake and the hydraulic gradient. Nevertheless, it should be recognized that the water-budget method of determining evaporation will prove most successful when applied to relatively 'tight' lakes in which the amount of seepage is relatively small in comparison to the amount of evaporation.

4. Determination of the change in storage, ΔS, requires that measurements of the water level (stage) are taken and an accurate area-capacity curve for the lake is available. Even with these records, the bank storage component may cause error in the calculation if the budget is applied over a short time period. However, bank storage can usually be neglected if the budget is used on an annual basis. Also, because of the lag effect, it is difficult to associate changes in inflow and outflow measurements with the change in storage over small time periods; Linsley et al. (1958, p. 94) also suggest that in large reservoirs all changes in the volume of water due to expansion and contraction caused by large temperature changes must be taken into account.

Also, it is pointed out in the Lake Hefner studies:
> Although it is theoretically possible to use the water budget method for determination of evaporation from any lake or reservoir it is usually impractical to do so because of the effects of errors in measuring various items. Evaporation as determined by this method is a residual, and therefore may be subject to considerable error if it is small relative to other items.

Because of these facts it is not recommended that the method be applied to time periods less than one month in duration if the estimate of evaporation is expected to be within ∓ 5 per cent of the actual amount which occurs.

III.2.1.4 Empirical Methods

The energy-budget and mass-transfer methods used to estimate evaporation amounts, while theoretically correct, require data which, for many studies, are not readily available. Also, in many cases it is highly questionable whether it is economically feasible to instrument a lake sufficiently to acquire these data. Frequently therefore, under such circumstances, resort is made to the use of empirical formulae to obtain estimates of evaporation. Many formulae have been developed using either the energy-balance or the mass-transfer method as a basis which may be used in these situations.

Most of the empirical equations are based on the simple aerodynamic equation:

$$E = K \; f(u) \; (e_O - e_a) \quad\quad\quad \text{III.31}$$

in which $f(u)$ = wind speed function,
e_O = vapour pressure of saturated air at
   the temperature of the surface,
e_a = actual vapour pressure of air at some
   height above the surface, and
K = constant.

Some of the more common of these empirical formulae used to estimate evaporation from lake surfaces are the following:

Meyer (1915). . U.S.A. . . . Shallow Lakes
$$E \; (\text{in/mos}) = 11(1+0.10 \; u_8) \; (e_O - e_8) \quad\quad \text{III.32}$$

Rohwer (1931) . . . U.S.A.
$$E \; (\text{in/day}) = 0.771 \; (1.465 - 0.0186 \; p) \; (0.44 + 0.118 u_O) \; (e_O - e_a) \quad\quad \text{III.33}$$

where p = barometric pressure (in/Hg).

Penman (1948) . . . England . . . Small Tank
$$E \; (\text{in/day}) = 0.35 \; (1 + 0.24 u_2) \; (e_O - e_a) \quad\quad \text{III.34}$$

Lake Hefner – Marciano and Harbeck(1952) . . . . U.S.A. . . . reservoir, 2200 A
$$E \; (\text{in/day}) = 0.0578 u_8 \; (e_O - e_8) \quad\quad \text{III.35}$$
and
$$E \; (\text{in/day}) = 0.0728 u_4 \; (e_O - e_2) \quad\quad \text{III.36}$$

Kuzmin (1957) . . . U.S.S.R. . . Reservoirs with surface >20–100m
$$E \; (\text{in/mos}) = 6.0 \; (1 + 0.21 u_8) \; (e_O - e_a) \quad\quad \text{III.37}$$

In the equations above it should be noted that the wind speed is expressed in mph and vapour pressure in in. Hg. Further, the subscripts attached to the terms refer to the height in meters at which the measurements are taken. Also, the vapour pressure term, e_O, is frequently taken as the saturated vapour pressure at the mean air temperature during the interval of measurement.

Probably the greatest appeal of empirical formulae is their simplicity and the fact that they permit estimates of evaporation to be made from standard meteorological data.

Nevertheless, it is essential that users of these equations recognize their limitations. The equations require a measure of the surface temperature of the body of water. This measurement is, of course, very difficult to obtain. If the mean air temperature is used in place of the surface temperature, then the effects of advected energy to the lake on evaporation are not considered. This may cause considerable error in the calculated amounts of evaporation, because small errors in temperature induce large errors in the calculations. Also, measurements of the wind speed and vapour pressure should be taken at the heights specified by the equation which is used. Usually, it is difficult to adjust data collected at different heights because neither a rigorous wind law, nor laws defining the variation in humidity with height, are presently available.

Kohler *et al.* (1955) developed a graphical technique for evaluating the evaporation from shallow lakes, based on the energy-budget and aerodynamic approach in a similar manner to that used by Penman (see Equation III.49). This graph is shown in Fig. III.11. In application of these data for computing evaporation from deep lakes the reader should use Equation III.39.

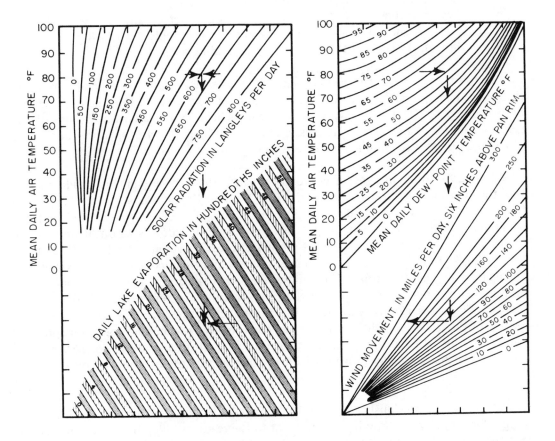

Fig. III.11 **Graphical Technique for Evaluating Evaporation from Shallow Lakes (Kohler *et al.*, 1955)**

III.2.1.5   Evaporation Pans

For many years, measurements taken on evaporation pans have been used to provide estimates of the amounts of evaporation from lakes and reservoirs. The popularity of pans for this purpose stems from the fact that they are inexpensive, simple to instrument, and generally, the annual ratio of lake to pan evaporation remains reasonably constant from year to year, and to a lesser extent, from region to region.

Various types of pan exposures may be used. These include sunken pan, floating pan, and surface pan.

The sunken pan is placed below ground level; it has the advantages that certain boundary effects such as direct solar radiation on (and thus the consequent heat exchange through) the side walls are negligible. On the other hand, these pans offer several disadvantages because

(a)   they tend to collect trash and are difficult to maintain,
(b)   they are difficult to install and leaks are hard to detect, and
(c)   it is difficult to evaluate the amount of heat which is transferred directly through the walls of the pan to the surrounding soil.

In the floating type, the pan is floated in a lake. Although it is generally conceded that these pans probably give the best estimates of lake evaporation, their use has not been widespread. Obviously, many operational difficulties prevail with the use of this type of pan; for example, their inaccessibility, the problem of water splashing into or out of pan because of wave action, and others.

The most common type of evaporation pan is the surface pan. In Canada and the U.S. the standard surface pan used is the U.S. Weather Bureau Class A pan. This pan is 4 ft. in diameter and 10 in. deep. It is constructed of galvanized steel and water is maintained within 2-3 in. of the top. When properly installed, it is set on timbers so that the bottom is 6 in. above the adjacent ground surface. This eliminates difficulties caused by drifting soil and snow. Earth is banked up over the timbers leaving approximately a one-inch space at the top to permit some air circulation under the pan. Wind measurements should also be taken at the same location. The temperature of the water in the pan fluctuates closely with the air temperature. Evaporation is determined from records of the water level changes in the pan corrected for the amounts of water added by rainfall and by artificial filling. In cases when heavy and intense rains have occurred, the correction of readings may prove difficult because of the loss of water by overflow and splash effects.

As indicated previously, in many instances lake evaporation is determined from pan evaporation simply by applying a coefficient to the readings taken from the pan. The pan coefficient is taken equal to the ratio of the lake evaporation to the pan evaporation. Table III.8 lists a few of these coefficients for a Class A pan which have been found by different investigators.

Discussion

The data reported in Table III.8 show that the annual pan coefficient is less than unity. This is to be expected because of differences which exist in factors affecting the heat exchange, such as exposure, mass, and areas of the evaporating surfaces of two different bodies of water. In the case of a pan, the effect of a particular weather event on evaporation is relatively independent of antecedent weather largely because of its limited heat storage capacity. For a lake, however, the evaporation occurring one day is not independent of the energy exchange which possibly occurs over a considerable time

Table III.8
Class 'A' Pan Coefficients

Reference	Location	Season	Years of Record	Coefficient
Sleight (1917) 12 ft pan	Denver, Colo.	annual	1915-16	0.67
Rowher (1931)	U.S.A.	annual		0.69
Rowher (1934) 85 ft reservoir	Ft. Collins, Colo.	Apr.-Nov.	1926-28	0.70
White (1932) 12 ft pan	Utah	May-Oct.	1926-27	0.67
Hall (1934)	U.S.A.	annual		0.69
Subcommittee (ASCE) (1934)	U.S.A.	annual		0.70
U.S. Dept. of Interior (1952)	Texas	annual	1939-47	0.68
U.S. Dept. of Interior (1952)	Florida	annual	1940-46	0.81
U.S. Dept. of Interior (1952)	Oklahoma	annual	1950-51	0.69
U.S. Dept. of Interior (1958)	Ariz., Nev.	annual	1952-53	0.60
Young (1947)	California	annual		0.77
Young (1947) 12 ft pan	California	annual		0.77

period preceding the time of measurement. Thus large differences between the heat capacities of lake and pan may produce large variations in the pan coefficients if only short time periods are considered. For example, during time intervals of one month or less, the pan coefficients may be quite erratic, and may show a pronounced seasonal march due to the temperature, resulting from the different capacities of the bodies of water to store heat. It should be recognized that, when comparing evaporation measurements from pans of different sizes, the heat storage terms should not be neglected, and the data should be adjusted by energy-budget calculations. Linsley *et al.* (1949) have developed a graphical solution of this calculation which employs air temperature, solar radiation, dew point and wind speed.

Correction for Advected Energy

Also, in the experimental determination of pan coefficients, some of the variation in the values may be attributed to differences in amounts of heat flow across the bottom and sides of the pan. Kohler *et al.* (1955) suggested a method which may be used to determine evaporation from shallow lakes from pan evaporation measurements which have been corrected for advected energy. The equation used for this calculation is

$$E_L = 0.70 \left[E_p + 0.00051 p \alpha_p \, (0.37 + 0.0041 \, u) \, (T_o - T_a)^{0.88} \right] \quad \ldots\ldots\ldots\ldots \text{III.38}$$

in which E_L = lake evaporation (in/day),
E_p = pan evaporation (in/day),
p = atmospheric pressure (in Hg),
α_p = proportion of advected energy to
the pan used for evaporation (see Fig. III.13)
u = wind speed (mpd),
T_o = surface water temperature ($^\circ$F), and
T_a = air temperature ($^\circ$F).

The solution of Equation III.38 can be obtained using Figs. III.13 and III.14.

Length of Evaporating Surface

The length of evaporating surface influences the evaporation rate because, as a unit of air containing a large concentration of vapour moves over an adjacent square unit of water lying in its path, it reduces the pressure gradient on that area, and hence, the evaporation potential. Experimentally it has been found that the evaporation from an area, whose axis is a distance 'x' downwind, is proportional to x to the power 0.67. Fig. III.12 shows the ratio of pan evaporation from different sizes of pans to evaporation from an 1800-acre reservoir.

Evaporation from Shallow and Deep Lakes

In the case of a deep lake it is necessary to consider the amount of heat energy that is stored and the net energy that is advected into the lake from inflow and outflow. Kohler *et al.* (1955) suggest the following equation may be used to obtain evaporation from a deep lake when using estimated amounts from shallow lakes:

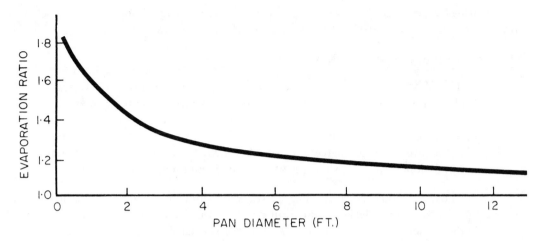

Fig. III.12   Ratio Pan Evaporation to Reservoir Evaporation

$$E_L' = E_L + \alpha_L (R_v - R_\theta) \dots\dots\dots\dots\dots\dots\dots\dots\dots\dots\dots\dots\dots\dots III.39$$

where E_L' and E_L = evaporation from deep and shallow lakes, respectively,

α_L = proportion of advected energy into the lake used for evaporation (see Fig. III.l5),

R_v = net energy advected to the lake as measured from inflow and outflow, and

R_θ = change in stored energy in the lake (measured).

Summary of the Use of Pan Measurements for Determining Lake Evaporation.

In manner of summary regarding the use of pan coefficients for determining lake evaporation the following points should be noted:

1. Daily pan evaporation can be estimated from energy-budget considerations.
2. Pan coefficients will probably provide estimates of annual lake evaporation withing 10-15 per cent (on the average) of actual evaporation amounts.
3. When pan measurements are used to estimate lake evaporation, corrections must be made to these measurements to account for advected energy.

III.2.1.6   Summary

In order to summarize the Section on methods of computing evaporation from a free water surface it is appropriate to refer to the study conducted by McKay and Stichling (1962) on the Weyburn Reservoir located in Southern Saskatchewan. In this study, McKay determined the evaporation from the reservoir by eight different methods: (a) Water Budget; (b) Kohler, Nordenson and Fox (1955); (c) Penman (1948); (d) Meyer (1915); (e) Lake Hefner and Lake Mead (U.S. Dept. of Interior, 1952);

(f) Floating Evaporation Pan; (g) Class 'A' pan; and (h) Energy Budget. The evaporation amounts calculated for 1961 by the different methods are shown plotted in Fig. III.16.

As a result of this study McKay concluded:

The Meyer formula, that of Kohler, Nordenson and Fox, the Energy Budget and 0.7 x the Class 'A' pan evaporation gave estimates of *seasonal* evaporation which were within 2 per cent of that determined by the water budget. Because of the heat-storage effects of the reservoir, these methods did not give the same agreement on a *monthly* basis. The energy-budget method alone gave consistent results seasonally and monthly.

Those methods which duplicated the water-budget evaporation are not equally suited to design, allocation or operational purposes. The energy-budget method, precisely used, requires special instrumentation and observations, and it is therefore most suitable for operational problems where high accuracy is required. For reservoirs with characteristics and exposures similar to the Weyburn reservoir, the 'Class A' pan and Meyer formula appear to give acceptable estimates of evaporation. However, methods using 'Class A' pan data are not suitable for design purposes since there are few acceptable historical records. The Meyer formula may be used for both purposes if reliable measurements or estimates of air-and-water-surface temperatures are available.

A water-surface-temperature network is required so that water-temperature—air-temperature relationships may be known more precisely. Few water-temperature records are available, therefore empirical relationships must be used to estimate this value for design studies. Small errors in temperature estimates may introduce large errors in computed evaporation. An observing network would give the high quality measurements required for current use and information from which more exact relationships may be determined.

The aerodynamic formulae also are dependent on accurate wind measurements or estimates. Since prairie anemovanes range in height from 35 to 90 ft. above ground the wind estimates must be reduced to the conditions under which the formulae were developed. Meyer (1915) uses a reduction method which may result in underestimates of prairie evaporation. More information is required on the wind structure on the prairies so that more accurate reductions of wind data may be made.

The floating pan may yield good quality estimates of seasonal evaporation when its coefficient is established. However, in view of the servicing difficulties and the reasonable accuracy of other methods, the pan is not likely to gain general favor.

The experimental period is still too short to make very positive claims. Also the experiment is based on a water budget which has a probable error of about 10 per cent. It is hoped that the energy-budget method as used at Weyburn may be placed on a more rigorous basis, thereby gaining an additional reference.

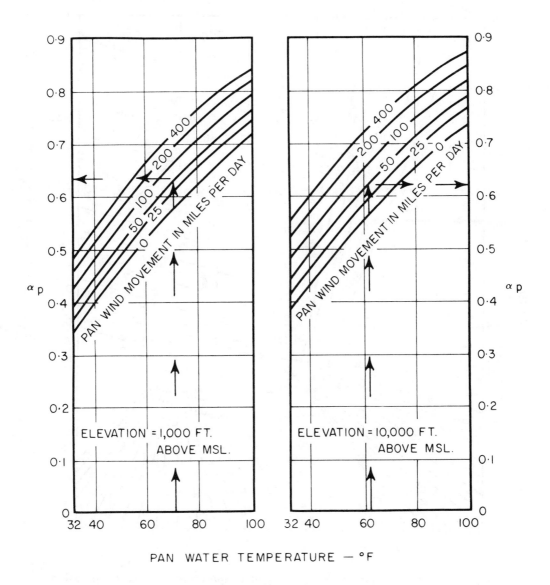

Fig. III.13   Proportion of Advected Energy (Into a Class 'A' Pan) Utilized for Evaporation

As an additional reference concerning the use of different methods of calculating lake evaporation, the reader is referred to the work of Bruce and Rodgers (1962), which summarizes evaporation studies of the Great Lakes System.

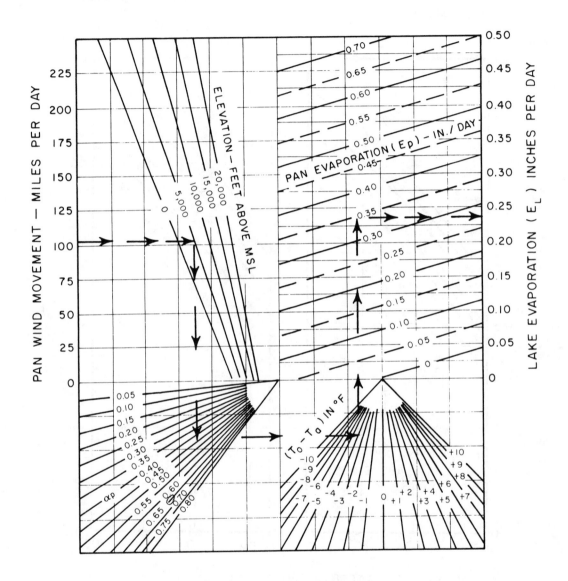

Fig. III.14   Conversion of Class 'A' Pan Evaporation to Lake Evaporation

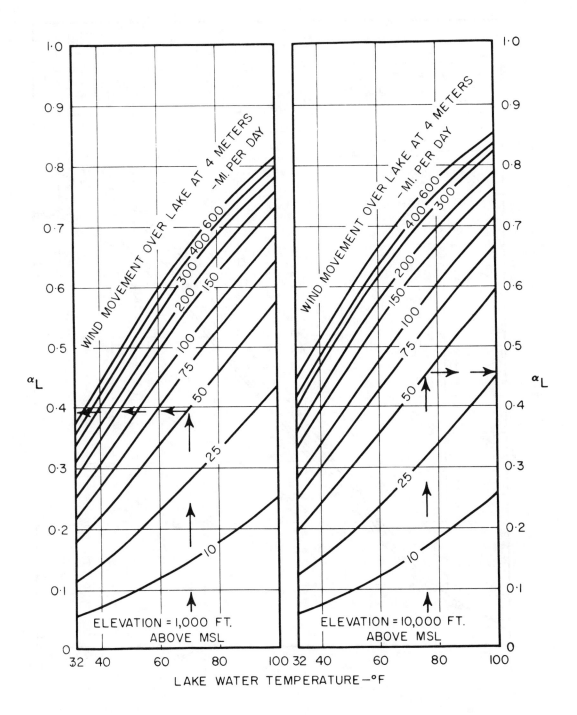

Fig. III.15 Proportion of Advected Energy (Into a Lake) Utilized for Evaporation

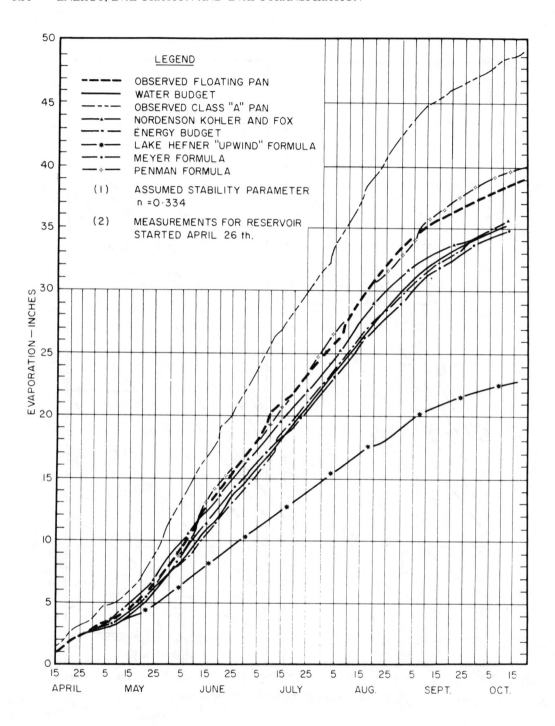

Fig. III.16 Weyburn Reservoir 1961 – Cumulative Measured and Computed Evaporation

III.2.2   Evaporation from Snow

Evaporation from snow may be treated fundamentally as evaporation from any other surface. There are, however, certain properties, peculiar to snow which affect the net energy exchange and consequent evaporation from these surfaces.

1. Snow surfaces have high and variable reflectivity coefficients whose magnitude depends on the age, water content, depth, etc. (refer to tabulated values given in Sections II and III).

2. Evaporation from snow involves a 3-phase change of state from a solid-liquid-gas. In general it requires approximately 675-680 cal/gm to sublimate snow, whereas the heat of fusion is only approximately 80 cal/gm. Frequently, because of these properties, the mass of water melted is substantially greater than the amount evaporated.

3. Snow has a limiting upper boundary of temperature ($0°C$). Under normal conditions the saturated vapour pressure over a snow surface at this temperature is approximately 6.11 mb. The fact that the saturated vapour pressure is low is extremely important to the evaporation from snow, inasmuch as evaporation can only occur when there is a vapour pressure gradient in the direction from snow to air. As pointed out by Diamond (1953), since the vapour pressure of air increases with increasing temperature, warm air must have a lower relative humidity than cold air when at the same vapour pressure. For example, the maximum relative humidities, which air can have at various temperatures and still have a vapour pressure of less than 6.11 mb, are

$0°C - 99.9\%, 5°C - 70\%, 10°C - 49.7\%, 15°C - 35.7\%,$ and $20°C - 26\%$

Thus, even though the air temperature may rise and more heat is transferred to a snowpack, if the relative humidity remains the same, the evaporation potential of the air decreases and more heat becomes available for melting. The occurrence of evaporation or condensation from snowcover can be determined when the temperature and relative humidity of the air are known.

4. Under conditions where the vapour pressure gradient is favourable for evaporation, yet there is insufficient heat to satisfy the heat of vapourization requirements, evaporation will continue at the expense of a decrease in the surface temperature of the snowpack.

5. The thermal conductivity of snow is very low, and thus the amounts of heat transferred to or from the evaporating surface by conduction are usually very small. Values of the diffusivities for various snowpacks are listed by Williams (1961).

III.2.2.1   Estimation of Evaporation from Snow

As indicated above, the amount of evaporation from snow can be determined by employing methods similar to those used to estimate free water surface evaporation. In these calculations, the primary components of the energy exchange include the net short-wave and net long-wave radiation and the turbulent or convective transfer of sensible heat. In winter, at our latitudes (the Prairies), because of the low solar elevation, the amounts of incident radiation received on a horizontal surface are very small (see Figs. III.4 and III.9). Because only small amounts of short-wave radiation are received and a large percentage of this is reflected, the net short-wave exchange is small (probably on the average less than 100 ly/day). Further, during the winter, the net long-wave loss may offset the short-wave gain. Under these conditions, when the mean air temperature is above the snow surface temperature, the major source of energy originates from turbulent transport, and the amount of evaporation can be calculated from Sverdrup's (1936) mass-transfer equation (Equation III.3) using $k_O = 0.42$ and $z_O = 0.25$ cm. On the average, as shown by Williams (1961), in northern latitudes during the winter the amount of evaporation is negligible (where chinooks do not prevail). In fact, in local regions adjacent to cities there may be a net gain of water due to condensation.

In the spring, the net short-wave radiation received by the earth's surface increases rapidly with time, and therefore becomes an important factor to both the evaporation and snowmelt process. Complete details concerning the snowmelt process are given in the text Snow Hydrology (U.S. Dept. of Commerce, 1956). This phenomenon will be discussed in a subsequent Section of these notes.

Two other equations which may be used to estimate evaporation from snow are

Kuzmin . . . U.S.S.R.

$$E = (0.18 + 0.098\, u_{10})\, (e_O - e_2) \quad\dots\dots\dots\dots\dots\dots\dots\dots\dots\dots\dots \text{III.40}$$

where E = evaporation from snow (mm/day),
u_{10} = average daily wind speed, at a height of 10 m (m/sec),
e_2 = vapour pressure at 2 m (mb), and
e_O = saturated vapour at snow surface temperature (mb).

Central Sierra Snow Laboratory . . . U.S.A. (Snow Hydrology)

$$E = 0.0063\, (z_a\, z_b)^{-1/6} (e_O - e_a) u_b \quad\dots\dots\dots\dots\dots\dots\dots\dots\dots \text{III.41}$$

where E = evaporation (in/day),
u_b = wind speed in mph at a height above ground,
z_b, in ft.,
e_O = saturated vapour pressure at snow surface
temperature (mb), and
e_a = vapour pressure (mb) at a height z_a,
in ft above ground.

III.3   EVAPOTRANSPIRATION

Evapotranspiration[1] of water by plants means, broadly, the amount of water required to produce the mature crop. The implication is that such water has been 'used' or is no longer directly available to man. It has been either stored in the plant tissue, or converted to vapour and discharged into the atmosphere where it re-enters the familiar hydrologic cycle and eventually appears again somewhere on the earth as precipitation.

III.3.1.   Factors Affecting Evapotranspiration

Inasmuch as evapotranspiration includes the sum of volumes of water used by both evaporation and transpiration processes, it is obvious that many of the factors, primarily climatic factors, which influence the amount of evaporation from a free water surface, also affect the amount of evapotranspiration. For example, solar radiation intensity and duration, wind conditions, relative humidity, cloud cover, atmospheric pressure and others. In addition to the climatic factors, however, both soil and vegetative factors govern the evapotranspiration from an area.

III.3.1.1 Vegetational Factors

There is a common conception that most plants, when providing full shading of the soil surface, have very close to the same evapotranspiration. It should be realized, however, that quantitatively at least there should be differences in the evapotranspiration rates of different species of plants when full cover is realized. The type, colour, density and stage of growth of a plant will affect its exposure, sensible heat and insulating properties, reflective power, etc., and thereby affect the proportioning of incoming solar radiation to the various components of the net exchange. Similarly, the stage of growth, density and shape of plants affect air turbulence in the overlying air and exchange of water between the evaporating surface and the atmosphere.

In addition, light, wind and other factors influence the stomatal opening and closing of different plants in different ways. These factors affect the ability of a plant to transmit water from its root system to the leaves. Further, different plants have different lengths of growing season, and therefore it can be expected that their seasonal water requirements will differ.

III.3.1.2 Soil Factors

The predominant soil factors affecting evapotranspiration are those that affect the amounts of water that are available at the soil surface and to the plants. When the surface layer of a soil is wet, evaporation is governed primarily by atmospheric conditions. However, as this layer dries out, the rate of evaporation decreases very rapidly and is greatly influenced by soil properties such as the relative humidity of soil air, the diffusion

---

[1] In this Section, the terms evapotranspiration and consumptive use are taken to be synonymous and will be used interchangeably throughout the text.

coefficient, the capillary conductivity and the hydraulic conductivity of the surface layer. These soil properties govern the rate at which water, either in liquid or vapour form, is transmitted from lower depths in the profile to the surface.

The diffusion coefficient of a soil, which is a measure of the diffusion of water vapour under a unit vapour pressure gradient, depends primarily on the number, size and distribution of air-filled pores in the soil. That is, the nature, texture, granulation and moisture content of a soil, which influence the diffusion process, are important only to the extent that they alter the air-filled pore space. On the other hand, the capillary conductivity, which characterizes the rate of capillary flow, is largely a function of the soil moisture content, the size, shape and distribution of the pores, and fluid properties.

Transpiration by a plant normally creates a diffusion pressure deficit in the plant roots, so that a potential gradient develops across the root surface which is in contact with the moist soil. This process results in a potential gradient in the soil that causes water to move to the plant. Again, as discussed previously, all factors such as texture, structure, pore space, etc., which affect the capillary conductivity, the diffusion coefficient or potential gradients, and thus the rate of movement of water in unsaturated soils, also affect transpiration rates.

The influence of soil moisture content or soil moisture tension on the evapotranspiration rate is imperfectly understood, and subject to considerable controversy. Veihmeyer and Hendrickson (1955) have indicated that soil moisture tension has little effect on the transpiration rate of plants until the permanent wilting point is reached (~ 15 atmosphere). Conversely, Taylor and Haddock (1956), and others, have demonstrated that soil moisture tension affects the availability of water and thus the rate at which it can be removed. In general, it is agreed that transpiration is dependent on both moisture tension and the water-conducting properties of the soil; however, the importance of this relationship may vary widely under different soil and plant conditions.

The preceding discussion attempts to point out a few of the complexities of the evapotranspiration process and the many factors which affect this phenomenon. In effect, we can consider evapotranspiration to depend primarily on three major factors: the capillary movement of water through the unsaturated soil; the transmissibility of the stomata; and the evaporative capacity of the atmosphere.

III.3.2   Methods of Estimating Evapotranspiration

Many methods are available which may be used to estimate the evapotranspiration requirements of plants. These methods may be grouped under the following headings:

1.   Soil moisture depletion studies on small plots.
2.   Tank and lysimeter experiments.
3.   Study of groundwater fluctuations.
4.   Differences of inflow-outflow measurements.
5.   Empirical constants applied to tank evaporation measurements.

6. Theoretical methods based on either the physics of vapour transfer or of heat energy.
7. Soil moisture budget.
8. Correlations of environmental climatic factors with irrigation and precipitation records.
9. Effective heat or Day degree method.

III.3.2.1 Field Plots

Determination of consumptive use from field plots simply involves an inventory procedure. The amounts of water added to the plot by precipitation and irrigation are measured in addition to surface runoff. Consumptive use is calculated as the difference between input and surface runoff, and adjusted for changes in the soil moisture storage (determined by direct soil moisture measurement).

Generally, no attempt is made in the method to evaluate deep percolation losses, hence water is necessarily added to the plot only in small quantities so as to minimize this loss. Further, the water table at the location of the plot should be deep, so that the plants do not derive any of their supply from this source.

The primary advantage of this method is that the determinations of consumptive use are made under field conditions.

III.3.2.2 Tanks and Lysimeters

As indicated by Kohnke, Dreibelbis and Davidson (1940), lysimeters have been used for water studies since 1688. However, their use as an instrument primarily for measurement of evapotranspiration did not develop until late in the 18th century.

Evapotranspiration rates are determined from lysimeters by measuring the water loss from tanks (lysimeters) on which plants are grown. As pointed out by Pelton (1961), lysimeters are classified according to their method of construction as 'Ebermayer', 'Filled-In', and 'Monolith'. In the 'Ebermayer' type, a percolate funnel is placed under a block of soil left *in situ*. The lysimeter has no side walls and therefore cannot be used in itself for evapotranspiration studies, but is limited to qualitative percolation investigations. The 'Filled-In' lysimeter consists of a container with vertical side walls and open top and bottom, that provides for percolation. Generally these lysimeters are filled with soil which has been stripped from the area in structural layers. Every attempt is made to fill the lysimeters in such a manner that they approach natural conditions as near as possible. The 'Monolith' or 'Soil-Block' lysimeter is composed of a casing built around a natural block of soil and provided with an open bottom. Percolation is collected in receiving tanks.

Evapotranspiration is determined on lysimeters from measurements of the amount of water applied, outflow and changes in the soil moisture in the tank. Changes in the soil moisture level may be obtained by direct moisture sampling (gravimetric samples, neutron probe, etc.), or by using weighing or hydraulic methods. In the weighing-type lysimeter, the tank is mounted on a self-recording scale. Perhaps the most famous of these is the installation at Coshocton, Ohio, which weighs a 65-ton mass within ∓ 5 lb. With the hydraulic method, the tank is floated and changes in weight are recorded as pressure changes by a manometer. King, at Guelph, has utilized this technique with considerable success.

Some controversy has arisen over the use of lysimetric data for determining the amount of evapotranspiration. The arguments against the use of these installations for this purpose are predicated on the basis that differences may exist between the lysimeter and natural conditions in soil profile, soil moisture regimen, plant rooting characteristics, methods of water application and the net energy exchange. Nevertheless, it is generally conceded that if the installations meet certain minimum standards, they will provide reasonably reliable information on evapotranspiration of plants over short time periods (Linsley *et al.*, 1949, and Pelton, 1961). For further reading on the subject of installation of lysimeters, the reader is referred to the article by Pelton. Other articles on the subject that should prove useful are those by Harrold and Dreibelbis (1958), Gilbert and Van Bavel (1954), King *et al.* (1956), and Visser (1962).

III.3.2.3   Inflow-Outflow Measurements

This method involves the application of the water-balance principle to large watersheds. The amount of water received by the area (inflow or precipitation) is measured in addition to the outflow from the area. The difference between the inflow and outflow measurements, adjusted for changes in groundwater storage, is taken as evapotranspiration.

III.3.2.4   Study of Groundwater Fluctuations

In situations where vegetation obtains most of its moisture for growth from the capillary fringe above a water table, then amount of use is reflected in changes in the position of the water table. For shallow water-table conditions, White (1932) suggests that consumptive use may be estimated by the expression:

$$CU = Y_s (24a + b)$$

in which CU = consumptive use (in/day),
Y_s = specific yield of the soil,
a = rate of rise of the water table from midnight to 4 am (in/hr), and
b = net change in water table elevation during day (in).

Values of 'a' and 'b' can be determined from water-stage recorder charts for wells in the area of interest. Linsley *et al.* (1949) suggest this method of determining CU is applicable only in areas where the water table is near the surface. It has been used quite widely to estimate the use of water by phreatophytes around small potholes or sloughs.

III.3.2.5   Evaporimeters—Pans and Atmometers

Numerous investigations have been undertaken to try to relate the amounts of evaporation from free water surfaces, or potential evapotranspiration, with measurements taken from various types of pans and atmometers. At this point, the reader is referred to previous discussions of the use of evaporation pans for estimating evaporation from lakes.

A wide variety of atmometers or small-plate evaporimeters, such as the Wilde, Piché, Bellani and Alundum disc, have been used throughout the world for measuring water loss by evaporation. A description of all the types is beyond the scope of these notes, however reference is given to two evaporimeters most commonly used, the Bellani Plate and the Piché. The Bellani Plate consists of a thin, porous, black ceramic disc, 7.5 cm in diameter, fused to the large end of a glazed, ceramic funnel. Water is conducted through the lower open end of the funnel from a burette which acts as a reservoir and a measuring device. The Piché evaporimeter consists of an 11-mm diameter glass tube open at one end and graduated to read directly the depth of water evaporated. The evaporating surface is a 3-cm diameter piece of blotting paper, giving a total evaporating surface of approximately 11 cm^2. It is obvious that losses of water from atmometers provide an index of the evaporative ability of the atmosphere.

The use of atmometers has increased materially during recent years, particularly by workers in the field of agriculture. The primary reasons for this increased use may be attributed to the fact that the instruments are portable, inexpensive, easily installed, maintained and standardized, and that they do not require large amounts of water. Nevertheless, there is considerable indecision among workers as to the value of these instruments for estimating evaporation from free water surfaces. Mukammal (1961) in his paper presents an excellent discussion of the use of evaporimeters for their intended purpose. A summary of his findings is given below:

1. Atmometer observations are difficult to interpret, and it is inadvisable to rely on their results until there is greater understanding of conditions created by the instrument. For example, Mukammal and Bruce (1960) report that the Bellani Plate is very responsive to wind, but less so to net radiation which is the dominant factor influencing water loss due to evaporation from extensive natural water surfaces. In comparing the relative importance of the three major factors involved in evaporation—net radiation, humidity and wind—they found the ratio of components for the pan was 80:6:14, and for the Bellani Plate 41:7:52.

2. Different conversion factors may be required with atmometers for different ranges of evaporation amounts as well as for different locations.

3.   A class 'A' type or larger pan would appear preferable to atmometers for use in estimating evaporation from lake surfaces.

One of the most intensive studies in Western Canada concerning the use of evaporimeters for estimating the consumptive use requirements of different crops, is that reported by Sonmor (1963). The values of seasonal coefficients to be applied to various evaporimeters to estimate the consumptive use of crops for maximum yield in southern Alberta, as listed by Sonmor, are given in Table III.9.

Table III.9   Coefficients to be Applied to Various Evaporimeter Readings to Estimate Seasonal Consumptive Use of Different Crops in Western Canada.

Crop	Black Bellani Plate (in/cc)	Class 'A' Pan (in/in)
Perennials	in/cc	in/in
alfalfa	0.0030	0.66
grass, pasture	0.0031	0.68
Annuals		
soft wheat	0.0030	0.66
hard wheat	0.0029	0.66
oats	0.0028	0.59
barley	0.0028	0.64
flax	0.0026	
peas	0.0031	
Row Crops		
sugar beets	0.0025	0.54
potatoes	0.0025	0.56
corn	0.0022	0.50
tomatoes	0.0024	

Coefficients to be applied over the length of the growing season.

III:3.2.6   Theoretical Methods—The Energy Budget

Considerable attention was given in previous Sections on the use of the energy-budget approach for evaluating the amount of evaporation from free water surfaces. This approach can also be used, however, in estimating the potential evapotranspiration[1] requirements of plants. It should be recognized that the main

---

[1] Potential Evapotranspiration, PE, is the amount of water a given plant, in given condition, will use in evaporation and transpiration if sufficient water is available in the soil to meet the demand. It is dependent on energy supply. Actual evapotranspiration refers to the actual loss of water by the processes as influenced by the combined effects of demand and available water supply.

differences in the net radiation exchange between a vegetative surface, a wet soil and a free water surface, are in the amount of radiation absorbed and the storage of heat during the day that causes evaporation at night. Further, as also indicated in previous discussions, other factors to be considered include the effect of stomatal opening, the roughness of the surface which affects the turbulent transport of water vapour, and the sensible heat requirements.

King (1961) summarizes the components of the net energy exchange for a land surface in illustrative form as shown in Fig. III.17. In the figure, if only the energy flux in the vertical direction is considered (neglecting the horizontal divergence of heat flux terms), a simple heat budget can be written in terms of gains and losses in energy.

That is

$$H = R_N = S + ET + K + N + \text{Storage Terms} \dots\dots\dots\dots\dots\dots\dots\text{III.42}$$

in which   H = heat budget,
R_N = net radiation,
S = energy to soil heat,
ET = energy used for evapotranspiration,
K = sensible heat to air, and
N = energy used by plant in photosynthesis.

King (1961) relates that the divergence terms are most significant near the borders of cropped fields. Horizontal divergence of energy usually shows up as more sensible heat, or less latent heat, entering the upwind side than being released downwind.

Equation III.42 can be simplified further if consideration is given to the relative importance of the various terms. For example, Penman (1954) indicates that, during the summer months, if the percentage of net radiation is considered to be 100, then the

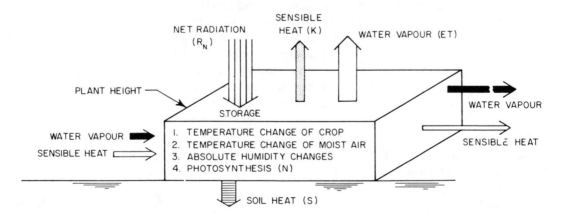

Fig. III.17   Energy Balance Over a Vegetational Surface

disposition of this energy would be evapotranspiration–85, heating air–9, soil heating–4 and plant growth–1. Measurements made by Lemon (1956) indicate that photosynthesis may require up to 10 per cent of the net radiation during early morning and late afternoon hours, and 5 per cent at midday. Thus, it would appear that photosynthesis should be considered in the budget for short-term studies, although it probably can be neglected when considering the balance over longer periods such as the growing season.

Similarly, Tanner (1960) suggests that for an alfalfa-brome crop, the storage terms were 2 per cent of the net radiation at night, and negligible during the day. For cereal crops, the storage factor would be smaller than for alfalfa; conversely, for a forest, the factor may be significant. Generally, however, dependent on the type, stand and density of the crop, it is believed that storage is insignificant if periods of 24 hours or longer are considered.

In view of the preceding comments, if the smaller terms of the budget are neglected, the heat balance on a daily basis may be written as

$$H = K + ET \dotfill \text{III.43}$$

Penman Method

Penman (1948) solved Equation III.43 by considering evaporation from a free water surface, in which case the free surface evaporation, E, replaces the evapotranspiration term, ET, of the equation. The basic difficulty in obtaining a solution to the equation is in determining the proportion of the net energy utilized by each of the components, E and K. Penman showed that this could be accomplished by utilizing the aerodynamic equations in the form

$$E = f(u)(e_0 - e_a) \dotfill \text{III.44}$$
$$K = \gamma f'(u)(T_0 - T_a) \dotfill \text{III.45}$$

in which $f(u), f'(u)$ = functions of wind speed,
e_0 = saturated vapour pressure of water at the surface temperature, T_0,
e_a = actual vapour pressure of the air,
T_a = mean air temperature, and
γ = constant to keep the units consistent (equal to 0.27 mm Hg/°F).

In practice the mean surface temperature, T_0, is extremely difficult to determine. However, if it is assumed that the temperature difference, $(T_0 - T_a)$, is small, then it can also be assumed without serious error that

$$\Delta(T_0 - T_a) = e_0 - e_s \dotfill \text{III.46}$$

in which Δ = slope of the saturation vapour pressure-temperature curve at,

T_a, and

e_s = saturated vapour pressure at T_a.

Substituting the relationships given by Equations III.44, III.45 and III.46 into Equation III.43 one obtains,

$$H = \frac{\gamma f'(u)(e_0 - e_s)}{\Delta} + f(u)(e_0 - e_a) \quad \dots\dots\dots\dots\dots\dots\dots \quad \text{III.47}$$

If it is also assumed that $f'(u) = f(u)$ (Penman suggests there is experimental evidence to confirm this, and also, only large differences in the function will cause important errors in the results) then Equation III.47 can be written

$$H = \frac{\gamma f(u)(e_0 - e_s)}{\Delta} + f(u)(e_0 - e_a) \quad \dots\dots\dots\dots\dots\dots\dots \quad \text{III.48}$$

Substituting the relationship given by Equation III.44 into Equation III.48 one obtains

$$\frac{\Delta(H-E)}{\gamma} = f(u)(e_0 - e_s)$$
$$= f(u)(e_0 - e_s + e_a - e_a)$$
$$= f(u)(e_0 - e_a) - f(u)(e_s - e_a)$$
$$= E - f(u)(e_s - e_a)$$
$$= E - E_a$$

in which $E_a = f(u)(e_s - e_a)$

Therefore $$E = \frac{\Delta H + \gamma E_a}{\Delta + \gamma} \quad \dots\dots\dots\dots\dots\dots\dots \quad \text{III.49}$$

The following equations may be used to evaluate the terms of Equation III.49:

$$H = R_{si}(1-r) - R_b \quad \dots\dots\dots\dots\dots \quad \text{III.50}$$

Penman

$$H = (1-r) R_A \left[0.18 + 0.55\, n/N\right] - \sigma T_a^4 \left[0.56 - 0.09\sqrt{e_a}\right]\left[0.10 + 0.90\, n/N\right] \quad \dots \quad \text{III.51}$$

Mateer; Canadian Conditions

$$H = (1-r) R_O \left[0.355 + 0.68\, n/N\right] - \sigma T_a^4 \left[0.56 - 0.09\sqrt{e_a}\right]\left[0.10 + 0.90\, n/N\right] \quad \dots \quad \text{III.52}$$

Note: The two equations above are but two alternative expressions which may be used to determine 'H'. Reference should be made to earlier discussions on the application

$$0.95 \ R_A\left(0.18 + 0.55 \ \frac{n}{N}\right) (mm \ day^{-1})$$

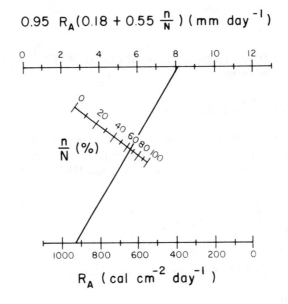

$$R_{si}(1-r) = 0.95 \ R_A\left(0.18 + 0.55 \ \frac{n}{N}\right)$$

Fig. III.18 Nomogram I

$$0.95 \ R_0\left(0.355 + 0.68 \ \frac{n}{N}\right) (mm \ day^{-1})$$

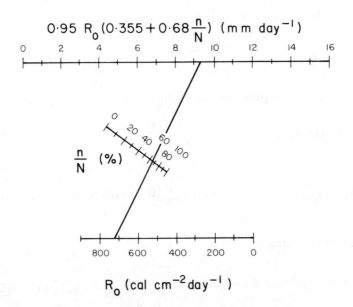

$$R_{si}(1-r) = R_0\left(0.355 + 0.68 \ \frac{n}{N}\right)(1-r)$$

Fig. III.19 Nomogram Ia

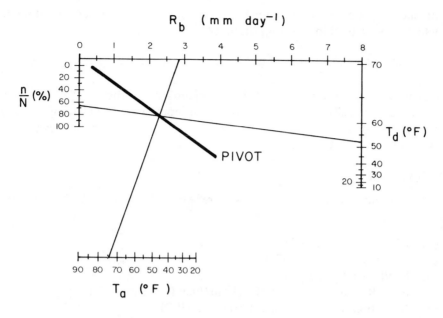

$$R_b = \sigma T_a^4 (0.56 - 0.092 \sqrt{e_a}) (0.10 + 0.90 \frac{n}{N})$$

Fig. III.20 Nomogram II

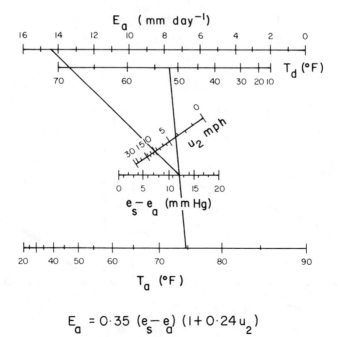

$$E_a = 0.35 (e_s - e_a) (1 + 0.24 u_2)$$

Fig. III.21 Nomogram III

of energy budget in determining the evaporation from lakes for other formulae which may be used in the computations.

$$E_a = 0.35 \, (e_s - e_a) \, (1 + 0.24 \, u_2) \dots\dots\dots\dots\dots\dots\dots\dots \text{III.53}$$

Nomogram solutions for these equations prepared by Turner (1957) are given in Figs. III.18 – III.21 inclusive. In the use of these nomograms, radiation is expressed in cal/cm²/day, sunshine ratio in per cent, temperatures in °F, vapour pressures in mm Hg wind speed in mph, and r = 0.05.

Example Calculation

Estimate the free water surface evaporation at Saskatoon by the Penman Method, given:

(1)   Air Temperature = 74.5°F
(2)   Relative Humidity = 47.5 %
(3)   Wind Speed = 10 mph
(4)   Solar Radiation, R_A, = 940 cal/cm²/day (see Table III.1)
(5)   Insolation, R_O, = 720 ly/day (see Fig. III.7)
(6)   Reflection coefficient = 0.05
(7)   Sunshine n/N = 65 %

From Nomogram I (Fig. III.18)

$$0.95 \, R_A \, (0.18 + 0.55 \, n/N) = 8.2 \text{ mm/day}$$

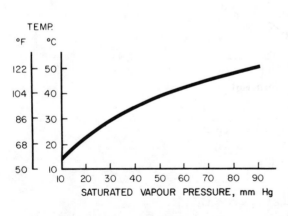

Fig. III.22 Saturated Vapour Pressure — Temperature Curve for Water

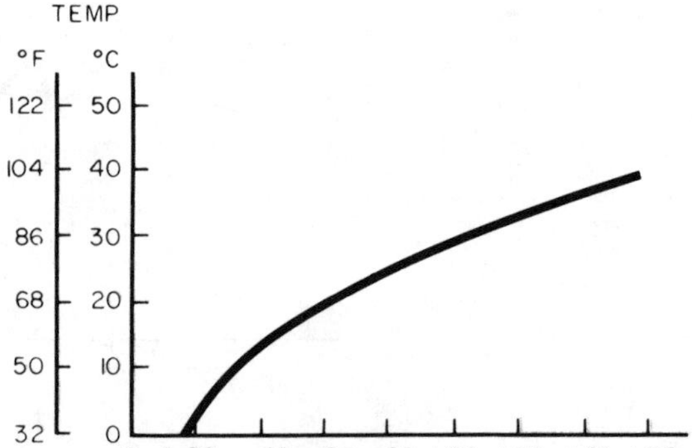

Fig. III.23 Slope of the Saturation Vapour Pressure Curve, Δ

From Nomogram III (Fig. III.21) and (Fig. III.22)

$$e_s = 23 \text{ mm Hg}$$
$$e_a = 0.475 \text{x} (23) = 10.9 \text{ mm Hg}$$
$$e_s - e_a = 12.1 \text{ mm Hg}$$
$$T_d = 52\,^{\circ}F$$
$$E_a = 14.3 \text{ mm/day}$$

From Nomogram II (Fig. III.20) using $T_d = 52^{\circ}F$; $R_b = 2.9$ mm/day

Thus $H = 8.2 - 2.9 = 5.3$ mm/day

$$E = \frac{(0.74)\ (5.3) + (0.27)\ (14.3)}{0.74 + 0.27} = 7.7 \text{ mm/day} = 0.30 \text{ in/day}$$

If Mateer's relationship is used

$$E = \frac{(0.74)\ (6.4) + (0.27)\ (14.3)}{0.74 + 0.27} = 8.5 \text{ mm/day} = 0.33 \text{ in/day}$$

In the above calculations, it is evident that if values of T_a, r, e_a, e_s and u_2 are measured (or selected) for a vegetative or soil surface, the evaporation amount calculated by the Penman Method estimates the potential evapotranspiration requirements of the crop; that is, $E \cong PE$.

Use of the Penman Method for Determining Evapotranspiration

Evaporation amounts determined by the Penman Method may be used to provide estimates of Potential Evapotranspiration, PE; that is, the maximum amount of water a plant will use assuming the supply to be non-limiting. By this definition it is assumed that the plant is lush-growing and completely covers the ground, and that the evaporative power of the atmosphere governs evapotranspiration.

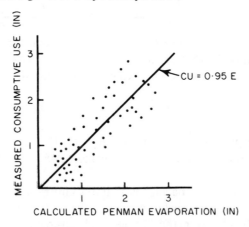

Fig. III.24   Relation Between Consumptive Use by Alfalfa Determined from Soil Moisture Measurements and Estimates Made from Penman's Heat Budget Equations

Nicholaichuk (1964) found the above concept to be applicable in defining the use rate of water by a heavy stand of alfalfa at Saskatoon over different time intervals (2, 6 and 14 days) under conditions where the moisture level of the soil was maintained above that which it would retain at 7 atmospheres pressure. In the study, the actual consumptive use (evapotranspiration) was determined in the field from direct readings of soil moisture taken with a neutron gauge, and for the calculated amounts, the reflection coefficient, r, was taken as 0.25. The findings of this study, shown in Fig. III.24, indicate that consumptive use can be related to evaporation as

$$CU = 0.95 \ E \ \dots\dots\dots\dots\dots\dots\dots\dots\dots\dots\dots\dots\dots\dots\dots\dots\dots\dots\dots III.54$$

These results provide evidence that when water is non-limiting, meteorological factors predominate in determining the use rate for the alfalfa crop. Also, there is evidence that the Penman equations provided estimates of consumptive use with comparative accuracy for all time intervals considered (2, 6 and 14 days).

When seasonal estimates of evapotranspiration are to be made from the evaporation amounts calculated by the Penman Method, Penman suggests that the coefficients listed in Table III.10 be used.

Table III.10   Crop Factors to be Applied to Free Surface Evaporation Estimates

Season	Coefficient
May—August (inclusive)	0.80
March, April, September, October	0.70
November—February (inclusive)	0.60

He states that it might be necessary to revise these constants for other latitudes to compensate for a change in the length of daylight, and that towards the equator the values should converge to a value of approximately 0.75 for all seasons.

Summary

In manner of summary regarding the use of the energy-budget approach for determining the evapotranspiration from land surfaces, King (1961) states:

> . . .the energy balance is an important means for estimating the evaporation from land surfaces. For relatively short periods of time and for non-uniform cover the storage and divergence terms may be of major importance. For many estimates of evaporation a knowledge of only the vertical energy balance is sufficient. The net radiation is the

principal energy source and for ample soil moisture is nearly equal to the energy used for evaporation. A sensible heat flux to the surface often is a source of energy for irrigated fields because of advection from dry and warmer surroundings.

III.3.2.7   Soil Moisture Budget

In the preceding discussions concerning the use of evaporimeter measurements and energy-budget calculations for estimates of evapotranspiration, consideration was given only to the case where water was non-limiting. In actual fact, however, plants cannot utilize the total available water (water held in the range between field capacity[1], FC, and the permanent wilting point[2], PWP, within the root zone) at a constant maximum rate. Holmes (1961) has shown that the ratio of actual evapotranspiration, AE, to potential evapotranspiration, PE, changes as the soil dries out, and that the shape of this curve differs both with the type of soil and the drying rate (see Fig. III.25).

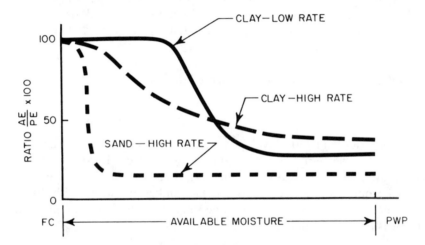

Fig. III.25   Schematic Drying Curves Showing Ratio of AE/PE Plotted with Soil Moisture Content for Different Soils and Drying Rates

Unfortunately, it is impossible to define quantitatively the shape of curve in terms of the factors affecting it. Undoubtedly, the shape reflects the influence of such factors as (a) the ability of the media to transmit water in the unsaturated state, (b) the ability of the plant to absorb water and convey it from the roots to the leaves, and (c) the root development and distribution pattern; and possibly, a host of other factors.

From the above discussion, it is obvious that an accurate description of the moisture status of a soil as it affects evapotranspiration is not simple; however, it is equally evident

---

[1]Field capacity: the amount of moisture retained by a soil initially at a high water content, which is permitted to drain by gravity, for a specified period of time (usually 2-3 days).

[2]Permanent wilting point: the soil moisture content when plants permanently wilt.

that potential rates must be modulated to account for such factors as soil moisture stress if they are to provide reasonable estimates of actual evapotranspiration rates and the soil moisture status. Various methods of modulating potential rates to predict soil moisture withdrawal are presented in the literature (Holmes 1961, Kerr 1963, Robertson and Holmes 1959, and Taylor and Haddock 1956). For example, the budget proposed by Holmes (1961) divides the soil root zone into two zones, an upper and lower zone, as to water availability. Holmes assumed: (a) all moisture from the upper zone is evapotranspired at the potential rate; (b) moisture in the lower zone is withdrawn at a decreasing rate depending on the amount of moisture in the zone (that is the PE values are modulated using a factor of less than 1, Holmes using the following factors—first 25% of soil moisture evapotranspires at 0.50 PE, second 25% at 0.20 PE, next 25% at 0.10 PE, and last 25% at 0.05 PE); and (c) available water is withdrawn from uppermost moist soil zone before extraction occurs from the lower zone.

Modulated values of evapotranspiration rates are generally used in conjunction with Soil Moisture Budget calculations which simply involve a book-keeping of the soil moisture. This procedure simply accounts for the moisture by taking into consideration the amounts of water added or depleted from the soil by precipitation, evapotranspiration and percolation.

III.3.2.8   Correlations of Environmental Climatic Factors

Blaney-Criddle Method

Blaney and Criddle (1950) developed a simplified formula for estimating consumptive use in the arid western regions of the United States. In this method, as in the Thornthwaite Method (discussed subsequently), it is assumed that the heat budget is shared in fixed proportion between heating the air and evaporation. With this assumption, it is evident that consumptive use should be related to the hours of sunshine and temperature (a measure of solar radiation).

In the Blaney- Criddle Method, monthly consumptive use, cu, is found by multiplying together the mean monthly temperature, T_m the monthly per cent of annual daytime hours, p, and a monthly crop coefficient, k. Expressed mathematically

$$cu = \frac{kT_m p}{100} = kf \qquad \dots \dots \dots \dots \dots \dots \dots \text{III.55}$$

where: $\dfrac{T_m p}{100}$ = monthly consumptive use factor, f.

The seasonal consumptive use, CU, is obtained by summing the monthly consumptive use amounts over the growing season.

$$CU = \Sigma kf = K\Sigma f = KF \dots \dots \dots \dots \dots \dots \dots \text{III.56}$$

Note: K is an average seasonal consumptive use coefficient selected for different crops and lengths of growing season, see Table III.11.

Table III.11   Consumptive Use Coefficients (K) for Irrigated Crops in Western U.S.A.

Item	Length of Growing Season or Period	Consumptive Use Coefficients Seasonal (K)	Maximum Monthly * (k)
Alfalfa	frost-free	0.85	0.95–1.25
Beans	3 months	0.65	0.75–0.85
Corn	4 months	0.75	0.80–1.20
Deciduous orchard	frost-free	0.65	0.70–0.75
Pasture, grass, hay annuals	frost-free	0.75	0.85–1.15
Potatoes	3 months	0.70	0.86–1.00
Small grains	3 months	0.75	0.85–1.00
Sorghum	5 months	0.70	0.85–1.10
Sugar beets	5-1/2 months		0.85–1.00

*Dependent upon mean monthly temperature and stage growth of crop.

Table III.12   Monthly Percentage of Annual Daytime Hours (p) for Different Latitudes.

Latitude °N	Jan.	Feb.	Mar.	Apr.	May	June	July	Aug.	Sept.	Oct.	Nov.	Dec.
40	6.76	6.72	8.33	8.95	10.02	10.08	10.22	9.54	8.29	7.75	6.72	7.52
42	6.63	6.65	8.31	9.00	10.14	10.22	10.35	9.62	8.40	7.69	6.62	6.37
44	6.49	6.58	8.30	9.06	10.26	10.38	10.49	9.70	8.41	7.63	6.49	6.21
46	6.34	6.50	8.29	9.12	10.39	10.54	10.64	9.79	8.42	7.57	6.36	6.04
48	6.17	6.41	8.27	9.18	10.53	10.71	10.80	9.89	8.44	7.51	6.23	5.86
50	5.98	6.30	8.24	9.24	10.68	10.91	10.99	10.00	8.46	7.45	6.10	5.65
52	5.77	6.19	8.21	9.29	10.85	11.13	11.20	10.12	8.49	7.39	5.93	5.43
54	5.55	6.08	8.18	9.36	11.03	11.38	11.43	10.26	8.51	7.30	5.74	5.18
56	5.30	5.95	8.15	9.45	11.22	11.67	11.69	10.40	8.53	7.21	5.54	5.89
58	5.01	5.81	8.12	9.55	11.46	12.00	11.98	10.55	8.55	7.10	5.04	4.56
60	4.67	5.65	8.08	9.65	11.74	12.39	12.31	10.70	8.57	6.98	4.31	4.22

Example Calculation

Problem: Estimate the Seasonal Consumptive Use of Water for an Alfalfa Crop at Saskatoon during 1961. Growing Season May 15 – Sept. 15.

See Table III.13 for Computations.

Table III.13   Consumptive Use Calculations – Blaney - Criddle Method

Month	Mean Monthly Temp. (°F) T_m	% Annual Daytime Hrs p	Monthly Consumptive Use Factor $f = \dfrac{T_m p}{100}$	Crop Coefficient k	Monthly Consumptive Use cu in.
May	52.8	10.85	5.73	0.70	4.02
June	67.4	11.13	7.50	0.80	6.00
July	66.8	11.20	7.50	1.00	7.50
August	70.3	10.12	7.10	0.95	6.75
Sept.	47.7	8.49	4.05	0.80	3.24

Seasonal Consumptive Use: $CU = \Sigma kf$

$CU = 16/31\,(4.02) + 6.00 + 7.50 + 6.75 + 15/30\,(3.24)$

$= 24$ inches

Note: In the example calculation, an attempt has been made to adjust consumptive use for different stage of growth of crop by varying the monthly crop coefficient (or crop factor) 'k'.

Thornthwaite Method

In 1948, Thornthwaite presented a method which may be used to determine evapotranspiration from climatic data. Like the Blaney-Criddle Method, the parameters used in the method are temperature and length of day, and a fixed sharing of the heat budget is assumed. Unlike the Blaney-Criddle Method, however, no crop constant is introduced to the final result. The formulae developed by Thornthwaite are,

$$E = cT_m^a \dotfill \text{III.57}$$

in which E = evaporation or potential evapotranspiration (water unlimited), (cm),

c = coefficient,

T_m = mean monthly temperature (°C), and

a = exponent.

Both constants, a and c, depend on the location. The exponent 'a' can be evaluated in terms of the annual Heat Index, I, as

$$a = 67.5 \times 10^{-8} I^3 - 77.1 \times 10^{-6} I^2 + 0.0179 I + 0.492 \dotfill \text{III.58}$$

in which $I = \text{Heat Index} = \sum_{m=1}^{12} \left[\dfrac{T_m}{5} \right]^{1.51}$ III.59

If each month has 12 hrs. of sunshine each day and 30 days a month, the basic equation reduces to

$$E = 1.62 \left[\frac{10\,T_m}{I} \right]^a$$ III.60

Although the calculations of this method appear to be somewhat cumbersome, most of this difficulty is eliminated by making use of a simple nomogram. Obviously, Equation III.60 is but a simple power equation which reduces to a simple linear equation (in log terms) when exponentiated. That is

$\ln E = \ln 1.62 + a(\ln 10 + \ln T_m - \ln I)$.

When $E = 1.62$, $\ln 10 T_m = \ln I$ or $I = 10 T_m$.

Also, as shown by Thornthwaite, all lines following this equation have a common point of convergence at $T_m = 26.5°C$ and $E = 13.5$ cms.

Construction of Nomogram (See Figure with Example Problem).

1.   On a sheet of 2 or 3 cycle logarithmic paper mark off ordinate scale, T_m, in °C and abscissa, E, in cms.

2.   At E = 1.62 cm, mark off heat index scale, I. Remember the I scale should correspond to 10 times the T_m scale.

3.   Calculate the heat index, I, for the year

$$I = \sum_{m=1}^{12} \left[\frac{T_m}{5} \right]^{1.51}$$

4.   Pass a straight line through calculated Heat Index, I, and the point of convergence (26.5,13.5).

5.   The potential evapotranspiration corresponding to a given temperature can now be read directly from the graph.

Example Calculation

Potential Evapotranspiration for Alfalfa Crop at Saskatoon, 1961.

Month	T_m °F	T^* °C	i	E cms	Daylight † Factor	Adjusted ‡ E cms	Adjusted E ins
Jan.							
Feb.							
Mar.							
April	35.2	1.8	0.21				
May	52.8	11.6	3.56	6.00	1.33	7.98	3.14
June	67.4	19.7	7.92	10.00	1.36	13.60	5.35
July	66.8	19.3	7.68	9.00	1.37	12.33	4.86
Aug.	70.3	21.3	8.90	11.00	1.25	13.75	5.42
Sept.	47.7	8.7	2.32	4.45	1.06	4.72	1.86
Oct.	40.2	4.6	0.89				
Nov.							
Dec.			I = 31.5				

* °C = 5/9 (°F - 32).

† Correction Factor for Daylight Hrs. taken from Table III.2.

‡ Columns 6 x 7.

Estimated Potential Evapotranspiration for Alfalfa

$$= 16/31 \, (3.14) + 5.35 + 4.86 + 5.42 + 15/20 \, (1.86)$$
$$= 18.6 \text{ inches}$$

Comments on Penman, Blaney-Criddle and Thornthwaite Methods of Estimating Consumptive Use

1. The Blaney-Criddle and Thornthwaite Methods are similar in that an estimate of consumptive use is arrived at by using the same data. These methods were developed primarily for arriving at an annual or seasonal consumptive use value. They tend to give an overestimate of use for the emergence period and an underestimate of use for midseason, unless an appropriate crop factor is used.

2. The Blaney-Criddle Method appears to give a more reliable estimate of seasonal use than the Thornthwaite Method for arid regions.

3. The Blaney-Criddle and Thornthwaite Methods should not be applied for estimations of use for short periods of time inasmuch as no allowance is made in the methods for variations in wind and relative humidity.

4. The Penman Method will provide a reliable estimate of evapotranspiration. This method, however, is somewhat limited in application because of the many

meteorological observations required. A great degree of caution should be used in application of the method to arid regions unless some attempt is made to account for advective energy.

Turc

All the preceding methods attempt to estimate the rate of water use when the soil water supply is non-limiting. In nature, this condition is seldom realized and actual evapotranspiration is usually less than the potential evapotranspiration. Usually, the general experience of watershed behaviour is that evaporation is greater in wetter years than in the drier years. Turc, in Europe, developed formulae, based on a statistical study of data collected from 254 watersheds located in all parts of the world, to relate evaporation, rainfall and temperature over watersheds. He suggests that annual evaporation or evapotranspiration may be estimated as

$$E = \frac{P}{\left[0.90 + \left(\frac{P}{I_T}\right)^2 \right]^{1/2}} \quad \dots\dots\dots\dots\dots\dots\dots\dots\dots\dots\dots\dots \text{III.61}$$

in which E = annual evaporation (mm),
P = annual precipitation (mm),
I_T = 300 + 25T + 0.05T³, and
T = mean air temperature (°C).

Turc also suggests a more complex formula to give evaporation over short periods of time. In this formula he attempts to take into account the affect on evapotranspiration of different levels of soil moisture by different crops:

$$E = \frac{P + E_{10} + K}{\left[1 + \left(\frac{P+E}{I_T} + \frac{K}{2I_T}\right)^2 \right]^{1/2}} \quad \dots\dots\dots\dots\dots\dots\dots\dots\dots\dots\dots \text{III.62}$$

where E = evaporation in 10 day period (mm),
P = precipitation in 10 day period (mm),
E_{10} = estimated evaporation (in 10 day period) from bare soil assuming no precipitation, and is not greater than 10 mm,
K = a crop factor expressible in two ways, one of which is
K = 25 (Mc/G)$^{1/2}$,
where 100M = the final yield of dry matter (kg./ha.),
10G = the length of the growing season (days)
c = a crop constant,

I_T = the "evaporation capacity" of the air, calculated from

$$I_T = \frac{(T + 2) \, R_{si}^{1/2}}{16}$$

where T = mean air temperature over the the 10-day period ($^{\circ}$C), and

R_{si} = incoming radiant energy (cal/cm^2/day).

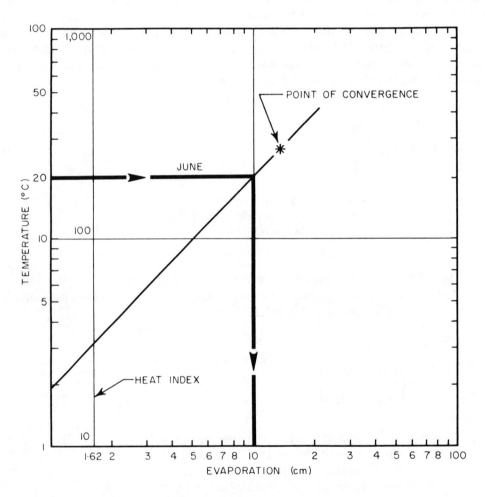

Fig. III.26 Nomograph Solution of Thornthwaite Equations

Values of c, given by Turc, are

Maize and beet 0.67
Potatoes 0.83
Cereals, flax, carrots 1.00

Peas, clover, legumes except
   lucerne 1.17
Lucerne, meadow grasses, and
   mustard . . . . 1.33.

When soil moisture is not limiting, in summer weather when $I_T > 10$, Turc reduces the equation to

$$ E = \frac{P + E_{10} + 70}{\left[1 + \left(\frac{P+E}{I_T} + \frac{70}{2I_T} \right)^2 \right]^{1/2}} \text{ mm./10 days} \quad \ldots \ldots \ldots \ldots \ldots \ldots \text{III.63} $$

III.3.2.9   Effective Heat-Day Degree

Lowry-Johnson Method

Lowry and Johnson (1942) developed a procedure for estimating water requirements for irrigation projects which has been used by the United States Bureau of Reclamation. This method applies to a valley, not to an individual farm, and has been widely used with good results by the Bureau in the arid western portions of the United States. It is essentially an empirical procedure.

A linear relationship is assumed between 'effective heat' and consumptive use. Effective heat is defined as the accumulation, in day-degrees, of maximum daily growing season temperatures above $32°F$.

The approximate relationship

$$ CU = 0.8 + 0.156 \, I_F \ldots \ldots \ldots \ldots \ldots \ldots \ldots \ldots \ldots \ldots \ldots \ldots \ldots \text{III.64} $$

is used in estimating the valley consumptive use by the Lowry-Johnson Method,

where CU = consumptive use in acre-feet per acre, and
   I_F = effective heat in thousands of day-degrees.

III.4   LITERATURE CITED

Anderson, E.R., 1954. Energy-budget studies. U.S. Geol. Survey Professional Paper 269.

Anderson, E.R., Anderson, L.J. and Marciano, J.J., 1950. A review of evaporation theory and development of instrumentation, Lake Mead Water Loss Investigation. U.S. Navy Electronics Laboratory Report 159, NE 121215 San Diego, Calif.

Blaney, H.F. and Criddle, W.D.,1950. Determining water requirements in irrigated areas from climatological and irrigation data. U.S.D.A. - SCS TP 96. Washington, D.C.

Bowen, I.S., 1926. The ratio of heat losses by conduction and by evaporation from any water surface. Physics Rev. Vol. 27.

Bruce, J.P. and Rodgers, G.K., 1962. Water balance of the Great Lakes system. American Assoc. Advancement of Science, No. 71, pp. 41-70.

Brunt, D., 1944. Physical and dynamical meteorology. Cambridge University Press.

Criddle, W.D.,1958. Methods of computing consumptive use of water. J. of Irrigation and Drainage Div. Amer. Soc. Civil Engr. Proc. Paper 1507-26.

Diamond, M., 1953. Evaporation or melt of snow cover. SIPRE, U.S. Corps of Engrs., Research Paper No. 6. November.

Elsasser, W.M. and Culbertson, M.F., 1960. Atmospheric radiation tables. Meteorological Monographs, Amer. Meteor. Soc. 23:1-43. August.

Evans, N.A., 1962. Methods of estimating evapotranspiration of water by crops, in Water Requirements of Crops Special Publication SP-SW-0162 of Amer. Soc. Agr. Engrs. St. Joseph, Mich.

Geiger, R., 1950. The climate near the ground. Harvard University Press. Cambridge, Mass. (Translation by M.N. Stewart and others).

Gilbert, M.J. and Van Bavel, C.H.M., 1954. A simple field installation for measuring evapotranspiration. Trans. Amer. Geophys. Union 35:937-942.

Hall, L.S., 1934. Evaporation from water surfaces. Trans. Amer. Soc. Civil Engrs. 99:715-724.

Harrold, L.L. and Dreibelbis, F.R., 1958. Evaluation of agricultural hydrology by monolith lysimeters. USDA Tech. Bull. 1179.

Holmes, R.M., 1961. Estimation of soil moisture content using evaporation data. Proc. of Hydrology Symposium, No. 2. Evaporation. Queen's Printer, Ottawa, pp. 184-196.

Israelsen, O.W. and Hansen, V.E., 1962. Irrigation Practices. 3rd ed. John Wiley and Sons, New York, N.Y.

Kerr, H.A., 1963. The development of an irrigation budget. Unpublished M.Sc. Thesis. College of Graduate Studies, Univ. of Saskatchewan, Saskatoon.

Kimball, H.H., 1914. Monthly Weather Review 42:474.

King, K.M., 1961. Evaporation from land surfaces. Proc. of Hydrology Symposium No. 2, Evaporation. Queen's Printer, Ottawa, pp. 55-80.

King, K.M., Tanner, C.B. and Soumi, V.E., 1956. A floating lysimeter and its evaporation recorder. Trans. Amer. Geophys. Union 37:738-742. (And discussion in Trans. Amer. Geophys. Union 38:765-768. 1957).

Kohler, M.A., Nordenson, T.J. and Fox, W.E., 1955. Evaporation from Pans and Lakes. U.S. Weather Bureau Research, Paper 38.

Kohnke, H., Dreibelbis, F.R. and Davidson, J.M., 1940. A survey and discussion of lysimeters and a bibliography on their construction and performance. USDA Misc. Pub. 372.

Kuzmin, P.O. Hydrophysical investigations of land waters. Int. Assoc. Sci. Hydrology, Int. Union of Geodesy and Geophysics 3:468-478.

Lemon, E.R., 1960. Photosynthesis under field conditions II: An aerodynamic method for determining the turbulent CO_2 exchange between the atmosphere and a corn field. Agron. J. 152:697-703.

Linsley, R.K., Kohler, M.A. and Paulhus, J.L.H., 1949. Applied Hydrology. McGraw-Hill Book Co., Inc., New York.

Linsley, R.K., Kohler, M.A. and Paulhus, J.L.H., 1958. Hydrology for Engineers. McGraw-Hill Book Co., Inc. New York.

Lowry, R.L. and Johnson, A.F., 1942. Consumptive use of water for Agriculture. Trans. Amer. Soc. Civil Engr. 107:1252.

Mateer, C.L., 1955a. Average insolation in Canada during cloudless days. Can. J. of Technology 33:12-32.

Mateer, C.L., 1955b. A preliminary estimate of average insolation in Canada. Can. J.Agr. Sci. 35:579-594.

Marciano, J.J. and Harbeck, G.E., 1954. Mass transfer studies. U.S. Geol. Survey Professional Paper 269.

McKay, G.A., 1962. Evaporation from Weyburn Reservoir. Progress Rept. 1960-61, Canada Dept. of Agric. PFRA, March, 1962.

McKay, G.A. and Stichling, W., 1961. Evaporation computations for Prairie reservoirs. Proceedings of Hydrology Symposium No. 2 , Evaporation. Queen's Printer, Ottawa.

Meyer, A.F., 1915. Computing runoff from rainfall and other physical data. Trans. Amer. Soc. Civil Engr. 79:1056-1155.

Meyer, A.F., 1942. Evaporation from lakes and reservoirs. Minnesota Resources Commission, St. Paul, Minn. p. 56.

Ministry of Agriculture and Fisheries, 1954. The calculation of irrigation need. Tech. Bull. No. 4. 1-37. Her Majesty's Stationery Office.

Mukammal, E.I., 1961. Evaporation pans and atmometers. Proc. of Hydrology. Symposium No. 2, Evaporation. Queen's Printer, Ottawa. pp. 84 - 105.

Mukammal, E.I., and J.P. Bruce, 1960. Evaporation measurements of pan and atmometer. Meteor. Br., Can. CIR - 300, TEC - 315.

Munn, R.E., 1961. Energy-budget and mass transfer theories of evaporation. Proceedings of Hydrology Symposium No. 2, Evaporation. Queen's Printer, Ottawa.

Nicholaichuk, W., 1964. Frequency of occurrence of evaporation amounts. Unpublished M. Sc. Thesis. University of Saskatchewan Library, Saskatoon.

Pelton, W.L., 1961. The use of lysimetric methods to measure evapotranspiration. Proc. of Hydrology Symposium No. 2, Evaporation. Queen's Printer, Ottawa. pp. 106-127.

Penman, H.L., 1948. Natural evaporation from open water, bare soil and grass. Proc. Royal Soc. London. 193:120-145.

Penman, H.L., 1963. Vegetation and hydrology. Commonwealth Agricultural Bureau, Farnham Royal, Bucks., England. Commonwealth Bureau of Soils Technical Communication 53.

Prairie Prov. Water Board, 1952. Evaporation from lakes and reservoirs on the Canadian Prairies (1921-1950) Rept. No.5.

Prandtl, L., 1904. On fluid motions with very small friction. Proceedings of the third Mathematical Congress. Heidelberg 1904, Leipzig 1934.

Prandtl, L. and Tietjens, O.G., 1934. Applied hydro-and aeromechanics, McGraw-Hill, Book Co., Inc., New York.

Prescott, J.A., 1940. Trans. Royal Society S. Australia 64:114.

Rider, N.E., 1954. Eddy diffusion of momentum, water vapour and heat near the ground. Phil. Trans. Roy. Soc. London A246.

Roberston, G.W. and Holmes, R.M., 1959. A modulated soil moisture budget. Monthly Weather Review 87:101-106.

Rohwer, C., 1931. Evaporation from free water surface. USDA Tech. Bull. 271. U.S. Gov't.Printing Office, Washington, D.C.

Rohwer, C., 1934. Evaporation from different types of pans. Trans. Amer. Soc. Civil

Engrs. 99:673-677.

Schmidt, W., 1925. Der massenaustausch in freir luft und verwandte rescheinungen, Henri Grand,Hamburg.

Sleight, R.B., 1917. Evaporation from the surfaces of water and river-bed materials. J. Agr. Res. 209-261, July.

Sonmor, L.G., 1963. Seasonal consumptive use of water by crops grown in Southern Alberta and its relationship to evaporation. Can. J. Soil Sc. 43:287-297.

Subcommittee, 1934. Standard equipment for evaporation studies Rept. Subcommittee on Irrigation Hydraulics. Trans. Amer. Soc. Civil Engrs. 99:716-718.

Sutton, O.G., 1949. The application to micrometeorology of the theory of turbulent flow over rough surfaces. Royal Meteor. Soc. Quarterly Jour. 75:335-350.

Sverdrup,H.U., 1936. Eddy conductivity of air over a smooth snow field. Geofysiske Publikasjoner 11:1-69.

Sverdrup,H.U., 1946. The humidity gradient over the sea surface. Journal of Meteorology 3:1-8.

Tabata, S., 1958. Heat budget of the water in the vicinity of Triple Island, B.C. J. Fisheries Research Board Can. 15:429-451.

Tanner, C.B., 1960. Energy balance approach to evapotranspiration from crops. Soil Sci. Soc. Amer. Proc. 24:1-9.

Taylor, S.A. and Haddock, J.L., 1956. Soil moisture availability related to power to remove water. Soil Sci. Soc. Amer. Proc. 20:284-288.

Thornthwaite, C.W., 1948. An approach toward a rational classification of climate. Amer. Geographical Review. Vol. 38.

Thornthwaite, C.W.and Holzman, B., 1939. The determination of evaporation from land and water surfaces. Monthly Weather Review. 67:4-11.

Turner, J.A., 1957. A nomographic solution of Penman's equation for the computation of evaporation. Proc. 3rd National Mtg. of the Royal Meteor. Soc., Toronto.

U.S. Dept. of Commerce, Office of Technical Services, 1956. Snow Hydrology. Washington, D.C.

U.S. Dept. of Interior, 1952. Water loss investigations: Volume 1 Lake Hefner studies technical report. Geological Survey Circular 229. Washington.

Veihmeyer, F.J. and Hendrickson, A.H., 1955. Does transpiration decrease as the soil moisture decreases. Trans. Amer. Geophys. Union 36:425-428.

Visser, W.C., 1962. A method of determining evapotranspiration in soil monoliths. Institute for Land and Water Management Research, Wageningen, The Netherlands, Reprint 25. pp. 453-460.

White, W.N., 1932. A method of estimating ground water supplies based on discharge by plants and evaporation from soil. Results of investigation in Escalante Valley, Utah. U.S. Geol. Survey Water Supply Paper 659-A.

Williams, G.P., 1961. Evaporation from water, snow and ice. Proceedings of Hydrology Symposium No. 2 - Evaporation. Queen's Printer, Ottawa.

Young, A.A., 1947. Evaporation from water surface in California. Summary of pan records and coefficients, 1881-1946. Bull. Public Works Dept., Calif. 54.

Section IV

INTERCEPTION

by

John M. Wigham

TABLE OF CONTENTS

LIST OF TABLES

Section IV

INTERCEPTION

INTRODUCTION

A portion of the precipitation falling to the earth's surface may be stored or collected by vegetal cover and subsequently evaporated; that volume of water caught by vegetation is referred to as interception, and the portion thereof retained and evaporated is called the interception loss. In studies of major storm events and floods, the component of interception is frequently neglected, though depending on the nature, type and density, etc., of the cover, and precipitation characteristics, it may be a very significant factor in water balance studies. For example, Helvey and Patric(1965) suggest that in humid, forested regions, the annual interception may be of the order of magnitude of 10 in. Linsley *et al.* (1949) indicate that under similar conditions, interception losses may account for as much as 25 per cent of the annual precipitation.

Numerous data have been collected by investigators on the magnitude of interception losses. Unfortunately, in many cases, because of difficulties—in measuring the absolute interception and in the differences of opinion which exist in defining this component— it is hard to separate the actual loss attributable to interception; thus, frequently, it is erroneously considered as part of the loss from some other process such as infiltration.

IV.2   PHYSICAL ASPECTS OF INTERCEPTION

If there is a vegetal cover over the soil surface, then any precipitation will be caught and redistributed as throughfall, stemflow and evaporation (interception loss) from the vegetation. Horton (1919) described the process as follows:

> When rain begins, drops striking leaves are mostly retained, spreading
> over the leaf surfaces in a thin layer or collecting in drops or blotches at
> points, edges, or on ridges or in depressions of the leaf surface. Only a
> meager spattered fall reaches the ground, until the leaf surfaces have
> retained a certain volume of water, dependent on the position of the leaf
> surface, whether horizontal or inclined, on the form of the leaf, and on

the surface tension relations between the water and the leaf surface, on the wind velocity, the intensity of the rainfall, and the size and impact of the falling drops. When the maximum surface storage capacity for a given leaf is reached, added water striking the leaf causes one after another of the drops to accumulate on the leaf edges at the lower points. Each drop grows in size (the air being still) until the weight of the drop overbalances the surface tension between the drop and the leaf film, when it falls, perhaps to the ground, perhaps to a lower leaf hitherto more sheltered. These drops may also be shaken off by wind or by impact of rain on the leaf. The leaf system temporarily stores the precipitation, transforming the original rain drops usually into larger drops. In the meantime the films and drops on the leaves are freely exposed to evaporation.

It is evident that the amount of interception in a given shower comprises two elements. The first may be called interception storage. If the shower continues, and its volume is sufficient, the leaves and branches will reach a state where no more water can be stored on their surfaces. Thereafter, if there is not wind, the rain would drop off as fast as it fell, were it not for the fact that even during rain there is a considerable evaporation loss from the enormous wet surface exposed by the tree and its foliage. As long as this evaporation loss continues and after the interception storage is filled, the amount of rain reaching the ground is measured by the difference between the rate of rainfall and the evaporation loss. When the rain ceases, the interception storage still remains on the tree and is subsequently lost by evaporation. If there is wind accompanying the rain, then, owing to motion of the leaves and branches, it is probable that the maximum interception storage capacity for the given tree is materially reduced, as compared with still air conditions. Furthermore, in such a case, after the rain has ceased, a part of the interception storage remaining on the tree may be shaken off by the wind, and the storage loss in such a case is measured only by the portion of the interception storage which is lost by evaporation and is not shaken off the tree after the rain has ceased. One effect of wind is, therefore, to reduce materially the interception storage. As regards evaporation loss during rain, the effect of wind is, of course, to increase it materially.

It may be deduced from this discussion that the effect of wind may either increase or decrease the interception loss depending on its velocity, the duration and the amount of rain, and the humidity of the air. In general, wind will increase the total interception loss for a long storm and decrease it for a storm of short duration. The rate of evaporation per unit surface area is normally quite small during storms because of the low vapor pressure differences which persist. Nevertheless, when this small amount is multiplied by the large area of vegetal surface, the total evaporation loss which occurs may be appreciable.

The total interception loss during and after a storm may be related to the storage capacity of the vegetation and the evaporation rate, by the following equation (Horton 1919):

$$I_{ri} = S_v + REt_R \quad \dots \dots \dots \dots \dots \dots \dots \dots \dots \text{IV.1}$$

where I_{ri} = the total interception loss for the projected
area of the canopy (in),
S_v = the storage capacity of the vegetation for
the projected area of the canopy (in),
R = the ratio of the vegetal surface area to its
projected area,
E = the evaporation rate from the vegetal surfaces (in/hr), and
t_R = the duration of rainfall (hr).

For using this equation, there is the disadvantage that the total storage capacity, S_v, of the foliage must be filled. This condition may or may not be fulfilled near the beginning of a storm depending on many conditions—for example, the volume of foliage, precipitation rate, time of prior precipitation, etc. In addition, the amount of rainfall is not considered in the calculations. Linsley *et al.* (1949) suggested that the equation should be modified to include an exponential term to consider rainfall quantities. Meriam (1960) used this concept and suggested the following equations:

$$I_{ri} = S_v (1 - e^{-P/S_v}) + REt_R \dots \dots \dots \dots \dots \dots \text{IV.2}$$

$$I_{ri} = S_v (1 - e^{-P/S_v}) + KP \dots \dots \dots \dots \dots \text{IV.3}$$

where P = the precipitation (in),
e = the base of natural logarithms, and
$K = \dfrac{REt_R}{P}$ is assumed constant.

These equations describe the process of interception fairly well because they indicate an exponential increase in the storage as the amount of precipitation increases. This trend to an increase in storage with increased precipitation has been observed in practice. The terms R and Et_R are difficult to evaluate separately, so Equation IV.3 has found favor in many instances. The assumption that K is a constant implies a constant relationship between the evaporation, E, and the precipitation, P, and this may not be theoretically justified (see Kohler, 1961) but practically the assumption may be acceptable in many cases. It should be noted that if the precipitation, P, is large, the equation essentially reduces to

$$I_{ri} = S_v + KP \quad \dots \dots \dots \dots \dots \dots \dots \dots \dots \dots \dots \text{IV.4}$$

IV.3   MEASUREMENT OF INTERCEPTION

Measurements of interception loss usually involve the evaluation of various terms in the following equation:

$$I_{ri} = P - T_h - S_f \dots\dots\dots\dots\dots\dots\dots\dots\dots\dots\dots IV.5$$

where I_{ri} = the interception loss,
   P = the amount of precipitation over the vegetal cover,
   T_h = the throughfall or the precipitation reaching the
     ground through the cover, and
   S_f = the amount of water reaching the ground from flow
     down the trunks or stems of the vegetation (stemflow).

IV.3.1   Precipitation Measurement

The measurement of the precipitation available over the vegetal cover is normally accomplished by using standard rain gauges and standard operating procedures. The sites chosen for placement of the gauges should be in large open areas near the interception study plots. The gauges should be properly shielded, and the sites selected should have the same slope and aspect as for the interception plots. Helvey and Patric (1965) in their paper stress the importance of methods which may be used to determine the required number of gauges to provide the desired accuracy. They suggest that two to four gauges may prove satisfactory for most studies.

When large open sites are not available for gauge placement, small clearings may be used provided there is an unobstructed sky view at an angle of $45°$ from the horizontal surrounding the gauge orifice. The value of placing gauges at tree-top level to measure the total areal precipitation is currently under study and development.

IV.3.2   Measurement of Throughfall

The method used to measure throughfall is governed, to a degree, by the type of vegetal cover. For low bushes and shrubs the precipitation gauges are placed at low level. In these cases, simple sheet-metal troughs are frequently used as gauges. Since a trough is a non-standard type of gauge, it must be designed to suit each particular type of cover under study. In all cases, however the sides of the troughs should be sloped inward to prevent splash. Since throughfall amounts within any given forest are highly variable, it is advisable that a network of troughs are used to obtain an estimate of this variation. Care should be taken in the placement of troughs to minimize wind effects. Note that catch measurements taken from troughs placed with their long axes, parallel to the main direction of wind flow will differ markedly with those taken from troughs placed at right angles to the direction of wind flow.

If the vegetal canopy is tall, then standard rain gauges may be used to measure throughfall (Helvey and Patric, 1965). One advantage of using standard gauges is that the data obtained may be compared directly with standard precipitation measurements taken in open areas. The effects of variation in throughfall can be reduced by periodically moving the gauges to new, randomly chosen locations, and by using a greater number of gauges than those used just to obtain measurements of precipitation. Helvey and Patric (1965) indicate that nine or ten times more throughfall gauges should be used than precipitation gauges. They present equations and graphs that may be used to calculate these gauge requirements.

The total interception loss includes losses from the primary and secondary vegetal covers, so throughfall from the primary cover must be adjusted for losses in the secondary cover. The secondary cover may be low shrubs, grasses or litter. If the cover consists of low shrubs or grasses, a low trough system may be required to determine the final throughfall. The interception loss for a litter cover may be calculated by determining the moisture content of the litter cover, immediately after the storm event, and periodically during periods of no precipitation.

IV.3.3   Measurement of Stemflow

It is practically impossible to measure stemflow for vegetal covers with numerous, small stems. Fortunately, interception studies have indicated that stemflow constitutes only a small proportion of the interception of such covers. Stemflow can be measured, however, for forest cover studies. In these cases, a ring or collar is attached and sealed to the tree trunk. The stemflow enters the collar through a small opening along the top of the collar and it is transported through a pipe to a container which is weighed periodically to determine the stemflow. The collar opening must be small to reduce the possibility of throughfall entering the collar. The container should be enclosed, in some manner, to prevent evaporation loss. Thompson (1964) describes in his paper a simple and economical method of stemflow measurement.

The amount of stemflow depends primarily on the roughness of the bark of the tree, and it may be from 0.01– 0.15 P for some smooth-bark species, and from 0.02 – 0.03 P for some rough-bark trees (Chow, 1964). Stemflow is, therefore, rather small but quite variable, thus indicating that it should be collected from several trees in a plot to provide better estimates of average conditions. Helvey and Patric (1965) indicate that several plots are required to obtain accurate results, and they suggest that five to ten randomly located plots should provide reliable samples.

IV.4   EMPIRICAL EQUATIONS AND METHODS

The interception loss for a particular storm is determined by the type and density of vegetal cover, provided there is no water in storage on the foliage prior to the rain. One of the most important vegetal factors that influence the interception loss is the surface area of the foliage. In addition, the geometry and physical properties of the vegetation are important. The seasonal change in surface area of the foliage results in variations in interception. Climatic variations from region to region also affect evaporation rates, and therefore interception. Empirical equations or methods of evaluating interception should include all of the above factors for universal application. Most methods do not attempt to include the effects of all these factors, particularly with regard to regional changes, so that most results from experimental studies must be extrapolated with care.

Many of the existing empirical equations used to estimate the interception during a particular storm event follow the form

$$I_{ri} = a + bP^n \dots\dots\dots\dots\dots\dots\dots\dots\dots\dots \quad \text{IV.6}$$

where I_{ri} = the interception loss,
P = the storm precipitation, and a,b and n are constants.

Equation IV.6 is similar in form to Equation IV.4, particularly since 'n' is often equal to unity. Some values of the constants, a, b and n, as given by Horton (1919) are listed in Table IV.1.

Table IV.1   Evaluation of Constants, a, b and n, in an
Interception Equation

Vegetal Cover	Interception = a + bP^n			Projection Factor
	a	b	n	
Orchards	0.04	0.18	1.00	
Ash, in woods	0.02	0.18	1.00	
Beech, in woods	0.04	0.18	1.00	
Oak, in woods	0.05	0.18	1.00	
Maple, in woods	0.04	0.18	1.00	
Willow, shrubs	0.02	0.40	1.00	
Hemlock and pine woods	0.05	0.20	0.50	
Beans, potatoes, cabbage and other small hilled crops	0.02h	0.15h	1.00	0.25h
Clover and meadow grass	0.005h	0.08h	1.00	1.00
Forage, alfalfa, vetch, millet, etc.	0.01h	0.10h	1.00	1.00
Small grains, rye, wheat, barley	0.005h	0.05h	1.00	1.00
Corn	0.005h	0.005h	1.00	0.10h

Note: Interception is in inches for P in inches. The symbol 'h' refers to the height of plant in feet.

The value of a, b and n given in Table IV.1 apply to individual storm events rather than to monthly or annual periods. The average interception over an area is determined by applying a projection factor to the computed interception to adjust for that portion of the area not covered by vegetation. Where no projection factor is listed it must be estimated on the basis of cover density for existing conditions.

The constants were evaluated from limited experimental data from relatively small storms and are not applicable for large storms. They should be used with caution, but they are valuable in that they allow initial estimates of interception values.

Other equations used to predict the interception loss are

$$I_{ri} = 0.083 \, (1 - \bar{e}^{P/0.083}) + 0.062P \ldots \ldots \ldots \ldots \ldots \ldots \ldots \text{IV.7}$$

$$I_{ri} = P - 0.09P^{1.25} - 0.87P^{1.16} \ldots \ldots \ldots \ldots \ldots \ldots \ldots \text{IV.8}$$

for $P \leqslant 0.5$ in.

Table IV.2   Interception under Natural Rainfall for
One Square Meter of Area

Vegetation	Precipitation		Interception
	Inches	Character	Percentage
Wheat	0.02	one very light shower	90
	0.06	one very light shower	80
	0.07	one very light shower	72
	0.07	one very light shower	76
	0.24	two light showers	74
	0.32	one short shower	52
	0.35	one short shower	64
	0.46	one hard shower	46
	0.80	three showers	51
	1.48	heavy rain followed by showers and mist	33
Oats	0.11	one light shower	72
	0.15	several light showers	57
	0.74	heavy rain followed by light showers	45
Slough grass	0.02	very light shower	80
	0.06	light shower	80
	0.07	light shower	66
	0.07	light shower	76
	0.38	hard shower	78
	0.39	hard shower	67
	0.45	hard shower	73

Equation IV.7 presented by Meriam (1960) is applicable to stands of evergreen and deciduous chaparral and ponderosa pine. Equation IV.8 was derived by Collins (1966) from his work on juniper and pinyon woodlands in Arizona, and only applies for rains in which the amount of precipitation is less than 0.5 in.

Clark (1940) performed a number of interception studies on Prairie grasses and crops. The data were not reduced to equational form, but precipitation amounts and the percentage losses through interception were tabulated. Interception losses for natural rainfall and artificial sprinkling were obtained for alfalfa, slough grass, wheat, oats, and bluestem. Some of these values are listed in Table IV.2.

Unfortunately, most of the data are for rather small amounts of precipitation. The comparison of these data with data from the sprinkled plots does allow some extrapolation, and the results could be used for design estimates.

IV.5   INTERCEPTION OF SNOW

The preceding discussion has primarily dealt with interception of rainfall. Snow particles are delivered to and stored in the vegetal canopy somewhat differently than rain drops, hence differences occur in the amounts of interception. The interception of snow by short vegetation is not important to flood runoff prediction, and will not be considered herein in detail. But interception of snow by forests is important in such studies, and thus will be discussed briefly in this context.

In forests, snow particles move almost horizontally (about 4° from the horizontal) at low speeds. This has the effect of closing an incompletely closed (in the vertical direction) canopy, thus increasing the surface area. This should have the effect of increasing interception. However, higher wind speeds may decrease interception, though higher temperatures will increase it due to greater adhesion effects.

In many cases though, the interception loss is of more interest than the amounts trapped by the cover. The loss must occur through evaporation and sublimation from the snow surface on the trees. These processes require considerably more energy than does the melting process. The radiation budget for a forested area does provide larger amounts of energy than for an open snowfield in winter, due to the high absorptivity of the tree crowns. This energy can be used in the evaporation process. Using an energy budget approach, Miller (1962) suggested that daily evaporation from intercepted snow would probably not exceed 0.7 in. in the Sierra Nevada. This he compared to a similar figure obtained by Müller for maximum daily evaporation for the Allegheny Plateau. These figures, although not exact, do give indications of possible losses through evaporation. The amount of evaporation does not necessarily indicate the water loss, however, because some of the vapour may be transferred downward to condense on the lower surfaces. Probably, the figures above are too high for the more northerly Canadian conditions.

More could be said in general about the interception processes and losses; but unfortunately only minor conclusions can be drawn of the importance of the interception component, because data, which would substantiate some of the existing concepts, is lacking.

IV.6   LITERATURE CITED

Chow, V.T. 1964. Handbook of Applied Hydrology. McGraw-Hill Book Co., New York, N.Y.

Clark, O.R. 1940. Interception of rainfall by Prairie grasses. Weeds and Certain Crop Plants, Ecological Monographs.Vol. 10. April.

Helvey, J.D. and Patric, J.H. 1965. Design criteria for interception studies, Symposium on Design of Hydrometeorological Networks. WMO/IASH, Laval Univ., Quebec City.

Horton, R.E. 1919. Rainfall interception. U.S. Monthly Weather Rev. Vol. 47.

Kohler, M.A. 1961. Discussion of Paper by R.A. Meriam "A note on the interception loss equation." Journal of Geophysical Research. Vol. 66, No. 61.

Linsley, R.K., Kohler, M.A., and Paulhus, J.L.H. 1949. Applied Hydrology, McGraw-Hill Book Co., New York, N.Y.

Meriam, R.A. 1960. A note on the interception loss equation. Jour. of Geophys. Res. Vol. 65. No. 11.

Miller, D.H. 1962. Snow in the trees – where does it go? Western Snow Conference Proceedings.

Thompson. T.A. 1964. Instrumentation for stemflow measurement in the Marmot Creek Project.Forest Research Branch Publication, Canada Department of Forestry, Calgary, Alberta.

Section V

INFILTRATION AND THE PHYSICS OF FLOW OF WATER THROUGH POROUS MEDIA

by

Donald M. Gray, Donald I. Norum
and John M. Wigham

TABLE OF CONTENTS

LIST OF TABLES

LIST OF FIGURES

Section V

INFILTRATION AND THE PHYSICS OF FLOW OF WATER THROUGH POROUS MEDIA

V.1   INFILTRATION

V.1.1   Introduction

The phenomenon of infiltration deserves special attention and study inasmuch as a thorough understanding of this phenomenon will enable the hydrologist to more effectively estimate amounts of runoff originating from precipitation and thus to more confidently and competently apply the results to design problems.

Horton (1933) introduced the concept of infiltration in the hydrologic cycle and defined the infiltration capacity, f_m as the maximum rate at which a given soil in a given condition can absorb water. Obviously the actual infiltration rate, f, of a soil will be equal to f_m when the rainfall intensity, i, or rate of snowmelt equals or exceeds the magnitude of f_m. Further, when $i < f_m$ then f=i. Inasmuch, therefore, as the term infiltration capacity is somewhat redundant, in the remaining discussions we will simply utilize the term infiltration rate to refer to the rate at which water enters the surface layer of the soil.

Depending on the rainfall intensity or the snowmelt rate relative to the infiltration rate, water may be entirely absorbed by the soil or may accumulate and flow from the area as surface runoff (see Figs. V.1 and V.2).

V.1.2   Mechanics of Infiltration

Water enters the soil surface due to the combined influence of gravity and capillary forces. Both forces act in the vertical direction to cause percolation downward, whereas capillary forces also act to divert water laterally from larger pores (feeder canals) to capillary pore spaces that are much smaller in dimension, but may be very numerous. As the process continues, the capillary pore spaces become filled, and with percolation to greater depths, the gravitational water normally encounters increased resistance to flow due to a reduction in the extent or dimensions of the flow channels, an increase in length of the channels, or the presence of an impermeable barrier such as rock or clay. At the same time there may be increased resistance to inflow of water at the soil surface due to a surface sealing effect as a result of the mechanical action of raindrops in breaking down the soil aggregates and subsequent inwash of the finer soil particles. The result is a rapid reduction of infiltration rate in the first few hours of a storm, after which the rate remains nearly constant for the remainder of the period of storm rainfall excess.

From this qualitative description of the infiltration process, it should be recognized that the phenomenon includes the separate functions of water storage and transmission by a soil. Further, the infiltration rate of a given soil may be governed by either of three separate processes:

1. Entry of the water into the surface layer of soil.
2. The downward movement or percolation of water through the soil profile.
3. Flow through deep cracks in the profile.

In general, therefore, the infiltration rate depends on many factors, including precipitation intensity and type; the condition of the soil surface; the density, type and stand of vegetation; the temperature and chemical composition of the water; the physical properties of the soil (porosity, grain and pore size, moisture content, etc.); and others.

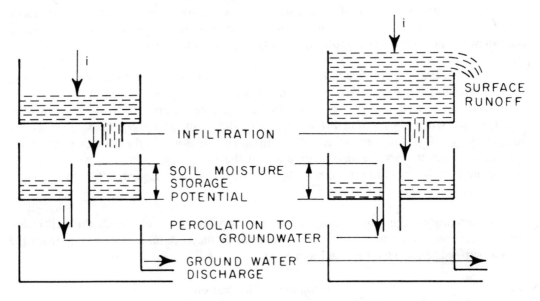

Fig. V.1   Low Intensity Rainfall   Fig. V.2   High Intensity Rainfall

V.1.2.1   Entry of Water into the Surface Layer

The rate of water entry into the surface layer of a soil depends not only on the total porosity but also on the size of the pores. As shown by the Hagen-Poiseuille equation for saturated, laminar flow through capillary tubes (see Subsection V.2.1.2.) the quantity of flow per unit time through small tubes is directly proportional to the fourth power of the tube diameter. Hence, any factor which reduces either the number or size of pores reduces the rate at which the soil will absorb and transmit water.

The surface entry of water is greatly affected by the nature, type and amount of surface cover. By its canopy effect, vegetation absorbs the energy of raindrops and protects the soil from dispersion. Wischmeier and Smith (1958) have shown that the kinetic energy expended by raindrops falling at their terminal velocity and striking the soil surface may be calculated from the equation:

$$KE = 916 + 331 \log i \dots\dots\dots\dots\dots\dots\dots\dots\dots\dots\dots\dots\dots V.1$$

where KE = kinetic energy in units of ft-tons/acre-inch of rain, and
i = rainfall intensity (in/hr).

It is sufficient to assume that the breakdown and dispersion of soil aggregates vary directly with the amount of kinetic energy. Thus, one may expect a decrease in the rate of surface entry of water with higher intensity storms resulting from the plugging of pores by soil particles and compaction.

Vegetation also influences the infiltration rate of a soil through its effect on soil aggregation. A well-aggregated, stable surface layer increases the size of the pores and permits rapid entry of water.

Another factor, important to the surface entry of water, is the dispersive qualities of the surface layer. If the layer contains a high percentage of exchangeable sodium, the soil system will likely be highly unstable. On wetting, a puddled condition can rapidly result from the impact of raindrops, thereby causing a decrease in the infiltration rate.

V.1.2.2   Downward Movement or Percolation Through the Profile

In cases where the rate of entry of water through the soil surface is not the limiting value, then the infiltration rate is governed by the downward movement through the profile. To comprehend fully the interrelationship between the percolation rate and the infiltration rate, it is necessary to review briefly the distribution of soil moisture within the profile during the downward movement of water from a ponded water surface. Bodman and Colman (1943) found that the moisture distribution pattern within the profile may be divided into four distinct moisture zones:

(a)   saturation and transition zone,
(b)   transmission zone,
(c)   wetting zone, and
(d)   the wet front (see Fig. V.3)

Zone of Saturation

The zone in which the soil is saturated.

Zone of Transmission

This zone is an ever-lengthening unsaturated zone of fairly uniform water content. Norum and Gray (1964) measured the moisture contents in the transmission zone of some Saskatchewan soils during irrigation trials and found these levels to be between 60-80 per cent of pore saturation in heavy glacial tills. These results agree with the findings reported by Moore (1939), Bodman and Colman (1943), and Kirkham and Feng (1949). Further, several investigators—for example, Miller and Richard (1952), Marshall and Stirk (1949), and Taylor and Heuser (1953)—found that only a small tension gradient existed in the zone, and that when water has penetrated sufficiently deeply into the soil, its movement is principally due to gravity. Under these conditions, in a uniform soil profile the infiltration rate will approach a constant value. Of course, these conditions will change in stratified or layered soils containing horizons of different hydraulic conductivities.

Wetting Zone

Zone which joins the wet front and transmission zone. The wetting zone terminates abruptly at the wet front and the moisture content of the zone increases as infiltration proceeds.

Wet Front

A sharp line of demarkation in which the soil moisture changes from wet to dry.

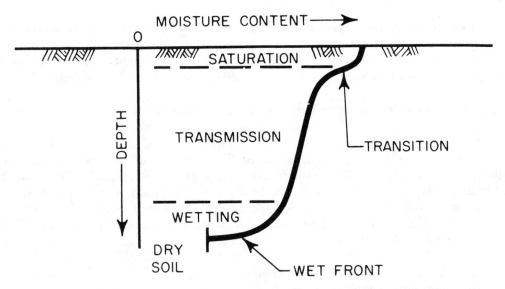

Fig. V.3   Moisture Distribution Pattern Through Soil Profile

Calculation of Precipitation Excess from Consideration of the Soil Moisture Profile

Acceptance of the principle that infiltration involves storage and transmission permits a rational approach to calculation of the mass infiltration, or similarly, the amount of precipitation excess from a given storm. As suggested by Ayers (1959), the storage capacity of the soil will result in an initial abstraction from precipitation, and thereafter, with time, the transmission capacity of the soil accounts for the subsequent reduction of the amount of precipitation which is available for surface runoff. Thus, in equational form

$$P_e = P - A_i - f_c t \dots\dots\dots\dots\dots\dots\dots\dots\dots\dots\dots\dots\dots\dots V.2$$

where P_e = precipitation excess (in),
P = total precipitation (in),
A_i = initial abstraction (in),
f_c = transmission rate (in/hr), and
t = time after A_i is satisfied or the moisture profile
is developed (hr).

Initial abstraction, A_i, is composed of water necessary to replace the soil moisture deficit (field capacity-soil moisture content) plus that required to fill the non-capillary pores to near saturation within the depth of soil that it takes to establish the moisture profile. Ayers found it convenient to evaluate A_i as the water required to raise the moisture level of the upper 1 ft. of the soil profile to saturation. He suggests the following values for the transmission rates, f_c, for different soil profiles.

**Table V.1 Net Infiltration Rates for Unfrozen Soils, f_c,
in Inches per Hour**

Soil Profile Category	Ground Cover Condition					
	Bare Soil	Row Crops	Poor Pasture	Small Grains	Good Pasture	Forested
I	0.3	0.5	0.6	0.7	1.0	3.0
II	0.1	0.2	0.3	0.4	0.5	0.6
III	0.05	0.07	0.10	0.15	0.20	0.25
IV	0.02	0.02	0.02	0.02	0.02	0.02

Category I
    Course and medium textured soils over, sand and gravel glacial outwash materials or, coarse open till or, coarse alluvial deposits.
Category II
    Medium textured soils over medium textured till.
Category III
    Medium and fine textured soils over fine textured clay till.
Category IV
    Soil over shallow bedrock (two ft. or less).

Holtan (1961) has reported a similar approach to define the infiltration rate of a soil as a function of the exhaustion of soil moisture storage. The expression used is

$$f = a(S - M_f)^n + f_c \quad \dots\dots\dots\dots\dots\dots\dots\dots\dots\dots\dots\text{V.3}$$

in which S = potential soil moisture storage volume or the volumetric diff-erence between pore saturation and the 15-bar or permanent wilting percentage for the soil zone above the control layer,

M_f = mass infiltration,

f_c = final constant rate of infiltration through the control horizon, and

a, n = constants for a particular soil in given condition (according to Philip, n = -½).

In Equation V.3, the effect of increasing the mass infiltration, M_f, is analogous to increasing the (initial) soil moisture content, which will cause a decrease in the infiltration rate. As pointed out by Holtan, one important aspect of Equation V.3, as applied to hydrologic analyses, is that by subdividing the storage potential into the free or gravitational water volume and the capillary water volume, the infiltration recovery between rain periods can be computed. In this calculation it is usually assumed that the free water is removed at the rate of gravity flow (perhaps f_c), and that the available water capacity is depleted at a slower rate of evapotranspiration. Another feature shown in the equation is that when the mass infiltration, M_f, equals the moisture storage, S, the infiltration rate is equal to the transmission rate through the control layer. Hanks and Bowers (1962) substantiated this result. They concluded that infiltration was governed by transmission through the least permeable layer, once the wetting front extended into that layer.

V.1.2.3   Flow Through Deep Cracks

Many soils that contain high percentages of colloidal clay shrink and crack during prolonged periods of drought. In the cracked condition, these soils will generally absorb greater quantities of water at a faster rate than under normal conditions. The increase in amount and rate of entry of water depends on the degree of cracking and the continuity of the cracks to the deeper subsoil layers. It has been observed that the infiltration rates of heavily-cracked clays tend to be very high, near a constant rate, until the cracks are filled with water. After this time the rate of entry becomes very low.

The extent or degree of cracking of a given soil is very transient, and appears to be directly related to moisture changes within the profile. As such, the process is governed by the climatic conditions which prevail over the area. It has been the writer's experience that severe cracking can develop over a very small range of soil moisture (4-5 per cent by weight). Further work is needed to describe the process.

V.1.3   Time Variation of Infiltration Rate

In hydrologic analysis, it is frequently necessary to know the time variation of the infiltration rate to be able to compute the rate at which the net storm rain is generated.

Normally, when excess water is applied to a soil, the infiltration rate decreases with time from the commencement of the trial; and the rate of decrease also decreases with time with a tendency for the infiltration rate to approach a final minimum value. Variations in the shape of some standard infiltration curves for different conditions are shown in Figs. V.4 and V.5.

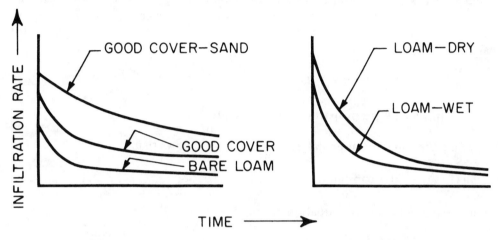

Fig. V.4   **Fig. V.5**
Variations in the Shape of Standard Infiltration Curves

Often, it is more practical to present infiltration data in the form of a cumulative curve of the total amount of water intake, M_f, in (ft^3/ft^2) which has entered the soil in a given time, t, in hr. This curve is called a mass infiltration curve. The slope of curve at any time is the infiltration rate at that time. Numerous investigators have proposed mathematical expressions to describe the shape of the mass infiltration curve. Some of the most commonly-used expressions are the following:

V.1.3.1   Kostiakov (1932) and Lewis (1937)

$$M_f = at^n \dots\dots\dots\dots\dots\dots\dots\dots\dots\dots\dots\dots\dots\dots\dots\dots\dots\dots\dots\dots\dots\dots\dots\text{V.4}$$

The two parameters, a and n are evaluated from a graph by fitting a straight line to the points when log M_f is plotted against log t.

V.1.3.2   Gardner and Widstoe (1921), and Horton (1940)

$$M_f = f_c t + de^{-kt} \dots\dots\dots\dots\dots\dots\dots\dots\dots\dots\dots\dots\dots\dots\dots\dots\dots\dots\dots\dots\text{V.5}$$

where f_c = infiltration rate which represents the fairly steady
rate of water absorption reached after water has been
applied continuously for a long period of time,

e = base of natural logarithms, and

d, k = constants.

The form of the equation given by Equation V.5 can be derived by considering the processes involved in the reduction of the infiltration rate as rain continues; these are similar in nature to an exhaustion process—that is a process in which the rate of performing work is proportional to the amount of work remaining to be performed. In the infiltration process, the work remaining to be performed at a given time, t, is that required to change the infiltration rate from its existing value, f, to its final constant value, f_c. The rate of performing work is df/dt, and since the infiltration process f decreases with time, the differential is negative. Using the above assumptions, it follows that

$$-\frac{df}{dt} = k(f - f_c) \dotfill \text{V.6}$$

in which k is a proportionality constant.

Integrating the equation, as an indefinite integral, it follows that

$$\ln(f - f_c) = -kt + \text{constant} \dotfill \text{V.7}$$

But at t = 0, f = f_0, and the equation becomes

$$f = f_c + (f_0 - f_c)\, e^{-kt} \dotfill \text{V.8}$$

But

$$\frac{dM_f}{dt} = f = f_c + (f_0 - f_c)\, e^{-kt} \dotfill \text{V.9}$$

Integrating Equation V.9 with the boundary conditions, that is t = 0 when M_f = 0 it follows that

$$M_f = f_c t + \frac{1}{k}(f_0 - f_c)(1 - e^{-kt}) \dotfill \text{V.10}$$

Generally, Equation V.5 is used in watershed studies in preference to Equation V.4, because it accounts for minimun infiltration rates which occur at large values of time. Obviously, according to Equation V.4, as $t \to \infty$, $f \to 0$. This equation has wider usage in irrigation studies to which water applications are not continued over long periods.

V.1.3.3   Kirkham and Feng (1949)

$$M_f = ct^{1/2} + a \dotfill \text{V.11}$$

where c and a are constants, a being of small magnitude.

The authors found that Equation V.11 could be used to describe the volume of water imbibed horizontally into a column of air-dry soil as a result of capillary forces when the core was placed in contact with a free water surface. That is, in cases where water entry is relatively unaffected by gravity (horizontal flow or conditions in which a high tension gradient may exist), M_f can be assumed to be approximately proportional to $t^{1/2}$. In wet and sandy soils, where the effect of gravity is more important to flow, M_f tends to be more proportional to t.

V.1.3.4   Philip (1957)

$$M_f = St^{1/2} + At \dots\dots\dots\dots\dots\dots\dots\dots\dots\dots\dots\dots\dots\dots\dots V.12$$

where S and A are coefficients to be evaluated.

In essence, Equation V.12 proposed by Philip includes sufficient terms to account for both gravitational and capillary forces. The relative magnitudes of the coefficients will depend on which force predominates as affected by soil texture, soil moisture, etc.

V.1.3.5   Summary

In the preceding discussions, it has been shown that, frequently, the variation in the infiltration rate can be assumed to decrease inversely with the square root of time. Wiegand and Taylor (1961) reported the results of several studies which show the yield of water, Q, for several different flow geometries can be expressed as a simple power function of the square root of time analogous to Equation V.4 (see Table V.2).

Nevertheless, to apply the infiltration equations given above, it is necessary to evaluate the coefficients and exponents by either field or laboratory techniques.

**Table V.2 Examples from the Literature of Moisture Yield
by Soil which Fit the Equation $Q = at^n$.**

Reference	Flow Geometry	Flow Induced by (Driving Force)	Initial Moisture Condition
Ubell (1956)	Radial, but some well drawdown	Pumping a well	Saturated
Read (1958)[1]	Radial (disc of soil)	Vacuum pump suction on filter cone	Saturated; unsaturated
Vasquez and Taylor (1958)	Small Tensiometers normal to axis of large cylinder	Vacuum pump suction on filter cone	Unsaturated
Richards and Weeks (1953)	Linear, horizontal	Vacuum pump suction on a ceramic plate	Unsaturated
Richards et al. (1956)	Linear, vertical	Atmospheric conditions	Field Capacity

[1] By the nature of the method used, the soil moisture tension could at no time exceed the equivalent of about 1.0 bar suction.

V.1.4   Evaluation of Infiltration Rates

Because of the complexity of the infiltration phenomenon and the fact that many factors affect the process, the measurements of infiltration rates and volumes should be accomplished under field conditions. Two methods of evaluating infiltration rates are in common use; (a) the analysis of hydrographs of runoff produced by natural rainfall occurring on plots or watersheds; and (b) the use of an infiltrometer in which water is applied artificially to the sample area.

V.1.4.1   Hydrograph Analysis

Determination of infiltration rates from analysis of hydrographs of runoff has the advantage that field data are used, and thus, quantitative results can be obtained for the watershed under study. On large watersheds, however, the separation of a hydrograph into its different components of flow is difficult to accomplish, and in addition, the analysis is usually complicated by variations in precipitation patterns (distribution and intensities) and storage effects (see Section VII). In such cases, only an average value of infiltration can be obtained. Details of the methodology which may be employed in such analysis, may be found in the works of Sharp and Holtan (1940, 1942).

V.1.4.2   Infiltrometers

Infiltrometers may be conveniently classified as either of two types: sprinkler or flooding. Both types require only a small area of soil surface on which to conduct a test. Further, both the amount of water supplied to the plot and the volume infiltrated can be measured or calculated fairly accurately. The disadvantages of infiltrometers are (a) the results are representative only of the soil and vegetal conditions of the plot, (b) the natural action of raindrops cannot be reproduced exactly, and (c) great care must be taken in the installation to reduce boundary effects.

Because of the rather artificial conditions, the results of infiltrometer tests can only be considered as qualitative until they are confirmed by hydrograph analysis. Such data are extremely useful, however, in that results of tests for different soils, vegetation conditions and initial moisture content can be compared. In addition, tests on plots in different soil and/or vegetation regions in a watershed can be used to evaluate the overall infiltration characteristics of the watershed. Areas of equal infiltration characteristics can be defined, and these can be considered in conjunction with the rainfall conditions over these subareas, resulting in more precision in analyzing the flow conditions of the basin.

Flooding Type

The flooding type infiltrometers normally consist of tubes or concentric rings which are driven into the ground to depths of 15-21 in. These tubes are usually about 9 in. in diameter and 18-24 in. in length. Water is applied to the tube from graduated burettes and maintained at a constant head ($\sim \frac{1}{4}$ in.) over the soil surface. Readings of the water level in the burettes taken at successive time intervals permit direct determination of infiltration volumes and rates.

Concentric-ring or double-ring infiltrometers usually consist of an inner ring, 9 in. in diameter, surrounded by a 14-in. diameter outer ring. The rings are only forced into the soil to depths necessary to prevent leakage under their edges. The water level is maintained at the same level within both rings, but measurements of applied volumes and time are only taken on the inner ring. The outer ring serves only to reduce the boundary effects on the inner ring (that is lateral flow).

The flooding type infiltrometers normally yield values of infiltration about twice those obtained from sprinkler-type infiltrometers. This is because there is no dynamic action of falling water drops. The results therefore should only be considered to apply for similar field conditions—for example, flood-irrigation.

Sprinkler Type

There are many different models of sprinkler type infiltrometers, but the most common types are designated type F and type FA. The type F infiltrometer utilizes a plot 6 ft. wide by 12 ft. in length; the FA type is normally used on plots only 1 ft. x 2 1/2 ft. in size. Each type contains two rows of special nozzles mounted along each side of the plot. The nozzles used in both types are the same; but the type FA infiltrometer is operated at lower nozzle pressures than on type F, hence the discharge and dynamic action of the water from the nozzles is reduced. The nozzles direct their spray upward and slightly inward to cover the plot and surrounding area with simulated, uniform rainfall. The water drops from the nozzle spray reach a height of 6-7 ft. above the plot surface. However, this height is insufficient to permit the drops to reach their terminal velocity by the time they strike the ground (Chow, 1964).

The intensities of simulated rainfall that may be obtained with the nozzles are governed by the number of nozzles used. Normally, in type F infiltrometers the range of intensities are multiples of 1.75 in/hr, whereas in the type FA they are of multiples of 1.5 in/hr. The actual intensity of rain provided by the nozzles is determined by placing a plastic or metal sheet over the plot, and by measuring the runoff from the plot. The 'test' run commences when the plastic sheet is removed. Water is applied to the plot through the nozzles, and runoff is measured continuously until the discharge rate from the plot becomes essentially constant (see Fig. V.6). The nozzles are then shut off but the discharge measurements are continued until there is no further runoff. As soon as all free water disappears from the plot the nozzles are turned on again and remain on until the discharge rate is once again constant. They are then shut off but the measurements of discharge continue until runoff ceases (see Fig. V.6). This latter portion of the test may be called the 'analytical' run.

Method of Analysis

The analysis of the results of the sprinkler tests is based on the following equations

$$M_f = P - SRO - D_A - S_d \dots\dots\dots\dots\dots\dots\dots\dots\dots\dots\dots\dots V.13$$

where M_f = depth of water which has been infiltrated (in),
P = depth of rainfall which has fallen (in),

SRO = total depth of runoff from the plot (in),
D_A = depth of water on the soil surface (in),
and
S_d = depression storage in terms of depth (in).

The equation applies at any time during the test and analytical runs. The terms P and SRO are measured quantities for the run, but D_A and S_d must be calculated.

The depth of water, D_A, on the soil surface is determined from the recession curves of runoff (those following the cessation of rainfall). The recession runoff, representing the reduction of surface detention storage, is equal to D_A times the area of the plot. In addition, of course, there is continuous infiltration during the recession period, which also reduces detention storage. The infiltration rate during this period may be estimated by the equation

$$f_r = \frac{f_c}{Q_c} Q_r \dots\dots\dots\dots\dots\dots\dots\dots\dots\dots\dots\dots\dots\dots\dots\dots\dots\dots\dots\dots\text{V.14}$$

where f_r = the infiltration rate at any time during
the recession runoff (in/hr),
f_c = the infiltration rate at the cessation
of rainfall (in/hr),
Q_c = the discharge rate at the cessation of
rainfall (in/hr), and
Q_r = the measured discharge rate at any time
during the recession (in/hr).

The calculated values of f_r are added to the observed Q_r values during the recession period. The area under the curve of $f_r + Q_r$ versus time is, at any chosen time, the volume of detention storage per unit area of the plot, which is the term D_A. A curve of D_A versus Q, the runoff rate may therefore be obtained. This is often a straight line on logarithmic graph paper (see Fig. V.6; from Chow 1964).

The term S_d is determined from the 'analytical' run. Because of the wet condition of the soil during this run, the infiltration rate will be at a relatively constant low value equal in magnitude to the rainfall rate minus the discharge rate at the time when the rainfall ceases. Equation V.13 can therefore be solved for S_d, since P and SRO are measured, D_A can be determined at any time from the plot of Q versus D_A, and M_f can be calculated as the infiltration rate during the analytical run times the time from the beginning of the analytical run. The various terms and their relation to each other are also shown in Fig. V.6.

Since S_d and D_A can be calculated for any discharge outflow from the plot, the test run data can now be used with Equation V.13 to give values of M_f versus time from the beginning of the test (see Fig. V.6). It is assumed that depression storage fills very quickly in the initial portion of the test run. The infiltration rate versus time curve may also be obtained, as the infiltration rate is simply the slope of the M_f versus time curve at any time.

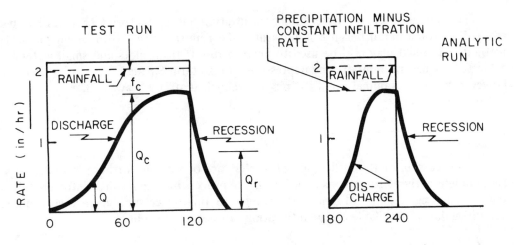

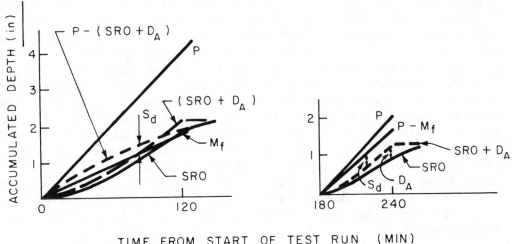

TIME FROM START OF TEST RUN (MIN)

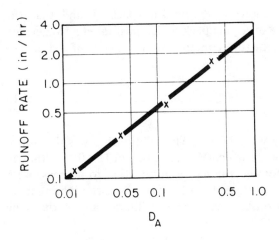

Fig. V.6 Precipitation-Infiltration-
Runoff Relationships
for Hydrograph Analysis of Small Plots

The methodology for determining infiltration rates for sprinkler plots has been covered in some detail because it illustrates the runoff process from small plots. The methods described may also be used to analyze runoff from plots and small watersheds subjected to natural rainfall. The depression storage term S_d may have to be evaluated, however, from an examination of the plot or watershed after a heavy rain.

V.1.5   Infiltration into Frozen Soils

No discussion of the infiltration characteristics of soils in northern latitudes would be complete without some consideration of the process in frozen soils. Unfortunately, in spite of the significance of Snow Hydrology to our water supply, only very limited quantitive data are available on the infiltration rates of frozen soils.

One of the basic parameters which govern the infiltration rate of a frozen soil is the quantity and size of the ice-free pores. Hence, it is obvious that the moisture content of a soil at the time of freezing is an important factor. Several Russian workers, Larkin (1962); Kuznik and Bezmenov (1963); and others, (Post and Dreibelbis, 1942) report that if a soil is frozen when its moisture content is greater than the field capacity, its infiltration rate will be very low, and if saturated, the intake rate is virtually zero. Similarly, Willis *et al.* (1961) in their studies on small plots in North Dakota report that as much as 90 per cent of the snowpack water is lost as surface runoff when these plots were frozen at high moisture levels. Other experiments have also shown that whenever an extremely wet layer within the soil profile is frozen, the downward movement of water through this layer is impeded until the zone is thawed. The existence of these ice shields within the profile may be caused either by the result of autumn precipitation prior to freezing, or perhaps by moisture migration induced by thermal gradients which are set up within the profile during the freezing process. The mechanism and magnitude of moisture transport during freezing in unsaturated soils in the absence of water tables is still a subject of considerable study. Willis *et al.* in their studies found no measurable accumulation of moisture at the freezing depth.

Gillies (1968) in a study of infiltration to Prairie snowpacks found that the percentage of the total water content of the overlying pack which infiltrated a frozen soil, d (%), could be related to the initial soil moisture content, prior to melt, of the surface layer (0-2 in.) MC as

$$d(\%) = 53.5 - 0.65 \, MC \dots\dots\dots\dots\dots\dots\dots V.15$$

in which MC is expressed on an oven-dry weight basis. The relationship was developed from tests conducted on small plots in 1966 on which MC ranged from 15-60%. Although the data for 1966 showed a linear trend for the volumetric infiltration to decrease with moisture content, these results should be applied with extreme care, inasmuch as additional tests indicated that the relationship may be curvilinear, and perhaps be different for different years.

In addition, Gillies also observed, that in the infiltration process to frozen soils, two distinct zones of moisture migration could be delineated: (a) an upper, thin, saturated zone of thawed soil, and (b) downward movement of water (unsaturated) into the underlying subzero, frozen soil. Extension of the thawed zone occurred as the heat supply to the ground increased. In these regards, he observed that for the Prairie snowpack the amount of heat transmitted to the ground followed a simple extinction curve depending on the density and mass of the overlying pack. Thus, these results suggest that the penetration of light rays through shallow Prairie snowpacks may be extremely important to the energy exchange at the ground surface.

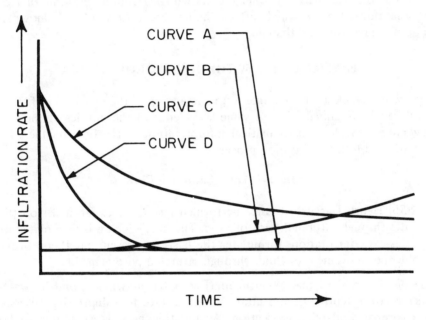

Fig. V.7   Schematic Diagram of Infiltration Rates of Frozen Soils

Curve A: This curve demonstrates the infiltration characteristics which may prevail when the soil is frozen when saturated, or when an impervious ice layer develops on the surface during the melting period. For these conditions, the rate of entry of water is very low, and consequently, the runoff coefficient is high.

Curve B: This curve exemplifies the conditions which may exist when a soil is frozen at a high moisture content (70-80% field capacity). In this case, some of the melt water is able to penetrate the soil, and thus transfer heat to melt the ice-filled pores. Progressively, as the soil warms and the pores melt, the infiltration rate increases. Zavodchikov (1962) cited examples in which the infiltration rate of the soil increased 6-8 times its initial rate during the melting period. In time, of course, the intake rate will again decrease due to its high moisture content.

Curve C: This curve characterizes the infiltration rate when the soil is frozen at a low moisture content, and the soil temperature is near or above freezing. In this case, only the small pores are filled with ice, and these are thawed rapidly with the downward movement of water. Under these conditions infiltration proceeds as under normal unfrozen conditions.

Curve D: This curve represents the condition in which the soil is frozen at a low moisture content, but the soil temperature at the time of snowmelt is well below freezing. Water entering the soil is frozen in the pores and movement is inhibited.

The infiltration rate curves of a frozen soil may adopt three distinct shapes dependent on conditions which prevail at the time of freezing or thawing. These shapes are (a) an intake rate which is reasonably constant with time at a low value, (b) an increase in the infiltration rate with time, and (c) a decrease in the infiltration rate with time (see Fig. V.7).

Inasmuch as the infiltration curves for frozen soils may adopt a multitude of shapes, to a certain degree it is therefore unrealistic that predictions of the yields from accumulated snowfall depths are made on a single criterion involving only climatic or meteorological conditions at the time of melt. Such estimates should be based also on a knowledge of both climatic and other factors which prevail at the time of freezing and melting, and those factors which affect the metamorphism of the snowpack in the interval between freezing and thawing.

V.2   PHYSICS OF FLOW THROUGH POROUS MEDIA

It is inconceivable that a complete understanding of the mechanics of flow of water in the soil can be obtained without some knowledge of the physics of the system. The following notes provide a mathematical interpretation of the flow regime with specific examples directed to the infiltration process.

V.2.1   Darcy's Law—Saturated Flow

In 1856, Henry Darcy in France performed his classic experiments on the vertical flow of water through saturated beds of sand. The law, which was formulated during the course of this experimentation, formulates the basis for present day theoretical solutions to the problem of movement of fluids through saturated porous media.

Consider Darcy's original experiment (Fig. V.8) in which *ponded* water, at the upper surface to a depth, h_1, and at the lower surface to a depth, h_2, filters vertically through a core of sand of cross section, A, and thickness, l. According to Darcy, the quantity of water, Vol, flowing through the core in time, t, can be expressed in equational form as

$$Vol = kA (h_1 + l - h_2) t/l$$

or

$$Vol/At = V = ki \dots\dots\dots\dots\dots\dots\dots\dots\dots\dots\dots\dots\dots\dots\dots\dots\dots V.16$$

where V = velocity of flow,
   k = proportionality constant (hydraulic conductivity), and
   i = hydraulic gradient.

It should be noted that Darcy's law gives the macroscopic velocity (average velocity) of flow through a core of cross section, A. In reality, flow through the medium only occurs through the interstices or pore spaces, and hence the velocity of flow as calculated, is somewhat fictitious. The actual velocity of flow in the pores is equal to V/n, where n is the porosity.

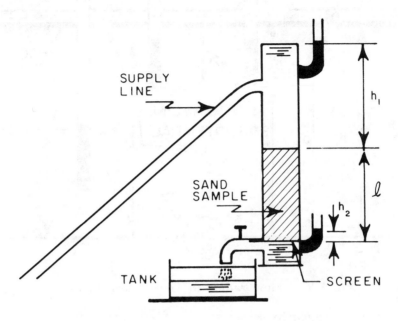

Fig. V.8   Darcy's Apparatus

V.2.1.1   Hydraulic Gradient, i

The hydraulic gradient, i, represents the change in hydraulic head per unit length, or the force per unit volume, of water acting through the soil core. Hydraulic head, h, is the potential energy per unit weight of fluid and includes the two components; pressure head, p/γ and elevation head, z. Hence

$$h = p/\gamma + z \quad \dots\dots\dots\dots\dots\dots\dots\dots\dots\dots\dots\dots\dots\dots\dots\dots V.17$$

By definition, h is equal to the depth to which water will stand in an open piezometer, as measured from a *fixed datum*. To illustrate, consider the flow to a tile line above which water is ponded at the surface (as shown in Fig. V.9).

In the presence of an energy difference, water will move from a position of higher energy to that of lower energy. Therefore, in the figure, flow is in the direction from the surface to the tile; $h_A > h_B$ or h_D. Piezometers B and C show points in the system at which the potential energy is equal; hence no flow occurs between B and C. Lines joining points of equal potential energy are referred to as equipotential lines. Further, it should be noted that the direction of flow will be at right-angles to equipotential lines, since this is the direction of maximum hydraulic gradient. ($\Delta h/l$ is large because l is small). The complex, or family, of equipotential lines and lines defining the direction along which water particles move—streamlines—is called a flow net.

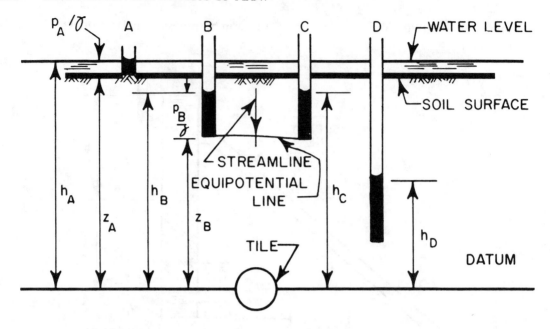

Fig. V.9   Flow Pattern to Tile-Ponded Water Case

V.2.1.2   Hydraulic Conductivity, k

The proportionality coefficient, k, of Darcy's law, commonly referred to as the hydraulic conductivity, expresses the interaction between the fluid and the media. Its dimensions are those of velocity (LT^{-1}) as ft/sec, cm/sec, in/hr, etc.

The Hagen-Poiseuille Equation Describing Laminar Flow Through Capillaries

Possibly, the best method of gaining a thorough physical interpretation of the coefficient is to make the comparison with viscous, laminar flow in a pipe.

Consider the streamtube (see Fig. V.10). For steady uniform flow, the forces are summed:

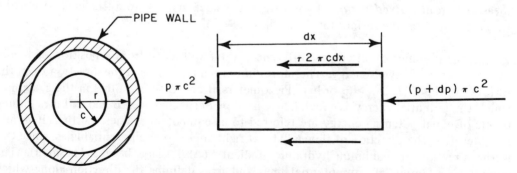

Fig. V.10   Free Body Diagram of Streamtube

$$\Sigma F_x = 0 = -(p+dp)\pi c^2 + p\pi c^2 - \tau 2\pi cdx \quad \text{or} \quad -cdp = 2\tau dx \dots\dots\dots\dots\dots\text{V.18}$$

where   p = pressure,
c = radius of streamtube
dx = length of streamtube, and
τ = shear stress.

however

$$\tau = \frac{-\mu dV}{dc} \dots\dots\dots\dots\dots\dots\dots\dots\dots\dots\dots\dots\dots\text{V.19}$$

where μ = dynamic viscosity, and
dV/dc = velocity gradient.

therefore

$$cdp = 2dx\mu\frac{dV}{dc} \dots\dots\dots\dots\dots\dots\dots\dots\dots\dots\dots\dots\text{V.20}$$

Integrating the equation along the streamtube between two points a distance L apart, at which the pressures are p_1 and p_2 respectively, one obtains

$$\int_{p_1}^{p_2} cdp = \int_0^L 2\mu \frac{dV}{dc} \, dx$$

$$(p_1-p_2)\, c = -2\mu\frac{dV}{dc} L \dots\dots\dots\dots\dots\dots\dots\dots\dots\dots\dots\dots\text{V.21}$$

The velocity, V_c, at any distance, c, from the centre may be found in terms of the velocity, V_0, at the centre by integration of Equation V.21. That is

$$\int_0^c (p_1 - p_2)\, cdc = -\int_{V_0}^{V_c} 2\mu L dV$$

$$(p_1-p_2)\frac{c^2}{2} = 2\mu L(V_0 - V_c) \dots\dots\dots\dots\dots\dots\dots\dots\dots\dots\dots\text{V.22}$$

For boundary conditions,

when c = r, V_c = 0, therefore

$$V_0 = (p_1 - p_2)\frac{r^2}{4\mu L}$$

$$V_c = \frac{(p_1 - p_2)(r^2 - c^2)}{4\mu L} \dots\dots\dots\dots\dots\dots\dots\dots\dots\dots\dots\dots\text{V.23}$$

Equation V.22 shows that change in velocity is proportional to the square of the distance from the center of the pipe—i.e., the velocity distribution is parabolic (laminar flow). For this distribution, the mean velocity, V, is related to V_0, as

$$V_0 = 2V$$

Substituting the equality into Equation V.23, one obtains

$$V = \frac{r^2}{8\mu} \cdot \frac{p_1 - p_2}{L} \dots\dots\dots\dots\dots\dots\dots\dots\dots\dots\dots\dots\dots\dots\dots\dots\dots\dots \text{V.24}$$

But, $\left[\dfrac{p_1 - p_2}{L}\right]$ the pressure gradient, can be written in the form of a hydraulic gradient as

$\rho g \left[\dfrac{h_1 - h_2}{L}\right]$. Thus Equation V.24 becomes

$$V = \frac{\rho g r^2}{8\mu}\left(\frac{h_1 - h_2}{L}\right) \frac{\rho g d^2}{32\mu}\left(\frac{h_1 - h_2}{L}\right) \dots\dots\dots\dots\dots\dots\dots\dots\dots\dots\dots\dots \text{V.25}$$

where ρ = mass density of fluid.

Equation V.25 is referred to as the Hagen-Poiseuille equation for laminar flow through small tubes (capillaries). In comparing this equation with Equation V.16, it is obvious the two are identical when

$$k = \frac{\rho g d^2}{32\mu} \dots\dots\dots\dots\dots\dots\dots\dots\dots\dots\dots\dots\dots\dots\dots\dots\dots\dots\dots\dots \text{V.26}$$

Factors Affecting the Magnitude of k

The analogy provided by Equations V.16 and V.25 implies that the hydraulic conductivity of a soil is a function of both soil and fluid properties. Specifically, such properties include the size, shape, number and continuity of pores, the mineralogy of the medium and the density and viscosity of the fluid.

Several investigators have presented formulae that may be used to evaluate k by considering the properties of the media. Slichter (1900) studied the flow of water through an idealized, homogeneous medium consisting of spheres of uniform size. He found the conductivities through tortuous capillaries of triangular cross section could be expressed by

$$k = \frac{\rho g d^2}{96 C\mu} \cdot \dots\dots\dots\dots\dots\dots\dots\dots\dots\dots\dots\dots\dots\dots\dots\dots \text{V.27}$$

in which d = diameter of sphere, and
C = dimensionless function of porosity.

Another relationship developed for this purpose is the Kozeny-Carman formula:

$$k = \frac{\rho g}{C'\mu S_v^2} \cdot \frac{n^2}{(1-n)^2} \quad \dots\dots\dots\dots\dots\dots\dots\dots\dots\dots\dots\dots\dots\dots\dots\dots \text{V.28}$$

in which C' = constant whose magnitude is 5 for
unconsolidated media
S_v = surface area/unit volume, and
n = porosity

Frequently, the hydraulic conductivity, k, is written in the more generalized form as

$$k = \frac{k'\rho g}{\mu} \dots\dots\dots\dots\dots\dots\dots\dots\dots\dots\dots\dots\dots\dots\dots\dots\dots\dots\dots \text{V.29}$$

in which k' = constant which characterizes the
flow properties of the medium (size
of pores etc.) having the dimensions of
L^2. Theoretically, k' is a constant
for a given medium and is independent
of fluid properties. The constant
is frequently referred to as the in-
trinsic hydraulic conductivity or permeability.

Using Equation V.29, one can express Darcy's law in terms of a pressure gradient as

$$V = \frac{k\Delta h}{\Delta L} = \frac{k'\rho g}{\mu}\frac{\Delta h}{\Delta L} = \frac{k'}{\mu}\frac{\Delta p}{\Delta L}\left[\begin{array}{c}\text{Horizontal}\\\text{Flow}\end{array}\right] \quad \dots\dots\dots\dots\dots\dots\dots\dots \text{V.30}$$

Needless-to-say, the magnitude of k varies widely for a given medium. Hence, it is somewhat meaningless to quote specific values for a given soil to cover all conditions. In manner of reference, the values for k of different subsoils, as reported by O'Neal (1949) and Smith and Browning (1946), may serve as useful guides (see Tables V.3 and V.4).

Table V.3 Classes of Hydraulic Conductivity or Percolation Rates
for Saturated Subsoils (after O'Neal 1949).

Class	Hydraulic Conductivity, k in/hr
Very Slow	<0.05
Slow	0.05 - 0.2
Moderately Slow	0.2 - 0.8
Moderate	0.8 - 2.5
Moderately Rapid	2.5 - 5.0
Rapid	5.0 - 10.0
Very Rapid	>10.0

Changes in Hydraulic Conductivity with Time

It should also be recognized that the magnitude of the hydraulic conductivity is not independent of time. The effects of factors such as soil mineralogy, ionic exchange, the chemical composition of soil and water, the presence of entrapped air and microorganism activity as they influence pore size and distribution by swelling, blockage, migration and rearrangement of clay particles, etc., may change with time and thus affect the magnitude of k.

Effect of Viscosity and/or Temperature Changes on the Hydraulic Conductivity

As shown by Equations V.26–V.29, any change in viscosity brought about by either a change in the type of fluid or a large temperature change, will necessarily change the magnitude of k. Hence, when evaluating the hydraulic conductivity of a soil *in situ,* all readings should be adjusted to standard conditions. In general, the effect of density

Table V.4 Classes of Permeability or Percolation Rates for Saturated Subsoils (after Smith & Browning 1946)

Class	Hydraulic Conductivity: or Percolation Rate in/hr	Comments
Extremely Slow	<0.001	So nearly impervious that leaching process is insignificant.
Very Slow	0.001 - 0.01	Poor drainage results in staining; too slow for artificial drainage.
Slow	0.01 - 0.1	Too slow for favourable air-water relations and for deep root development.
Moderate	0.1 - 1	Adequate permeability (conductivity)
Rapid	1 -10	Excellent water holding relations as well as excellent permeability (conductivity).
Very Rapid	> 10	Associated with poor water holding conditions.

changes in the water due to the presence of dissolved salts is usually neglected. The effect of temperature changes in the water on the value of k can be made by the following equation:

$$k_s = k_m \, \mu_m / \mu_s \; \dots\dots\dots\dots\dots\dots\dots\dots\dots\dots\dots\dots\dots\dots\dots\dots \text{ V.31}$$

> where k_s = hydraulic conductivity value adjusted
> to standard base of viscosity, μ_s, and
> k_m = measured hydraulic conductivity with
> water of viscosity, μ_m.

Measurement of Hydraulic Conductivity

Several methods are available for determination of the hydraulic conductivity of a soil. These may be divided into two broad categories: laboratory methods and field methods. A list of the more important of these methods is given below.

Laboratory methods

> (1) Constant head permeameter, or
> (2) Variable head permeameter.

Field Methods:

> (1) Auger-hole,
> (2) Two-well,
> (3) Four-well,
> (4) Piezometer,
> (5) Tube, and
> (6) Pumping test.

Because of the number of these methods, it is impossible within the limits of these notes to give a complete description of each. Each method has individual advantages and disadvantages which should be evaluated in terms of the physical requirements of the investigation, for which the measurement of 'k' is needed. For a complete description of each method listed above, the reader is referred to the following references: Luthin, *Advances in Agronomy*, pp. 414-445 (1957); Muskat (1946); and Leliavsky (1955).

V.2.1.3 Range of Validity of Darcy's Law

A general equation describing the flow of fluids through porous material may be written as

$$V = ki^n \dots\dots\dots\dots\dots\dots\dots\dots\dots\dots\dots\dots\dots\dots\dots\dots\dots\dots\dots \text{ V.32}$$

in which n is an exponent whose magnitude varies unexplainably with the flow regime. According to Darcy's law, n = 1 for a given range of conditions. In the Darcy relationship, viscous or resistive forces predominate.

One criterion which is used to establish the range of validity of Darcy's law is the Reynolds Number, N_R, which gives the relative magnitude of viscous to inertial forces in a system. N_R is defined

$$N_R = \frac{\rho V d}{\mu} \dotfill V.33$$

where ρ = mass density of the fluid,
       V = average velocity of flow,
       μ = dynamic viscosity, and
       d = some average diameter of the soil grains.
            For example, the mean weight diameter obtained
            from a particle-size distribution curve.

In general, it is conceded by most investigators that Darcy's law will apply at values of $N_R < 10$, (laminar flow). Nevertheless, one should only use these limits of N_R with discretion, inasmuch as the Reynolds Number is not a particularly sensitive criterion and flow which obey   Darcy's law can exist in a porous medium at values of $N_R > 10$.

Frequently, the limits of Darcy's law are fixed with reference to the size of the grains and the velocity and hydraulic gradient. Several investigators have shown that the magnitude of the exponent, n, of Equation V.32 tends to become less than unity as the size of the material increases. For example, experiments conducted in Germany indicated n = 0.65 for gravel containing particles of diameter 0.33-2 in. and n = 1.00-1.15 for sand having 86 per cent of the grains smaller than 0.08 in. Hazen suggests that the velocity is proportional to the gradient if the effective size of the material is not greater than 0.118

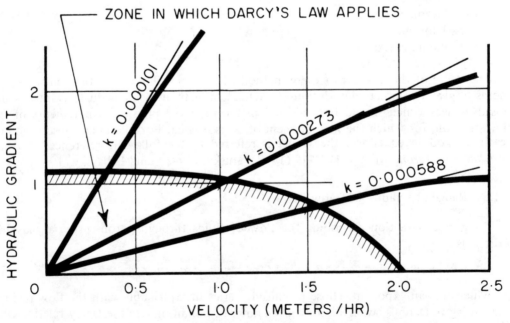

Fig. V.11 Limits of Darcy's Law According to Prinz

in. In summary, it would appear that Darcy's law can be safely applied to materials whose texture is not coarser than that of a fine gravel.

On the other hand, other investigations have shown that Darcy's law may also be limited by consideration of the velocity of filtration or hydraulic gradient. These effects are best illustrated in graphical form by the works of Prinz as taken from p. 24 of Leliavsky (1955)—shown in Fig. V.11.

In the foregoing discussions, no attempt has been made to give physical reasons why the exponent 'n' varies under different conditions. Some writers attribute this variation to factors such as the manner of energy dissipation due to turbulence in eddies or to pressure exercised by the filtering water on the grains. Yet, others submit that the deviation is principally the result of the increased importance of inertial forces compared to viscous forces. In effect, the phenomenon has not been clearly explained. Nevertheless, it is an established fact that beyond a given limit Darcy's law ceases to be valid.

V.2.2   Generalized Equation for Flow of Fluids Through Porous Media

To completely describe the motion of a fluid, three relationships need to be taken into consideration: (a) the conditions imposed on fluid motion by the media; (b) the state of the fluid; and (c) conservation of mass. In previous discussions of this Section we have presented a law, Darcy's law, which gives a dynamic description of flow as imposed by the media, and hence suffices condition (a). To complete the requirements, therefore, it remains to consider conditions (b) and (c).

V.2.2.1   State of the Fluid

For any fluid, we can write a general equation of the state as

$$f(p, \rho, T) = 0 \dots\dots\dots\dots\dots\dots\dots\dots\dots\dots\dots\dots\dots V.34$$

where p = pressure,
ρ = density, and
T = temperature.

In the case of an ideal gas, the equation takes the form $p V_{ol} - mRT = 0$ where V_{ol} = volume of the gas, m = mass, and R is the universal gas constant. For other conditions other laws apply. Similarly, for an incompressible fluid such as water, in flow problems in porous media other than those which involve such anomalies as salt intrusion or diffusion problems, we can write ρ = constant. It has to be remembered that information must be given regarding the equation of the state to solve the flow problem.

V.2.2.2   Conservation of Mass

According to the basic law of physics, barring subatomic reaction, mass can neither be created nor destroyed.

V.2.2.3   Derivation of Continuity Equation for Steady-State, Incompressible Flow

Let us now proceed to develop a generalized mathematical expression to describe the flow of fluids through a porous medium that incorporates all conditions mentioned above. To do this, consider a small cube whose sides are of differential lengths, dx, dy and dz, taken from a point in a saturated porous medium.

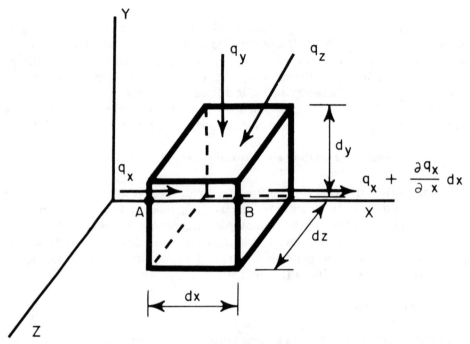

Fig. V.12   Basic Factors in Calculation of Continuity Equation

Let ρ = mass density of fluid entering cube in three principal directions (assumed constant)

q_x, q_y, q_z = flow rate per unit time per unit cross-section entering the cube in the x, y and z directions.

From the Law of Conservation of Mass:

Net Mass Flux of water in the x - direction to the cube $\Big\}$ = Mass Entering - Mass Leaving

$$= \rho q_x \, dydz - \left[\rho q_x + \frac{\partial}{\partial x} (\rho q_x) \, dx \right] dydz$$

$$= - \frac{\partial}{\partial x} (\rho q_x) \, dxdydz \dots\dots\dots\dots\dots \text{V.35}$$

Note: in this development because we are dealing with more than one variable we make use of the partial derivative (Lebnitz dee, ∂) to define the change in mass flow over the distance dx. The physical interpretation of this procedure is shown below.

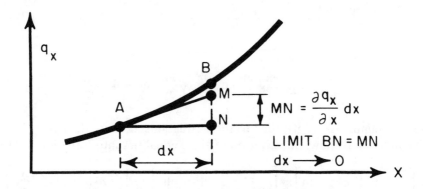

Fig. V.13   Change in Flow Over Distance

Similarly, it can be shown that the net mass of fluid entering or leaving the cube per unit time in the y and z directions are respectively

$$-\frac{\partial}{\partial y}(\rho q_y)\, dxdydz \dots\dots\dots\dots\dots\dots\dots\dots V.36$$

$$-\frac{\partial}{\partial z}(\rho q_z)\, dxdydz \dots\dots\dots\dots\dots\dots\dots\dots V.37$$

Hence, the net mass flux of fluid entering or leaving the cube per unit time can be obtained by summing Equations V.35, V.36 and V.37 to obtain

$$-\left[\frac{\partial}{\partial x}(\rho q_x)+\frac{\partial}{\partial y}(\rho q_y)+\frac{\partial}{\partial z}(\rho q_z)\right] dxdydz \dots\dots\dots\dots\dots V.38$$

However, according to the conservation of mass, this mass flux must be equal to the time rate of change of fluid mass within the cube itself. That is

the mass of fluid in cube $= \rho F dxdydz$

where   F = porosity, and

the rate of change $=\frac{\partial}{\partial t}(\rho F)\, dxdydz$ $\dots\dots\dots\dots\dots\dots\dots\dots$ V.39

Combining Equations V.38 and V.39 one obtains

$$-\left[\frac{\partial}{\partial x}(\rho q_x)+\frac{\partial}{\partial y}(\rho q_y)+\frac{\partial}{\partial z}(\rho q_z)\right]dxdydz=\frac{\partial}{\partial t}(\rho F)\, dxdydz \dots\dots\dots V.40$$

Equation V.40 is the continuity equation of fluid flow. Now, if we assume that the fluid is incompressible and its density remains constant, and the medium is incompressible so that the porosity is constant, then Equation V.40 reduces to

$$\left[\frac{\partial q_x}{\partial x} + \frac{\partial q_y}{\partial y} + \frac{\partial q_z}{\partial z}\right] = 0 \quad \cdots\cdots\cdots\cdots\cdots\cdots\cdots\cdots\cdots\cdots\cdots \text{V.41}$$

But, we also know from Darcy's law

$$q_x = -k_x \frac{\partial h}{\partial x}, \ q_y = -k_y \frac{\partial h}{\partial y}, \ q_z = -k_z \frac{\partial h}{\partial z}$$

Note, the minus sign is attached to these equations, because the hydraulic head decreases in the direction of flow. Substitution of these equalities into Equation V.41 gives.

$$\left[k_x \frac{\partial^2 h}{\partial x^2} + \frac{\partial k_x}{\partial x}\frac{\partial h}{\partial x}\right] + \left[k_y \frac{\partial^2 h}{\partial y^2} + \frac{\partial k_y}{\partial y}\frac{\partial h}{\partial y}\right] + \left[k_z \frac{\partial^2 h}{\partial z^2} + \frac{\partial k_z}{\partial z}\frac{\partial h}{\partial z}\right] = 0 \ldots \text{V.42}$$

If we further assume the medium is homogeneous and isotropic, such that $k_x = k_y = k_z$ = constant, then Equation V.42 can be written as

$$\frac{\partial^2 h}{\partial x^2} + \frac{\partial^2 h}{\partial y^2} + \frac{\partial^2 h}{\partial z^2} = 0$$

$$\nabla^2 h = 0 \ldots\ldots\ldots\ldots\ldots\ldots\ldots\ldots\ldots\ldots\ldots\ldots\ldots\ldots \text{V.43}$$

Equation V.43 is the Laplace equation for steady-state flow. It should also be mentioned that this equation is applicable to electrical, magnetic, thermal and other systems, and to the field of celestial mechanics. As applied to fluid flow systems, the components of Equation V.43 represent the curvature of the water table in the different directions. If either term of the equation is positive, this indicates that the shape of the water table is concave up; if negative, then concave down. Further, it should be noted, if the total curvature is either positive or negative, then according to Equation V.42 there must be a gradient in the hydraulic conductivity and thus the derivatives

$$\frac{\partial k_x}{\partial x}, \frac{\partial k_y}{\partial y}, \frac{\partial k_z}{\partial z}$$

are not zero to produce steady-state flow conditions.

V.2.2.4   Solving the Equations of Flow

It is well known in the theory of differential equations that the Laplace equation, which we have developed to describe the flow through porous media, has an infinite number of solutions — so how may we choose one that applies specifically to a particular problem? Since all the problems we are considering at present are the same dynamically in character, the solution selected must necessarily be based on differences in the boundaries defining the system, and the physical conditions imposed on these boundaries at the initial instant when these boundary conditions are introduced.

These boundaries are not necessarily impermeable walls which confine the fluid to a given region or space, but can be thought of as geometrical conditions over which the fluid velocity, or velocity potential (head distribution), are known. For example, along the interface of an impermeable layer we know that the normal velocity equals zero, or $\partial h/\partial n = 0$.

For *steady-state* conditions (Equation V.43), pre-assigned values of the velocity potential and normal velocities, or a combination of the two, at all points on the boundaries of the system suffice to determine uniquely the pressure or hydraulic distribution in the interior of the regions defined by the boundaries. For *unsteady-state* conditions (Equation V.40) one must specify the initial conditions (the initial density distribution with which the system begins its history) and follow the change in this pattern with time. An incompressible fluid, with the boundary changing with time, will go through a continuous succession of steady-state distributions, each corresponding to the instantaneous conditions at the boundaries. As time progresses, these boundary conditions tend to fixed boundaries—a steady-state distribution—independent of the initial conditions and determined only by the limiting values of the initial boundary conditions.

Thus, the physical problem in the flow of fluids through a porous medium is in general defined by

(1) a geometrical statement of the boundaries of the region in space for which a solution is desired,

(2) specification of the 'boundary conditions' to be satisfied over the boundaries, and

(3) assignment of the density and hence pressure distribution at initial instant.

When a function has been selected such as to satisfy the boundary conditions and the equation of continuity, this is the unique solution to the flow problem.

V.2.2.5 Practical Applications of the Laplace Equation.

Problem

Determine the seepage loss in ft³/day/ft of width through a peat layer under an earth-filled dam when

$$k = 0.10 \text{ ft./day}$$
$$H = 80 \text{ ft.}$$
$$L = 100 \text{ ft.}$$
$$b = 50 \text{ ft.}$$

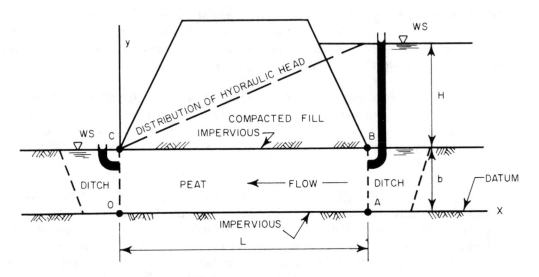

Fig. V.14   Application of Laplace Equation to the Problem
of Flow Under an Impervious Embankment

Boundary conditions:

Along face   OC $h = b$ $0 \leqslant y \leqslant OC, x = 0$

"   "   CB, OA: $k_y \, \partial h/\partial y = 0$ (no flow across) : $k_y = 0$

$y = 0, \; y = b, 0 \leqslant x \leqslant L$

"   "   AB $h = b + H$ $0 \leqslant y \leqslant AB, x = L$

Perpendicular to paper: $\partial h/\partial z = 0$

Assume that the hydraulic head is dissipated linearly through the peat.

Thus $h = ax + c$.

For the given boundary conditions

(1) When $x = 0, h = b$ (2) When $x = L, h = b + H$
$\therefore b = a(0) + c$ $\therefore b + H = a(L) + b$
$b = c$ $a = H/L$
$h = ax + b$

Thus $h = Hx/L + b$

Notice this function also satisfies the other boundary conditions, namely $\partial h/\partial z = 0$
Checking by Laplace $\partial h/\partial x = H/L$ and $\partial^2 h/\partial x^2 = 0$
Hence $\nabla^2 h = 0$

Since our function satisfied both the boundary conditions and the Laplace equation,

$h = Hx/L + b$ describes the distribution of hydraulic head in our flow system.

According to Darcy:

$$Q = kiA = k \frac{\partial h}{\partial x} A$$

$$= k \ HA/L$$

$$= 0.10 \times 80 \times 50/100 = 4.0 \ \text{ft}^3/\text{day/ft}.$$

Problem

Determine the steady-state flow rate from a well of radius $r_w = 1$ ft which fully penetrates a confined aquifer of thickness of $b = 100$ ft and $k = 1$ ft/day. Under steady-state pumping the water levels measured in piezometers show that when $r = r_w = 1$ ft, $h_w = 130$ ft, $r = r_i = 1000$ ft and $h_i = 150$ ft (see Fig. V.15).

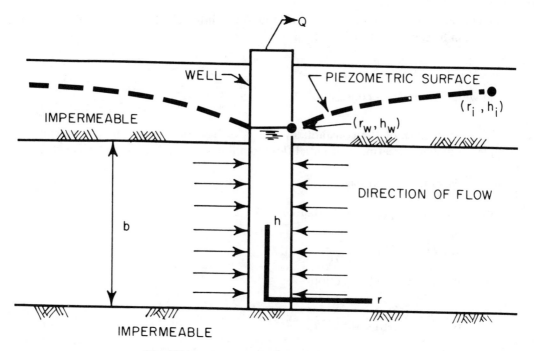

Fig. V.15   Confined Aquifer Flow System

Solution:

Inasmuch as the flow system represents a case of horizontal, radial flow, the solution to our problem is obtained more easily if we consider the continuity equation for steady-state flow in the cylindrical coordinate system. That is

$$\frac{1}{r} \frac{\partial}{\partial r} \left[r \frac{\partial h}{\partial r} \right] + \frac{1}{r^2} \frac{\partial^2 h}{\partial \theta^2} + \frac{\partial^2 h}{\partial z^2} = 0 \quad \dots\dots\dots\dots\dots\dots\dots\dots\dots \text{V.44}$$

in which $r = \sqrt{x^2+y^2}$, $\theta = \tan^{-1}\frac{y}{x}$, $z = z$

$x = r\cos\theta$, $y = r\sin\theta$, $z = z$

For the flow system described in the problem, Equation V.44 can be reduced by recognizing that

$$\frac{\partial h}{\partial \theta} = 0 \text{ (Flow is radial ... no circular motion)}$$

$$\frac{\partial h}{\partial z} = 0 \text{ (Horizontal flow ... streamlines horizontal)}$$

Therefore, the equation becomes

$$\frac{1}{r}\frac{\partial}{\partial r}\left[r\frac{\partial h}{\partial r}\right] = 0 \quad \dots\dots\dots\dots\dots\dots\dots\dots\dots\text{V.45}$$

By integrating Equation V.45 twice as an indefinite integral, the general solution of the equation can be shown to be of the form

$$h = c_1 \ln r + c_2 \dots\dots\dots\dots\dots\dots\dots\dots\dots\dots\dots\text{V.46}$$

in which c_1 and c_2 = constants.

The significance of the constants c_1 and c_2 of Equation V.46 can be appreciated by a simple geometrical treatment of radial flow to a line sink (see Fig. V.16) of length, b.

From the diagrams on next page:

Flow rate through concentric rings

$$Q = 2\pi rb\, V_R \text{ or } V_R = \frac{Q}{2\pi rb} \quad \dots\dots\dots\dots\dots\dots\dots\text{V.47}$$

Velocity vector in 'x' direction

$$u = V_R \cos\theta = \frac{Q}{2\pi rb}\left[\frac{x}{\sqrt{x^2 + y^2}}\right] \quad \dots\dots\dots\dots\dots\dots\text{V.48}$$

But by definition of a velocity potential, ϕ

$$u = \frac{\partial \phi}{\partial x} = \frac{Q}{2\pi rb}\left[\frac{x}{\sqrt{x^2 + y^2}}\right] = \frac{Q}{2\pi b}\left[\frac{x}{x^2 + y^2}\right] \dots\dots\dots\dots\text{V.49}$$

or $\phi = \frac{Q}{4\pi b} \ln\left[x^2 + y^2\right] + \text{constant} \quad \dots\dots\dots\dots\dots\text{V.50}$

$$= \frac{Q}{4\pi b} \ln\left[r^2\right] + \text{constant}$$

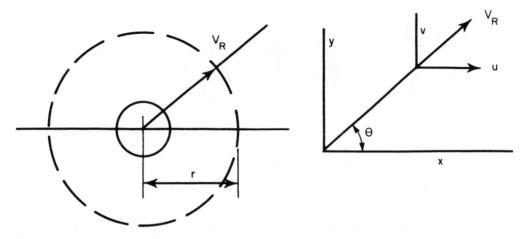

Fig. V.16   Geometrical Treatment of Radial Flow to a Line Sink

$$= \frac{Q}{2\pi b} \ln r + \text{constant} \dots\dots\dots\dots\dots\dots\dots\dots\dots\dots\dots\dots V.51$$

and for saturated flow $h = \phi$. Therefore

$$h = \frac{Q}{2\pi b} \ln r + \text{constant} \dots\dots\dots\dots\dots\dots\dots\dots\dots\dots\dots V.52$$

It follows that Equation V.46 will enable us to solve our problem when the constants, c_1 and c_2, are evaluated in terms of the boundary conditions of the system. This can be accomplished by solving simultaneously two equations which utilize the known boundary conditions.

$$\text{when } r = r_w, \ h = h_w \text{ and } h_w = c_1 \ln r_w + c_2 \dots\dots\dots\dots\dots\dots V.53(a)$$

$$\text{and when } r = r_i, \ h = h_i \text{ and } h_i = c_1 \ln r_i + c_2 \dots\dots\dots\dots\dots\dots V.53(b)$$

Solving

$$c_1 = \frac{h_i - h_w}{\ln\left(\dfrac{r_i}{r_w}\right)} \dots\dots\dots\dots\dots\dots\dots\dots\dots\dots\dots\dots\dots\dots\dots\dots V.54$$

and

$$c_2 = h_w - \frac{(h_i - h_w) \ln r_w}{\ln\left(\dfrac{r_i}{r_w}\right)} \dots\dots\dots\dots\dots\dots\dots\dots\dots\dots\dots\dots V.55$$

Therefore, the head distribution around the well may be defined by substituting Equations V.54 and V.55 into Equation V.46 to obtain

$$h = \left[\frac{h_i - h_w}{\ln\left(\frac{r_i}{r_w}\right)}\right] \ln r + h_w - \left[\frac{h_i - h_w}{\ln\left(\frac{r_i}{r_w}\right)}\right] \ln r_w \quad \dots\dots\dots\dots \text{V.56}$$

From the continuity equation, $Q = AV$, knowing from Darcy's law $V = ki$, the flow rate, perpendicular to concentric rings surrounding the well, can be expressed as

$$Q = 2\pi r b k \frac{\partial h}{\partial r} \quad \dots\dots\dots\dots\dots\dots\dots\dots\dots\dots\dots\dots\dots\dots\dots\dots\dots\dots\dots\dots\dots\dots\dots\dots\dots\dots \text{V.57}$$

Utilizing Equations V.56 and V.57, the steady state discharge to the well, Q_w, $(r = r_w)$, can be expressed as

$$Q_w = 2\pi r_w b k \frac{\partial}{\partial r}\left[\left[\frac{h_i - h_w}{\ln\left(\frac{r_i}{r_w}\right)}\right] \ln r + h_w - \left[\frac{h_i - h_w}{\ln\left(\frac{r_i}{r_w}\right)}\right]\ln r_w\right]_{r = r_w}$$

$$Q_w = 2\pi r_w b k \left[\frac{h_i - h_w}{\ln\left(\frac{r_i}{r_w}\right)}\right] \frac{1}{r_w}$$

$$Q_w = 2\pi k b \left[\frac{h_i - h_w}{\ln\left(\frac{r_i}{r_w}\right)}\right] \quad \dots\dots\dots\dots\dots\dots\dots\dots\dots\dots\dots\dots \text{V.58}$$

Substituting the known values for k, b, h_i, h_w, r_i and r_w into Equation V.58 one can obtain the discharge as

$$Q = 2\pi(100)\ (1)\ \frac{(150 - 130)}{\ln \frac{1000}{1}} = 1825 \text{ ft}^3/\text{day}$$

V.2.2.6   Generalized Equation of Nonsteady or Transient Flow Through Porous Media

Jacob (1949) suggests that a generalized form of continuity equation for nonsteady state flow, which considers the compressibility of the media and water, can be written as

$$-\left[\frac{\partial(\rho q_x)}{\partial x} + \frac{\partial(\rho q_y)}{\partial y} + \frac{\partial(\rho q_z)}{\partial z}\right] = \rho F\left[\beta + \frac{\alpha}{F}\right]\frac{\partial p}{\partial t} \quad \dots\dots\dots\dots \text{V.59}$$

in which β = compressibility index of water
or the reciprocal of its bulk
modulus of elasticity (1/300,000 psi),
α = vertical compressibility index of
the medium (α may be equal to up
to 10β depending on crystalline
structure of the medium) and
p = pressure.

If we assume that the density of the fluid can be considered reasonably constant, and the medium is isotropic, then Equation V.59 can be reduced to a less formidable form as

$$\frac{\partial^2 h}{\partial x^2} + \frac{\partial^2 h}{\partial y^2} + \frac{\partial^2 h}{\partial z^2} = \frac{F\gamma}{k} \left[\beta + \frac{\alpha}{F} \right] \frac{\partial h}{\partial t} \quad \dots\dots\dots\dots\dots\dots\dots \text{V.60}$$

For the special case of flow in a confined aquifer system of thickness, b, Equation V.60 can be reduced to

$$\frac{\partial^2 h}{\partial x^2} + \frac{\partial^2 h}{\partial y^2} + \frac{\partial^2 h}{\partial z^2} = \frac{S}{T} \frac{\partial h}{\partial t} \quad \dots\dots\dots\dots\dots\dots\dots \text{V.61}$$

in which   S = storativity or storage coefficient
equal to $F\gamma b/\left[\beta + (\alpha/F)\right]$ and,
T = transmissibility equal to the product, kb.

Further discussions of these equations and their application to groundwater flow systems will be given in Section VI.

V.2.3   Movement of Water in the Unsaturated Phase

So far, we have considered only the movement of water through saturated soils in which the effective conducting pore space is the total pore space. In unsaturated soils, there is insufficient water present to fill completely all voids, and thus some of the pores are filled with air or water vapour. It should be recognized that unsaturated soils are much more prevalent in nature than saturated soils. Further, the theoretical equations describing the movement in the unsaturated phase are extremely complex, and correspondingly, we have less knowledge of this flow phenomenon.

In unsaturated soils, the movement of water can occur in either liquid or vapour phase, or both. The relative importance of the two types of flow depends on many factors, one being the degree of saturation of the medium. In the balance of these notes we will concern ourselves only with movement in the liquid phase.

V.2.3.1   Capillary-Tube Hypothesis

Movement of water in the liquid phase in unsaturated soils takes place through soil moisture films, or in small pores through the combined action of gravitational and capillary forces. The basis of capillary movement of water in soils can be shown through the analogy of the capillary-tube hypothesis. Suppose a small glass tube is inserted into a vessel of water which is open to the atmosphere (see Fig. V.17), water will rise in this tube and come to equilibrium at some definite height above the surface.

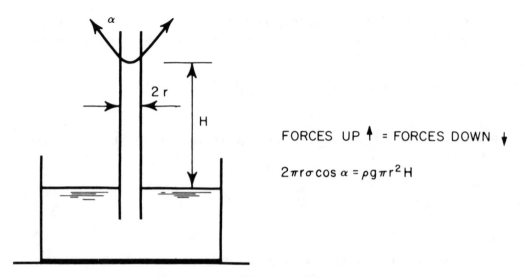

FORCES UP $\uparrow$ = FORCES DOWN $\downarrow$

$$2\pi r\sigma \cos\alpha = \rho g \pi r^2 H$$

Fig. V.17   Capillary-Tube Hypothesis

$$H = \frac{2\sigma}{r\rho g} \cos\alpha \quad \dots\dots\dots\dots\dots\dots\dots\dots\dots\dots\dots\dots\dots\dots\dots\dots \text{V.62}$$

where H = height of rise,
        σ = surface tension coefficient,
        α = contact angle,
        r = radius of the tube,
        ρ = density of the fluid, and
        g = acceleration of gravity.

Interpreting Equation V.62 in terms of the soil, we see that the height of rise is inversely related to the radius of the pore. Hence, the presence of many small pores in the soil would suggest considerable capillary movement. Keen (1922) calculated the theoretical heights of rise of water in different textured soils as fine gravel−1/3 ft, fine sand−7.5 ft, silt−31.25 ft, and clay−150 ft.

Another important concept, which the reader should gain from these discussions, is that the soil water in unsaturated soils is under a tension or suction. As a soil dries out, it develops a greater affinity or attraction for water. This fact is evident from Equation V.62, inasmuch as smaller pores dry, the radius, r, decreases and the tension increases. Further, it follows that if two soils, say a clay and sand, are at the same moisture content, the soil water in the clay will be at a higher suction than that in the sand. Fig. V.18 shows the relative positioning of soil moisture-tension curves for different soil types.

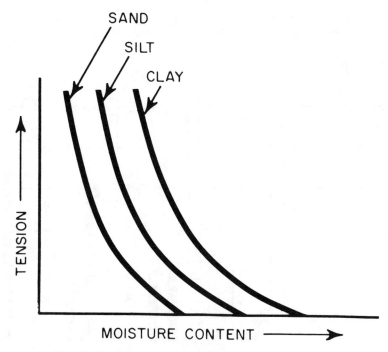

Fig. V.18   Moisture Desorption Curves

V.2.3.2   Concept of Total Potential

As indicated in previous discussions, water moves as a liquid in unsaturated soils under the combined influence of gravity and suction. Obviously, the direction of movement will be from a position of high total potential to a point of lower potential. It is often difficult to visualize how both gravitational and capillary potential enter into an unsaturated flow problem. To illustrate, consider a bank of tensiometers inserted in a soil as shown in Fig. V.19.

If we consider the gravitational potential, ϕ and the capillary potential ψ, then the total potential at any point in the system is the sum of the two terms; that is

$$\Phi = \phi + \psi \qquad \text{(where } \psi \text{ is negative)} \dots\dots\dots\dots\dots\dots\dots\dots\dots\text{V.63}$$

The concept of potential is introduced here to infer potential energy. Buckingham defined the capillary potential of soil water as the amount of work required to pull a unit mass of water from under a soil film to a place of the same gravitational potential where the curvature was zero.

In the figure, for locations A and B note that $\psi_A = \psi_B$. Thus the soil moisture is of equal dryness at A and B. But this does not mean that moisture will not move between A and B. On the contrary, since the total potential Φ_B is greater than Φ_A, moisture will move from B to A. Note again, because ψ_C is more negative than ψ_B, the soil is drier at C.

However, because the total head Φ_B is equal to Φ_C moisture neither moves from B to C, nor vice versa. Note finally that moisture will move upward from C to D, since Φ_C is greater than Φ_D; but the soil at D has to be considerably drier than at C (to cause movement upward).

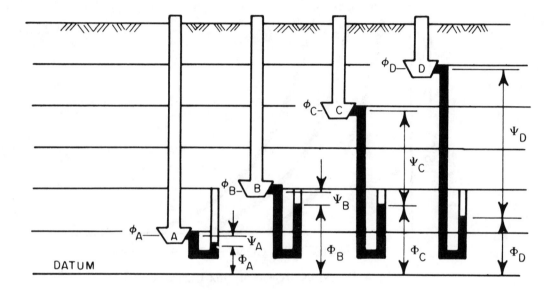

Fig. V.19   Concept of Total Potential

V.2.3.3   Equations of Movement

Generally, it is assumed that Darcy's law also applies to describe the rate of movement in an unsaturated soil. Accordingly, we can write

$$V = -k(\theta) \frac{d\Phi}{d l} = -k(\theta) \frac{d}{dl} (\phi + \psi) \dots\dots\dots\dots\dots\dots\dots\dots\dots V.64$$

in which   V = velocity of flow (LT^{-1}),
$k(\theta)$ = unsaturated hydraulic conductivity
or capillary conductivity (LT^{-1}),
Φ = total potential composed of the
gravitational potential, ϕ, and
the capillary potential, ψ expressed
in units of length (L), and
l = length (L)

Further, if we use a procedure similar to that used in deriving the Laplace equation for saturated flow, we can derive the continuity equation for unsaturated flow as

$$\frac{\partial}{\partial x}\left[k(\theta)_x \frac{\partial \psi}{\partial x}\right] + \frac{\partial}{\partial y}\left[k(\theta)_y \frac{\partial \psi}{\partial y}\right] + \frac{\partial}{\partial z}\left[k(\theta)_z\right] + \frac{\partial}{\partial z}\left[k(\theta)_z \frac{\partial \psi}{\partial z}\right] = \frac{\partial \theta}{\partial t} \quad .. \text{V.65}$$

If it is further assumed that both $k(\theta)$ and ψ are single-valued functions of moisture content, then Equation V.65 can be reduced to the form

$$\frac{\partial}{\partial x}\left[D(\theta)_x \frac{\partial \theta}{\partial x}\right] + \frac{\partial}{\partial y}\left[D(\theta)_y \frac{\partial \theta}{\partial y}\right] + \frac{\partial}{\partial z}\left[k(\theta)_z\right] + \frac{\partial}{\partial z}\left[D(\theta)_z \frac{\partial \theta}{\partial z}\right] = \frac{\partial \theta}{\partial t} \quad .... \text{V.66}$$

in which $D(\theta)_x$, $D(\theta)_y$ and $D(\theta)_z$ are so-called diffusivities in the three principal directions. The magnitudes of these are

$$D(\theta)_x = k(\theta)_x \frac{\partial \psi}{\partial \theta} \dots\dots\dots\dots\dots\dots\dots\dots\dots\dots\dots\dots\dots .\text{V.67}$$

$$D(\theta)_y = k(\theta)_y \frac{\partial \psi}{\partial \theta} \dots\dots\dots\dots\dots\dots\dots\dots\dots\dots\dots\dots\dots .\text{V.68}$$

$$D(\theta)_z = k(\theta)_z \frac{\partial \psi}{\partial \theta} \dots\dots\dots\dots\dots\dots\dots\dots\dots\dots\dots\dots\dots .\text{V.69}$$

and θ is bulk volume of water/unit volume of soil.

Variations of $k(\theta)$ and ψ with Moisture Content

In the above, it is obvious that the theoretical equations for unsaturated movement are exceedingly more complex than those for the saturated case. One should realize that with unsaturated flow, movement of water at any point usually results in a change in water content at that point. This causes a change in $k(\theta)$ and ψ at that point. As indicated in the development of Equation V.66, it is frequently assumed that the unsaturated conductivity and capillary potential are unique functions of the soil moisture content. When this is true, $k(\theta)$ can be related to ψ in the form

$$k(\theta) = a/(\psi^n + b)$$

With $k(\theta)$ in cms/day and ψ in cms, n has been found to be ≈ 2 for clay and ≈ 4 in sand. This uniqueness can never be exact in soils however, because of the effect of hysteresis (see Figs. V.20 and V.21). As shown in the Fig. V.20, the curves showing ψ plotted against θ of a given soil are different under wetting and drying conditions. Nevertheless, if the investigator knows the point on the curve that defines the energy conditions of the soil with which he is working, the relationship likewise can be defined.

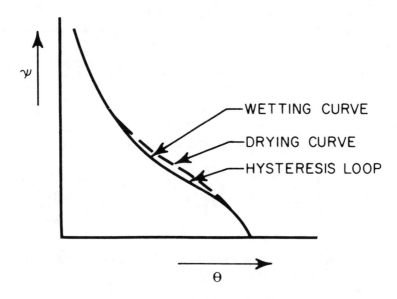

Fig. V.20   Hysteresis Loop

The unsaturated hydraulic conductivity (capillary conductivity) is dependent on many of the same factors that affect the hydraulic conductivity (for example: soil type, density, compaction; fluid properties–density, viscosity, surface tension; and others- chemical constituents, back air pressure, continuity of films, etc.) and $k(\theta)$ is near a maximum at or near saturation; it then decreases rapidly with moisture content to a very low level, where it remains practically constant with any further decreases in moisture (see Fig. V.22). Childs (1957) suggests that the reason for this reduction may be attributed to four separate effects which operate in the same direction:

1.   A reduction of moisture content reduces the effective porosity.
2.   The larger pores empty first, and hence the more effectively conducting pores are put out of action in the early stages of unsaturation.
3.   The pores that are emptied have to be avoided by the remaining paths of flow, which therefore become more tortuous in nature as water removal proceeds.
4.   In soils that shrink, the increased suction reduces the size of pores that remain full.

It is generally accepted that at moisture contents less than the *moisture equivalent,* the water films interconnecting adjacent pores become discontinuous, so that the conductivity approaches zero, and thus the rate of capillary flow is very low. In general, the magnitude of $k(\theta)$ is in reverse order of the magnitude of the saturated 'k' for different soil types. That is $k(\theta)$ of sand $<$ $k(\theta)$ of fine sandy loam $<$ $k(\theta)$ of light clay $<$ $k(\theta)$ of clay.

Measurement of $k(\theta)$

Youngs (1964) suggested a simple method of determining the unsaturated hydraulic

RISING OR WETTING CURVE — ABSORPTION

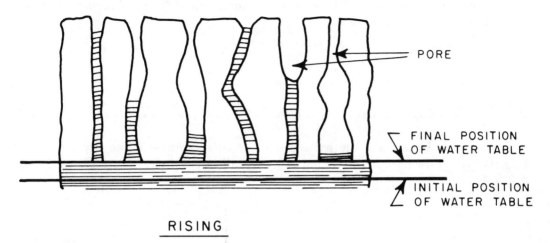

PORE

FINAL POSITION
OF WATER TABLE

INITIAL POSITION
OF WATER TABLE

RISING

FALLING OR DRYING CURVE — DESORPTION

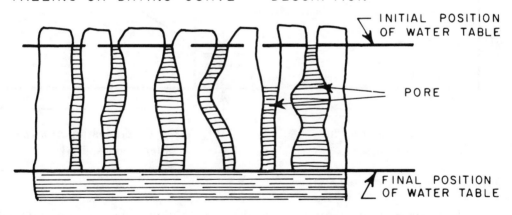

INITIAL POSITION
OF WATER TABLE

PORE

FINAL POSITION
OF WATER TABLE

FALLING

Fig. V.21 Capillary Fringe Above Rising and Falling Water Table

conductivity from infiltration measurements. This method is based on the theory that the total volume of water, Vol, of infiltration to any time, t, is

$$Vol = \int_{0}^{t} k(\theta)_z \left[\frac{d\psi}{dz} + 1 \right]_{z=0} dt \quad \dots\dots\dots\dots\dots\dots\dots\dots\dots\dots\dots V.70$$

in which $k(\theta)_z$ is the conductivity at a given suction and z is the distance measured vertically from the surface. When $z = 0$, then $d\psi/dz = 0$ (near saturation), and the first term of the integral (Equation V.70) approaches some constant in keeping with the infiltration theory. Thus, as $t \to \infty$

$$\text{Vol} = a + k(\theta)_z t$$

or

$$\frac{\text{Vol}}{t} = \frac{a}{t} + k(\theta)_z \quad \dots\dots\dots\dots\dots\dots\dots\dots\dots\dots\dots\dots\dots\dots\text{V.71}$$

It follows from Equation V.71 that the value of $k(\theta)_z$ can be conveniently determined from the asymptotic slope of the volume of water infiltrated with time (see Figs. V.22 and V.23).

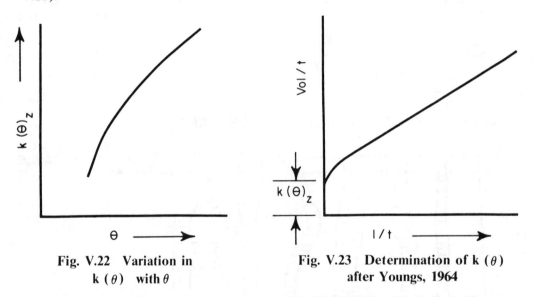

Fig. V.22   Variation in
k (θ)   with θ

Fig. V.23   Determination of k (θ)
after Youngs, 1964

Solutions of Unsaturated Flow Systems

The following example solutions of the basic continuity equation for unsaturated flow Equations V.65 and V.66 were selected from the literature because of their applicability to the infiltration process. For further details of these solutions, the reader should consult the specified references.

Horizontal Flow Systems

In the case where flow is horizontal (unidirectional), the effect of gravity and components of flow in the y and z directions can be neglected. Thus the differential equation of flow (Equation V.66) reduces to

$$\frac{\partial}{\partial x}\left[D(\theta)_x \frac{\partial\theta}{\partial x} \right] = \frac{\partial\theta}{\partial t} \quad \dots\dots\dots\dots\dots\dots\dots\dots\dots\dots\dots\dots\text{V.72}$$

This non-linear differential equation can be solved by use of the Boltzmann transformation if the medium is uniform and semi-infinite; for example, a long uniform soil column of constant initial moisture content.

The Boltzmann transformation assumes that there is a unique function $\lambda(\theta)$, that has the form

$$\lambda(\theta) = xt^{-\frac{1}{2}} \quad \dots\dots\dots\dots\dots\dots\dots\dots\dots\dots\dots\dots \text{V.73}$$

By substitution of the equality given by Equation V.73 into Equation V.72, one obtains the ordinary differential equation

$$-\frac{\lambda}{2}\frac{d\theta}{d\lambda} = \frac{d}{d\lambda}\left[D(\theta)_x \frac{d\theta}{d\lambda}\right] \quad \dots\dots\dots\dots\dots\dots\dots\dots\dots\dots\text{V.74}$$

For infiltration into a semi-infinite system with horizontal flow, the boundary and initial conditions to be satisfied are

$$\theta = \theta_i \qquad \text{when } x > 0, \ t = 0$$
$$\theta = \theta_f \qquad \text{when } x = 0, \ t > 0$$

where θ_i = initial constant moisture content, and
θ_f = constant moisture content at inlet to the system

Bruce and Klute (1956) have used a solution of Equation V.74 to determine the diffusivity $D(\theta)$. Their procedure is as follows. Integrating Equation V.74 with respect to λ within the limits imposed by the boundary conditions

$$\int_{\theta_i}^{\theta_x} -\frac{\lambda}{2}\, d\theta = \int_{\lambda_i}^{\lambda_x}\left[\frac{d}{d\lambda}\left(D(\theta)_x \frac{d\theta}{d\lambda}\right)\right]d\lambda$$

$$= D(\theta)_x \left(\frac{d\theta}{d\lambda}\right)_{\theta_x} - D(\theta)_x \left(\frac{d\theta}{d\lambda}\right)_{\theta_i} \quad \dots\dots\dots\dots\dots\dots\dots\dots \text{V.75}$$

in which $(d\theta/d\lambda)_\theta$ is the value of the derivative at a given value of θ. It should be recognized that the last term of Equation V.75 is zero; that is,

$$(d\theta/d\lambda)_{\theta_i} = 0$$

Thus

$$\int_{\theta_i}^{\theta_x} -\frac{\lambda}{2}\, d\theta = D(\theta)_x \left(\frac{d\theta}{d\lambda}\right)_{\theta_x}$$

Solving for the diffusivity

$$D(\theta)_x = -\frac{1}{2t} \left(\frac{dx}{d\theta} \right) \int_{\theta_i}^{\theta_x} x \, d\theta \quad \dots\dots\dots\dots\dots\dots\dots\dots\dots\dots\text{V.76}$$

To evaluate the integral:

1.   Obtain a moisture content-distance curve from the flow system at a constant value of time.
2.   From a plot of θ vs x, evaluate the integral and the derivative at a series of values of θ_x.
3.   Calculate the value of $D(\theta)_x$, given values of θ_x.

Experimental data from Bruce and Klute (1956) indicate that the maximum diffusivity occurs at a moisture content 75-80% of pore saturation. This occurs because the curve of θ vs x is concave upward in the transmission zone. It happens that the behaviour of $dx/d\theta$, as a function of θ, strongly influences the nature of the function $D(\theta)_x$; and if there is a maximum value of $dx/d\theta$ at $\theta < \theta_f$, then there will be a maximum value of diffusivity at $\theta < \theta_f$.

Application of Diffusion Theory

Kirkham and Feng (1949) applied the diffusion concept to the horizontal flow case. They assumed a constant diffusivity, resulting in a solution that is analogous to the heat flow problem in which a constant temperature is suddenly applied to, and maintained at, one end of a long bar of metal, initially at a uniform temperature throughout and having a constant thermal diffusivity. This solution gave rise to the so-called *Laws of Times* which govern movement horizontally in unsaturated cores; that is

"free water is imbibed under capillary forces into uniform drier soils, in accordance with the law

$$M_f = ct^{1/2} + a \dots\dots\dots\dots\dots\dots\dots\dots\dots\dots\dots\text{V.77}$$

and the movement of the wetted front follows the law

$$x = Bt^{1/2} + b \text{ "} \dots\dots\dots\dots\dots\dots\dots\dots\dots\dots\text{V.78}$$

The constants a and b were used so that a major portion of the experimental data would fit the linear relationship between x and $t^{1/2}$; however, these constants obviously invalidate the equations at very small times.

However, it is not necessary to assume constant diffusivity to obtain linear relationships between the volume of water infiltrated M_f and $t^{1/2}$, and distance to the wetting front and $t^{1/2}$. From Equation V.73 which is valid for variable diffusivity

$$x(\theta) = \lambda (\theta) t^{1/2} \quad \ldots\ldots\ldots\ldots\ldots\ldots\ldots\ldots\ldots\ldots\ldots\ldots\ldots\ldots\text{V.79}$$

where $x(\theta)$ is the distance to the section in the column which is at moisture content θ, at time t.

If it is assumed that the moisture content at the wetting front is a constant, then because (θ) is a unique function of θ, $\lambda (\theta)$ at the wet front will be a constant, and

$$x = Bt^{1/2} \quad \ldots\ldots\ldots\ldots\ldots\ldots\ldots\ldots\ldots\ldots\ldots\ldots\ldots\ldots\ldots\ldots\text{V.80}$$

where x is the distance to the wet front.

In a similar manner

$$Q = \int_{\theta_i}^{\theta_f} x \, d\theta$$

$$= t^{1/2} \int_{\theta_i}^{\theta_f} \lambda(\theta) \, d\theta$$

or

$$Q = ct^{1/2} \quad \ldots\ldots\ldots\ldots\ldots\ldots\ldots\ldots\ldots\ldots\ldots\text{V.81}$$

Work by Neilsen et al. (1962) has shown that Equation V.81 is valid if θ_f (the constant moisture content at the wetting end) is very close to saturation. However, if water is applied under a sufficient tension to make θ_f less than the saturated moisture content, then the relationship is no longer linear.

Vertical System

Water movement in a vertical column of soil is described by the differential equation

$$\frac{\partial \theta}{\partial t} = \frac{\partial}{\partial z} \left(D(\theta)_z \frac{\partial \theta}{\partial z} \right) + \frac{\partial}{\partial z} k(\theta)_z \quad \ldots\ldots\ldots\ldots\ldots\ldots\text{V.82}$$

Because of the complex nature of the solutions to this equation, an attempt will not be made here to present these in detail. For further references on the subject, the reader is referred to the works of Philip (1957). The solution proposed by Philip for vertical flow is in the form of a power series of the form

$$z = at^{1/2} + bt + ct^{3/2} \ldots\ldots\ldots n_m(\theta)t^{m/2} \ldots\ldots\ldots\ldots\ldots\ldots\text{V.83}$$

where a, b, c, etc. are coefficients which are functions of θ. The equation is generally solved by analytic processes or use of an analogue. The simplified form for mass infiltration is given as

$$M_f = St^{1/2} + At \dots\dots\dots\dots\dots\dots\dots\dots\dots\dots\dots\dots .V.84$$

> where S = sorptivity or a measure of uptake
> of water by the soil and,
> A = a parameter which arises from the analysis.
> (See Equation V.12)

Taylor and Heuser (1953) presented a somewhat different approach to solution of the differential form of the continuity equation

$$\frac{\partial}{\partial z}(\rho V_z) = \frac{\partial \rho_v}{\partial t} \dots\dots\dots\dots\dots\dots\dots\dots\dots\dots .V.85$$

> where ρ_v = mass of water per bulk volume of soil
> (ρ x water filled pores),
> ρ = mass density of water, and
> V_z = flow rate per unit cross sectional area
> per unit time in z direction or
> $V_z = k(\theta)_z \, \partial\Phi/\partial z$

Therefore

$$\frac{\partial}{\partial z}\left(\rho k(\theta)_z \frac{\partial \Phi}{\partial z}\right) = \frac{\partial \rho_v}{\partial t} \dots\dots\dots\dots\dots\dots\dots .V.86$$

or

$$\rho k(\theta)_z \frac{\partial^2 \Phi}{\partial z^2} + \frac{\partial \Phi}{\partial z}\left[\frac{\rho \partial k(\theta)_z}{\partial z} + k(\theta)_z \frac{\partial \rho}{\partial z}\right] \dots\dots\dots\dots .V.87$$

The authors did several measurements of the potential gradients in the transmission zone and wetting zone. Their findings indicated that the moisture potential in the transmission zone increased more rapidly than depth alone, such that the gradient of flow was greater than unity (1.06–1.25). If the flow is caused by the gradient in the transmission zone then the conductivity through the pores must be less than saturated conductivity. Similarly, they indicated that the potential gradient increased in the wetting zone. They suggest that the potential in this zone may be defined by the relationship,

$$\Phi = \log a + b \log z$$

where z is the distance to the wetting zone.

Thus, if the gradient in this zone is responsible for movement it can be defined as

$$\frac{d\Phi}{dz} = abz^{b-1} \quad \cdots\cdots\cdots\cdots\cdots\cdots\cdots\cdots\cdots\cdots \text{.V.88}$$

In summary, they concluded that because of the different gradients in the different zones, it will not suffice to solve the flow problem with the use of a simple, average gradient over the wetted area. Moreover, it is necessary to know the change in gradient throughout the flow column.

V.3   APPLICATION TO THE INFILTRATION PROCESS

V.3.1   Interrelation Between Infiltration and Soil Moisture Movement

Taking into consideration the preceding subject matter, it is immediately obvious that the infiltration rate of a soil can be evaluated in terms of the fundamental laws governing the flow of fluids through saturated or unsaturated porous media. It follows from this that any factor in the soil system that affects the magnitude of the hydraulic, or capillary conductivity and the potential gradient, will likewise affect the infiltration rate. The interrelationships of many factors influencing infiltration, and these factors, are shown in schematic form in Table. V.5.

V.3.2   Infiltration Potential of a Watershed,

Several techniques have been presented that may be used to evaluate the effect of the initial soil moisture content on the infiltration rate. To apply these methods, certain microhydrologic and physical characteristics of a soil must be known—such as (a) the capillary conductivity-moisture content curve, (b) the moisture-tension relationship, or (c) an experimental relation between the soil moisture content and the infiltration rate. If these properties for all soils in a watershed have been measured, then the infiltration potential of the watershed at any time can be evaluated from soil moisture measurements. Needless-to-say, the work involved in this computation would be reduced appreciably if soils could be grouped on their infiltration potential based on soil-moisture retention and transmission characteristics. Although it has been found that soils are amenable to grouping in accordance with their water-intake capacities in the wet condition, it has not yet been established if a system could be derived to group soils on their infiltration rates to include the full range of moisture content.

Though evaluation of the infiltration potential of a watershed based on infiltration characteristics of individual soils of the basin may be feasible for small experimental catchments on which necessary measurements are taken, and for which laboratory and analytic facilities are available, this approach is impractical for large watersheds—and particularly so for the practising hydrologist. The use of antecedent precipitation or groundwater indices to reflect the degree-of-wetness or the infiltration potential of a basin is thus frequently resorted to. In using these indices it is commonly assumed that the degree-of-wetness prior to the storm is closely related to the soil moisture, which is the controlling factor of runoff.

Probably, the most common form of antecedent precipitation index (API) in current use is

$$API = \sum_{t=1}^{n} b_t P_t \quad \dots\dots\dots\dots\dots\dots\dots\dots\dots\dots\dots\dots\dots\dots V.89$$

in which b_t and P_t respectively are a constant and an amount of precipitation occurring at selected times preceding the storm event (days). Usually, the constant, b_t, is assumed to be some function of time, t, as $b_t = 1/t$ or $b_t = K^t$. According to the latter expression, the effect of precipitation on wetness of the basin is decreased exponentially with time. Values for the constant, K, are usually assumed to be in the range 0.80-0.98; however, the choice of the constant is not usually critical, inasmuch as the calculation is used as an index of moisture deficiency. The final computation of the API for a given storm is obtained by calculating the cumulative effect of all precipitation amounts in the series. For example

$$API = KP_1 + K^2 P_2 + K^3 P_3 + \dots\dots\ K^n P_n \dots\dots\dots\dots\dots V.90$$

where P_1, P_2, $P_3 \dots\dots P_n$ are the amounts of precipitation at different times (days) preceding the storm event. Anywhere from 20 to 60 terms may be used in the series.

Frequent use is made of indices of the type mentioned above in multiple regression, graphical or correlation analyses where attempts are made to predict basin yield from storm and watershed characteristics. For individual studies, these indices may prove extremely valuable; however, their general applicability to watersheds (other than those studied) is questionable.

V.4 SOIL MOISTURE MEASUREMENT

Many methods that may be used to measure the moisture content of a soil are available. These may be categorized as those that measure the amount of water in a given mass or volume of soil, and those that measure the energy conditions of the soil water.

V.4.1. Measurement of the Amount of Water in a Given Mass or Volume of Soil

V.4.1.1 Gravimetric Method

In determining the soil moisture content, the gravimetric method is standard. A sample of soil obtained from the field is taken to the laboratory and dried in an oven at $110°C$ for approximately 24 hr. The moisture content is calculated as

$$MC = \frac{W_w - W_d}{W_d} \times 100 \quad \dots\dots\dots\dots\dots\dots\dots\dots\dots\dots\dots\dots\dots V.91$$

Table V.5 Factors Affecting the Infiltration Rate Into Unfrozen Soil

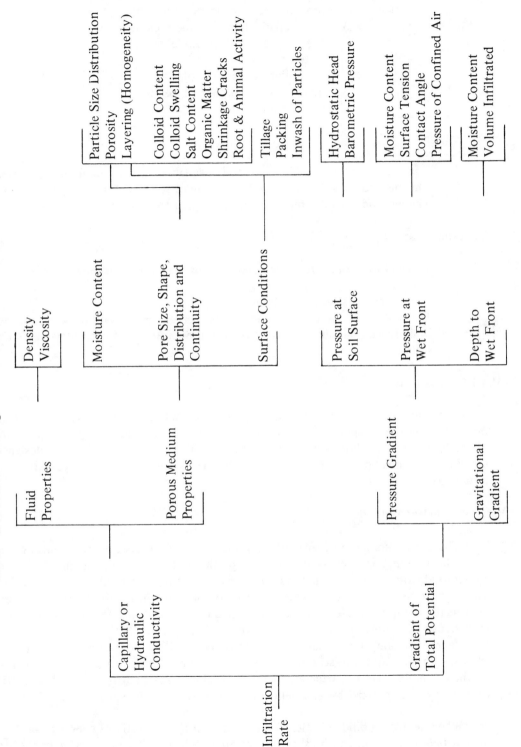

Infiltration Rate

Capillary or Hydraulic Conductivity
- Fluid Properties
  - Density
  - Viscosity
- Porous Medium Properties
  - Moisture Content
  - Pore Size, Shape, Distribution and Continuity
    - Particle Size Distribution
    - Porosity
    - Layering (Homogeneity)
    - Colloid Content
    - Colloid Swelling
    - Salt Content
    - Organic Matter
    - Shrinkage Cracks
    - Root & Animal Activity
  - Surface Conditions
    - Tillage
    - Packing
    - Inwash of Particles

Gradient of Total Potential
- Pressure Gradient
  - Pressure at Soil Surface
    - Hydrostatic Head
    - Barometric Pressure
  - Pressure at Wet Front
    - Moisture Content
    - Surface Tension
    - Contact Angle
    - Pressure of Confined Air
- Gravitational Gradient
  - Depth to Wet Front
    - Moisture Content
    - Volume Infiltrated

in which   MC = moisture content in % by weight,
           W_w = wet weight of the sample including
                the weight of soil grains plus
                water, and
           W_d = dry weight of the sample; this
                includes the soil grains plus some
                hygroscopic water.

Advantages:

1.   The equipment used is simple, readily available and reasonably inexpensive.
2.   The method provides an accurate measure of soil moisture of a given sample.
3.   The method is applicable over the entire moisture range.

Disadvantages:

1.   The method cannot be performed on soils *in situ,* therefore the sampling area is disturbed.
2.   The method is slow and tedious.
3.   The method provides a measure of the percentage of soil moisture by weight. Bulk density data are needed to convert these values to a percentage moisture by volume.

V.4.1.2   Lysimetry

The lysimetric method represents a special gravimetric technique whereby changes in moisture content of a specific block of soil are determined by recording changes in weight of the block. The method is accurate for the soil mass measured, and can be designed for continuous recording. The principal disadvantage is the high cost of construction for installation.

V.4.1.3   Nuclear Methods

The most common nuclear method of measuring soil moisture is by neutrons. In this method, neutrons, or uncharged subatomic particles of about the same mass as hydrogen, are emitted from a source, usually radium-beryllium at high thermal energies into the soil. These neutrons—called 'fast' neutrons—cannot be counted by the scintillation counter used in the equipment. However, as the neutrons move out into the soil mass, they react with nuclei of atoms and are either captured or scattered. If they are captured, either they form unstable nuclei that emit gamma rays or they disintegrate. More probably, however, the neutrons are scattered by elastic collision with atoms, simply giving up part of their kinetic energy. After a series of collisions, the neutrons have lost sufficient energy to be categorized as slow neutrons (low thermal energy) and they can be counted.

When neutrons collide elastically with other atoms, the ratio of energy after collision to that before collision can be shown to be an inverse logarithmic function of the atomic weight of the atom; that is, the lower the atomic weight of the atom, the greater the energy loss. In a soil, the most prevalent atom of low atomic weight is hydrogen, being bound in mineral constituents, humus and organic matter, and soil water. The major source of hydrogen in most soils is, of course, the soil water; this contains 11% hydrogen by weight.

It follows that the larger the number of hydrogen atoms in the soil, the greater will be the number of slow neutrons that return to the counter. Therefore, the number of counts that are recorded in a specified time can be related to the hydrogen content of the soil (primarily soil moisture content).

The most common type of neutron equipment in use today is the single probe system (Nuclear-Chicago, Troxler) in which the radioactive source and counter are contained in the same unit. In using this type of equipment, a hole is bored into the soil and cased with thin-walled plastic, steel, stainless steel or aluminum tubing of a diameter slightly larger than the probe; measurements are taken by setting the probe at various depths in the tube.

Double tube systems may also be used. In these systems the source and counter are inserted into different tubes, placed a short distance apart. The advantage of this system over the single tube is that the source and counter tubes can be collimated, and so moisture contents over small intervals of depth can be measured. Frequently, in this method a gamma source (Cesium 137) is used. Changes in moisture are thus associated with changes in the attenuation of the gamma wave caused by changes in the mass of material contained between the source and counter; this results from soil moisture movement.

Advantages:

1. The method gives a measurement of the volume of water in the soil.
2. In inorganic soils, the measurement can be expected to be within ± 1% of the gravimetric estimate.
3. The sampling is non-destructive.
4. A comparatively large volume of soil is sampled. For a single probe, the diameter of the sphere of soil sampled is calculable from

$$\text{dia (in.)} = 12 \sqrt[3]{\frac{100}{\theta(\% \text{ by volume})}} \qquad \ldots\ldots\ldots\ldots\ldots\ldots\ldots\ldots\text{V.92}$$

Disadvantages:

1. The equipment is expensive and usually quite delicate.
2. If measurements of moisture are to be made at shallow depths, a surface probe is required.
3. New calibrations of the equipment may be required in highly organic soils.

V.4.2   Measurements of the Energy Conditions of Soil Water

These are indirect methods of soil moisture measurements in which measurements are made of the physical (soil moisture tension), electrical, chemical, thermal or other properties of the soil mass, and are related to the soil moisture content. In most of these methods, calibration curves must be derived to relate the parameter measured to the moisture content, and therefore, the methods are subject to error because of the hysteresis effect. Further, the readings may be influenced by temperature, salt concentrations, degree-of-contact, etc.

V.4.2.1   Tensiometers

The use of tensiometers for soil moisture measurement is based on the principle that, as the soil dries out, the tension in the soil water increases (smaller pores and the total curvature of menisci increases—see previous discussions). Thus the amount of moisture in a soil is inversely related to the tension in the soil water.

In this method, a tensiometer containing a small porous ceramic cup (usually gypsum) and equipped with a manometer or pressure gauge, is placed in contact with the soil. Movement of water out of or into the cup (depending on whether the soil is drying or wetting) is reflected by an increase or decrease in tension readings on the gauge. These readings are used in conjunction with a calibration curve (previously obtained) to determine the soil moisture content.

Advantages:

1.   The method is simple and direct, and the equipment used is inexpensive.
2.   Measurements may be automatically recorded.

Disadvantages:

1.   The method can only be employed to measure soil moisture content within a limited range of tensions (saturation to 800 cm H_2O)
2.   Readings obtained from tensiometers usually lag actual soil moisture changes.
3.   A calibration curve is required.
4.   It is difficult to maintain contact between the cup and soil in soil moisture regimes which undergo wide changes in the moisture content (surface layers of soil).

V.4.2.2   Electrical Methods

Resistance Units

Electrical resistance units have been widely used by agronomists, engineers and foresters to measure indirectly the soil moisture content. These electrical units, or resistance blocks, consist of two electrodes that are spaced a small distance apart and imbedded in some type of porous block (gypsum, plaster of paris, nylon or fibreglass). In operation, the block is utilized as one of the resistors of a Wheatstone Bridge circuit, and the resistance between the two electrodes is measured (see Fig. V.24). As water moves into or out of a block which is placed in contact with the soil, its resistance changes; and these readings are used in conjunction with the calibration curve for the soil to obtain the soil moisture content.

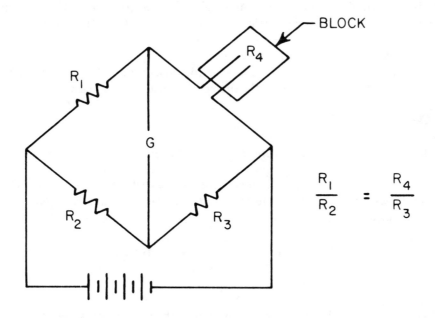

$$\frac{R_1}{R_2} = \frac{R_4}{R_3}$$

Fig. V.24 Circuit Diagram of Resistance Block

It is recommended in installing these blocks that the unit is placed in close contact with the soil and that the back fill is tamped down firmly. The leads of the units upon entering the soil surface should first dip down and then come up again to run under the surface for about three feet; this prevents water from running directly into the block. Further, because of the small sampling volume of the block, it is recommended that several units be used at one site.

Advantages:

1.   The units can be used to estimate soil moisture over a wide range of moisture contents.
2.   The units are reasonably inexpensive.
3.   The units are simple to operate and reasonably easy to install.
4.   Measurements are made *in situ.*

Disadvantages:

The problems encountered with resistance units are similar to those discussed earlier concerning the use of tensiometers. That is, these difficulties arise because of lag and hysteresis effects, difficulties in maintaining contact, and because of the fact that a calibration curve is required.

Capacitance

In this method, a sample of soil is placed in a condenser and its capacitance measured. Changes in the dielectric constant—the ratio of the capacitance of the condenser with the material between its plates to the capacitance of the condenser in a vacuum—reflects the amount of moisture in the sample. For example, the dielectric constant of water is 80.36 at 20°C, and that for quartz 4.3. Obviously, as the moisture content of the soil mass is increased, the dielectric constant is increased.

V.4.2.3   Chemical Methods

Reaction Between a Substance and Water in the Soil Sample

The most common reaction method is probably the carbide method of soil moisture measurement in which calcium carbide is mixed with the soil sample. Calcium carbide reacts with the moisture in the soil as

$$CaC_2 + 2H_2O \rightarrow C_2H_2 + Ca(OH)_2$$
$$\text{Calcium} \qquad\qquad \text{Acetylene}$$
$$\text{carbide} \qquad\qquad \text{gas}$$

If the mixture is placed in a tight container, the amount of water in the sample can be evaluated in terms of either the pressure-temperature increase of the gas, or the loss of weight of the sample as calculated by measuring the weight of acetylene gas produced.

Dilution of Substances in Water

The soil sample is 'washed' with a chemical substance (usually methyl alcohol) of known specific gravity. By measuring the change in specific gravity of the chemical, the amount of water present in the sample can be calculated.

The results obtained by the 'alcohol' method can be expected to fall within ±1% of those obtained by the gravimetric method when used in coarse-textured soils; though it should be noted that the method is affected by the presence of dissolved salts, hence it is not recommended for use in alkali soils. And further, because the success of this method of soil moisture determination depends on the fact that all pores of the sample must be washed by the chemical, results are usually unsatisfactory when it is applied to fine-textured soils.

V.4.2.4   Thermal Methods

The thermal methods of measuring soil moisture are based in principle on the fact that the thermal conductivity and the specific heat of a soil mass increase with moisture content. Using this knowledge, several thermal, or thermo-electric methods of moisture measurement, have been proposed. For example, one of these methods employs a mercury thermometer with a portion of its bulb wrapped with wire. The bulb is heated by passing a constant electric current through the wire and a recording is taken of the time for the temperature to rise 3-5 degrees. A relationship is then developed between the recorded time and moisture content. Another technique that is used relates the moisture content to the amount of current passing through a resistor for a given time. This method employs the concept that the resistance of a wire conductor increases with temperature, and when placed in a soil, the temperature increase is dependent on the thermal conductivity of the soil and the moisture content.

In general, thermal or thermo-electric methods are not used to any extent for the field measurement of soil moisture. The primary advantages of these methods are that they are unaffected by the presence of electrolytes and changes in temperature.

For further reference on methods of soil moisture determination, the reader is referred to the works of Shaw and Arbell (1959).

V.5   LITERATURE CITED

American Society Civil Engineering. 1949. Hydrology handbook. ASCE Manuals of Engineering Practice.

Ayers, H.D. 1959. Influence of soil profile and vegetation characteristics on net supply to runoff. Proceedings of Symposium No. 1 - Spillway Design Floods. Queen's Printer and Controller of Stationery, Ottawa. pp. 198-205.

Baver, L.D. 1948. Soil physics. John Wiley and Sons Inc., New York.

Bodman, G.B. and Colman, E.A. 1943. Moisture and energy conditions during the downward entry of water into soils. Proc. Soil Soc. Amer. 8:116-122.

Bruce, R.R. and Klute, A. 1956. The measurement of soil moisture diffusivity. Proc. Soil Sci. Soc. Amer. 20:458-462.

Childs, E.C. 1957. In Drainage of Agricultural Lands, Pt. 1. The physics of land drainage. Amer. Soc. of Agronomy, Madison, Wis.

Gardner, W. and Widstoe, J.A. 1921. The movement of soil moisture, Soil Sci. 11:215-232.

Hanks, R.J. and Bowers, S.A. 1962. Numerical solutions of the moisture flow equations into layered soils. Proc. Soil Sci. Soc. Amer. 26:530-534.

Holtan, H.N. 1961. A concept of infiltration estimates in watershed engineering. USDA-ARS. 41-51, 25 pp.

Horton, R.E. 1933. The role of infiltration in the hydrologic cycle. Trans. Amer. Geophys. Un. 14:446-460.

Horton, R.E. 1940. An approach to the physical interpretation of infiltration capacity. Proc. Soil Sci. Soc. Amer. 5:399-417.

Horton, R.E. 1945. Erosional development of streams and their drainage basins; hydrophysical approach to quantitive morphology. Bull. Geol. Soc. Amer. 56:275-370.

Jacob, C.E. 1949. Flow of groundwater. Chpt. 5. Engineering Hydraulics. H. Rouse ed. John Wiley and Sons Inc., New York. N.Y.

Keen, B.A. 1922. The system soil-soil moisture. Trans. Faraday Soc. 17:228-243.

Kirkham, D. and Feng. C.L. 1949. Some tests of the diffusion theory and laws of capillary flow in soils. Soil Sci. 67:29-40.

Klute, A., Bruce R. and Russell, M.B. 1956. The application of the diffusivity concept to soil moisture movement. Proc. 6th Int. Congress Soil Sci. B:345-354.

Kostiakov, A.N. 1932. On the dynamics of the coefficient of water percolation in soils and the necessity of studying it from dynamic point of view for purposes of amelioration. Trans. 6th Comm. Int. Soc. Soil Sci. Russian Pt. A15-21.

Kuznik, I.A. and Bezmenov, A.I. 1963. Infiltration of meltwater into frozen soils. Soviet Soil Sci. No. 7, pp. 665-670. (Translation Scripta Technica, Inc.).

Larkin, P.A. 1962. Permeability of frozen soils as a function of their moisture content and fall tillage. Soviet Hydrology: Selected Papers. No. 4, p.p. 445-460. (Amer. Geophys. Union publishers).

Leliavsky, S. 1955. Irrigation and Hydraulic Design: Vol. 1. General Principles of Hydraulic Design. Chapman and Hall, London.

Lewis, M.R. 1937. The rate of infiltration of water in irrigation practice. Trans. Amer. Geophys. Un. (18th meeting) Pt. II Sect. Hydro., 361-368.

Luthin, J.N. 1957. Ed. Drainage of agricultural lands. American Society of Agronomy, Madison, Wis.

Marshall, T.J. 1959. Relations between water and soil. Tech. Comm. 50, Commonwealth Bureau of Soils, Harpenden. Commonwealth Agric. Bureau, Farnham Royal, Bucks, England.

Marshall, T.J. and Stirk, G.B. 1949. Pressure potential of water moving downward into soil. Soil Sci. 68:359-370.

Miller, R.O. and Richard F. 1952. Hydraulic gradients during infiltration in soils. Proc. Soil Sci. Soc. Amer. 16:33-38.

Moore, R.E. 1939. Water conduction from shallow water table. Hilgardia 12:383-401.

Muskat, M. 1946. The flow of homogeneous fluids through porous media, J.W. Edwards, Inc. Ann Arbor, Michigan.

Neilson, D.R., Biggar, J.W. and Davidson, J.M. 1962. Experimental consideration of diffusion analysis in unsaturated flow problems. Proc. Soil Sci. Soc. Amer. 26:107-111.

Norum, D.I. and Gray, D.M. 1964. Unlined mole lines for irrigation. Unpublished paper presented to Amer. Soc. Agric. Engr. Mt., Fort Collins, Colo.

O'Neal, A.M. 1949. Soil characteristics significant in evaluating permeability. Soil Sci. 67:403-409.

Philip, J.R. 1957. The theory of infiltration: the infiltration equation and its solution. Soil Science 83:345-357.

Post, F.A. and Dreibelbis, F.R. 1942. Some influences of frost penetration and micro-climate on the water relationships of woodland, pasture and cultivated soils. Proc. Soil Sci. Soc. Amer. 7:95-104.

Read, D.W.L. 1958. Horizontal movement of moisture in soil. Utah State Univ. M.S. Thesis, Logan, Utah.

Richards, L.A., Gardner, W.R. and Ogata, G. 1956. Physical processes determining water loss from soil. Proc. Soil Sci. Soc. Amer. 20:310-314.

Richards, S.J. and Weeks, L.V. 1953. Capillary conductivity values from moisture yield and tension measurements of soil columns. Proc. Soil Sci. Soc. Amer. 17:206-209.

Sharp, A.L. and Holtan, H.N. 1940. A graphical method of analysis of sprinkler-plot hydrographs. Trans. Amer. Geophys. Un. Part II.

Sharp, A.L. and Holtan, H.N. 1942. Extension of graphic method of analysis of sprinkled-plot hydrographs to the analysis of hydrographs of control-plots and small homogeneous watersheds. Trans. Amer. Geophys. Un. 23:587 Part II.

Shaw, M.D. and Arbell, W.C. 1959. Bibliography on methods for determining soil moisture. Engr. Res. Bull. B-78, Coll. of Engr. and Arch. Penn. State Univ. Unversity Park. 152 p.

Smith, R.M. and Browning, D.R. 1946. Some suggested laboratory standards of subsoil permeability. Proc. Soil Sci. Soc. Amer. 11:21-26.

Taylor, S. and Heuser, N.C. 1953. Water entry and downwards movement in undisturbed soil cores. Proc. Soil Sci. Soc. Amer. 17:195-201.

Ubell, K. 1956. Unsteady flow of groundwater caused by well drawdown. Vol. II, Groundwater Inter. Union Geol. and Geophys. Assoc. Sci. Hydrol. Publ. 41:129-132.

U.S. Dept. of Commerce, Office of Technical Services, 1956. Snow Hydrology, Washington, D.C.

Vasquez, R. and Taylor, S.A. Simulated root distribution and water removal rates from moist soil. Proc. Soil Sci. Soc. Amer. 22:106-110.

Wiegand, C.L. and Taylor, S.A. 1961. Evaporative drying of porous media; influential factors, physical phenomena, mathematical analyses. Utah State Univ. Agric. Expt. Sta. Special Rept. 15.

Willis, W.O., Carlson, C.W., Alessi, J. and Haas, H.J. 1961. Depth of freezing and spring runoff as related to fall soil-moisture level. Can. Jour. Soil Sci. 41:115-124.

Wischmeier, W.H. and Smith, D.D. 1958. Rainfall energy and its relationship to soil loss. Trans. Amer. Geophys. Union 39:285-291.

Youngs, E.G. 1964. An infiltration method of measuring the hydraulic conductivity of unsaturated porous materials. Soil Sci. 97:307-311.

Zavodchikov, A.B. 1962. Snowmelt losses to infiltration and retention on drainage basins during snow melting period in Northern Kazakhstan. Soviet Hydrology: Selected Papers No. 1, pp. 37-42. (Amer. Geophys. Union Publishers).

Section VI

GROUNDWATER HYDROLOGY

by

James M. Murray

TABLE OF CONTENTS

LIST OF TABLES

LIST OF FIGURES

Section VI

GROUNDWATER HYDROLOGY

VI.1   INTRODUCTION

In the past groundwater hydrology has been the subject of study by specialists from a number of disciplines who have tended to use different names to describe terms in relation to their particular interests. Thus there are terms such as groundwater hydrology and geo-hydrology—synonymous names for the study of occurrence and movement of groundwater—as well as groundwater geology and hydro-geology—terms stressing the importance of geology.

The term groundwater hydrology will be used in this Section in its broadest sense, to include geologic, hydrologic, hydraulic and mechanical aspects of water occurring in the zone of saturation below the earth's surface. Water in unsaturated zones has already been discussed and will be given no further consideration here, but one should be careful not to divorce the study of groundwater from the rest of the Hydrologic Cycle of which it is a vital part.

VI.2   OCCURRENCE OF GROUNDWATER

Groundwater occurs in both consolidated and unconsolidated formations, any of which may be an aquifer if sufficiently porous and permeable. Sedimentary materials make up the largest percentage of the world's aquifers, including coarse unconsolidated materials and hard sedimentary rock such as limestone and dolomite. The openings in sub-surface formations are of three general classes:

1. Openings between individual particles as in sand and gravel, referred to as original interstices.
2. Crevices, joints or fractures in hard rock which have developed from breaking of the rock, called secondary interstices.
3. Solution channels and caverns in limestone and openings, resulting from shrinkage and from gas bubbles in lava.

The porosity of a formation is that part of its volume which consists of openings or pores. Porosity is an index of how much water can be stored in an aquifer but not of its ability to yield water. The yield rate of an aquifer is a function of pore size and interconnection and this characteristic will be considered in detail subsequently. Porosity is expressed as a percentage of the bulk volume of the material. When water is drained from a saturated material by gravity only a part of the total volume stored is released. This portion of the stored water released is called the specific yield. The water not drained by gravity is called the specific retention, and the sum of specific yield and specific retention, expressed as volume percentages, is equal to the porosity. In fine-grained materials, the pores are naturally small and tend to retain most of the water in the pores. Thus fine-grained materials have low specific yields and coarse-grained materials have high specific yields. Fig. VI.1 indicates the relationship between specific yield and particle size.

An aquifer can be defined as any formation that has sufficient porosity and water-yielding ability to permit the removal of water at reasonably useful rates. Obviously, aquifers are the product of geologic processes. The locations and properties of water-bearing formations are related to the geologic history of an area. The hydrologist is concerned not only with the thickness and depth of the water bearing formations, but also with the details of the kind of material that make up the formations. The sizes, shapes and uniformity of the particles are largely determined by the way in which the material was transported and laid down.

VI.3   GEOLOGIC FORMATIONS AS AQUIFERS

VI.3.1   Bedrock Aquifers

For those involved in groundwater development in regions where bedrock aquifers are the rule, the search for groundwater supplies can be more effective if several basic ideas are considered in the evaluation of available groundwater.

Sandstone formations, because they are made up of innumerable tiny sand grains, might be expected to be the type of bedrock unit which forms the best aquifer. This would seem to be true because it is reasonable that the permeability[1] of sandstone in general is higher than other types of sedimentary bedrock units, such as limestone and shale. While this concept is undoubtedly true when the availability of water in bedrock aquifers throughout the whole world is considered, there are notable exceptions where limestone and dolomite formations may be much more effective as aquifers. In general, geologic units composed entirely of shale may be considered as the least desirable type of bedrock aquifer. On the other hand, there are areas throughout the world, where shale units are the most productive, if not the only, usable aquifers.

There are bedrock units that are mixtures of sandstone and shale. In mixed formations such as this, the water-yielding capability generally varies in direct proportion to the amount of sandstone present. Where limestone and shale are interbedded together

[1]Because of the general acceptance of the term 'permeability' in Geo-hydrology in preference to 'hydraulic conductivity', the former is used extensively throughout this Section. It should be remembered, however, that the terms, as used, are synonymous and equal in magnitude to the Darcy "k" having dimensions of velocity.

in a geologic formation, the formation may generally be considered a usable aquifer only where the limestone beds are much thicker than the shale beds.

In comparing the relative ability of different types of rock units to yield water, it should be kept in mind just what a sandstone is. As stated above, it is composed of countless small grains of sand packed tightly together and consolidated into hard rock. What differentiates sandstone from loose sand is the fact that the grains are cemented together. What this means is that the void spaces, which normally occur between grains in loose sand, are partially filled with the cementing material which holds the grains together. This cement may be composed of calcium carbonate; silica, which is the same material as the grains in most sandstone; iron oxide; or other chemically-precipitated material which has been carried into the sand by percolating groundwater. Therefore, the permeability of unfractured sandstones of a given gradation is always less than that of an unconsolidated sand of the same gradation. In sandstone, as in limestone, dolomite and shale, much of the permeability, and thus the ability of the rock units to yield water, is dependent upon the existence of cracks, crevices and other voids distributed throughout the rock.

Sandstones that are tightly cemented have little permeability between the grains themselves. However, the tight cementation causes the sandstone to act as a solid body, and therefore, it is more susceptible to cracking when earth movements or other forces act upon it. In other words, tightly-cemented sandstone will break between the grains when subjected to force, whereas in a loose sand the grains will change their position.

Limestone and dolomites can be considered together because of the similarity between the two types of rock. The intergranular permeability of dolomite and limestone is much less than that of sandstone. Both can be thought of as practically impervious to the movement of water between the various crystalline particles of which they are composed. However, there is some evidence to suggest that dolomite in general has a slightly greater between-the-grains permeability than limestone. Because dolomite is primarily magnesium carbonate in contrast to limestone which is primarily calcium carbonate, the shapes of the crystalline particles that make up the two types of rock are somewhat different. In some cases, percolating groundwater has carried magnesium-bearing waters into a limestone and caused the formation of dolomite within the limestone. When this occurs, the original crystal structure of the limestone is changed, and the newly-formed dolomite particles are shaped differently and occupy less space than the limestone particles did previously. What this means is that some intergranular permeability can exist in limestones that have partially been converted to dolomite.

In both types of rock, however, the ability to yield usable quantities of water is a function of the number of fractures, crevices and other openings which have been enlarged by solution.

Shale has the ability to transmit water to wells only after it has been fractured or weathered. Fracturing provides openings through which water can move, and weathering processes tend to split some shales into thin, platy particles which then have void spaces between them. As a general rule, however, the weathering of shale progresses rapidly to a point where the platy structure of the rock is destroyed; the shale then becomes claylike, and thus, impermeable.

Wells several hundred feet deep that obtain water from shale units provide the evidence that fractures and crevices in shale can exist at these depths. At first glance this might seem illogical, since it might be supposed that the weight of rock overlying a relatively soft material like shale would tend to squeeze shut any openings occurring. However, these deep shale wells obtain water from hard shales which, because of this hardness and resultant strength, seem to fracture somewhat like soft limestones.

There is evidence that minor earth tremors, which occur throughout the world at all times, contribute to the original cracking of bedrock units. These movements, coupled with the slight expansion of rock units that takes place when the weight of overlying rocks is removed by erosion, probably are the prime causes of original hairline cracks forming in limestone, sandstone and shale.

The formation of these small cracks provides void spaces through which ground-water can move. In the case of limestones and dolomites, a great deal of the original rock can be removed by the dissolving action of groundwater. Because the original source of nearly all groundwater is precipitation, groundwater contains dissolved gases picked up by the water while it was falling as rain or snow. These gases make water a weak acid, dissolving carbonate rock relatively easily.

Even in sandstone, some dissolving of the previously-deposited cement between the grains may occur by percolating water. This in turn can cause enlargement of the original cracks, and the permeability of sandstones can thus be increased. By the same token, a slight chemical solution of shale can also occur, but probably to a lesser extent than in limestone and sandstone.

When geologic conditions are such that bedrock lies close to the surface of the ground and is mantled with only a thin layer of soil, deep-seated weathering of bedrock units can occur. This upper weathered zone is usually the source for water obtained in shallow, dug wells.

VI.3.2   Alluvial Aquifers

Alluvial formations include all the detrital material deposited by the action of modern rivers. The alluvium in a river valley consists of fine to coarse material. It is stratified and in some respects similar to a glacial valley train. Sometimes several cycles of erosion and deposition are preserved as terraces in a river valley. The alluvial fill may be similar under each terrace or the material may range greatly in sorting, size and color. Fig. VI.2 shows a valley with two cycles of down-cutting and deposition. The yield of wells constructed in alluvium ranges widely depending on the local texture and thickness of the material.

Where a heavily-loaded stream emerges from hills or mountains into a lowland, the sudden change in gradient results in deposition and formation of an alluvial cone or fan. The material comprising a fan varies in texture from boulders and pebbles at its head, to gravel and sand beyond, and to silt and clay at the end of the slope. As the mountains are worn down, the water deposits successively finer material. Therefore, an alluvial fan, similar to a valley deposit, usually has fine material overlying coarse deposits. If the mountain area has been subjected to repeated uplift, there will likely be several generations of fans. This succession of layers would be stacked up—like a deck of cards arranged in a fan-like shape—at the edge of the mountains. Fig. VI.3 is a cross-section of an alluvial fan showing two generations of fan building. In such an area, the performance of wells varies greatly with location and depth. Alluvial fans are an important source of groundwater in the western part of North America.

The static water level in the upper sand is shown to be well below the ground surface in Fig.VI.3, this level resulting from the natural discharge of groundwater from the formation at lower elevations to the right. In the lower sand, however, no natural discharge from the aquifer occurs because the sand body pinches out at the lower elevation and is confined between other impervious strata.

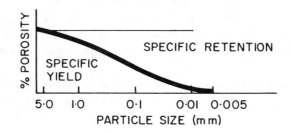

Fig. VI. 1   The Variation in Specific Yield Related to Soil Particle Size

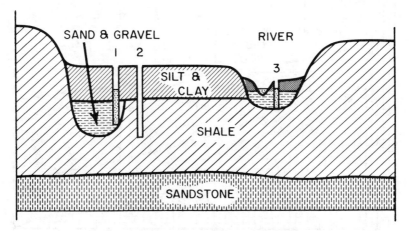

Fig. VI. 2   River Valley Cross Section Showing Two Levels of Filling (Johnson, 1957)

The sands in marine and lake deposits are generally more uniform in particle size and of finer texture than the alluvial deposits already described, and because of environmental conditions being more constant, they are more extensive and uniform in thickness–especially marine deposits. Many factors control the texture, thickness and extent of material that is deposited: these include the type and size of material available (transported by streams or eroded by ocean waves); the distance from shore and the depth of ocean. The best marine formations for groundwater are sands and sandstones that were former beach or nearshore deposits. The presence of fossil remains of sea organisms provides evidence of marine origin.

VI.3.3   Glacial Aquifers

Many forms of glacial deposits have been recognized and named, but only the major ones related to the occurrence of groundwater are described here.

Glacial till may be recognized by its lack of sorting and stratification. Most of the material is of clay, silt or sand size, but pebbles and huge boulders may be present. Where this material has been plastered by ice in a sheet over the pre-existing topography, it is referred to as ground moraine. The deposits range in thickness from a few feet to as much as several hundred feet over pre-glacial valleys. Ground moraine is not a good aquifer because it is a mixture of all sorts of material. However, thin, discontinuous, sandy layers may be scattered through the till where local washing may have taken place. These small deposits will frequently supply sufficient water for a domestic or farm well.

Other types of moraines include both terminal and recessional. Terminal or end moraine is formed at the outermost edge of a continental glacier. Recessional moraine

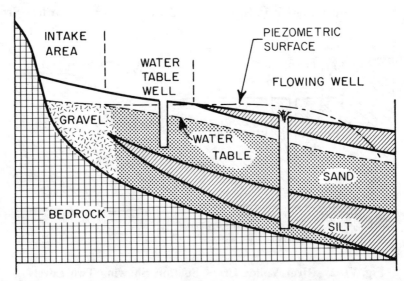

Fig. VI. 3   Typical Section Through an Alluvial Fan (Johnson, 1957)

may be deposited at the time of a slight re-advance of a retreating ice front, or when the front is halted temporarily during its retreat. These deposits are dumped at the margins in a mixed-up mass. However, the melting of the ice at the front can wash and sort some of the material, resulting in the accumulation of mixtures of sand and gravel near these types of moraine. Therefore, there is a better possibility of locating water-bearing sand and gravel in these deposits than in ground moraine areas.

Proglacial or fluvial-glacial deposits are those built by streams carrying material away from a melting glacier. As long as the land slopes sufficiently away from the ice, the streams continue to carry away the debris. Anything that decreases the water velocity causes the coarser material to drop out, but the fine material may be carried for miles beyond. When large quantities of material are continually washed out by the streams over considerable time, outwash fans or aprons merge into outwash plains that cover many square miles of area.

When outwash deposits form narrow fillings in the bottom of pre-existing valleys, they are called valley trains. The major streams draining an ice sheet have produced valley trains several hundred miles in length. Some of these trains are preserved as terrace remnants some distance above the present streams, because the streams have deepened the valleys. Many valley trains still remain buried under overlying materials. The outwash becomes progressively finer down the valley and grades from sand and gravel into silt and clay.

Additional detail of the more important glacial aquifers is presented in subsequent paragraphs.

VI.3.3.1   Glacial Fans

Glacial fans are formed where melt water emerges as a stream at the face of a glacier and spreads out over adjacent plains. The deposit formed will be coarse in the upper reaches, grading to fine at the extremities of the fan, and will often be extensive enough to provide large volumes of water. This type of fan will normally be found adjacent to terminal moraines where the ice has remained relatively stationary for periods of time, long enough to allow building of substantial fan deposits (see Fig. VI.4).

VI.3.3.2   Glacial Lake Deltas

This type of delta is found bordering the flat glacial lake clay deposits where the water flowing from the glacial front has had sufficient velocity to carry sand and coarser particles into the lake waters. Coarser particles settle out near the shore line producing fairly good aquifers in some locations. In general, the delta deposits are likely to be of finer material and less extensive than the fans described above; however, very shallow lake deposits may, in some cases, be indistinguishable from fans (see Fig. VI.5).

VI.3.3.3   Interlobate Deposits

Sediments deposited in melt water channels extending back from the ice front between lobes of melting ice are referred to as interlobate deposits. Such channels are

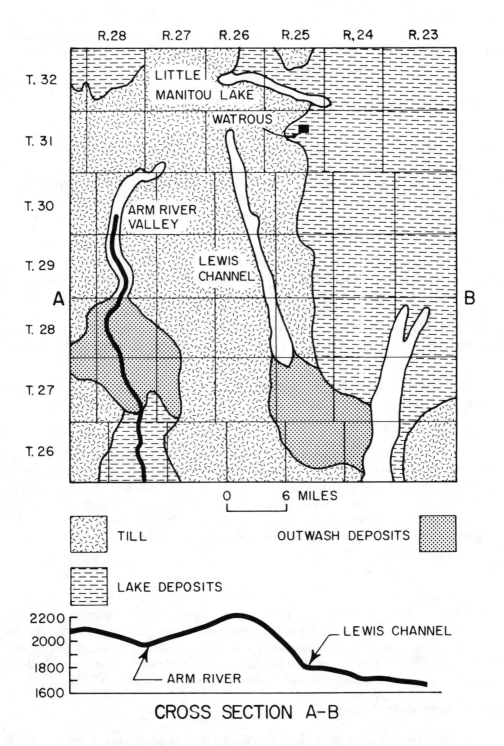

Fig. VI. 4   Glacial Outwash Fans (Christiansen, 1962)

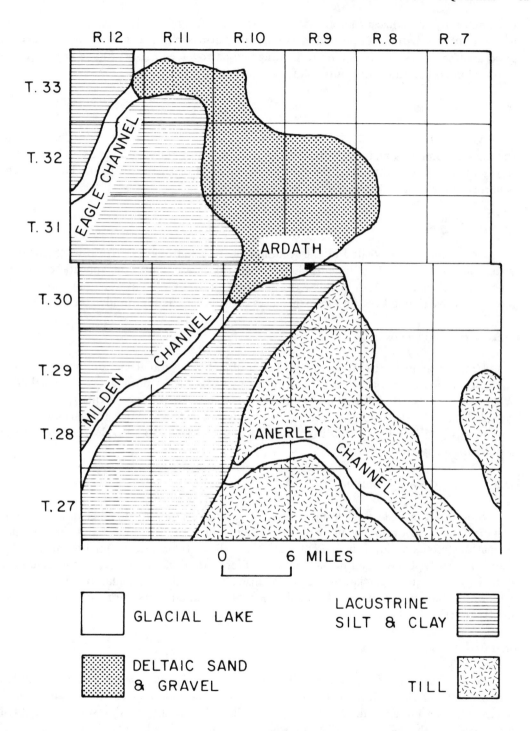

Fig. VI. 5 Deltaic Sands and Gravel (Christiansen, 1962)

constricted by the adjacent ice and serve as drainage for relatively large areas of melting ice so that the material deposited is primarily coarse sands and gravels which become good aquifers. These aquifers tend to be long and narrow (see Fig. VI.6) and are frequently found in conjunction with outwash fans as indicated.

VI.3.3.4   Eskers

Where narrow channels develop within the ice, coarse material will be deposited in the channel and this material, after the disappearance of the ice, remains as long gravel ridges. Such ridges tend to be at right angles to the position of the ice front. The sediment is normally very coarse, although not too extensive.

VI.3.3.5   Kames

Gravel may be deposited at the leading edge of a stationary ice front when the melt water carries away finer sediment. Such gravel ridges are similar in appearance to the eskers described above, but parallel the position of the ice front. They are frequently more extensive than eskers, and in many cases provide excellent groundwater supplies.

VI.3.3.6   Intertill Deposits

Many areas in North America have been subject to several glaciations with the result that formations as previously described occur beneath one or more layers of glacial drift. Wind and water sorting during interglacial periods may also produce formations of coarse granular material that has been subsequently covered with till. Such formations are not continuous and are therefore difficult to locate, but they do provide adequate water supplies in many areas.

VI.3.3.7   Buried Valley Aquifers

Some of the highest yielding and most extensive aquifers that occur in glaciated regions are those occurring in the river valleys which carried melt water. Where such formations are exposed, as in presently existing valleys, the coarse sediments called valley trains are easily located and developed as aquifers. Aquifers of this type that have been buried by subsequent glaciations occur extensively in Western Canada, and are of major importance as a water supply. Two sub-types of buried aquifers are recognized, and are briefly described below.

Buried Bedrock Valleys

Where glacial melt water has followed a pre-glacial valley or has cut into bedrock formations, and the valley has been subsequently filled with glacial till, the situation is referred to as a buried bedrock valley. Such valleys frequently contain small amounts of pre-glacial gravels below extensive deposits of glacial gravel. Where the bedrock material has some porosity, it can serve to recharge the coarser material in the valley, thus increasing the yield ability of such an aquifer. Buried bedrock valleys are not evident

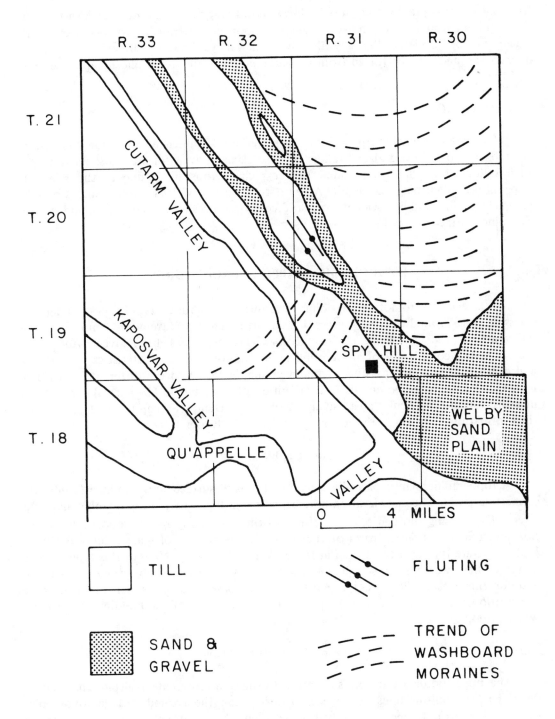

R. 33   R. 32   R. 31   R. 30

T. 21

T. 20

T. 19

T. 18

CUTARM VALLEY

KAPOSVAR VALLEY

QU'APPELLE

VALLEY

SPY HILL

WELBY
SAND
PLAIN

0   4   MILES

TILL

SAND &
GRAVEL

FLUTING

TREND OF
WASHBOARD
MORAINES

Fig. VI. 6   Interlobate Deposits of Sand and Gravel
(Christiansen, 1962)

from surface topography and must be located and mapped by sub-surface investigation. The aquifers which occur, however, are so important that all known occurrences are usually mapped in detail. Partially-filled bedrock valleys, however, are evident as surface depressions, and may be traced from topographic maps and aerial photographs (see Fig. VI.7).

Buried Glacial Valleys

Melt water channels formed in glacial material and buried by subsequent glaciation are referred to as buried glacial valleys. Coarse glacial gravels form extensive aquifers in such valleys, but there is a much slower recharge possibility due to the relatively impermeable, confining glacial till. As with the bedrock valleys, glacial valleys may be completely filled and not evident from surface appearances, or partially filled and evident as distinct surface depressions.

VI.4   HYDROLOGIC MAPPING PROCEDURES

As is evident from the preceding description of aquifer types, the problem of assessing the groundwater resources of an area can be extremely complex. Some indication of the variability due to geologic processes has been given. Additional factors in the complexity of any groundwater study include characteristics such as size and depth of formations, existing hydraulic gradients, flow between different formations, recharge and discharge potentials, and definition of boundaries. The normal approach to investigation of a particular area usually involves an attempt to map the geologically-distinct formations that comprise the regime of the groundwater system.

VI.4.1   Topographic Maps

A topographic map is basically a map of the land surface on which lines (contours) connecting points of equal elevation have been drawn. Fig. VI.8 is a topographic map of a river valley passing through a region of glacial till. The gently rolling topography in the north-west and south-east parts of the map is characteristic of glacial deposits. Steep bluffs separate that area from the flat flood plain of the valley. Maps of this type provide a surface base which is most valuable in locating the depths to sub-surface formations, and for this reason they are frequently used for base maps on which other geologic information is to be shown. Topographic maps are also one of the building blocks from which geologic cross-sections can be constructed.

VI.4.2   Bedrock Topographic Maps

The upper boundary of any sub-surface formation represents a surface that can be described by contour lines in the same way as can the ground surface topography (described above). In most cases bedrock topographic contours are superimposed on surface topographic base maps so that depths to the particular formation at any point will be clearly evident. An example of contour lines representing the upper surface of a shale deposit is shown in the south-east portion of Fig. VI.8. It should be noted that contour lines of this type are constructed on the basis of data from well logs and therefore are likely to be the result of a limited amount of data; thus the accuracy of such maps is frequently subject to question.

In Fig. VI.9 the small symbols similar in shape to the capital letter "T" with a number beside them define the dip and strike of the bedrock. The short stem of the symbol indicates the direction in which the rock layer dips, and the number is the value of the dip in degrees below a horizontal plane. The longer arm of the symbol represents the strike of the formation or the direction of a horizontal line on the formation surface.

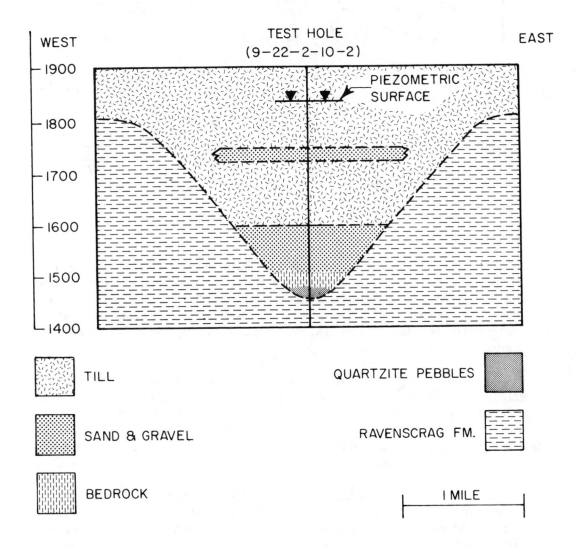

Fig. VI. 7   Cross Section of Preglacial Valley Deposits
(Christiansen, 1962)

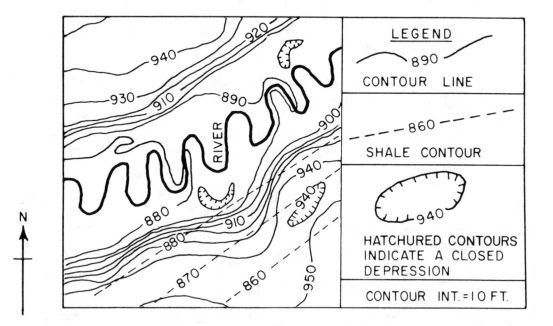

Fig. VI. 8   Topographic Map of River Valley Area (Johnson, 1966a)

VI.4.3   Geologic Maps

Fig. VI.9 is a geologic map of the same area shown in Fig. VI.8. It can be seen how closely the patterns representing the different rock units correspond to the different types of topography. This correlation will not be so evident where most of the materials are similar, or where deformation has occurred. However, in regions where the formations are relatively flat and intersected by streams, a topographic map can be a guide to where the various geologic boundaries occur.

Fig. VI.9 shows a geologic cross-section, which has been constructed from the topographic map, the geologic map and test-hole data. Sections of this type are used in conjunction with geologic maps to show clearly the nature of the formations in the vertical direction. One should be careful in extending information shown on sections to adjacent regions, as considerable variability may occur within short distances. This would be particularly true of the heterogeneous types of deposit that occur in river channels.

Geologic maps, similar to that shown in Fig. VI.9, are frequently used to show the various sub-surface formations that underlay a surface mantle. This technique is particularly important in glaciated regions where a glacial till mantle covers the previously-exposed bedrock material. Such a map, superimposed on a surface topographic map, and in many cases including a set of bedrock contours, will show a wealth of information in a single figure. In any map showing sub-surface details it is important to include the

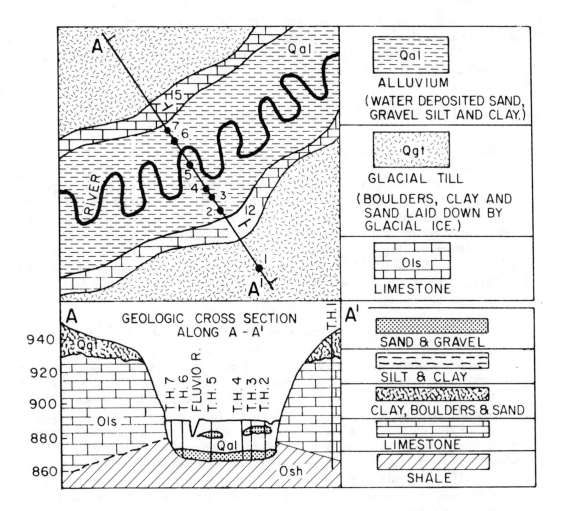

Fig. VI. 9   Geologic Map and Cross Section of River Valley (Johnson, 1966a)

locations of all test holes from which the data was obtained to indicate the extent of extrapolation that was used in preparing the map.

VI.4.4   Isopach Maps

In Fig. VI.10 is shown a fourth type of illustration commonly found in the geological literature dealing with groundwater geology. This is the isopach map, or thickness map. The contour lines shown are similar to those on a topographic map except that on an isopach map the lines connect points of equal thickness of a rock unit, or group of rock units.

In effect, the isopach map can be used to determine, with some accuracy, the approximate shape and lateral extent of sand and gravel formations. By knowing the dimensions and extent of a sand and gravel body in a river valley or in a glacial outwash terrain, an evaluation of the availability of groundwater can be made. An isopach map constructed from test-hole data can be valuable in pre-determining whether or not hydrologic boundaries exist in the proposed area of a well field, or around the location of an individual well.

VI.4.5   Surface Soil Maps

The surface soils of any area are the result of the weathering and development of natural material, which is most likely to be present in its natural state immediately below the developed soil mantle. For this reason surface soil maps are an excellent guide to character of the sub-surface materials, and can be of great value in determining the nature and extent of the shallower sub-surface deposits. Since agricultural soils maps are readily available for most regions, they represent an excellent starting point in any hydrologic study. Sandy areas that may represent shallow unconfined aquifers and/or regions of recharge are readily apparent. Textural variation will often coincide with changes in bedrock formations, and can thus be used in extrapolating data from test holes to construct geologic maps of the area.

VI.4.6   Groundwater Flow Maps

Contour maps of water tables or piezometric surfaces are used to depict the flow of water through waterbearing formations. Flow lines, drawn at right angles to the contour lines (see Fig. VI.11), can be sketched to indicate the direction of flow. Although maps of this type are valuable in showing both regional and local flow patterns, it may be desirable to provide additional information on vertical flow patterns through the use of vertical cross-sections, which show variations in piezometric head with depth and any tendency to upward or downward flow. This procedure would be most important in assessing recharge and discharge characteristics of any formation. Further information relative to construction and use of groundwater flow maps and profiles will be found in the following Section dealing more specifically with water movement through formations.

VI.5   WATER MOVEMENT

Groundwater as a part of the hydrologic cycle is subject to continuous movement— the rate of which depends on characteristics of the formations involved, and the hydraulic gradients which topography superimposes on the system. Thus there are regions of recharge where percolating water from rainfall, snow melt or surface bodies of water, moves into the groundwater system; regions of discharge where water is lost from the groundwater system to surface stream flow, evapotranspiration or directly to the oceans; and areas of groundwater transmission. Most hydrologic studies attempt to define more or less complete and independent flow systems. Where the flow system is essentially isolated by boundaries that prohibit the flow of water, the system may be so defined; but in most cases boundaries are not perfect aquicludes so that very complex and widespread flow

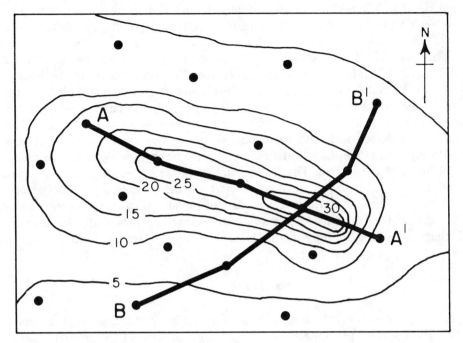

LEGEND :

● – TEST HOLE.   CONTOUR INTERVAL = 5 FEET

Fig. VI. 10 Isopach Map of Sand Deposits (Johnson, 1967)

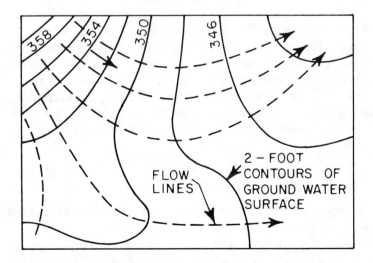

Fig. VI. 11 Contour Map of Groundwater Surface (Todd, 1959)

systems develop. Where flows across assumed boundaries are reasonably low, analysis of the system on the basis of the smaller regional system may still yield useful information.

The basis of all flow analyses is the Darcy equation, which has been discussed in detail in Section V. The reader is referred to that Section for a complete description of flow equations, and for definition of commonly-used terms such as hydraulic conductivity, hydraulic gradient and transmissibility. In general, the basic flow equations serve adequately to define flow characteristics where hydraulic gradients and hydraulic conductivities can be determined with reasonable accuracy. One of the major problems facing the hydrologist in attempting to define a complete basin flow system is to establish the regional limits of the system. That is, to establish areas of recharge, discharge and transmission, and to recognize such areas as part of the preliminary investigation of any groundwater system. Several classical models of possible flow systems can be recognized, and as such will be of considerable value in establishing procedures for investigational work that is designed to fully assess the flow system. Three such models are described below.

VI.5.1   Water-Table Aquifers

The typical water-table aquifer is shown in Fig. VI.12. Recharge occurs through percolation of surface precipitation over the major portion of the aquifer, and discharge occurs to a stream channel, lake or low depressional area when evapotranspiration is high. Transmission occurs at varying rates over the entire area of the basin and consists of a base flow over the lower confining boundary. The limits of the flow system can be established from studies of geologic and topographic conditions, discharge areas are easily recognized, and flow systems are not complicated.

VI.5.2   Artesian Aquifers

The classical artesian aquifer model is also shown in Fig. VI.12. It consists of an exposed permeable recharge area leading to a completely confined aquifer. Natural discharge areas may occur in aquifers of this type where the aquifer formation is partially exposed at lower elevations. Under this latter condition, portions of the aquifer become constricted zones of transmission. Both the water-table aquifers referred to previously, and the artesian aquifers, can become extremely complicated flow systems when the confining beds are not completely impermeable and leakage across the confining formations must be considered.

VI.5.3   The Prairie Profile

This profile shown in Fig. VI.13 is described by Meyboom (1962) as follows:

By definition the Prairie Profile consists of a central topographic high bounded at either side by an area of lower elevation. Geologically the profile is made up of two layers of different permeability, the upper layer having the lower permeability. Through the profile is a steady

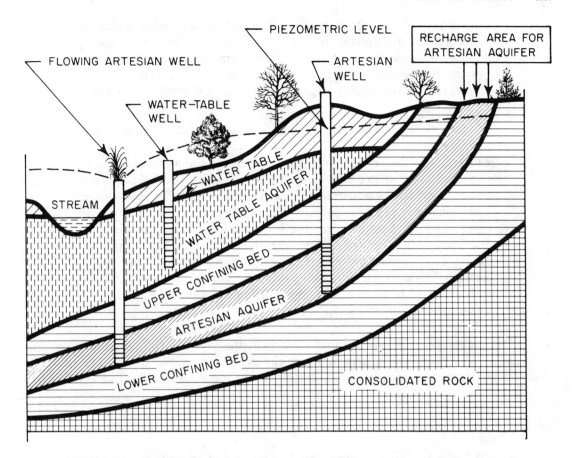

Fig. VI. 12 Classical Confined and Water-Table Aquifers (Johnson, 1966a)

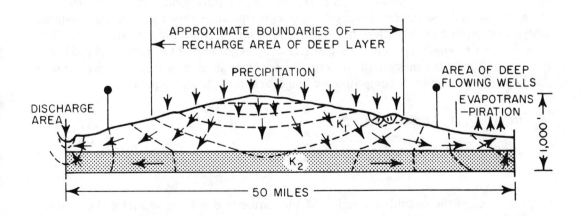

Fig. VI. 13 The Prairie Profile as a Groundwater System Model (Meyboom, 1962)

flow of groundwater from the area of recharge to the area of discharge. The ratio of permeabilities is such that groundwater flow is essentially downward through the material of low permeability, and lateral and upward through the underlying more permeable layer. The potential distribution in the profile is governed by the differential equation of Laplace.

By definition, areas of recharge overlie those parts of the flow system that are characterized by a decrease in head with increasing depth, whereas the flow system underneath areas of discharge exhibits an increase in head with depth. It should be noted that artesian conditions are not necessarily restricted to discharge areas but can exist underneath those areas where the total fluid potential in the deeper layer exceeds the fluid potential at the water table, although the potential distribution in the intervening slightly permeable layer prohibits spontaneous upward movement of water from the lower layer. Those parts of the flow system in which there is virtually no change of head with depth, as is the case in a large portion of the lower layer, may be called transmission zones.

This model is much more complicated than the previous typical aquifers, but it probably represents a flow system of which the occurrence is widespread. The reader is referred to Meyboom's original paper for details on recognizing recharge and discharge areas.

VI.5.4   Flow Nets

A field technique, used to ascertain the direction of groundwater movement, is to construct a two-dimensional flow net from piezometer data (see Section V.1.5 and Fig. V.7). Knowing the hydraulic conductivity and boundary conditions, Harr (1962) outlines the method of constructing a quantitative flow net. Other field techniques which may be used to study groundwater movement, are the use of tracers, and more recently, geochemical methods for large flow systems. Geochemical methods involve the chemical analysis of groundwater within the basin and their interpretation to determine flow direction. An example method is that of using the hypothesis of Chebotarev (1955), who stated that during the movement of groundwater from an area of recharge to an area of discharge, the groundwater composition changes according to the sequence:

$$HCO_3^- \rightarrow HCO_3^- + Cl^- \rightarrow Cl^- + HCO_3^- \dots\dots\dots\dots\dots\dots\dots\dots\dots\dots \text{VI.1}$$

$$\rightarrow (Cl^- + SO_4^= \text{ or } SO_4^= + Cl^-) \rightarrow Cl^- \dots\dots\dots\dots\dots\dots\dots\dots \text{VI.2}$$

Keeping in mind the possible alteration of the sequence due to local rocks, the method holds promise of being a useful tool for large flow systems. It is generally agreed that, in principle, water tends to degrade in quality as it moves towards the discharge area. In glacial regions water flowing through till formations shows an increase in Ca^{++} and Mg^{++} and that through bedrock formations shows an increase in $Na+$ content.

Reference to man's interference leads to the practical consideration of the 'safe yield' of the basin. Safe yield is a rather nebulous term as it means different things to different people depending on the individual philosophy of use and management of the basin. Probably the suggestion by Freeze (1966) is a reasonable approach; here the natural basin yield (discharge) is to be used as an estimate of the safe yield during the initial stages of development. Continual appraisal can be made as development progresses to avoid undesirable results. On the other hand, some people may look upon stored groundwater in the basin as a water resource that is the most economical source to develop for man's needs during the next 20 to 30 years. Both ideas deserve consideration, for with our technology and need, various sources of supply must be sought and developed in the future.

VI.6   GROUNDWATER EXPLORATION

VI.6.1   Drilling

The best information on the character of sub-surface formations is obtained by exploratory drilling for accurate samples from which a common well log can be prepared. The common well log is the drillers' description of the materials encountered in the different formations penetrated. Ideally the driller should collect representative samples at regular intervals of depth and at all changes in the formations to show the complete character of the materials. Core samples from hard rock drilling are the most accurate type of samples obtainable. Drive core samples from soft or unconsolidated materials will also be representative. However, as it is frequently impossible to obtain drive core samples, then other methods of sampling have to be used that do not produce as accurate samples. Cuttings from rotary drilling, carried to the surface in drilling mud, are frequently collected as samples, and the representativeness of such samples will depend to a great extent on the skill and experience of the driller. In general, some type of geophysical logging, as described subsequently, should be used in conjunction with rotary drilling to supplement and verify the log obtained through sampling. In cable tool drilling, samples are normally obtained directly from a baler, and as such are likely to be more representative and accurate than the washed-up samples which may experience various levels of separation. The importance of accurate sampling in groundwater exploration and development cannot be over stressed. Considerable detail on sampling procedures will be found in the reference *Groundwater and Wells,* Johnson (1966).

VI.6.2   Electric Logging

An electric log is a record of the resistivity of sub-surface formations and the spontaneous potentials generated in the well. Both characteristics are related to formation materials and water quality, and can be measured only in mud-filled, uncased boreholes. Dry materials show high resistivities, and saturated materials show lower resistivities. Since the mineral content of water in a formation will be related to the type of formation, resistivity values will be similarly related, and will provide a good indication of changes in formation. Samples are, of course, required to establish a relationship between the resistance and the type of formation. As a general guide, sand deposits saturated with fresh water will show high resistance, but clay formations usually contain more mineralized water and therefore show lower resistance. The self potential curve of an electro log gives an indication of the relative quality or mineralization of the groundwater and the drilling mud. Table VI.1 is a further guide to interpretation of electro logs.

Table VI. 1 Electric Log Interpretation
(Fresh Water Drilling Mud)

SP	Resistivity	Formation	Water Yielding Capacity
1. No response	Low	Shale or clay	None
2. No response	Very high	Tight sandstone or other dense imper- meable rock	None
3. Moderate to low negative or low positive	Moderate to high	Sand and gravel or sand	Fresh water good aquifer
4. Low	Low to moderate	Silty sand or gravel	Fresh water fair to poor aquifer
5. About the same as item 3	More than Item 3	Sandstone with a lower porosity than sand and gravel	Fair to good aquifer fresh water
6. Moderate to high	Low to moderate	Sand, sandstone or sand and gravel	Slightly salty water
7. High	Low, possibly less than shale	Sandstone or sand and gravel or sand	Salt water

Note: If salty mud is used, the resistivity of all formations is reduced from that indicated with fresh mud. If the mud is much saltier than the water in the aquifer, the potential is positive in respect to clay.

VI.6.3   Gamma Ray Logging

Gamma Ray logging is simply the measurement of the natural radioactivity of the various formations in a borehole. The curve obtained is similar in appearance to the resistivity curve of an electric log. In most cases, clays and shales contain a higher concentration of radioactive materials than limestones, sandstones or sands; thus high readings are associated with clays or shales, and low readings with the better aquifer materials.

VI.6.4   **Electrical Resistivity Surveying**

Electrical resistivity surveying is a procedure designed to measure the resistance of sub-surface materials from ground surface tests, and the relative resistances measured can be interpreted as an indication of sub-surface formation characteristics on the same basis as indicated in Table VI.1. Measurement of resistance is made using four equally-spaced electrodes (see Fig. VI.14). The apparent resistivity is calculated from the potential drop, the applied current and the electrode spacing. The value thus obtained is a measure of resistance of all material down to a depth which is proportional to the spacing of the potential electrodes. Electrical resistivity surveys may be made in two ways: 'depth profile' and 'step traverse' procedures.

Depth profile information is obtained by taking a series of readings at increasing electrode spacings and by plotting the resistance readings against the potential electrode spacing to produce a depth profile resistivity curve. The shape of the curve will approximate the type of curve obtained from electro logging, but may be more or less compressed where near-surface formations of greatly differing resistivity cause a reduced penetration of electric current. A number of methods of analyzing depth profile curves to predict the sub-surface strata conditions have been proposed. Most procedures involve a curve fitting technique with theoretical curves developed for a variety of possible layering conditions. Such procedures are useful for conditions that are not too complex, but the problem of handling multi-layered situations is not easily solved. Lennox (1962) gives additional information on the analysis of electrical resistivity curves.

The step traverse procedure consists of making measurements along established lines traversing the area of interest, the same electrode spacings being maintained. Resistivity profiles may then be plotted along the lines to indicate changes that are caused by lateral variation in the sub-surface formations. This type of survey offers excellent possibilities in tracing laterally the extent of various formations that have been encountered in a test hole. A combination of depth profiling and step traversing may be used to provide valuable preliminary information for an area. With some geologic information from test holes, the depth interval (and electrode spacing) that appears most favorable may be selected for step traversing of the area. The resistivities may then be plotted as contours on a map of the area, which will tend to show the extent and variability of the formation of interest. The area pattern will also indicate where additional test holes may be required to fill in required information, and will result in detailed exploration of an area with a minimum number of test holes.

VI.6.5   **Seismic Refraction Surveying**

When energy is generated by an explosion or concussion at or near the ground surface, a seismic wave will spread through the earth in all directions. The paths, and speed of the seismic wave, depend on the nature of the material being passed through.

The direct wave travels through the earth parallel to the ground surface. When a penetrating wave encounters an interface between two formations, it may be reflected or transmitted with a change in direction. The latter process is called refraction. As a refracted wave travels along the interface, it generates waves in the upper formation that return to the surface. Both direct waves and refracted waves can be detected at the surface, and if the time of arrival at various distances from the source is recorded, it is possible to measure the velocity of the seismic wave in each formation and the depth to the refracting interface.

Fig. VI.15 shows a typical plot of the time of arrival of waves at various distances. Each straight portion of the plot represents the arrival of waves through a different combination of earth strata, and the slope of the straight segments represent seismic velocity for the different wave paths. The thickness of the upper layer or depth to the first change in formation is given by

$$D = \frac{L}{2} \sqrt{\frac{V_2 - V_1}{V_2 + V_1}} \quad \dots\dots\dots\dots\dots\dots\dots\dots\dots\dots\dots\dots\dots\dots\dots\dots \quad VI.3$$

where D = depth, (ft)
V_1 = velocity of shock wave in the upper layer (fps),
V_2 = velocity of shock wave in second layer, (fps), and
L = distance to point of velocity change, (ft).

Where three layered conditions exist, the preceding equation can be expanded to determine the depth to the second change in formation.

VI.7   **WELL DESIGN**

Requirements to be considered in well construction include consideration of depth, diameter, water intake, design, development and use of suitable materials—each of which will be considered subsequently in some detail. Consideration will be given, in the main, to well construction in unconsolidated materials. The same general principles of well design will apply to well construction in consolidated or rock aquifers; however, such wells can normally be developed without the use of screens and are therefore of simpler construction. Previous discussion has indicated the importance of accurate and representative sampling in the drilling of test holes. The same principles will obviously apply in the drilling of production wells.

VI.7.1   **Well Diameter**

Well diameter should be selected to satisfy two major requirements:

1.   The well bore must be large enough to accommodate the pump that will be required to pump the anticipated capacity. Sufficient clearance between pump bowls and casing must be provided to allow for the possibility of misalignment

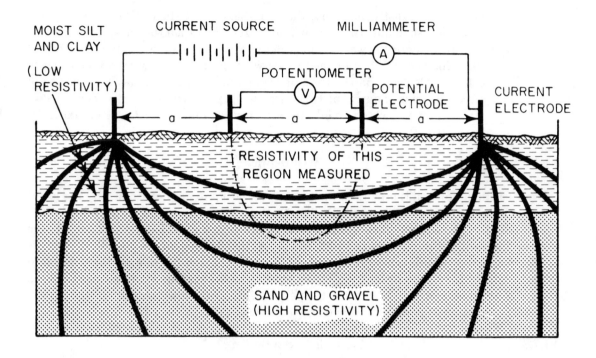

Fig. VI. 14 Electrical Resistivity Surveying (Johnson, 1966a)

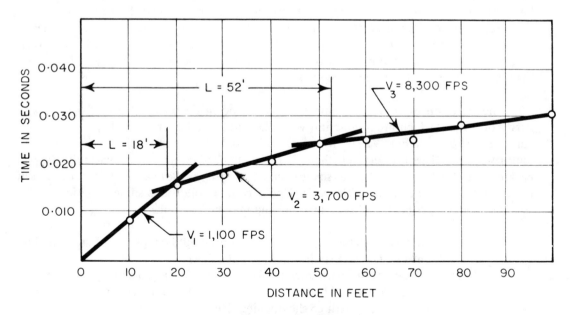

**Fig. VI. 15 Typical Graph Showing Seismic Velocities
(Johnson, 1966a)**

in the casing that could cause the pump shaft not to be vertical. Generally, the casing size should be two nominal pipe sizes larger than the pump bowl size.

2. The diameter of the well screen is chosen to assure good hydraulic efficiency of the well. This will be obtained if the water entrance velocities are kept below 0.1 fps, calculated on the basis of maximum production and the percentage of screen opening. Somewhat larger screens may be required if it is necessary to set the pump bowls down into the screened section, in which case clearance for the vertical flow of water in the well bore must be considered.

VI.7.2   Well Depth

The depth of well is usually to the bottom of the aquifer as determined by the log of a test hole or the production well. Drilling to the bottom of the aquifer provides for full utilization of the aquifer thickness as a well intake area, and will also provide maximum drawdown with resulting greater yield. Where less than the maximum possible capacity is required, screen length can be reduced, and the depth of well can similarly be reduced to that required for screen installation in the aquifer. Other reasons for reducing the depth of well in an aquifer may be because the permeability of the lower portion of the aquifer is low, or because of poor quality water in the lower regions of the aquifer.

VI.7.3   Well Screen Design

For maximum water production, the length of screen to be used depends on the type of aquifer encountered. For confined or artesian conditions, approximately 80 per cent of the aquifer depth should be screened. For water table aquifers, screening the lower one third of the aquifer depth will produce maximum yields. Some adjustment to these recommendations may be desirable where less permeable materials occur at various locations within an aquifer. Where yields of less than optimum are required, the screen length can be estimated on the basis of a permissible water entry velocity of not more than 0.1 fps. through the screen openings.

The selection of screen opening size should be made on the basis of a grain size curve, or curves, which are representative of the aquifer to be screened. Typical grain size curves are shown in Fig. VI.16. Where the effective size is greater than .01 in. and the uniformity coefficient is greater than 2, the well may be developed naturally (gravel packing will not be required); the size of screen opening or slot size may then be chosen to retain about 40 per cent of the sand. Under certain conditions some variation in slot size from the '40 per cent size' is permissible and may be desirable:

1. Where water is expected to be corrosive, one should choose a slot size equal to 50 per cent retained size (reduced screen opening).
2. For coarse formations with large uniformity coefficients, a slot size equal to 30 per cent retained size will be stable. This may be desirable where encrustation is expected to gradually reduce the opening size.
3. In stratified aquifers where different screen openings are used opposite the various strata, fine screen openings should extend down into coarser formations

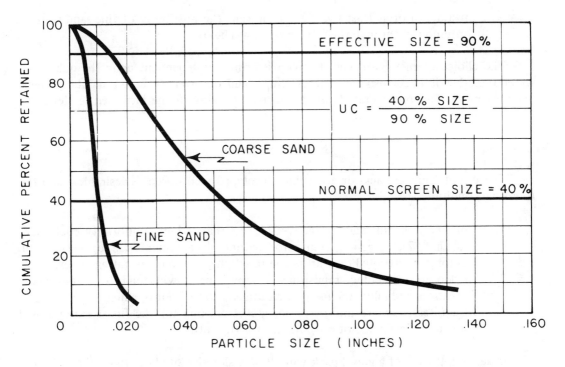

Fig. VI.16 Typical Grain Size Analysis Curves (Johnson, 1959)

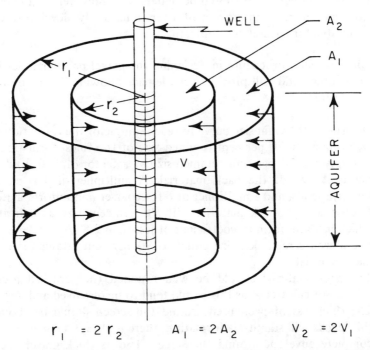

$$r_1 = 2r_2 \qquad A_1 = 2A_2 \qquad V_2 = 2V_1$$

Fig. VI. 17 Schematic Drawing of an Artesian Well (Johnson, 1966a)

for a depth of at least two feet to prevent the possibility of fine sand flowing continuously through the coarser openings below.

The choice of metal for a well screen is primarily dependent upon the quality of water (whether it is corrosive or encrustating), and on strength requirements. Screens made of more than one metal should be avoided because of the increased tendency to corrosion from bimetallic action.

VI.7.4   Gravel Pack Design

The use of a gravel pack, or artificially-developed well screen, is desirable under the following conditions:

1. For fine uniform sands (effective size less than .01 in., uniformity coefficient less than 2), screen openings for natural development will be very small, resulting in restricted flow areas and rapid plugging in encrustating water.
2. In loosely cemented sandstone, a gravel pack will tend to stabilize the formation, and allow the use of a reasonable size of slot opening.
3. In aquifers that are multi-layered, gravel packing for the finest material will allow use of a single slot opening size for the entire screen.

In general, the use of a gravel pack will allow a slot size of about twice that required for natural development, and as such should be used where, for any reason, it is desirable to use a larger screen opening. It should be understood, however, that gravel-packed wells will not show any marked improvement over a naturally developed well where the formation is suitable for natural development.

The following step procedure in designing the gravel pack material is based on the theory of sand filters, and will provide a stable pack allowing the use of an optimum slot size for the well screen:

1. Determine the grain size curve for each sample of the formation to be screened.
2. Determine the 70 per cent retained size of the finest formation in the aquifer.
3. Multiply the 70 per cent retained size by a factor of 4–6 to determine the 70 per cent size of the pack material. A multiplier of 4 is used for very fine uniform sands, and a multiplier of 6 for coarser nonuniform materials.
4. Choose a sand pack material with 70 per cent size as indicated above, and which has a uniformity coefficient of less than 2.5.
5. Use a well screen slot size equal to 90 per cent retained size of the selected pack material.
6. The pack material should be well rounded (not crushed rock) of siliceous origin, not calcareous as this would tend to be dissolved and lost.
7. The thickness of pack used around the screen should be about 3 in. Though thin packs will stabilize a formation, there is always a danger of not obtaining a complete envelope around the screen. Use of thick gravel packs will present problems in well development, and this is discussed later.

With a well screen, or properly-designed perforated casing in place in a well, some additional work is required to develop maximum yields from the well. The correct size of screen openings will allow the finer sand particles to pass into the well. A rapid up and down motion of a surge block in the water of the well will give reversals of flow through the screen, and will draw fine sand particles into the well until a coarse sand or gravel envelope is formed around the screen. This will then hold back the finer sand particles, further out in the formation, and will allow a larger flow of water into the screen. The fine sand drawn through the screen can be removed by baling or flushing. This process of development is very important in wells where drilling mud has been used, as the mud must be removed from the water formation. In most cases the yield of a well will be increased several times by proper development work. Pumping tests at intervals throughout the developing process will indicate when no further increase in yield can be obtained with additional development. Work should be stopped at this point, as there is danger in over developing and producing a well that pumps sand, especially in gravel-packed wells.

VI.8   HYDRAULICS OF WELLS

VI.8.1   General

The hydraulic conductivity—or transmissibility, storage, and vertical conductivity—is the major hydraulic property of an aquifer and confining beds upon which the foundation of quantitative groundwater studies is based.

The rate of flow of groundwater, in response to a given hydraulic gradient, is dependent upon the conductivity of the aquifer. The field hydraulic conductivity, k, is defined as the rate of flow of water in gal/day, through a cross-sectional area of 1 sq. ft. of the aquifer under a hydraulic gradient of 1 ft./ft. at the prevailing temperature of the water. A related term, the coefficient of transmissibility T, indicates the capacity of an aquifer as a whole to transmit water, and is equal to the hydraulic conductivity multiplied by the saturated thickness of the aquifer, b, in ft. The coefficient of transmissiblity is defined as the rate of flow of water in gal/day through a vertical strip of the aquifer 1 ft. in width and extending the full saturated thickness under a hydraulic gradient of 1 ft./ft. at the prevailing temperature of the water.

The rate of vertical leakage of groundwater through a confining bed in response to a given vertical hydraulic gradient is dependent upon the vertical conductivity of the confining bed. The field coefficient of vertical conductivity k_z is defined as the rate of vertical flow of water, (usually in gal./day in groundwater work) through a horizontal cross-sectional area of 1 sq. ft. of the confining bed under a hydraulic gradient of 1 ft./ft. at the prevailing temperature of the water.

The storage properties of an aquifer are expressed by its coefficient of storage, S, which is defined as the volume of water the aquifer releases from or takes into storage per unit surface area of the aquifer per unit change in the component of head normal to that surface. Under artesian conditions, when the piezometric surface is lowered by pumping, water is derived from storage by the compaction of the aquifer and its associated beds and by expansion of the water itself, while the interstices remain saturated. Under water table conditions, when the water table is lowered by pumping, groundwater is derived from storage mainly by the gravity drainage of the interstices in the portion of the aquifer dewatered by the pumping. The coefficient of storage for water table conditions is normally about 0.05; for artesian conditions it would be nearer 0.0005.

Under artesian conditions and for granular or loosely cemented aquifers reasonably free from clay beds, the coefficient of storage is a function of the elasticity of water and the aquifer skeleton, as expressed in the following equation. (See also Equation V.61).

$$S = (F\gamma b\beta/144)(1 + \alpha/F\beta) \dots\dots\dots\dots\dots\dots\dots\dots\dots\dots\dots VI.4$$

where S = coefficient of storage, $(ft^3/ft^2/ft)$,
$\quad\quad\quad F$ = porosity, fraction,
$\quad\quad\quad b$ = saturated thickness of aquifer, (ft),
$\quad\quad\quad \beta$ = reciprocal of the bulk modulus of elasticity of water, $(in.^2/lb,)$
$\quad\quad\quad \alpha$ = reciprocal of the bulk modulus of elasticity of aquifer skeleton, $(in.^2/lb)$, and
$\quad\quad\quad \gamma$ = specific weight of water, (lb/cu ft).

Recognizing that for practical purposes $\gamma = 62.4$ lb/cu ft and $\beta = 3.3 \times 10^{-6}$ in.^{2}/lb the fraction of storage attributable to expansibility of the water S_W is given by the following equation:

$$S_W = 1.4 \times 10^{-6} (Fb) \dots\dots\dots\dots\dots\dots\dots\dots\dots\dots\dots\dots\dots VI.5$$

VI.8.2   **Theis Equation**

With reference to Fig. VI.17, it will be noted that the discharge at two adjacent cylindrical sections differs by the rate of release of storage water between the sections, so that

$$Q_2 = Q_1 + \frac{\Delta \text{ storage}}{\Delta \text{ time}}$$

where Δ storage is proportional to the drawdown, s. This equation can be expressed in terms of differentials and solved to give the Theis nonequilibrium equation:

$$s = \frac{Q}{4\pi T} W(u) \dots\dots\dots\dots\dots\dots\dots\dots\dots\dots\dots\dots\dots\dots\dots VI.6$$

where s = drawdown in feet at radial distance, r,
$\quad\quad\quad T$ = transmissibility (U.S. gal/day/ft),[1]
$\quad\quad\quad Q$ = well discharge, (U.S. gal/day), and

---

[1] All calculations in the remaining portion of this Section are based upon the use of U.S. gallons.

$$W(u) = \int_{u}^{\infty} \frac{e^{-u}}{u}\, du - \text{called the well function.}$$

$$u = \frac{7.48r^2 S}{4Tt} \quad \dots\dots\dots\dots\dots\dots\dots\dots\dots\dots\dots\dots\dots\dots\dots\dots\dots\dots \text{VI.7}$$

where r = radial distance in ft. from the pumped well,
       S = storage coefficient, and
       t = duration of pumping in days.

The Theis equation is based on the following assumptions:

(1) the release of water from storage occurs instantaneously upon lowering of the drawdown curve;
(2) all water comes from storage;
(3) the coefficient of storage is constant;
(4) the nonpumping piezometric surface is horizontal;
(5) when pumping, all stream lines are horizontal;
(6) the well casing is perforated over the full depth of saturated, confined, aquifer; and
(7) the aquifer is horizontal, homogeneous, of uniform thickness and infinite in extent.

VI.8.2.1   Graphical Solution

Values of the well function for different values of u are given in Table VI.2, and a plot of u vs. W(u) on logarithmic paper is included as Fig. VI.18. Where values of storage coefficient, S, and transmissibility, T, are known, the value of u can be calculated for any time, t, and radius, r, and the well function can be determined. Values of drawdown, s, can thus be obtained for any value of time and radius. The more common requirement in solving the Theis equation is to determine the actual values of T and S from pump test data. Solution for T is relatively difficult since it occurs both inside and outside of the integral function.

For a given pumping situation it will be noted that the drawdown, s, is proportional to the well function, W(u), and that u is proportional to r^2/t. The factors of proportionality (constants) determine the specific values of u and W(u), and therefore determine the lateral and vertical position of the u vs. W(u) curve, but not its shape or slope. If a logarithmic plot of s vs r^2/t is superimposed on a u vs. W(u) plot as shown in Fig. VI.19 and a suitable match point chosen, relative values of W(u) and s, and u and r^2/t will be obtained. Relative values of W(u) and s can be used in Equation VI.6 to determine the transmissibility, and the relative values of u and r^2/t along with transmissibility can be used in Equation VI.7 to determine the storage coefficient.

VI.8.3   Modified Theis Equations

The preceding graphical solution is generally applicable to artesian wells, and may be applicable for free aquifer wells if drawdown is sufficiently small. A better approach to

Table VI.2   Values of W(u)   (Wenzel, 1942)

$\dfrac{u}{N}$	$N\times10^{-15}$	$N\times10^{-14}$	$N\times10^{-13}$	$N\times10^{-12}$	$N\times10^{-11}$	$N\times10^{-10}$	$N\times10^{-9}$	$N\times10^{-8}$	$N\times10^{-7}$	$N\times10^{-6}$	$N\times10^{-5}$	$N\times10^{-4}$	$N\times10^{-3}$	$N\times10^{-2}$	$N\times10^{-1}$	N
1.0	33.9616	31.6590	29.3564	27.0538	24.7512	22.4486	20.1460	17.8435	15.5409	13.2383	10.9357	8.6332	6.3315	4.0379	1.8229	0.2194
1.1	33.8662	31.5637	29.2611	26.9585	24.6559	22.3533	20.0507	17.7482	15.4456	13.1430	10.8404	8.5379	6.2363	3.9436	1.7371	.1860
1.2	33.7792	31.4767	29.1741	26.8715	24.5689	22.2663	19.9637	17.6611	15.3586	13.0560	10.7534	8.4509	6.1494	3.8576	1.6595	.1584
1.3	33.6992	31.3966	29.0940	26.7914	24.4889	22.1863	19.8837	17.5811	15.2785	12.9759	10.6734	8.3709	6.0695	3.7785	1.5889	.1355
1.4	33.6251	31.3225	29.0199	26.7173	24.4147	22.1122	19.8096	17.5070	15.2044	12.9018	10.5993	8.2968	5.9955	3.7054	1.5241	.1162
1.5	33.5561	31.2535	28.9509	26.6483	24.3458	22.0432	19.7406	17.4380	15.1354	12.8328	10.5303	8.2278	5.9266	3.6374	1.4645	.1000
1.6	33.4916	31.1890	28.8864	26.5838	24.2812	21.9786	19.6760	17.3735	15.0709	12.7683	10.4657	8.1634	5.8621	3.5739	1.4092	.08631
1.7	33.4309	31.1283	28.8258	26.5232	24.2206	21.9180	19.6154	17.3128	15.0103	12.7077	10.4051	8.1027	5.8016	3.5143	1.3578	.07465
1.8	33.3738	31.0712	28.7686	26.4660	24.1634	21.8608	19.5583	17.2557	14.9531	12.6505	10.3479	8.0455	5.7446	3.4581	1.3098	.06471
1.9	33.3197	31.0171	28.7145	26.4119	24.1094	21.8068	19.5042	17.2016	14.8990	12.5964	10.2939	7.9915	5.6906	3.4050	1.2649	.05620
2.0	33.2684	30.9658	28.6632	26.3607	24.0581	21.7555	19.4529	17.1503	14.8477	12.5451	10.2426	7.9402	5.6394	3.3547	1.2227	.04890
2.1	33.2196	30.9170	28.6145	26.3119	24.0093	21.7067	19.4041	17.1015	14.7989	12.4964	10.1938	7.8914	5.5907	3.3069	1.1829	.04261
2.2	33.1731	30.8705	28.5679	26.2653	23.9628	21.6602	19.3576	17.0550	14.7524	12.4498	10.1473	7.8449	5.5443	3.2614	1.1454	.03719
2.3	33.1286	30.8261	28.5235	26.2209	23.9183	21.6157	19.3131	17.0106	14.7080	12.4054	10.1028	7.8004	5.4999	3.2179	1.1099	.03250
2.4	33.0861	30.7835	28.4809	26.1783	23.8758	21.5732	19.2706	16.9680	14.6654	12.3628	10.0603	7.7579	5.4575	3.1763	1.0762	.02844
2.5	33.0453	30.7427	28.4401	26.1375	23.8349	21.5323	19.2298	16.9272	14.6246	12.3220	10.0194	7.7172	5.4167	3.1365	1.0443	.02491
2.6	33.0060	30.7035	28.4009	26.0983	23.7957	21.4931	19.1905	16.8880	14.5854	12.2828	9.9802	7.6779	5.3776	3.0983	1.0139	.02185
2.7	32.9683	30.6657	28.3631	26.0606	23.7580	21.4554	19.1528	16.8502	14.5476	12.2450	9.9425	7.6401	5.3400	3.0615	.9849	.01918
2.8	32.9319	30.6294	28.3268	26.0242	23.7216	21.4190	19.1164	16.8138	14.5113	12.2087	9.9061	7.6038	5.3037	3.0261	.9573	.01686
2.9	32.8968	30.5943	28.2917	25.9891	23.6865	21.3839	19.0813	16.7788	14.4762	12.1736	9.8710	7.5687	5.2687	2.9920	.9309	.01482
3.0	32.8629	30.5604	28.2578	25.9552	23.6526	21.3500	19.0474	16.7449	14.4423	12.1397	9.8371	7.5348	5.2349	2.9591	.9057	.01305
3.1	32.8302	30.5276	28.2250	25.9224	23.6198	21.3172	19.0146	16.7121	14.4095	12.1069	9.8043	7.5020	5.2022	2.9273	.8815	.01149
3.2	32.7984	30.4958	28.1932	25.8907	23.5881	21.2855	18.9829	16.6803	14.3777	12.0751	9.7726	7.4703	5.1706	2.8965	.8583	.01013
3.3	32.7676	30.4651	28.1625	25.8599	23.5573	21.2547	18.9521	16.6495	14.3470	12.0444	9.7418	7.4395	5.1399	2.8668	.8361	.008939
3.4	32.7378	30.4352	28.1326	25.8300	23.5274	21.2249	18.9223	16.6197	14.3171	12.0145	9.7120	7.4097	5.1102	2.8379	.8147	.007891
3.5	32.7088	30.4062	28.1036	25.8010	23.4985	21.1959	18.8933	16.5907	14.2881	11.9855	9.6830	7.3807	5.0813	2.8099	.7942	.006970
3.6	32.6806	30.3780	28.0755	25.7729	23.4703	21.1677	18.8651	16.5625	14.2599	11.9574	9.6548	7.3526	5.0532	2.7827	.7745	.006160
3.7	32.6532	30.3506	28.0481	25.7455	23.4429	21.1403	18.8377	16.5351	14.2325	11.9300	9.6274	7.3252	5.0259	2.7563	.7554	.005448
3.8	32.6266	30.3240	28.0214	25.7188	23.4162	21.1136	18.8110	16.5085	14.2059	11.9033	9.6007	7.2985	4.9993	2.7306	.7371	.004820
3.9	32.6006	30.2980	27.9954	25.6928	23.3902	21.0877	18.7851	16.4825	14.1799	11.8773	9.5748	7.2725	4.9735	2.7056	.7194	.004267
4.0	32.5753	30.2727	27.9701	25.6675	23.3649	21.0623	18.7598	16.4572	14.1546	11.8520	9.5495	7.2472	4.9482	2.6813	.7024	.003779
4.1	32.5506	30.2480	27.9454	25.6428	23.3402	21.0376	18.7351	16.4325	14.1299	11.8273	9.5248	7.2225	4.9236	2.6576	.6859	.003349
4.2	32.5265	30.2239	27.9213	25.6187	23.3161	21.0136	18.7110	16.4084	14.1058	11.8032	9.5007	7.1985	4.8997	2.6344	.6700	.002969
4.3	32.5029	30.2004	27.8978	25.5952	23.2926	20.9900	18.6874	16.3848	14.0823	11.7797	9.4771	7.1749	4.8762	2.6119	.6546	.002633
4.4	32.4800	30.1774	27.8748	25.5722	23.2696	20.9670	18.6644	16.3619	14.0593	11.7567	9.4541	7.1520	4.8533	2.5899	.6397	.002336
4.5	32.4575	30.1549	27.8523	25.5497	23.2471	20.9446	18.6420	16.3394	14.0368	11.7342	9.4317	7.1295	4.8310	2.5684	.6253	.002073
4.6	32.4355	30.1329	27.8303	25.5277	23.2252	20.9226	18.6200	16.3174	14.0148	11.7122	9.4097	7.1075	4.8091	2.5474	.6114	.001841
4.7	32.4140	30.1114	27.8088	25.5062	23.2037	20.9011	18.5985	16.2959	13.9933	11.6907	9.3882	7.0860	4.7877	2.5268	.5979	.001635
4.8	32.3929	30.0904	27.7878	25.4852	23.1826	20.8800	18.5774	16.2748	13.9723	11.6697	9.3671	7.0650	4.7667	2.5068	.5848	.001453
4.9	32.3723	30.0697	27.7672	25.4646	23.1620	20.8594	18.5568	16.2542	13.9516	11.6491	9.3465	7.0444	4.7462	2.4871	.5721	.001291
5.0	32.3521	30.0495	27.7470	25.4444	23.1418	20.8392	18.5366	16.2340	13.9314	11.6289	9.3263	7.0242	4.7261	2.4679	.5598	.001148
5.1	32.3323	30.0297	27.7271	25.4246	23.1220	20.8194	18.5168	16.2142	13.9116	11.6091	9.3065	7.0044	4.7064	2.4491	.5478	.001021
5.2	32.3129	30.0103	27.7077	25.4051	23.1026	20.8000	18.4974	16.1948	13.8922	11.5896	9.2871	6.9850	4.6871	2.4306	.5362	.0009086
5.3	32.2939	29.9913	27.6887	25.3861	23.0835	20.7809	18.4783	16.1758	13.8732	11.5706	9.2681	6.9659	4.6681	2.4126	.5250	.0008086
5.4	32.2752	29.9726	27.6700	25.3674	23.0648	20.7622	18.4596	16.1571	13.8545	11.5519	9.2494	6.9473	4.6495	2.3948	.5140	.0007198
5.5	32.2568	29.9542	27.6516	25.3491	23.0465	20.7439	18.4413	16.1387	13.8361	11.5336	9.2310	6.9289	4.6313	2.3775	.5034	.0006409
5.6	32.2388	29.9362	27.6336	25.3310	23.0285	20.7259	18.4233	16.1207	13.8181	11.5155	9.2130	6.9109	4.6134	2.3604	.4930	.0005708
5.7	32.2211	29.9185	27.6159	25.3133	23.0103	20.7082	18.4056	16.1030	13.8004	11.4978	9.1953	6.8932	4.5958	2.3437	.4830	.0005085
5.8	32.2037	29.9011	27.5985	25.2959	22.9934	20.6908	18.3882	16.0856	13.7830	11.4804	9.1779	6.8758	4.5785	2.3273	.4732	.0004532
5.9	32.1866	29.8840	27.5814	25.2789	22.9763	20.6737	18.3711	16.0685	13.7659	11.4633	9.1608	6.8588	4.5615	2.3111	.4637	.0004039
6.0	32.1698	29.8672	27.5646	25.2620	22.9595	20.6569	18.3543	16.0517	13.7491	11.4465	9.1440	6.8420	4.5448	2.2953	.4544	.0003601
6.1	32.1533	29.8507	27.5481	25.2455	22.9429	20.6403	18.3378	16.0352	13.7326	11.4300	9.1275	6.8254	4.5283	2.2797	.4454	.0003211
6.2	32.1370	29.8344	27.5318	25.2293	22.9267	20.6241	18.3215	16.0189	13.7163	11.4138	9.1112	6.8092	4.5122	2.2645	.4366	.0002864
6.3	32.1210	29.8184	27.5158	25.2133	22.9107	20.6081	18.3055	16.0029	13.7003	11.3978	9.0952	6.7932	4.4963	2.2494	.4280	.0002555
6.4	32.1053	29.8027	27.5001	25.1975	22.8949	20.5923	18.2898	15.9872	13.6846	11.3820	9.0795	6.7775	4.4806	2.2346	.4197	.0002279
6.5	32.0898	29.7872	27.4846	25.1820	22.8794	20.5768	18.2742	15.9717	13.6691	11.3665	9.0640	6.7620	4.4652	2.2201	.4115	.0002034
6.6	32.0745	29.7719	27.4693	25.1667	22.8641	20.5616	18.2590	15.9564	13.6538	11.3512	9.0487	6.7467	4.4501	2.2058	.4036	.0001816
6.7	32.0595	29.7569	27.4543	25.1517	22.8491	20.5465	18.2439	15.9414	13.6388	11.3362	9.0337	6.7317	4.4351	2.1917	.3959	.0001621
6.8	32.0446	29.7421	27.4395	25.1369	22.8343	20.5317	18.2291	15.9265	13.6240	11.3214	9.0189	6.7169	4.4204	2.1779	.3883	.0001448
6.9	32.0300	29.7275	27.4249	25.1223	22.8197	20.5171	18.2145	15.9119	13.6094	11.3068	9.0043	6.7023	4.4059	2.1643	.3810	.0001293
7.0	32.0156	29.7131	27.4105	25.1079	22.8053	20.5027	18.2001	15.8976	13.5950	11.2924	8.9899	6.6879	4.3916	2.1508	.3738	.0001155
7.1	32.0015	29.6989	27.3963	25.0937	22.7911	20.4885	18.1860	15.8834	13.5808	11.2782	8.9757	6.6737	4.3775	2.1376	.3668	.0001032
7.2	31.9875	29.6849	27.3823	25.0797	22.7771	20.4746	18.1720	15.8694	13.5668	11.2642	8.9617	6.6598	4.3636	2.1246	.3599	.00009219
7.3	31.9737	29.6711	27.3685	25.0659	22.7633	20.4608	18.1582	15.8556	13.5530	11.2504	8.9479	6.6460	4.3500	2.1118	.3532	.00008239
7.4	31.9601	29.6575	27.3549	25.0523	22.7497	20.4472	18.1446	15.8420	13.5394	11.2368	8.9343	6.6324	4.3364	2.0991	.3467	.00007364
7.5	31.9467	29.6441	27.3415	25.0389	22.7363	20.4337	18.1311	15.8286	13.5260	11.2234	8.9209	6.6190	4.3231	2.0867	.3403	.00006583
7.6	31.9334	29.6308	27.3282	25.0257	22.7231	23.4205	18.1179	15.8153	13.5127	11.2102	8.9076	6.6057	4.3100	2.0744	.3341	.00005886
7.7	31.9203	29.6178	27.3152	25.0126	22.7100	20.4074	18.1048	15.8022	13.4997	11.1971	8.8946	6.5927	4.2970	2.0623	.3280	.00005263
7.8	31.9074	29.6048	27.3023	24.9997	22.6971	20.3945	18.0919	15.7893	13.4868	11.1842	8.8817	6.5798	4.2842	2.0503	.3221	.00004707
7.9	31.8947	29.5921	27.2895	24.9869	22.6844	20.3818	18.0792	15.7766	13.4740	11.1714	8.8689	6.5671	4.2716	2.0386	.3163	.00004210
8.0	31.8821	29.5795	27.2769	24.9744	22.6718	20.3692	18.0666	15.7640	13.4614	11.1589	8.8563	6.5545	4.2591	2.0269	.3106	.00003767
8.1	31.8697	29.5671	27.2645	24.9619	22.6594	20.3568	18.0542	15.7516	13.4490	11.1464	8.8439	6.5421	4.2468	2.0155	.3050	.00003370
8.2	31.8574	29.5548	27.2523	24.9497	22.6471	20.3445	18.0419	15.7393	13.4367	11.1342	8.8317	6.5298	4.2346	2.0042	.2996	.00003015
8.3	31.8453	29.5427	27.2401	24.9375	22.6350	20.3324	18.0298	15.7272	13.4246	11.1220	8.8195	6.5177	4.2226	1.9930	.2943	.00002699
8.4	31.8333	29.5307	27.2282	24.9256	22.6230	20.3204	18.0178	15.7152	13.4126	11.1101	8.8076	6.5057	4.2107	1.9820	.2891	.00002415
8.5	31.8215	29.5189	27.2163	24.9137	22.6112	20.3086	18.0060	15.7034	13.4008	11.0982	8.7957	6.4939	4.1990	1.9711	.2840	.00002162
8.6	31.8098	29.5072	27.2046	24.9020	22.5995	20.2969	17.9943	15.6917	13.3891	11.0865	8.7840	6.4822	4.1874	1.9604	.2790	.00001936
8.7	31.7982	29.4957	27.1931	24.8905	22.5879	20.2853	17.9827	15.6801	13.3776	11.0750	8.7725	6.4707	4.1759	1.9498	.2742	.00001733
8.8	31.7868	29.4842	27.1816	24.8790	22.5765	20.2739	17.9713	15.6687	13.3661	11.0635	8.7610	6.4592	4.1646	1.9393	.2694	.00001552
8.9	31.7755	29.4729	27.1703	24.8678	22.5652	20.2626	17.9600	15.6574	13.3548	11.0523	8.7497	6.4480	4.1534	1.9290	.2647	.00001390
9.0	31.7643	29.4618	27.1592	24.8566	22.5540	20.2514	17.9488	15.6462	13.3437	11.0411	8.7386	6.4368	4.1423	1.9187	.2602	.00001245
9.1	31.7533	29.4507	27.1481	24.8455	22.5429	20.2404	17.9378	15.6352	13.3326	11.0300	8.7275	6.4258	4.1313	1.9087	.2557	.00001115
9.2	31.7424	29.4398	27.1372	24.8346	22.5320	20.2294	17.9268	15.6213	13.3217	11.0191	8.7166	6.4148	4.1205	1.8987	.2513	.000009988
9.3	31.7315	29.4290	27.1264	24.8238	22.5212	20.2186	17.9160	15.6135	13.3109	11.0083	8.7058	6.4040	4.1098	1.8888	.2470	.000008948
9.4	31.7208	29.4183	27.1157	24.8131	22.5105	20.2079	17.9053	15.6028	13.3002	10.9976	8.6951	6.3934	4.0992	1.8791	.2429	.000008018
9.5	31.7103	29.4077	27.1051	24.8025	22.4999	20.1973	17.8948	15.5922	13.2896	10.9870	8.6845	6.3828	4.0887	1.8695	.2387	.000007185
9.6	31.6998	29.3972	27.0946	24.7920	22.4895	20.1869	17.8843	15.5817	13.2791	10.9765	8.6740	6.3723	4.0784	1.8599	.2347	.000006439
9.7	31.6894	29.3868	27.0843	24.7817	22.4791	20.1765	17.8739	15.5713	13.2688	10.9662	8.6637	6.3620	4.0681	1.8505	.2308	.000005771
9.8	31.6792	29.3766	27.0740	24.7714	22.4688	20.1663	17.8637	15.5611	13.2585	10.9559	8.6534	6.3517	4.0579	1.8412	.2269	.000005173
9.9	31.6690	29.3664	27.0639	24.7613	22.4587	20.1561	17.8535	15.5509	13.2483	10.9458	8.6433	6.3416	4.0479	1.8320	.2231	.000004637

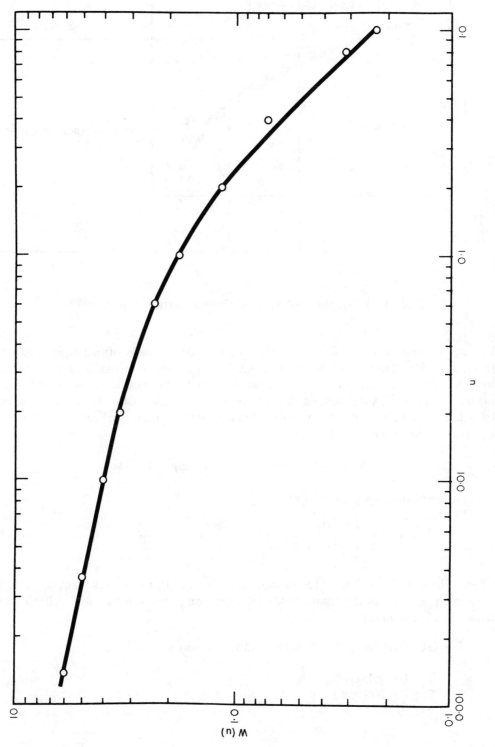

Fig. VI. 18 Type Curve—u vs. W(u)

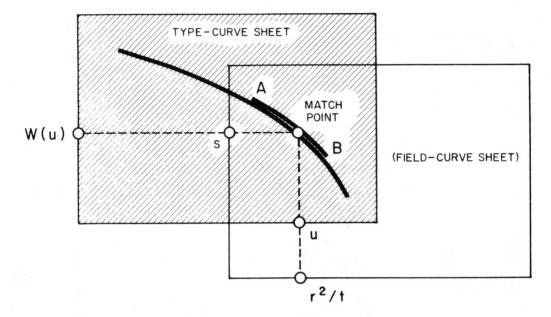

Fig. VI. 19 Graphic Solution of Theis Equation (Todd, 1959)

free aquifer conditions, and one which is very useful in the confined aquifer situation, involves a simplification of the nonequilibrium formula to a modified equation. The modification of the Theis equation is possible, due to the fact that a plot of W(u) versus log u (semi log plot) approaches a straight line for small values of u. Thus if one can pump for relatively long periods of time the value of u will become small, and one can use the equation of the asymptote, which is

$$W(u_2) - W(u_1) = -2.3 (\log u_2 - \log u_1)$$

or substituting for u and W(u)

$$s_2 - s_1 = \frac{2.3Q}{4\pi T} \log \left[\frac{r_1^2/t_1}{r_2^2/t_2} \right] \dots\dots\dots\dots\dots\dots\dots\dots\dots\dots\dots\dots VI.8$$

where *subscripts 1 and 2* refer to chronological times and/or distances from the well. This equation is known as the 'Basic Modified Equation', and is reasonably reliable for all values of u less than 0.1.

Typical values for a free aquifer condition might be

    T = 10³ gal/day/ft,
    S = 5 x 10⁻², and
    r = 50 ft.

in which case the value of u would be equal to

$$u = \frac{7.48 \ r^2 S}{4Tt} = \frac{0.23}{t}$$

and for u to be less than 0.1, the time of pumping would have to be about 2.3 days before the basic modified equation could be applied. It is obvious that much shorter pumping times are required to give reasonably small u values in artesian aquifers where the storage coefficient is much smaller.

VI.8.3.1 Time-Drawdown Relationships

By considering conditions in a single observation well, the radius r becomes a constant and the basic modified equation becomes

$$s_2 - s_1 = \frac{2.3Q}{4\pi T} \ \log (t_2/t_1) \ \dots\dots\dots\dots\dots\dots\dots\dots\dots\dots .VI.9 \ (a)$$

$$\text{or } \Delta s = \frac{2.3Q}{4\pi T}$$

where Δs is the drawdown over one log cycle of time.

Fig. VI.20 is a typical straight line plot of field data from a pumping test in which drawdown from a single observation well is plotted against time on semi log paper. The transmissibility, T, can be determined directly as shown. In order to determine the storage coefficient, S, one recalls from Equation VI.7 that

$$S = \frac{4uTt}{7.48r^2}$$

A plot of the asymptote to the u vs. W(u) semi log plot shows that it intersects the value u = 0.565 when W(u) = 0, or in other words, when the drawdown is zero. Thus extending the straight line section of the field data plot back to the point of zero drawdown gives an initial time, t_o, at which u has the value of 0.565 and the storage coefficient is then given by

$$S = \frac{4 \times 0.565 Tt_o}{7.48r^2} = \frac{0.3Tt_o}{r^2}$$

The calculation of storage coefficient from typical field data is also shown in Fig. VI.20.

VI.8.3.2 Distance-Drawdown Relationships

If one considers the Basic Modified Equation at one particular time, t, the equation may be written

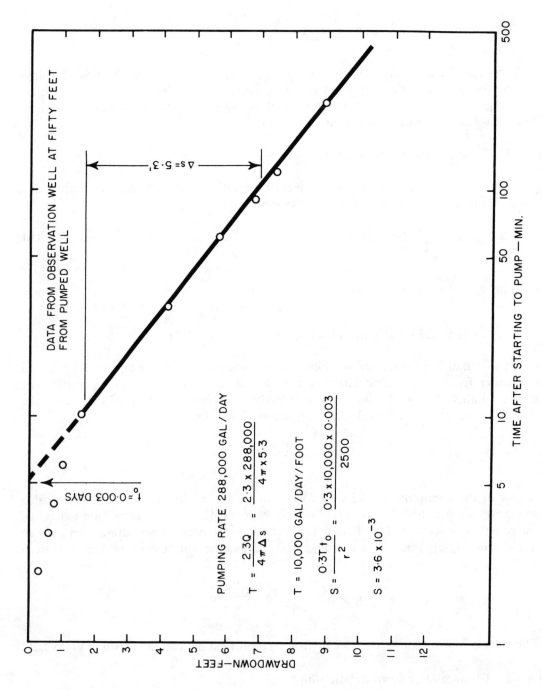

Fig. VI. 20 Time - Drawdown Plot—Field Data (Johnson, 1959)

$$s_2 - s_1 = \frac{2.3Q}{2 \pi T} \log (r_1/r_2) \quad \dots\dots\dots\dots\dots\dots\dots\dots\dots\dots \text{VI.9 (b)}$$

$$\text{or } \Delta s = \frac{2.3Q}{2\pi T}$$

where Δs is the drawdown over one log cycle of distance measured from the pumped well.

Fig. VI.21 shows typical field data from a number of observation wells at different distances from a pumped well, plotted as a straight line on semi log paper. The transmissibility and storage coefficient can be calculated directly as shown. It is interesting to note that the slope of the distance-drawdown curve is just double the slope of the time-drawdown curve discussed previously. Because this is true, it is obviously possible to construct distance-drawdown curves with the information from a single observation well for any time period desired. One has only to determine the slope of the distance drawdown curve, and construct a line at that slope through the co-ordinate represented by the observation well radius and the desired time.

Remembering that the drawdown at any time, and at any radius, is always proportional to the pumping rate, Q, one can construct distance-drawdown curves for any required pumping rate. For example, the following data are calculated from the original time-drawdown data shown in Fig. VI.20 for a time of 300 minutes.

Pumping Rate	Drawdown at 50 ft.	Δs
100 gpm	4.7 ft	2.6 ft
250 gpm	11.8 ft	6.3 ft
400 gpm	18.8 ft	10.6 ft

The curves represented by this table of data are shown in Fig. VI.22.

One of the main purposes in constructing distance-drawdown curves is to determine the interference effect between adjacent wells in an aquifer. Interference is actually a reduction in available drawdown at the site of a pumped well due to drawdown from a nearby well or wells. The reduction in yield due to interference can be calculated as the loss in available drawdown at the site multiplied by the specific capacity of the well, where specific capacity is the pumping rate divided by drawdown at the well.

VI.8.3.3   Modifications for Free Aquifer Conditions

To apply any of the preceding equations to free aquifer or water table conditions, the transmissibility must be replaced by hydraulic conductivity multiplied by the average thickness of saturated aquifer. Where s_1 and s_2 represent drawdowns at two different

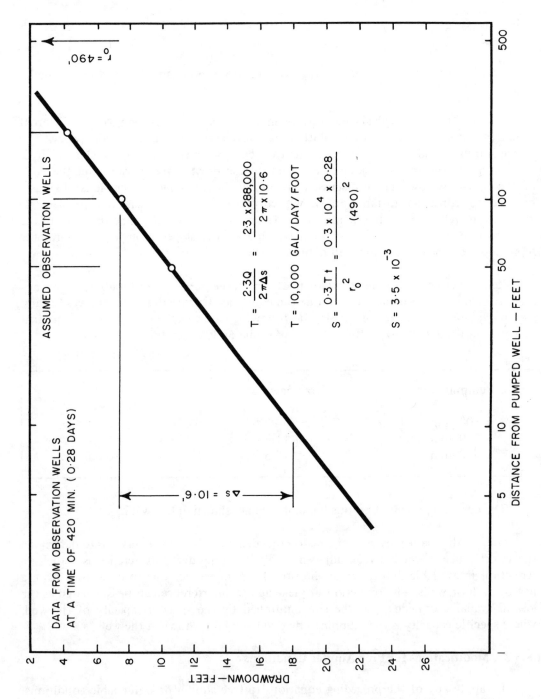

Fig. VI.21 Distance - Drawdown Plot—Field Data (Johnson, 1959)

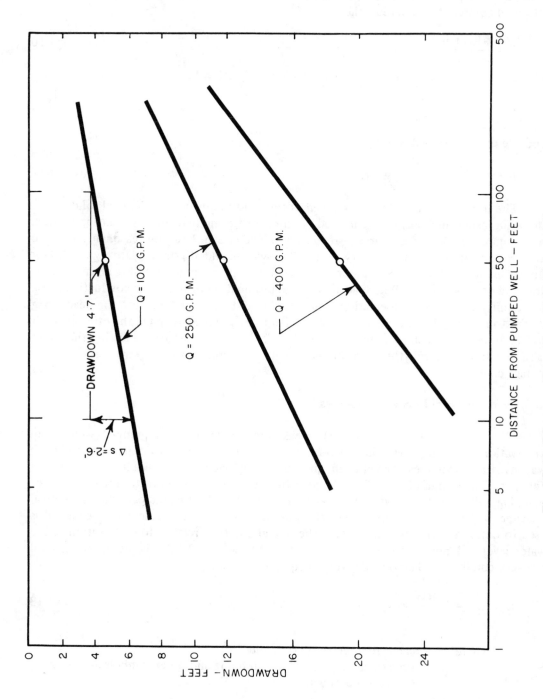

Fig. VI. 22 Distance - Drawdown Plot—Calculated

radii in the aquifer, and the symbols h_1 and h_2 are used to represent the depth of saturated aquifer at each radius, then

$$T = \frac{k(h_1 + h_2)}{2} \quad \dots\dots\dots\dots\dots\dots\dots\dots\dots\dots\dots\dots\dots\dots \text{VI.10}$$

where k is in gal/sq. ft./day. The time-drawdown equation becomes

$$h_1{}^2 - h_2{}^2 = \frac{2.3Q}{2\pi k} \log (t_2/t_1) \quad \dots\dots\dots\dots\dots\dots\dots\dots\dots\dots \text{VI.11}$$

and the distance-drawdown equation becomes

$$h_1{}^2 - h_2{}^2 = \frac{2.3Q}{\pi k} \log (r_1/r_2) \quad \dots\dots\dots\dots\dots\dots\dots\dots\dots\dots \text{VI.12}$$

The use of semi log plots of 'h²' versus time and radius permit identical solutions to those suggested previously for confined aquifers, providing that some precaution is taken to minimize the effect of curved stream lines. The stream lines are reasonably close to horizontal near the lower boundary of the flow system, and at points relatively far distant from the pumped well. The water surface in the pumped well is also an acceptable drawdown measurement. Where accuracy in determination of hydraulic conductivity of a free aquifer is required, observation wells should be designed to measure the head along the bottom stream lines, or at some distance from the pumped well. On the other hand, it will be noted that shallow observation wells will be most accurate in locating the water table position, and will therefore provide greater accuracy in the determination of the storage coefficient.

VI.8.3.4   Water-Level Recovery Analyses

The use of water-level recovery measurements made in a pumped well or an observation well can provide information on the aquifer characteristics in much the same way as drawdown measurements in an observation well were used in the preceding Sections. This analysis is based on the idea that recovery with time after the cessation of pumping will be identical with the changes in water level that would occur if an identical recharge well were superimposed on the pumped well. A term 'calculated recovery' is used to define the rise in water level in the well at any time relative to the position of the water level had pumping been continued to that time. Fig. VI.23 shows the water-level recovery conditions. The time drawdown equation becomes

$$s_2{}' - s_1{}' = \frac{2.3Q}{4\pi T} \log(t_2{}' / t_1{}') \quad \dots\dots\dots\dots\dots\dots\dots\dots\dots\dots \text{VI.13}$$

where s′ = calculated recovery, (ft), and
       t′ = time since pumping stopped, hours and other symbols as previously defined.

Obviously the calculated recovery will plot on semi log paper as a straight line, and calculation of transmissibility may be made as shown in Fig. VI.24. Adjustment for free

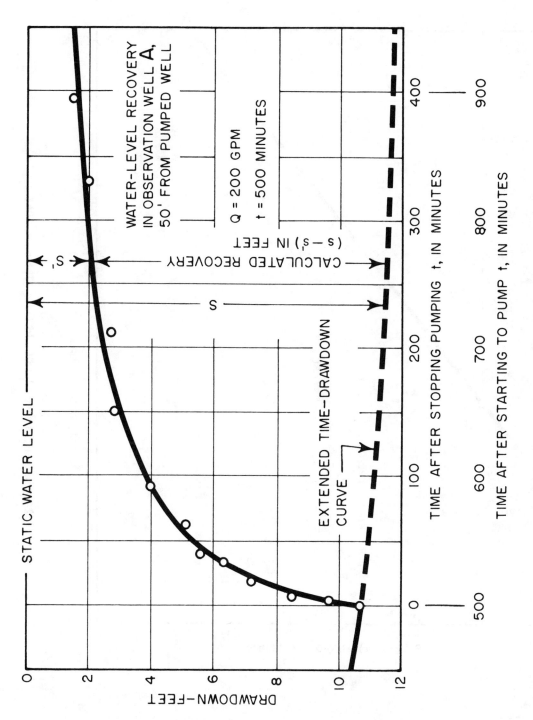

Fig. VI. 23 Water-Level Recovery Curve (Johnson, 1961a)

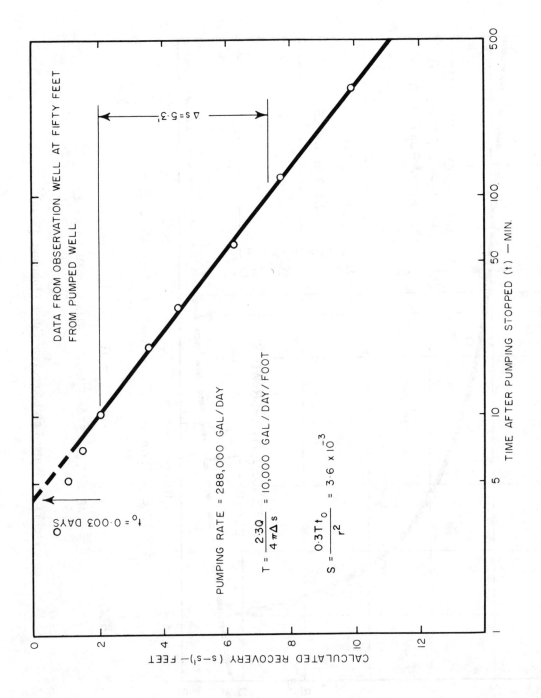

Fig. VI. 24 Time - Recovery Plot—Field Data (Johnson, 1961a)

aquifer conditions may be made as before by replacing transmissibility as

$$T = \frac{k\ (h_1 + h_2)}{2}$$

Further analyses of recovery data may be made by plotting residual drawdown versus t/t' (see Fig. VI.25). On this plot, time is increasing towards the right, and theoretically the straight line plotted should pass through the point of zero drawdown at $t/t' = 1$. Reasons for this not occurring are due to variations in aquifer characteristics as indicated in Fig. VI.26.

VI.8.3.5   Boundary and Recharge Effects

Boundaries of an aquifer, or recharge sources, within the range of well drawdown limits, are excluded from the previous analyses through the basic assumptions listed. The occurrence of such anomalies, however, have a predictable effect on the shape of measured time-drawdown curves from a pumped well or observation well. Where a source of recharge, such as a lake or stream channel is intercepted by the spreading radius of influence of a pumped well, the drawdown will become steady, that is the slope of the time-drawdown curve will change to zero. Thus it is evident that a decrease in the slope of a measured time-drawdown curve is evidence of a recharge source. As will be shown subsequently in the Section 'Boundary Delineation', the presence of an aquifer boundary, within the radius of influence of a pumped well, will result in a doubling of the slope of the time-drawdown curve. Thus any sudden increase in the slope of measured time-drawdown curves is evidence of the occurrence of an aquifer boundary or boundaries.

VI.8.4   Boundary Delineation

In developing formulae applicable to pumped aquifers, the idea of infinite areal extent has been adopted. Few major aquifers are so large that for practical purposes they can be considered as infinite in extent; therefore, for most aquifers some adjustment in procedure of analysis is required to account for the presence of geologic boundaries. It is recognized that most boundaries do not occur as straight line and abrupt changes, but as tapered or irregular terminals. However, since the area affected by a pumped well is likely to be relatively large in comparison with the area encompassed by a boundary, it is often possible to treat the boundary as an abrupt discontinuity. Two procedures, recognized as methods for handling boundary conditions, are the Method of Images and the Aquifer Limits Test.

VI.8.4.1   Method of Images

The Method of Images is based on the fact that a boundary can be simulated in an aquifer of infinite area extent by the use of image wells, as follows:

1.   For the case of a line recharge source, the real and bounded aquifer is replaced by an imaginary recharging well, which is placed equidistant to the real well on

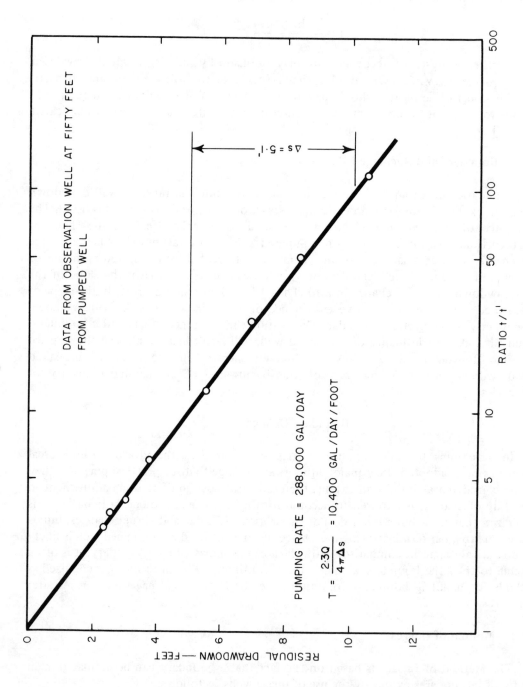

Fig. VI. 25 Residual Drawdown vs. t/t′–Field Data (Johnson, 1961b)

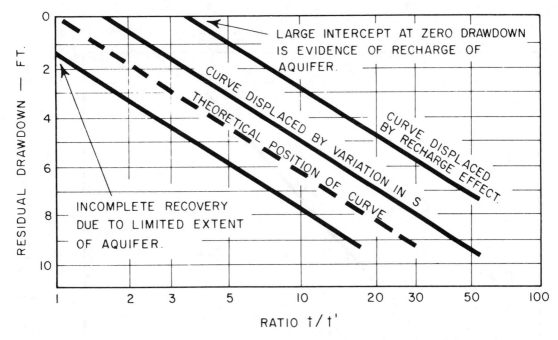

Fig. VI. 26 Displacement of Residual Drawdown (Johnson, 1961b)

the opposite side of the boundary. The recharge well returns water to the aquifer at the same rate as it is withdrawn, so that no drawdown occurs at the boundary. Therefore the limits of the real problem are met.

2.   For the case of an impermeable aquifer boundary, a pumped image well equidistant and on the opposite side of the boundary will result in equal drawdown at the boundary from each well. This means that there can be no flow across the boundary and the limits of the real problem are satisfied.

Fig. VI.27 shows a typical time-drawdown curve for an observation well in which the effect of an impermeable boundary has become evident. The cone of influence from the pumped well reaches the observation well at time t_o. The cone of influence from the image well reaches the observation well at time t'_o. Recalling that the storage coefficient is constant and is defined by

$$S = \frac{0.3T\, t_o}{r^2} \text{ for both wells, then}$$

$$\frac{t_o}{r^2_R} = \frac{t_o'}{r^2_I} \quad \dots\dots\dots\dots\dots\dots\dots\dots\dots\dots\dots\dots\dots\dots\dots\dots\dots\dots\dots\dots\dots\dots VI.14$$

where r^2_R = distance between the real well and observation well, (ft), and

   r^2_I = distance between the image well and observation well, (ft).

Thus one can determine the distance from the observation well to the image well. If more than one observation well is available, the actual location of the image well may be

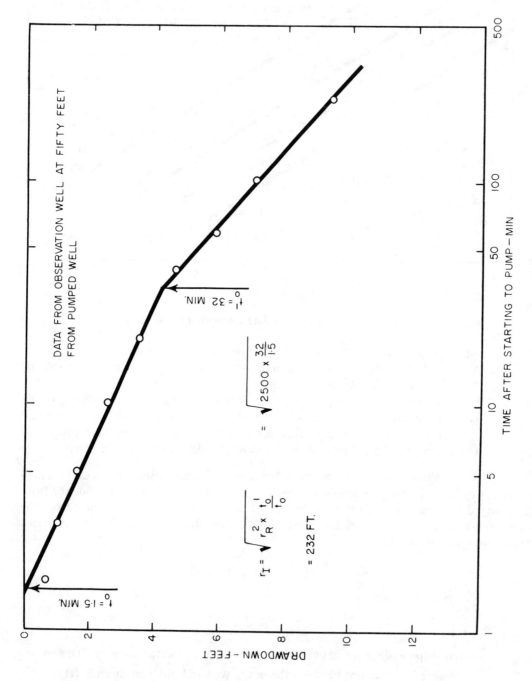

Fig. VI. 27 Time - Drawdown Plot—Boundary Effect

determined, and the position of the boundary is then established as midway between, and at right angles to, the line connecting the pumped well and the image well. The actual drawdown that occurs at any point in the aquifer affected by a single boundary can be determined by adding the drawdown from both the real and image wells at that point.

When more than one boundary condition affects the drawdown for a pumped well, the number of image wells required to simulate the actual conditions increase very rapidly, and the analysis becomes extremely complicated Any attempt to locate or define several boundaries on the basis of pumping tests is likely to be of limited practical value. A few multi-boundary conditions are discussed further in *Theory of Aquifer Tests*, Ferris (1962).

VI.8.4.2 Aquifer Limits Test

The Aquifer Limits Test is designed to provide information relative to aquifers of limited areal extent. The test is based upon the assumption that there will be three distinct periods related to flow conditions in pumping from an aquifer of limited size. These are

(1) an initial time during which the cone of influence of the pumped well is expanding and has not reached any boundary;

(2) a transitional period during which more and more boundary is contacted by the expanding cone of influence; and

(3) a semi-steady state period when all of the aquifer is contributing to the flow to the well.

A semi log plot of time versus drawdown will show a straight line section for the initial period with curvature occurring as soon as any boundary is contacted. An arithmetic plot of the same data will show continuous curvature until the semi-steady state condition occurs, and thereafter will be a straight line. Fig. VI.28 shows typical plots as described. The straight line section of the semi log plot can be used to determine transmissibility and storage coefficient.

During the semi-steady state period the production is

$$Q\Delta t = S \times A \times \Delta s \ldots\ldots\ldots\ldots\ldots\ldots\ldots\ldots\ldots\ldots\ldots VI.15$$

where   A = area of the aquifer (sq ft).
Δs = change in head during time, Δt, and
Q and S as previously defined.

Therefore the slope of the straight line section of the arithmetic plot is

$$\frac{\Delta s}{\Delta t} = \frac{Q}{SA}$$

and the areal extent of the aquifer, A, can be determined if the storage coefficient is known.

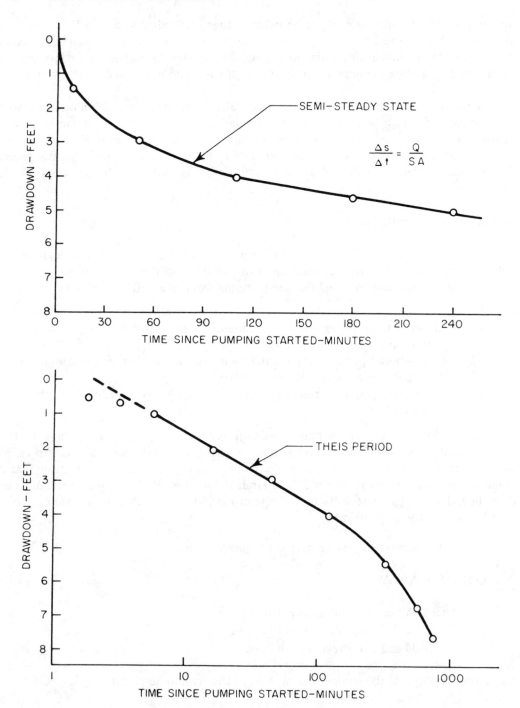

Fig. VI. 28 Arithmetic and Log Plots—Aquifer Limits Test.

VI.8.5   Leaky Artesian Aquifers

In many field situations the condition of a completely impermeable confining bed does not occur. Instead there is frequently percolation of water through the confining bed of sufficient magnitude to affect the hydraulics of flow toward pumped wells in the aquifer.

The data collected during aquifer tests may be analyzed by means of the leaky artesian formula (Hantush and Jacob, 1955). The leaky artesian formula may be written as

$$s = (114.6Q/T)\,W(u, r/B) \dots\dots\dots\dots\dots\dots\dots\dots\dots\dots VI.16$$

where

$$u = 7.48\ r^2 S/4Tt \dots\dots\dots\dots\dots\dots\dots\dots\dots\dots\dots VI.17$$

and

$$r/B = r/\sqrt{T/(k'_z/b')} \dots\dots\dots\dots\dots\dots\dots\dots\dots\dots VI.18$$

where s = drawdown in observation well, (ft),
 r = distance from pumped well to observation well, (ft),
 Q = discharge, (gpm),
 t = time after pumping started, (days),
 T = coefficient of transmissibility, (gpd/ft),
 S = coefficient of storage of aquifer,
 k'_z = coefficient of vertical hydraulic conductivity of the confining bed, (gpd/ft), and
 b' = thickness of confining bed through which leakage occurs, (ft), and
$W(u, r/B)$ is read as the well function for leaky artesian aquifers.

The leaky artesian formula was developed on the basis of the following assumptions that

(1)   the aquifer is infinite in areal extent and is of the same thickness throughout;
(2)   it is homogeneous and isotropic;
(3)   it is confined between an impermeable bed and a bed through which leakage can occur;
(4)   the coefficient of storage is constant;
(5)   water is released from storage instantaneously with a decline in head;
(6)   the well has an infinitesimal diameter and penetrates the entire thickness of the formation;
(7)   leakage through the confining bed into the aquifer is vertical and proportional to the drawdown;
(8)   the hydraulic head in the deposits supplying leakage remains more or less uniform;
(9)   the flow is vertical in the confining bed and horizontal in the aquifer; and
(10) the storage in the confining bed is neglected.

The leaky artesian formula may be solved by the following method which is a modification of the type curve graphical method devised by Theis. Values of W(u,r/B) in terms of the practical range of u and r/B are presented in tabular form in Table VI.3. Values of W(u,r/B) are plotted against values of l/u on logarithmic paper, and a family of leaky artesian type curves is constructed. Values of s plotted on logarithmic paper of the same scale as the type curves against values of t describe a time-drawdown field data curve that is analogous to one of the family of leaky artesian type curves.

The time-drawdown field data curve is superimposed on the family of leaky artesian type curves, keeping the W(u,r/B) axis parallel with the s axis and the l/u axis parallel with the t axis. In the matched position a point at the intersection of the major axes of the leaky artesian type curve is selected and marked on the time-drawdown field data curve (the point may also be selected anywhere on the type curve). The co-ordinates of this common point (match point) W(u,r/B), l/u, s and t, are substituted into the preceding equations to determine T, k'_z and S.

VI.9   GROUNDWATER INVENTORIES

Studies of groundwater basins are made to define the complete groundwater system in terms of the potential for development. This involves determining the location and extent of aquifers, aquifer characteristics, recharge potential and the effects of groundwater development. Some of the procedures that may be used to delineate the occurrence and character of groundwater supplies have been outlined in the preceding portion of this Section. Since the time involved and the cost of undertaking various components of a complete study on a groundwater basin can be prohibitive, the hydrologist often has to limit the extent of his investigation. Under these circumstances the aim of an investigation must be to recognize and record data relevant to factors that may change with development, and to limit the assessment of unchanging factors to those absolutely necessary to the proposed development of the groundwater supply. The following outline is thus presented as a general guide, or logical series of steps, that may be followed in building up a complete inventory of a groundwater basin, and within which the hydrologist may develop an investigational program as time and finances permit.

The term *groundwater basin* is used here to distinguish an area of particular interest as far as the groundwater regime is concerned. However, in most cases the idea of an isolated groundwater system is not likely to be found in nature; in fact, any groundwater study most likely involves consideration of a 'leaky' basin in which more or less flow occurs into and out of the designated area.

1.   *Topographic Mapping and Area Delineation*—the first step in any basin study will be to delineate the area to be studied on the basis of surface topography and flow systems. The groundwater basin may or may not be coincident with a surface basin; in fact, subsurface flows into and out of the area are most likely to occur, but in many cases the groundwater system below a surface basin is treated as a single groundwater unit and as such can be evaluated in detail.

2. *Inflow-Outflow Estimates*—valuable information on groundwater supplies may be obtained through consideration of the water balance within the defined basin. Records of surface flow and precipitation are normally available, estimates of evapotranspiration can be made, and fluctuations in groundwater levels may often be determined from existing wells or through installation of a limited number of simple observation wells. The balance of water indicated by such information is indicative of the recharge or discharge of the groundwater system. It is desirable that information of this type be obtained prior to groundwater development, which would mask the natural fluctuations in groundwater level. Many fluctuations in groundwater levels or piezometric heads, due to locally changing conditions, occur over relatively short periods of time. Such fluctuations may represent only redistribution of water on a temporary basis, and would be extremely difficult to analyze as well as having little meaning in terms of overall flow systems. Inflow-outflow measurement to be of any real value should be made over seasonal, or longer, periods of time.

3. *Geologic Mapping*—some mapping of geologic formations of an area can usually be done using available information from sources such as surface soil maps, aerial photographs, regional geologic maps and logs of existing wells. A test drilling program will likely be required to fill in some missing detail, and considerable care should be taken in laying out such a program to ensure that maximum results are obtained from a limited number of test holes. Consideration in the location of test holes should include the following:

    (a) To provide missing information on the event, depth and thickness of formations of interest to the study.
    (b) Locations should be such that test holes can be developed into observation wells to record groundwater fluctuations.
    (c) Test holes should provide samples of various formations that can be used to estimate aquifer characteristics.
    (d) The pattern of test holes should provide a broad covering network adaptable to the technique of contouring, which is used with both geologic mapping and mapping groundwater levels.
    (e) It is likely that additional test holes, to those planned in the initial stages, will be required eventually, so the initial installations should be spread as broadly as possible over the area.

4. *Mapping Groundwater Flow Systems*—the mapping of groundwater flow systems consists of the preparation of groundwater level contour maps and appropriate cross-sections showing hydraulic gradients. In general, the problem of locating suitable observation wells is similar to that of locating geologic test holes, and it is unlikely that more test holes than those required for geologic information would be required, particularly for deeper confined flow systems. Some additional information may be required on the water table position, but installations for this purpose will normally require relatively shallow and less costly installations. Considerations in the installation of piezometers for

Table VI.3   Values of W(u,r/B)   (Walton, 1962)

u \ r/B	0	0.001	0.002	0.003	0.004	0.005	0.006	0.007	0.008	0.009	0.01	0.015	0.02	0.025	0.03	0.035	0.04	0.045	0.05	0.055	0.06	0.065	0.07	0.075	0.08	0.085	0.09
0	∞	14.0474	12.6611	11.8502	11.2748	10.8286	10.4640	10.1557	9.8887	9.6532	9.4425	8.6319	8.0569	7.6111	7.2471	6.9394	6.6731	6.4383	6.2285	6.0388	5.8658	5.7067	5.5596	5.4228	5.2950	5.1750	5.0620
0.00001	13.2383	13.0051	12.4417	11.8153	11.2711	10.8283	10.4640	10.1557	9.8887																		
0.00002	12.5451	12.4240	12.1013	11.6716	11.2259	10.8174	10.4619	10.1554	9.8886	9.6532																	
0.00003	12.1397	12.0581	11.8322	11.5098	11.1462	10.7849	10.4509	10.1523	9.8879	9.6530	9.4425																
0.00004	11.8520	11.7905	11.6168	11.3597	11.0555	10.7374	10.4291	10.1436	9.8849	9.6521	9.4422																
0.00005	11.6289	11.5795	11.4384	11.2248	10.9642	10.6822	10.3993	10.1290	9.8786	9.6496	9.4413																
0.00006	11.4465	11.4053	11.2866	11.1040	10.8764	10.6240	10.3640	10.1094	9.8686	9.6450	9.4394																
0.00007	11.2924	11.2570	11.1545	10.9951	10.7933	10.5652	10.3255	10.0862	9.8555	9.6382	9.4361	8.6319															
0.00008	11.1589	11.1279	11.0377	10.8962	10.7151	10.5072	10.2854	10.0602	9.8398	9.6292	9.4313	8.6318															
0.00009	11.0411	11.0135	10.9330	10.8059	10.6416	10.4508	10.2446	10.0324	9.8219	9.6182	9.4251	8.6316															
0.0001	10.9357	10.9109	10.8382	10.7228	10.5725	10.3963	10.2038	10.0034	9.8024	9.6059	9.4176	8.6313	8.0569														
0.0002	10.2426	10.2301	10.1932	10.1332	10.0522	9.9530	9.8386	9.7126	9.5781	9.4383	9.2961	8.6152	8.0558	7.6111	7.2471												
0.0003	9.8371	9.8288	9.8041	9.7635	9.7081	9.6392	9.5583	9.4671	9.3674	9.2611	9.1499	8.5737	8.0483	7.6101	7.2470												
0.0004	9.5495	9.5432	9.5246	9.4940	9.4520	9.3992	9.3366	9.2653	9.1863	9.1009	9.0102	8.5168	8.0320	7.6069	7.2465	6.9394	6.6731										
0.0005	9.3263	9.3213	9.3064	9.2818	9.2480	9.2052	9.1542	9.0957	9.0304	8.9591	8.8827	8.4533	8.0080	7.6000	7.2450	6.9391	6.6730										
0.0006	9.1440	9.1398	9.1274	9.1069	9.0785	9.0426	8.9996	8.9500	8.8943	8.8332	8.7673	8.3880	7.9786	7.5894	7.2419	6.9384	6.6729	6.4383									
0.0007	8.9899	8.9863	8.9756	8.9580	8.9336	8.9027	8.8654	8.8224	8.7739	8.7204	8.6625	8.3233	7.9456	7.5754	7.2371	6.9370	6.6726	6.4382	6.2285								
0.0008	8.8563	8.8532	8.8439	8.8284	8.8070	8.7798	8.7470	8.7090	8.6661	8.6186	8.5669	8.2603	7.9105	7.5589	7.2305	6.9347	6.6719	6.4381	6.2284								
0.0009	8.7386	8.7358	8.7275	8.7138	8.6947	8.6703	8.6411	8.6071	8.5686	8.5258	8.4792	8.1996	7.8743	7.5402	7.2222	6.9316	6.6709	6.4378	6.2283								
0.001	8.6332	8.6308	8.6233	8.6109	8.5937	8.5717	8.5453	8.5145	8.4796	8.4407	8.3983	8.1414	7.8375	7.5199	7.2122	6.9273	6.6693	6.4372	6.2282	6.0388	5.8658	5.7067	5.5596	5.4228	5.2950		
0.002	7.9402	7.9390	7.9352	7.9290	7.9203	7.9092	7.8958	7.8800	7.8619	7.8416	7.8192	7.6780	7.4972	7.2898	7.0685	6.8439	6.6242	6.4143	6.2173	6.0338	5.8637	5.7059	5.5593	5.4227	5.2949	5.1750	5.0620
0.003	7.5348	7.5340	7.5315	7.5274	7.5216	7.5141	7.5051	7.4945	7.4823	7.4686	7.4534	7.3562	7.2281	7.0759	6.9068	6.7276	6.5444	6.3623	6.1848	6.0145	5.8527	5.6999	5.5562	5.4212	5.2942	5.1747	5.0619
0.004	7.2472	7.2466	7.2447	7.2416	7.2373	7.2317	7.2249	7.2169	7.2078	7.1974	7.1859	7.1119	7.0128	6.8929	6.7567	6.6088	6.4538	6.2955	6.1373	5.9818	5.8309	5.6860	5.5476	5.4160	5.2912	5.1730	5.0610
0.005	7.0242	7.0237	7.0222	7.0197	7.0163	7.0118	7.0063	6.9999	6.9926	6.9843	6.9750	6.9152	6.8346	6.7357	6.6219	6.4964	6.3626	6.2236	6.0821	5.9406	5.8011	5.6648	5.5330	5.4062	5.2848	5.1689	5.0585
0.006	6.8420	6.8416	6.8403	6.8383	6.8353	6.8316	6.8271	6.8218	6.8156	6.8086	6.8009	6.7508	6.6828	6.5988	6.5011	6.3923	6.2748	6.1512	6.0239	5.8948	5.7658	5.6383	5.5134	5.3921	5.2749	5.1621	5.0539
0.007	6.6879	6.6876	6.6865	6.6848	6.6823	6.6790	6.6752	6.6706	6.6653	6.6594	6.6527	6.6096	6.5508	6.4777	6.3923	6.2962	6.1917	6.0807	5.9652	5.8468	5.7274	5.6081	5.4902	5.3745	5.2618	5.1526	5.0471
0.008	6.5545	6.5542	6.5532	6.5517	6.5495	6.5467	6.5433	6.5393	6.5347	6.5295	6.5237	6.4858	6.4340	6.3695	6.2935	6.2076	6.1136	6.0129	5.9073	5.7982	5.6873	5.5755	5.4642	5.3542	5.2461	5.1406	5.0381
0.009	6.4368	6.4365	6.4357	6.4344	6.4324	6.4299	6.4269	6.4233	6.4192	6.4146	6.4094	6.3757	6.3294	6.2716	6.2032	6.1256	6.0401	5.9481	5.8509	5.7500	5.6465	5.5416	5.4364	5.3317	5.2282	5.1266	5.0272
0.01	6.3315	6.3313	6.3305	6.3293	6.3276	6.3253	6.3226	6.3194	6.3157	6.3115	6.3069	6.2765	6.2347	6.1823	6.1202	6.0494	5.9711	5.8864	5.7965	5.7026	5.6058	5.5071	5.4075	5.3078	5.2087	5.1109	5.0133
0.02	5.6394	5.6393	5.6389	5.6383	5.6374	5.6363	5.6350	5.6334	5.6315	5.6294	5.6271	5.6118	5.5907	5.5638	5.5314	5.4939	5.4516	5.4047	5.3538	5.2991	5.2411	5.1803	5.1170	5.0517	4.9848	4.9166	4.8475
0.03	5.2349	5.2348	5.2346	5.2342	5.2336	5.2329	5.2320	5.2310	5.2297	5.2283	5.2267	5.2166	5.2025	5.1845	5.1627	5.1373	5.1084	5.0762	5.0408	5.0025	4.9615	4.9180	4.8722	4.8243	4.7746	4.7234	4.6707
0.04	4.9482	4.9482	4.9480	4.9477	4.9472	4.9467	4.9460	4.9453	4.9443	4.9433	4.9421	4.9345	4.9240	4.9105	4.8941	4.8749	4.8530	4.8286	4.8016	4.7722	4.7406	4.7068	4.6710	4.6335	4.5942	4.5533	4.5111
0.05	4.7261	4.7260	4.7259	4.7256	4.7253	4.7249	4.7244	4.7237	4.7230	4.7222	4.7212	4.7152	4.7068	4.6960	4.6829	4.6675	4.6499	4.6302	4.6084	4.5846	4.5590	4.5314	4.5022	4.4713	4.4389	4.4050	4.3699
0.06	4.5448	4.5448	4.5447	4.5444	4.5441	4.5438	4.5433	4.5428	4.5422	4.5415	4.5407	4.5357	4.5287	4.5197	4.5088	4.4960	4.4814	4.4649	4.4467	4.4267	4.4051	4.3819	4.3573	4.3311	4.3036	4.2747	4.2446
0.07	4.3916	4.3916	4.3915	4.3913	4.3910	4.3908	4.3904	4.3899	4.3894	4.3888	4.3882	4.3839	4.3779	4.3702	4.3609	4.3500	4.3374	4.3233	4.3077	4.2905	4.2719	4.2518	4.2305	4.2078	4.1839	4.1588	4.1326
0.08	4.2591	4.2590	4.2590	4.2588	4.2586	4.2583	4.2580	4.2576	4.2572	4.2567	4.2561	4.2524	4.2471	4.2404	4.2323	4.2228	4.2118	4.1994	4.1857	4.1707	4.1544	4.1368	4.1180	4.0980	4.0769	4.0547	4.0315
0.09	4.1423	4.1423	4.1422	4.1420	4.1418	4.1416	4.1413	4.1410	4.1406	4.1401	4.1396	4.1363	4.1317	4.1258	4.1186	4.1101	4.1004	4.0894	4.0772	4.0638	4.0493	4.0336	4.0169	3.9991	3.9802	3.9603	3.9395
0.1	4.0379	4.0379	4.0378	4.0377	4.0375	4.0373	4.0371	4.0368	4.0364	4.0360	4.0356	4.0326	4.0285	4.0231	4.0167	4.0091	4.0003	3.9903	3.9795	3.9675	3.9544	3.9403	3.9252	3.9091	3.8920	3.8741	3.8552
0.2	3.3547	3.3547	3.3547	3.3546	3.3545	3.3544	3.3543	3.3542	3.3540	3.3538	3.3536	3.3521	3.3502	3.3476	3.3444	3.3408	3.3365	3.3317	3.3264	3.3205	3.3141	3.3071	3.2997	3.2917	3.2832	3.2742	3.2647
0.3	2.9591	2.9591	2.9591	2.9591	2.9590	2.9590	2.9589	2.9589	2.9588	2.9587	2.9584	2.9575	2.9562	2.9545	2.9523	2.9501	2.9474	2.9444	2.9409	2.9370	2.9329	2.9284	2.9235	2.9183	2.9127	2.9069	2.9007
0.4	2.6813	2.6812	2.6812	2.6812	2.6812	2.6811	2.6810	2.6810	2.6809	2.6808	2.6807	2.6800	2.6791	2.6779	2.6765	2.6747	2.6727	2.6705	2.6680	2.6652	2.6622	2.6589	2.6553	2.6515	2.6475	2.6432	2.6386
0.5	2.4679	2.4679	2.4679	2.4679	2.4678	2.4678	2.4678	2.4677	2.4676	2.4676	2.4675	2.4670	2.4662	2.4653	2.4642	2.4628	2.4613	2.4595	2.4576	2.4554	2.4531	2.4505	2.4478	2.4448	2.4416	2.4383	2.4347
0.6	2.2953	2.2953	2.2953	2.2953	2.2952	2.2952	2.2952	2.2952	2.2951	2.2950	2.2950	2.2945	2.2940	2.2932	2.2923	2.2912	2.2900	2.2885	2.2870	2.2853	2.2833	2.2812	2.2790	2.2766	2.2740	2.2713	2.2684
0.7	2.1508	2.1508	2.1508	2.1508	2.1508	2.1508	2.1507	2.1507	2.1507	2.1506	2.1506	2.1502	2.1497	2.1491	2.1483	2.1474	2.1464	2.1452	2.1439	2.1424	2.1408	2.1391	2.1372	2.1352	2.1331	2.1308	2.1284
0.8	2.0269	2.0269	2.0269	2.0269	2.0269	2.0269	2.0269	2.0268	2.0268	2.0267	2.0260	2.0255	2.0249	2.0240	2.0231	2.0221	2.0210	2.0198	2.0184	2.0169	2.0153	2.0136	2.0118	2.0099	2.0078		
0.9	1.9187	1.9187	1.9187	1.9187	1.9187	1.9187	1.9187	1.9186	1.9186	1.9186	1.9185	1.9183	1.9179	1.9174	1.9169	1.9162	1.9154	1.9146	1.9136	1.9125	1.9114	1.9101	1.9087	1.9072	1.9056	1.9040	1.9022
1	1.8229	1.8229	1.8229	1.8229	1.8229	1.8229	1.8229	1.8228	1.8228	1.8228	1.8227	1.8225	1.8222	1.8218	1.8213	1.8207	1.8200	1.8193	1.8184	1.8175	1.8164	1.8153	1.8141	1.8128	1.8114	1.8099	1.8084
2	1.2227	1.2226	1.2226	1.2226	1.2226	1.2226	1.2226	1.2226	1.2226	1.2226	1.2225	1.2224	1.2222	1.2220	1.2218	1.2215	1.2212	1.2209	1.2205	1.2201	1.2196	1.2192	1.2186	1.2181	1.2175	1.2168	
3	0.9057	0.9057	0.9057	0.9057	0.9057	0.9057	0.9057	0.9057	0.9056	0.9056	0.9056	0.9056	0.9055	0.9054	0.9053	0.9052	0.9050	0.9049	0.9047	0.9045	0.9043	0.9040	0.9038	0.9035	0.9032	0.9029	0.9025
4	7024	7024	7024	7024	7024	7024	7024	7024	7024	7024	7024	7023	7023	7022	7021	7020	7019	7018	7016	7015	7014	7012	7010	7008	7006	7004	
5	5598	5598	5598	5598	5598	5598	5598	5598	5598	5598	5598	5597	5597	5597	5596	5596	5595	5595	5594	5593	5592	5591	5590	5588	5587	5586	5584
6	4544	4544	4544	4544	4544	4544	4544	4544	4544	4544	4544	4543	4543	4543	4542	4542	4542	4541	4540	4540	4539	4538	4537	4536	4535	4534	
7	3738	3738	3738	3738	3738	3738	3738	3738	3738	3738	3738	3737	3737	3737	3736	3735	3735	3734	3733	3733	3732	3731					
8	3106	3106	3106	3106	3106	3106	3106	3106	3106	3106	3106	3106	3106	3105	3105	3105	3104	3104	3103	3103	3102	3102	3101	3101			
9	2602	2602	2602	2602	2602	2602	2602	2602	2602	2602	2602	2602	2601	2601	2601	2601	2601	2600	2600	2600	2599	2599	2599	2598	2598		
1.0	0.2194	0.2194	0.2194	0.2194	0.2194	0.2194	0.2194	0.2194	0.2194	0.2194	0.2194	0.2194	0.2194	0.2194	0.2194	0.2193	0.2193	0.2193	0.2193	0.2193	0.2192	0.2192	0.2192	0.2191	0.2191	0.2191	0.2191
2.0	489	489	489	489	489	489	489	489	489	489	489	489	489	489	489	489	489	489	489	489	489	489	489	489	489	489	
3.0	130	130	130	130	130	130	130	130	130	130	130	130	130	130	130	130	130	130	130	130	130	130	130	130	130	130	
4.0	38	38	38	38	38	38	38	38	38	38	38	38	38	38	38	38	38	38	38	38	38	38	38	38	38	38	
5.0	11	11	11	11	11	11	11	11	11	11	11	11	11	11	11	11	11	11	11	11	11	11	11	11	11	11	
6.0	4	4	4	4	4	4	4	4	4	4	4	4	4	4	4	4	4	4	4	4	4	4	4	4	4	4	
7.0	1	1	1	1	1	1	1	1	1	1	1	1	1	1	1	1	1	1	1	1	1	1	1	1	1	1	
8.0	0	0	0	0	0	0	0	0	0	0	0	0	0	0	0	0	0	0	0	0	0	0	0	0	0	0	

Table VI.3 (Cont'd)   Values of W(u,r/B)   (Walton, 1962)

	0.15	0.2	0.25	0.3	0.35	0.4	0.45	0.5	0.55	0.6	0.65	0.7	0.75	0.8	0.85	0.9	0.95	1.0	1.5	2.0	2.5	3.0	3.5	4.0	4.5	5.0	6.0	7.0	8.0	9.0
041	4.0601	3.5054	3.0830	2.7449	2.4654	2.2291	2.0258	1.8488	1.6981	1.5550	1.4317	1.3210	1.2212	1.1307	1.0485	0.9735	0.9049	0.8420	0.4276	0.2278	0.1247	0.0695	0.0392	0.0223	0.0128	0.0074	0.0025	0.0008	0.0003	0.0001
41																														
39																														
30																														
40	4.0601																													
78	4.0600																													
30	4.0599																													
68	4.0598																													
92	4.0595	3.5054																												
79	4.0435	3.5043	3.0830	2.7449																										
22	4.0092	3.4969	3.0821	2.7448																										
30	3.9551	3.4806	3.0788	2.7444	2.4654	2.2291																								
60	3.8821	3.4567	3.0719	2.7428	2.4651	2.2290																								
42	3.8384	3.4274	3.0614	2.7398	2.4644	2.2289	2.0258																							
71	3.7529	3.3947	3.0476	2.7350	2.4630	2.2286	2.0257																							
22	3.6903	3.3598	3.0311	2.7284	2.4608	2.2279	2.0256	1.8488																						
52	3.6302	3.3239	3.0126	2.7202	2.4576	2.2269	2.0253	1.8487																						
50	3.5725	3.2875	2.9925	2.7104	2.4534	2.2253	2.0248	1.8486	1.6931	1.5550	1.4317	1.3210	1.2212	1.1307	1.0485															
42	3.1158	2.9521	2.7658	2.5688	2.3713	2.1809	2.0023	1.8379	1.6883	1.5530	1.4309	1.3207	1.2210	1.1306	1.0484	0.9735	0.9049													
73	2.8017	2.6896	2.5571	2.4110	2.2578	2.1031	1.9515	1.8062	1.6695	1.5423	1.4251	1.3177	1.2195	1.1299	1.0481	9733	.9048	0.8420												
88	2.5655	2.4816	2.3802	2.2661	2.1431	2.0155	1.8869	1.7603	1.6379	1.5213	1.4117	1.3094	1.2146	1.1270	1.0465	9724	9044	8418												
71	2.3776	2.3110	2.2299	2.1371	2.0356	1.9283	1.8181	1.7075	1.5985	1.4927	1.3914	1.2955	1.2052	1.1210	1.0426	9700	9029	8409												
22	2.2218	2.1673	2.1002	2.0227	1.9369	1.8452	1.7497	1.6524	1.5551	1.4593	1.3663	1.2770	1.1919	1.1116	1.0362	9657	9001	8391												
32	2.0894	2.0435	1.9867	1.9206	1.8469	1.7673	1.6835	1.5973	1.5101	1.4232	1.3380	1.2551	1.1754	1.0993	1.0272	9593	8956	8360	0.4276											
34	1.9745	1.9351	1.8861	1.8290	1.7646	1.6947	1.6206	1.5436	1.4650	1.3860	1.3078	1.2310	1.1564	1.0847	1.0161	9510	8895	8316	4275											
83	1.8732	1.8389	1.7961	1.7460	1.6892	1.6272	1.5609	1.4918	1.4206	1.3486	1.2766	1.2054	1.1358	1.0682	1.0032	9411	8819	8259	4274											
50	1.7829	1.7527	1.7149	1.6704	1.6198	1.5644	1.5048	1.4422	1.3774	1.3115	1.2451	1.1791	1.1140	1.0505	0.9890	0.9297	0.8730	0.8190	0.4271	0.2278										
55	1.2066	1.1944	1.1789	1.1602	1.1387	1.1145	1.0879	1.0592	1.0286	0.9964	0.9629	0.9284	0.8932	0.8575	0.8216	7857	7501	7148	4135	2268	0.1247	0.0695								
18	0.8969	0.8902	0.8817	0.8713	0.8593	0.8457	0.8306	0.8142	0.7964	7775	7577	7369	7154	6932	6706	6476	6244	6010	3812	2211	1240	694								
00	6969	6927	6874	6809	6733	6647	6551	6446	6332	6209	6080	5943	5801	5653	5501	5345	5186	5024	3411	2096	1217	691	0.0192							
81	5561	5532	5496	5453	5402	5344	5278	5206	5128	5044	4955	4860	4761	4658	4550	4440	4326	4210	3007	1944	1174	681	390	0.0233						
32	4518	4498	4472	4441	4405	4364	4317	4266	4210	4150	4086	4018	3946	3871	3793	3712	3629	3543	2630	1774	1112	664	386	222	0.0128					
29	3719	3704	3685	3663	3636	3606	3572	3534	3493	3449	3401	3351	3297	3242	3183	3123	3060	2996	2292	1602	1040	639	379	221	127					
00	3092	3081	3067	3050	3030	3008	2982	2953	2922	2889	2853	2815	2774	2732	2687	2641	2592	2543	1994	1436	961	607	368	218	127	0.0074				
97	2591	2583	2572	2559	2544	2527	2507	2485	2461	2436	2408	2378	2347	2314	2280	2244	2207	2168	1734	1281	881	572	354	213	125	73				
90	0.2186	0.2179	0.2171	0.2161	0.2149	0.2135	0.2120	0.2103	0.2085	0.2065	0.2043	0.2020	0.1995	0.1970	0.1943	0.1914	0.1885	0.1855	0.1509	0.1139	0.0803	0.0534	0.0338	0.0207	0.0123	0.0073	0.0025			
88	488	487	486	485	484	482	480	477	475	473	470	467	463	460	456	452	448	444	394	335	271	210	156	112	77	51	21	0.0008	0.0003	
30	130	130	130	130	130	129	129	128	128	127	127	126	125	125	124	123	123	122	112	100	86	71	57	45	34	25	12	6	2	
38	38	38	38	38	38	38	37	37	37	37	37	37	37	36	36	36	34	31	27	24	20	16	13	10	6	3	2	0.0001		
11	11	11	11	11	11	11	11	11	11	11	11	11	11	11	11	11	11	10	10	9	8	7	6	5	4	2	1	1	0	
4	4	4	4	4	4	4	4	4	4	4	4	4	4	4	4	4	4	3	3	3	3	2	2	2	1	1	0			
1	1	1	1	1	1	1	1	1	1	1	1	1	1	1	1	1	1	1	1	1	1	1	1	1	0	0			
0	0	0	0	0	0	0	0	0	0	0	0	0	0	0	0	0	0	0	0	0	0	0	0	0	0	0			

mapping and evaluating flow systems within a basin might include the following:

(a)   Recharge areas—water table position and vertical gradients should be measured.

(b)   Discharge areas—water table position should be determined and additional piezometers will be required to measure vertical or horizontal gradients depending upon the type of discharge.

(c)   Basin outlet areas—where subsurface flow out of a basin is likely to occur, as for example in paralleling a surface outflow stream, additional piezometer installations will be required to evaluate the flow system. It should be noted that any surface flow-measuring structure will tend to pond water and to increase the subsurface flow.

(d)   A further consideration in the installation of piezometers or observation wells is the desirability of obtaining water samples for chemical analysis. As a general approach, one should attempt to obtain samples of water from all formations that occur at a given location, so that changes in water quality with depth can be assessed.

5.   *Determination of Aquifer Characteristics*—pump tests to determine aquifer characteristics are likely to be the most costly part of any groundwater investigation. Reasonable estimates of transmissibility and storage coefficient can be made from laboratory tests on representative samples of the various formations, and this approach should be used as fully as possible. Pump testing should then be considered only as a field check on the reliability of other aquifer characteristic determinations. As such, the extent of pump testing can be extremely limited; often one test will be sufficient, and this will still provide the type of information required to complete the study of a groundwater basin.

6.   *Safe Yield Determination*—in terms of the optimum development of a groundwater basin, the concept of *safe yield* will eventually become of prime importance. Safe yield is broadly defined as the maximum rate at which water can be removed from the basin without producing an undesirable result. For example, any lowering of the water table may be undesirable to a person who depends on a water-table well as a water supply. Conversely lowering the water table may tend to increase aquifer recharge and thereby increase the total available water supply for other users. A satisfactory balance between water withdrawal and groundwater levels can only be determined in the light of existing conditions on each individual groundwater basin.

VI.10   LITERATURE CITED

Brown, I.C. 1962. Chemical methods as an aid to hydrogeology. Proc. of Hydrology Symposium No. 3. N.R.C. Subcommittee on Hydrology. The Queen's Printer, Ottawa. pp 181-203.

Butler, S.S. 1957. Engineering Hydrology. Prentice-Hall Inc. Englewood Cliffs, New Jersey.

Chebotarev, I.I. 1955. Metamorphism of natural waters in the crust weathering. Geochemica et Cosmochimica Acta 8:22-48.

Christiansen, E.A. 1962. Hydrogeology of surficial and bedrock aquifers in Southern Saskatchewan. Proc. of Hydrology Symposium No. 3. N.R.C. Subcommittee on Hydrology. The Queen's Printer, Ottawa. pp. 49-66.

Ferris, J.G. *et al.,* 1962. Theory of aquifer tests. Geological Survey Water Supply Paper 1536 E. United States Govt. Printing Office, Washington, D.C.

Freeze, R.A. 1966. Theoretical analysis of regional groundwater flow. Scientific Series No. 3, Inland Waters Branch, Department of Energy, Mines and Resources, Ottawa.

Hantush, M.S. and Jacob, C.E. 1955. Non steady radial flow in an infinite leaky aquifer. Trans. Am. Geophys. Union 36:95-100.

Harr, M.E. 1962. Groundwater and seepage. McGraw-Hill Book Co., Inc., New York, N.Y.

Johnson, E.E. 1957. The Drillers Journal, *July-Aug.* E.E. Johnson Inc. St. Paul, Minnesota, U.S.A.

Johnson, E.E. 1959. "Factors Affecting Permeability", The Drillers Journal, *Jan - Feb.* E.E. Johnson Inc., St. Paul, Minnesota, U.S.A.

Johnson, E.E. 1961a "Analyzing Water Level Recovery Data". The Drillers Journal, *May-June. p. 9* E.E. Johnson Inc., St. Paul, Minnesota, U.S.A.

Johnson, E.E. 1961b. "Analyzing Water Level Recovery Data". The Drillers Journal, *July-Aug p. 10.* E.E. Johnson Inc., St. Paul, Minnesota, U.S.A.

Johnson, E.E. 1966a. "Some Basic Principles of Geology". The Drillers Journal, *July-Aug. p. 11.* E.E. Johnson Inc., St. Paul, Minnesota, U.S.A.

Johnson, E.E. 1966b. Groundwater and Wells. E.E. Johnson Inc., St. Paul, Minnesota, U.S.A.

Johnson, E.E. 1967. "Analyzing Water Level Recovery Data". The Drillers Journal, *Jan-Feb. p. 8.* E.E. Johnson, Inc., St. Paul, Minnesota, U.S.A.

Lennox, D.H. 1962. Geophysical methods of locating aquifers in the prairies. Proc. of Hydrology Symposium No. 3. N.R.C. Subcommittee on Hydrology. The Queen's Printer, Ottawa. pp. 5-20.

Meyboom, Peter. 1962. Patterns of groundwater flow in the prairie profile. Proc. of Hydrology Symposium No. 3. N.R.C. Subcommittee on Hydrology. The Queen's Printer, Ottawa. pp. 5-20.

Todd, D.K. 1959. Groundwater hydrology. John Wiley & Sons, New York.

Walton, W.C. 1962. Selected analytical methods for well and aquifer evaluation. Bulletin 49. State Water Survey Division, Urbana, Illinois.

Wenzel, L.K. 1942. Methods of determining permeability of water–bearing materials, with special reference to discharging-well methods. U.S. Geol. Survey Water Supply Paper 887.

Section VII

RUNOFF – RAINFALL – GENERAL

by

Donald M. Gray and John M. Wigham

TABLE OF CONTENTS

LIST OF FIGURES

Section VII

RUNOFF – RAINFALL – GENERAL

VII.1   **COMPONENTS OF HYDROGRAPH**

A hydrograph is the graphical representation of the instantaneous rate of discharge of a stream plotted with time. It includes the integrated contributions from surface runoff, interflow, groundwater flow and channel precipitation.

VII.1.1   **Surface Runoff Phenomena**

Depending upon the rate at which rain falls, water may either infiltrate into the soil or accumulate and flow from an area as surface runoff. If the intensity of rainfall—neglecting interception and evaporation losses—is less than the rate of infiltration, all water will enter the soil profile. Conversely, when the intensity of rainfall is greater than the rate of infiltration, a sequence of events occurs which ultimately will produce surface runoff.

Excess water produced by a high intensity of rainfall must first satisfy soil and vegetal storage, detention and interception requirements. When the surface depressions are filled, surface water then begins to move down the slopes in thin films and tiny streams. At this stage, the flow overland is influenced greatly by surface tension and friction forces. Horton (1945) shows that, as precipitation continues, the depth of surface detention increases, and it is distributed according to the distance from the outlet (refer to Fig. VII.1). With an increase in depth or volume of supply, there is a corresponding increase in the rate of discharge. Therefore, the rate of outflow is a function of the depth of water detained on the area.

The paths of the small streams being tortuous in nature, every small obstruction causes a delay until sufficient head is built up to overcome such resistance (Horton 1935); then upon release, the stream suddenly speeds on its way again. Each time there is a merging of two or more streams, the water is accelerated still more in its downhill path. The culmination of all these small contributions thus produces the ultimate hydrograph of surface runoff. After the excess rain ends, the water remaining on the area (surface detention) disappears progressively from the watershed as a result of the combined action of surface runoff and infiltration.

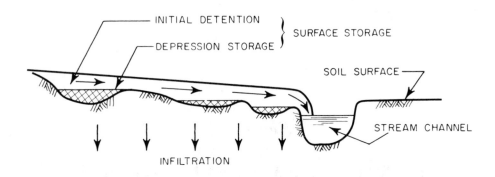

Fig. VII.1 Surface Runoff Phenomena

VII.1.1.1 Genetic Sequence of Surface Runoff

According to the preceding discussion, it is evident that the total time that elapses for an elemental volume of water, originating from a position in space on a watershed, to reach a specified point on the stream (for example, the gauging station) includes the sum of the times required for the water to flow over slopes (overland flow), through hollows, ravines and gullies, and finally along the stream channel. That is

$$t_t = \int_0^{l_1} \frac{dl}{V_o} + \int_{l_1}^{l_2} \frac{dl}{V_g} + \ldots \ldots + \int_{l_{n-1}}^{l_n} \frac{dl}{V_s} \quad \ldots \ldots \ldots \ldots \ldots \ldots \text{VII.1}$$

where t_t = total time elapsed,
$l_1, l_2 - l_1, l_n - l_{n-1}$ = lengths of flow paths in overland, gully,
  channel, etc., and
V_o, V_g, V_s = velocities of flow overland, in gullies and
  channels, etc.

It should be mentioned that, frequently, the relative times-of-travel in overland and channel flow are used for hydrologic categorization of the size of the watershed. That is, a *small watershed* is an area in which the time involved in overland flow is significant (and thus cannot be neglected); whereas a *large watershed* is an area in which the time of travel by surface runoff in channel flow predominates, being much greater than the travel time in overland flow.

Let us assume that a watershed can be divided into incremental areas, ΔA_j, of boundaries defined by lines connecting points in space which have equal travel times to the closing profile or gauging station (isochronal lines); see Fig. VII.2. If a storm of uniform intensity, i, (i>f) is distributed over the watershed area, then water begins to flow first from areas adjacent to the gauging station; and progressively with time, a larger percentage of the total area will contribute to the flow. That is

After "t_1"

$$Q_t = (i_1 - f_1) \, \Delta A_1 \dots \dots \dots \dots \dots \text{VII.2}$$

After "t_2"

$$Q_t = (i_2 - f_2) \, \Delta A_1 + (i_1 - f_1) \, \Delta A_2 \dots \dots \dots \text{VII.3}$$

After "t_j"

$$Q_t = (i_j - f_j) \, \Delta A_1 + (i_{j-1} - f_{j-1}) \, \Delta A_2 + \dots \dots + (i_1 - f_1) \, \Delta A_j \dots \dots \dots \text{VII.4}$$

in which i_1, i_2 and i_j = average precipitation rates
during the increments of
travel time respectively, and
f_1, f_2, and f_j = average loss rates during the
increments of travel time.

In effect, if the excess rate (i-f) remains constant over the watershed for a duration equal to the time of concentration of the basin, t_c,[1] then the shape of the surface runoff hydrograph (Equation VII.4) will be similar to that of the cumulative time-area curve shown in Fig. VII.3.

The volume of water originating on an elemental area of the watershed at any time, ΔV_{tj}, can be expressed by

$$\Delta V_{tj} = \Delta A_j \, (i_{tj} - f_{tj}) \, \Delta t \dots \dots \dots \dots \dots \text{VII.5}$$

However, it is convenient to express these volumes according to the travel-time criterion, or the time at which they appear at the gauging station. Velikanov suggests that the volume of water passing the station in an increment of time, dV, is given by

$$dV = \frac{\partial A}{\partial \tau} \, (i_{t-\tau} - f_{t-\tau}) \, d\tau \, dt \dots \dots \dots \dots \dots \text{VII.6}$$

and the discharge rate within a given isochronal period is

$$Q_t = \int_0^\tau \frac{\partial A}{\partial \tau} \, (i_{t-\tau} - f_{t-\tau}) \, d\tau \dots \dots \dots \dots \dots \text{VII.7}$$

The above method of calculating surface runoff rates from a watershed is referred to as the *isochronal method*. Although the method is fundamentally sound and provides an

[1]Time of concentration, t_c, is the time required for a particle of water to move from the most hydraulically remote point of the watershed to the outlet.

insight of surface runoff phenomena, its application is limited because of the difficulty of constructing an isochronal map of the watershed. In practice, isochrones will not appear as smooth lines, because of the presence of major and micro water divides. Further, the position of these lines in space will vary with time, because of such factors as variable infiltration rates and rainfall intensities, changes in storage and depths of flow, etc.

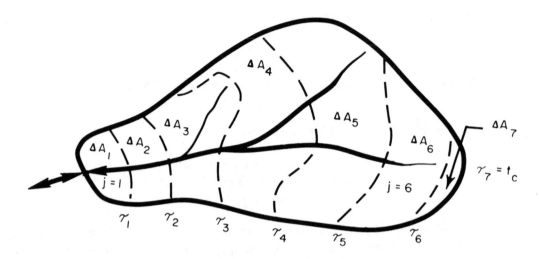

Fig. VII.2 Isochronal Map of Watershed

VII.1.2   **Interflow**

Water infiltrating the soil surface and moving laterally through the upper horizons of the soil until it returns to the surface at some point downslope (from its point of infiltration) to flow to the stream as surface runoff, or to be intercepted in its course by a stream channel, is known as interflow. This water does not become part of the characteristic groundwater flow system. Geologic conditions which favour interflow are those where the porous surface layers are underlain by relatively impervious strata. Under such conditions, the contribution of interflow to streamflow may be very significant.

The primary effect of the interflow component is that it tends to lengthen the time elements of the hydrograph; that is, the times of arrival of interflow contributions are delayed, and thus lag surface runoff contributions.

VII.1.3   **Groundwater Flow**

Groundwater flow is that component of streamflow which originates from flow occurring below the groundwater table, and this flow from accretions of water that have been built up during any storm event occurs over a long period of time. As a result, this component is extremely important to watershed yield (see Section X).

Streams may be categorized as being either *effluent* or *influent* depending on the direction of movement of water from the stream channel. During those periods when the water-surface level in the stream intersects the water table at the bank, such that groundwater flow occurs to the stream (because of the slope of the water table), the stream is considered to be effluent. On the other hand, if the water table falls below the bottom of the channel and the water level in the stream is higher than the groundwater level, or if either occur, then flow from the stream to groundwater begins, and the stream is considered to be influent.

VII.1.4   Channel Precipitation

The component of streamflow originating from precipitation that falls directly on the water surfaces of lakes and streams is known as channel precipitation. This amount can be computed by multiplying the average rainfall by the area of the basin covered by water surfaces which are connected with the stream system.

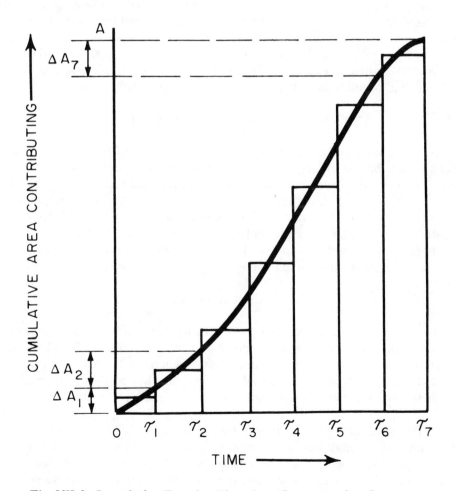

Fig. VII.3 Cumulative Travel – Time-Area Concentration Curve

Obviously, the percentage of channel precipitation varies from basin to basin, and from time to time within a given basin depending on the water level in the streams. As pointed out by Linsley *et al.* (1949), the water surface area for most basins does not exceed 5 per cent of the total area at fairly high stages.

Generally, channel precipitation is not considered as a separate component of runoff, since it is usually a relatively small amount and is thus included with surface runoff. In reservoirs, however, significant inflow may occur from precipitation falling directly on the lake surface.

VII.2   HYDROGRAPH SHAPE

The shape of a hydrograph of a single, short-duration storm occurring over the drainage area follows a general pattern. This pattern shows a period of rise, or a period of increasing discharge, that culminates in a peak or crest. Following is a period of decreasing discharge (recession limb) which may or may not decrease to zero discharge, depending on the amount of groundwater flow. A typical hydrograph, divided into three principal parts, is shown in Fig. VII.4. For small watershed areas, the total contributions to the runoff hydrograph by groundwater flow, channel precipitation and interflow are usually small in comparison to the amount received from surface runoff.

VII.2.1   Rising Limb or Concentration Curve

The rising limb extends from the time of beginning of surface runoff to the first inflection point on the hydrograph, and represents the increase in discharge produced by an increase in storage or detention on the watershed. Its geometry is characterized by the shape of the time-area histogram of the basin, and by the duration, intensity and uniformity of the rain. The initial portion is concave as a result of two factors: the greater concentration of area between adjacent isochrones within the middle and upper reaches of the basin; and the greater opportunity for infiltration, evaporation, surface detention and interception during the initial periods of the storm (Linsley *et al.* 1949, p. 390).

VII.2.2   Crest Segment

The crest segment includes that part of the hydrograph from the inflection point on the rising limb to a corresponding point on the recession limb. The peak of the hydrograph, or the maximum instantaneous discharge rate, occurs within this time interval. The peak represents the arrival of flow from that portion of the basin receiving the highest concentration of area-inches of runoff. Ramser (1927) states:

> The maximum rate of runoff from any watershed area for a given intensity rainfall occurs when all parts of the area are contributing to flow. That part of the watershed nearest the outlet must still be contributing to the flow when the water from the most remote point on the watershed reaches the outlet.

That is, the duration of rain must equal or exceed the time of concentration. As a measure of caution it should be reiterated that this concept is limited to small watersheds.

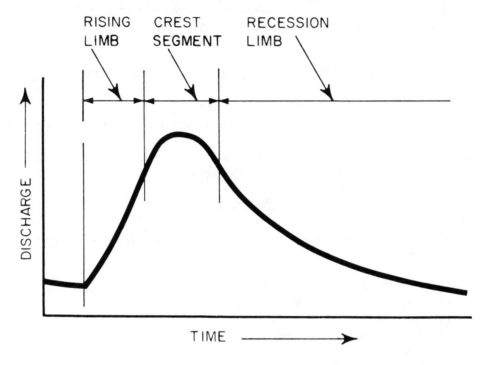

Fig. VII.4 Components of Hydrograph

VII.2.3   Recession Limb

The recession limb includes the remaining part of the hydrograph, which may or may not decrease to zero discharge depending on the amount of base flow or groundwater flow. It represents the withdrawal of water from storage after excess rainfall has ceased. Consequently, it may be considered as the natural decrease in the rate of discharge resulting from the draining-off process. The shape of the curve is independent of time variations in rainfall or infiltration, and is essentially dependent upon the physical features of the channel alone. Horner and Flynt (1936) and Barnes (1939) have stated that the general mathematical form of the equation defining this segment of the hydrograph is:

$$Q_2 = Q_1 \, K^{-\Delta t} \quad \dots\dots\dots\dots\dots\dots\dots\dots\dots\dots\dots\dots\dots\dots\dots\dots\dots\dots\dots\dots\dots VII.8$$

where Q_2 = instantaneous discharge rate at time, t_2,
       Q_1 = instantaneous discharge rate at time, t_1,
        K = recession constant, and
       Δt = elapsed time interval, $(t_2 - t_1)$.

This equation produces a straight line when plotted on semilogarithmic paper. The value of the recession constant, K, is generally not constant throughout all discharge rates. Frequently, the recession curve is broken into a series of line segments to obtain several values of K, with each value applicable within a given range of flows.

VII.3   FACTORS AFFECTING HYDROGRAPH SHAPE

The time distribution of runoff (the shape of the hydrograph) is influenced by climatic factors, and by the topographic and geologic features of the basin; thus the final hydrograph is affected by all three factors. However, it may be stated that climatic factors predominate in producing the rising limb, while the recession limb is largely independent of the storm characteristics producing the runoff.

VII.3.1   Climatic Factors

The climatic factors which influence the hydrograph shape, and of course, the volume of runoff are

(1)   rainfall intensity and duration;
(2)   distribution of rainfall on the basin;
(3)   direction of storm movement; and
(4)   type of precipitation and type of storm.

VII.3.1.1   Rainfall Intensity and Duration

Both govern the amount of runoff, the peak flow rate, and the duration of surface runoff for a given basin.

For a constant rainfall duration, an increase in intensity of rainfall will increase the peak discharge and volume of runoff, provided of course the infiltration rate of the soil is exceeded. However, under these conditions the duration of surface runoff remains essentially the same. Short time variations in intensity during a storm may affect the shape of the hydrograph for small basins, but generally they will have little noticeable effect on hydrographs from a large basin.

For a rain of given intensity, rainfall duration determines in part the peak flow and time period of surface runoff. If a storm lasts long enough, eventually almost all the precipitation will become runoff; consequently, the peak flow will approach a rate equal to the product 'iA', where 'i' is the intensity of rainfall and 'A' is the area of the basin. This situation is never reached on basins of large size, but it may occur on small areas and is frequently used as the criteria for design of storm sewers, airport drainage or small culverts. Theoretically, there will be a different surface runoff period for each different rainfall duration. Practically, however, it has been found that small variations in rainfall duration (up to 25 per cent of the storm duration) do not produce measurable effects on surface runoff duration, because of the modifying influence of storage.

VII.3.1.2   Distribution of Rainfall

The areal distribution of rainfall can cause variations in hydrograph shape. If an area of high rainfall is near the basin outlet, a rapid rise, sharp peak and rapid recession of the hydrograph usually result. If a larger amount of rainfall occurs in the upper reaches of a basin, the hydrograph exhibits a lower and broader peak.

VII.3.1.3   Direction of Storm Movement

The direction of storm movement with respect to orientation of the drainage net of the basin affects both the magnitude of the peak flow and the duration of surface runoff. Storm direction has the greatest effect on elongated basins. On these basins, storms that move upstream tend to produce lower peaks and a broader time base of surface runoff than storms that move downstream.

VII.3.1.4   Type of Precipitation and Storm

The hydrograph of runoff from snowmelt release differs greatly in shape from that caused by rainfall. Probably, the single factor which makes the runoff pattern different is the difference in the rate of generation of the excess volumes. In rain-flood problems it is customary to study situations in which direct runoff volumes are generated very rapidly. By comparison, the rate at which runoff is generated from snowmelt is usually rather sluggish, because of the lag effects in the snowpack, the local distribution of the source areas of runoff and diurnal fluctuations in temperature. As a result, the snowmelt hydrograph usually tends to exhibit a lower and broader runoff pattern than the rainfall hydrograph. In certain cases, where the rates of snowmelt may be low, subsurface flow may be very important to flood hydrographs.

The type of storm is important in that thunderstorms produce the peak flows on small basins, whereas large cyclonic or frontal-type storms are generally important in flood prediction on large basins.

VII.3.2   Topographic Factors and the Hydrograph

The surface-runoff hydrograph for a watershed represents the integrated effect of all physical characteristics of the basin and their modifying influence on the translation and storage of a rainfall-excess volume. The factors involved are numerous, some having a major bearing on the phenomena and others being of negligible consequence. Sherman (1932) suggests the following are the dominant factors:

1.   Drainage-area size and shape.
2.   Distribution of the watercourses.
3.   Slope of the valley sides or general land slope.
4.   Slope of the main stream.
5.   Pondage resulting from surface or channel obstructions forming natural detention reservoirs.

VII.3.2.1   Drainage-Area Size and Shape

The major effect of increasing the drainage - area size on the geometry of the sur-face-runoff hydrograph is that the time base of the hydrograph is lengthened (Wisler and Brater, 1959, p.40). Thus for a given rainfall excess, the peak ordinate, expressed in units of cubic feet per second (cfs) per square mile, will decrease with basin size.

Drainage-area shape is instrumental in governing the rate at which water is supplied to the main stream as it proceeds to the outlet. It is, therefore, a significant feature that influences the period of rise. For example, the hydrograph, produced by a semicircular basin on which flow converges from all points to the outlet, will have a shorter time to peak than a hydrograph produced on a long, narrow basin of equal size. Langbein and others (1947) summarized the effect:

> A drainage basin whose drainage tributaries are compactly organized so
> that water from all parts of the basin has a comparatively short distance
> to travel will discharge its runoff more quickly and reach greater flood
> crests than one in which the larger part of the basin is remote from the
> outlet.

Although drainage areas can be of a multiplicity of shapes, they are generally ovoid or pear-shaped. Dooge (1956) found that unless the shape of a watershed deviated appreciably from generally ovoid, the geometry of the hydrograph remained relatively constant.

VII.3.2.2   Distribution of Water Courses

The pattern and arrangement of the natural stream channels determine the efficiency of the drainage system. Other factors being constant, the time required for water to flow a given distance is directly proportional to length. Since a well-defined system reduces the distance water must move in overland flow, the corresponding reduction in time involved is reflected by an outflow hydrograph having a short, pronounced time of concentration of runoff.

VII.3.2.3   Slope of Main Stream

After reaching the main drainage way, the time necessary for a flood wave to pass the outlet is related to the length of travel and the slope of the waterway. The velocity of flow of water, V, in an open channel under steady, uniform flow may be expressed in the general form

$$V = CR^m S_c^n \dots\dots\dots\dots\dots\dots\dots\dots\dots\dots\dots\dots\dots\dots\dots\dots VII.9$$

where   C = constant whose magnitude depends on the
              roughness of the channel,
      R = hydraulic radius,

S_C = channel slope, and
m and n = exponents.

It follows from Equation VII.9 that the time, t, required for a particle of water to move a given distance, l, is inversely related to some power of the slope value. According to Manning, the values of the exponents are m = 2/3 and n = 1/2, respectively. Dooge (1956) shows that in loose boundary hydraulics, however, roughness and slope are not independent, and that the velocity relationship depends on the size of the bed material. He indicates that, for a channel in equilibrium, the travel time varies inversely with the cube root of the channel slope.

The influence of channel slope is reflected in the time elements of the hydrograph. Since the recession limb represents the withdrawal of water from channel storage, the effect of channel slope should be very influential in that portion of the hydrograph. Correspondingly, with increased channel slope, the slope of the recession limb increases, and the base time of the hydrograph decreases.

VII.3.2.4   Slope of Valley Sides or General Land Slope

The general land slope has a complex relationship to the surface runoff phenomena because of its influence on infiltration, soil moisture content and vegetal growth. The influence of land slope on hydrograph shape is manifested in the time of concentration of the runoff volumes to defined stream channels. On large watershed areas, the time involved in overland flow is small in comparison with the time of flow in the stream channel. Conversely, on smaller areas, the overland flow regime exerts a dominating effect on the time relationships and the peak of the hydrograph (Commons, 1942).

The velocity of overland flow is not readily computed because of variations in types of flow that may exist along the paths of transit. Overland flow over smooth slopes may range from purely laminar to purely turbulent. Horton (1935, 1938) describes an additional type of flow—subdivided flow—in which flow is subdivided by grass or vegetal matter, so producing a condition where the velocity is practically uniform over the depth of flow. Under this condition the resistance to flow is very great.

Theoretical and empirical considerations of the overland flow regime (expressed by Butler, 1957, p.316) are in the following relationship:

$$q = ad^b S_L{}^c \quad \dots\dots\dots\dots\dots\dots\dots\dots\dots\dots\dots\dots\dots\dots\dots\dots\dots \text{VII.10}$$

where   q = rate of outflow per unit width,
         d = average depth of surface storage,
       S_L = land slope, and
a, b and c = coefficient and exponents which vary
             with Reynold's number, raindrop impact
             and roughness.

Equation VII.10 indicates that the effect of land slope on the velocity of flow is similar to that of channel slope. With increasing land slope, the time elements of the hydrograph decrease.

VII.3.2.5   Pondage or Storage

Since storage must first be filled, then emptied, its delaying and modifying effect on the excess precipitation volumes is instrumental in determining hydrograph shape. Much of the variation, caused by differences in subintensity patterns and areal distribution of a rain, and by differences in the time of travel of runoff volumes from individual subbasins to the outlet, is evened out.

Storage effects exist in both overland and channel flow. Sherman (1932) summarizes the effect on the unit graph of differences in topography:

> Topography with steep slopes and few pondage pockets gives a unit graph with a high sharp peak and short time period. A flat country with large pondage pockets gives a graph with a flat rounded peak and a long time period.

During its passage through a watercourse, a flood wave may be considered to undergo a simple translation (uniformly progressive flow) and reservoir or pondage action (Langbein, 1949). The extent of modification of the flood wave can be ascertained by employing flood routing procedures if the flow characteristics and the geometric properties of the stream channel are known. In general, storage causes a decrease in the peak discharge and a lengthening of the time base of the hydrograph.

The foregoing discussion considers only the generalized influences of topographic factors on hydrograph shape. It is impossible within the bounds of these notes to cover the influence of each individual factor in detail. The effect of each factor may be obscured by the effect of another. The final hydrograph will depend on the cumulative effect of all of the factors as they act either alone or in combination with others.

VII.3.3   Geological Factors

The geological factors, which affect the shape of the runoff hydrograph, are primarily those which govern the flow of groundwater and interflow to a stream. For example, an impervious formation or layer close to the surface would affect the amount of interflow (flow through the surface soil layers), hence the resulting hydrograph. The hydraulic conductivity of the surface layers affects the infiltration to lower levels, and thus determines the groundwater and interflow contributions to runoff. It should be mentioned that subsurface formations can make the groundwater drainage area to a stream much larger or much smaller than the surface drainage area. That is, the phreatic divide need not correspond and in many cases does not conform, to the topographic divide; hence a stream may show a proportionately high or low groundwater contribution depending on the subsurface formations. It is also possible that the groundwater table is normally at such a level that the stream continually supplies water (influent stream) to subsurface aquifers; or the stream may be effluent, receiving a continuous supply of groundwater; or influent at high stages and effluent at low.

VII.4   HYDROGRAPH SEPARATION

For certain types of studies, it is necessary to separate the observed hydrograph to its component parts. Several techniques are used to accomplish this task.

The recession limb of a hydrograph—sometimes referred to as a depletion curve since it represents the depletion from channel, interflow and groundwater storage—can be described in general mathematical form (from Equation VII.8)

$$\ln Q_t = \ln Q_o - \Delta t \ln K \ldots\ldots\ldots\ldots\ldots\ldots\ldots\ldots\ldots\ldots VII.11$$

As mentioned previously, according to Equation VII.11 the recession limb should produce a straight line when plotted on semilogarithmic paper. In practice, however, this will usually occur only when the storage in the stream channel is mainly surface runoff (interflow and groundwater flow being negligible), or when only values are taken from the recession curve that represents groundwater flow for plotting (that is, values of 'Q' from the recession limb are taken only after sufficient time, so that storage from interflow and surface runoff have been released). Generally, however, because of the effects of the different components of storage, the plot will be curvilinear having a series of line segments of different slopes. Referring to Fig. VII.5, Barnes (1939) suggests the

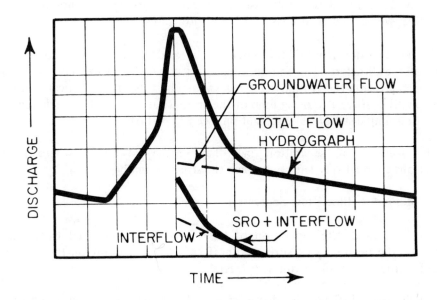

Fig. VII.5 Semilogarithmic Plotting of a Hydrograph Showing the Recession Analysis

following procedure to be followed in separating the hydrograph into various components:

1. Plot the hydrograph on semilogarithmic paper.
2. Approximate the groundwater recession by a straight line extended back under the hydrograph.
3. Plot the residuals (surface runoff + interflow).
4. Fit a straight line to the recession of this curve (item 3) and extend under the hydrograph.
5. Plot the residuals (surface runoff).

In this method, as in all others, the rising limb of the groundwater and interflow hydrographs must be approximated.

VII.4.1   Separation of Hydrograph into Two Components

Frequently, for engineering applications, it is unnecessary to divide the hydrograph into all its components because the procedures used in the separations are somewhat arbitrary and artificial. Usually, however, it is standard practice to separate the base flow from direct runoff (surface runoff and interflow). The three methods that are used to accomplish this separation are shown in Fig. VII.6.

VII.4.1.i   Method 1—Base Flow Separation

This method of separation consists of extending the base flow recession curve back under the peak of the hydrograph. To accomplish this, a composite base flow recession curve, derived from analyses of recessions for several storm events, should be used. These composite curves should encompass a range of flow rates.

The point of departure of the observed and calculated depletion curves is taken as the end of the direct runoff, and the depletion curve that is extended back from this point then represents the groundwater recession. The time of groundwater peak must be determined arbitrarily. The rising limb of the groundwater curve is constructed by using a template or a simple straight line separation (see Fig. VII.6).

This method of separation may have some advantage where groundwater contributions are relatively large and reach the stream quickly.

VII.4.1.2   Method 2—Base Flow Separation

In this method, the separation of base flow is accomplished by joining with a straight line the beginning of surface runoff (pt. A) to a point on the recession curve representing the end of direct runoff (pt. B).

Usually, little difficulty is encountered in determining pt. A, but the break between the base flow recession and direct runoff may be difficult to define. If pt. B is not well

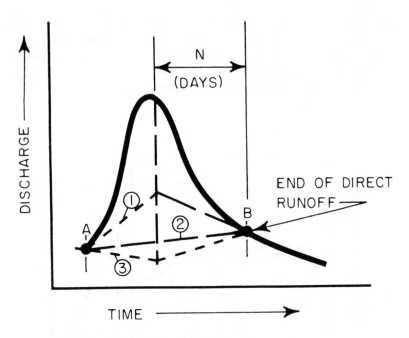

Fig. VII.6 Methods of Base Flow Separation

defined, its position may be determined using empirical equations defining N, the number of days after the peak at which direct runoff essentially ends. Linsley *et al.* (1958, p.156) approximate N by the following equation:

$$N = A^{0.2} \quad \dots\dots\dots\dots\dots\dots\dots\dots\dots\dots\dots\dots\dots\dots\dots\dots\dots\dots\dots \quad \text{VII.12}$$

where N is in days, and
        A is the drainage area in square miles.

However 'N' is better determined, probably, by inspection of a number of hydrographs for the basin under study, keeping in mind that the time base of direct runoff should not be too long, and the increase in groundwater flow should not be too great. Consideration of the topographic, soil and geologic conditions of the basin will aid in this selection (Butler, 1957, p.212).

VII.4.1.3   Method 3—Base Flow Separation

This method (see curve 3 in Fig. VII.6) consists of extending the base flow recession curve, which occurs before surface runoff, past pt. A to a point beneath the peak, and then of extending a straight line to pt. B, representing the end of direct runoff.

VII.4.2   Base Flow Separation of Complex Hydrographs

The previous discussion of hydrograph separation assumed an isolated streamflow event. More complex hydrographs caused by two or more closely-spaced rainfall events are quite common, and analyses of these are often necessary. For these cases, the effects of the rainfall events must be separated, in addition to separating base flow and direct runoff. If Methods 2 and 3 are used for separation, then the division between rainfall events may be accomplished using a total runoff recession curve.

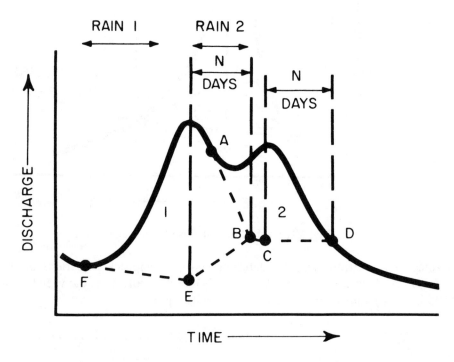

Fig. VII.7 Separation of Complex Hydrograph

The small portion of recession between the peaks is reconstructed from pts. A to B (pt. B being 'N' days after the occurrence of the first peak) using a composite total recession curve (surface runoff + interflow + groundwater flow). Lines FE and EB (see Fig. VII.7) can then be drawn as for Method 3. If the position of pt. B falls before the occurrence of the second peak, then the base flow recession curve is used to construct the line BC, and the straight line CD can be drawn. Usually pt. B will fall after the second peak, in which case a straight line is drawn from pt. D to a point directly below the second peak on the recession curve AB.

VII.4.3   Summary of Techniques of Base Flow Separation

Distinctions between methods of separating the three components of flow are somewhat arbitrary and artificial. In most cases, neither the quantity nor the time distribution of all the components is known. In any event, and for any given study, the investigator should employ a common method of separation in the analysis of all hydrographs.

As a check on the separation method, it should be remembered that the volume of surface runoff (area under the surface-runoff hydrograph) must be equal to the volume of net storm rain on the watershed. Further, the volume of infiltration over the basin should equal (neglecting evapotranspiration) the change in groundwater storage between the beginning and end of the hydrograph plus the interflow volume plus the increased base flow volume under the hydrograph. The amount of groundwater storage can be obtained by integration of Equation VII.8, that is

$$\text{Storage} = \int_{t=t}^{t=\infty} Q_t dt = \int_{t=t}^{t=\infty} Q_o K^{-t} dt = \left[\frac{-Q_o K^{-t}}{\ln K} \right]_t^{\infty}$$

$$\text{Storage} = Q_t/\ln K \dots\dots\dots\dots\dots\dots\dots\dots\dots\dots\dots\dots\dots\text{VII.13}$$

VII.5   NET STORM RAIN

Surface runoff phenomena have already been discussed (Section VII.1.1.); and the time sequence of events—infiltration, depressional storage, detention storage, overland flow and surface runoff—are illustrated in Fig. VII.8.

From the figure it can be deduced that the net storm rain (that portion of the total precipitation which will appear as direct surface runoff) can be obtained from consideration of the phenomena of retention, infiltration and overland flow. Essentially, in terms of watershed yield (surface runoff), one may write a simple continuity equation:

Inflow = Outflow + Change in Storage.

Precipitation = Depressional Storage + Evaporation + Infiltration + Interception + Surface Runoff.

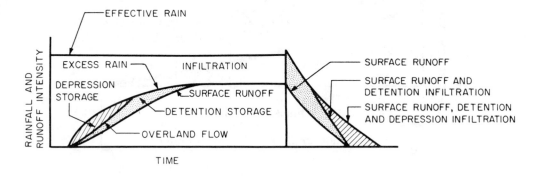

Fig. VII.8 Hydrographs Showing the Rates of Disposition of a Hypothetical, Uniform, Effective Rainfall

In other words, runoff is the residual of precipitation after accounting for the various component losses. Frequently, the object of analysis is to distribute these losses according to time so as to determine the time-distribution with which volumes of precipitation excess are generated.

VII.5.1   The Infiltration Approach

In computing the net storm rain by the infiltration approach, four basic steps are involved:

1.   Select the appropriate infiltration curve after consideration of factors such as soil type, soil moisture content, vegetation.
2.   Construct and superimpose the infiltration curve on the design rainfall histogram (see Fig. VII.9).
3.   Deduct a proportion of the rainfall to account for depressional storage. This deduction is made from the first part of the 'excess' rain, since it is during this part of the storm that depressions are filled (see Fig. VII.9). If the rain stops for an hour or two, allowing part or all of the depression storage to infiltrate to the ground, a new deduction for depression storage may be necessary when the storm resumes. In some cases, depression storage may be negligible and can be ignored. The amount to allow for can be estimated on the basis of the topography of the land, perhaps aided by an examination of the land immediately after a heavy rainfall to note the extent of depression storage.
4.   Deduct a part of the precipitation to consider the residual detention storage, which infiltrates during rainless intervals and after the end of the storm as

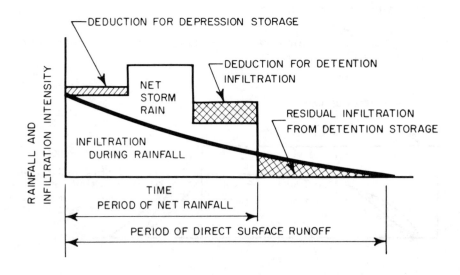

Fig. VII.9 The Infiltration Method of Computing Net Storm Rain

water drains to a stream channel. As shown in Fig. VII.9, this residual infiltration is deducted from the excess rain near the end of the storm period, as it is this rain which supplies the water for detention.

In the foregoing example, the losses due to interception and evaporation have been considered negligible—hence, neglected. The decision to include these terms in the analysis must be based on consideration of the purpose of the study, and whether or not these factors are of significant value relative to the magnitude of the storm event. Often, for large design storms, on watersheds in which depressional storage is also small, it is assumed that infiltration constitutes the major loss. Under such conditions it may be impractical to consider the losses of storage and evaporation, and the net storm rain is determined on the basis of the infiltration characteristics of the watershed alone.

VII.5.1.1   Adjustment of Infiltration Rate Curve To Account for Variable Rainfall Intensities

Perhaps the major difficulty arising in the use of the infiltration approach to determine the precipitation excess is in evaluating the effect of rainfall which falls at intensities less than the infiltration rate, f. Figure VII.10 illustrates the problem in simplified form, showing a rain with low intensity (less than the soil infiltration rate) preceding the principal burst. The question remains as to what value should the infiltration curve begin at the time of the main burst. Fig. VII.10 also demonstrates two diverse approaches to the problem:

1. Curve 1 – assumes infiltration occurs at capacity regardless of the rainfall intensity.
2. Curve 2 – assumes that the f-curve does not begin to decrease until the runoff-producing rain occurs.

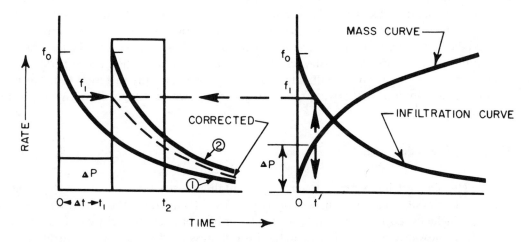

Fig. VII.10 Adjustment of Infiltration Rate Curve for Low Intensity Rain

It is quite obvious that application of these two techniques will give greatly different values of precipitation excess for the same storm. Arguments can be presented to show that neither approach is acceptable.

The commonly-used method of adjusting the infiltration curve is based on the assumption that the low intensity rain will infiltrate at its maximum rate, and that f should be adjusted according to the mass curve. If ΔP is the amount of water that falls in the time interval, $\Delta t'$ (t_1-0), it is assumed to be absorbed by the soil at the maximum rate, and the time required to absorb this water is t' (see Fig. VII.10). In this time interval, $\Delta t'$, the infiltration curve experiences a change from f_0 to f_1. Thus the infiltration curve (Curve 1) should be shifted to the right to intersect the runoff, producing rain at t_1 with a rate of f_1. Corrections for intermediate showers between runoff-producing parts of a storm utilize the same assumptions.

Although workable results are secured, it must be emphasized that accurate evaluation of the effects of initial rainfall requires additional research. Further, it should be recognized that there is no provision in the technique to account for some recovery of the infiltration capacity curve in the event rainfall ceases between runoff-producing storms. Generally, for lighter soils (well-drained), it is conceded that the infiltration capacity may be totally recovered if this interval is greater than 6 hr., in which case consecutive storms may be considered as individual events. For heavier soils, because of their poor drainability, the effects of preceding rains may last several days, and provision must be made to adjust the infiltration curve accordingly.

VII.5.2   Infiltration Indices

In many instances the design problem does not require the accuracy, nor warrant the time involved, to apply the more sophisticated infiltration approach to determine the net storm rain. For these cases, short-cut 'index methods' will give acceptable results with less work. The general procedure to follow in the use of these methods is to establish some easily-computed index factor for use in place of the infiltration curve, and to determine numerical values for the factor from natural storms which have occurred on the watershed. As these indices are computed from actual storm records and take into account such factors as intensity characteristics of a storm, they are superior to using mere ratios of runoff to rainfall. There are, of course, as many different kinds of indices as there are methods of deriving them. In general, however, they will follow one of three general types.

VII.5.2.1 ϕ Index

The simplest index, ϕ, is the average rainfall intensity over and above that when the mass of rainfall equals the mass runoff. Fig. VII.11 illustrates the nature of this index, which includes both surface retention and infiltration characteristics of the area. No separate allowance is made in the method for depression storage or for infiltration during rainless periods.

For long storms, when the volume of infiltration is large compared to that retained in surface storage, ϕ approaches an f-average value. At the other extreme are instances where short, high intensity rains occur over highly impermeable surfaces, and ϕ represents, almost wholly, the surface storage of the area. Also, it has been noted that the rainfall intensity and intensity pattern affect the magnitude of the index. For higher intensity rains, ϕ increases. This variation does not reflect an increased infiltration capacity due to the intensity of rain, but it is due to the fact that the proportion of area producing runoff throughout the rain becomes greater as the rainfall intensity increases. In general, values of ϕ derived from advanced storm patterns will be greater than those determined from equivalent storms having a delayed pattern. This variation reflects the increased loss to infiltration with advanced-type storms.

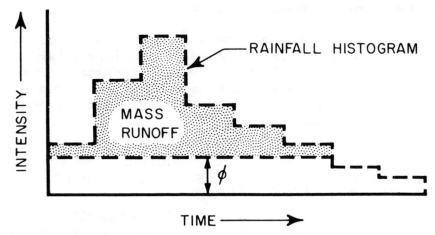

Fig. VII.11 Derivation and Meaning of the ϕ Index

VII.5.2.2   W Index

The W index is the average infiltration rate during the time when rainfall intensity exceeds the infiltration rate. That is

$$W = \frac{M_f}{t} = \frac{1}{t} (P - SRO - D) \dots\dots\dots\dots\dots\dots\dots\dots VII.14$$

where M_f = total infiltration,
        t = time during which the rainfall intensity exceeds the infiltration rate,
        P = the total precipitation corresponding to time, t,
    SRO = surface runoff, and
       D = the effective surface retention.

Values of the indices are easily derived from rainfall and discharge data, using an objective method of hydrograph separation. To apply the indices, the procedure is reversed; runoff is then computed from a selected index value.

As in the case for ϕ, the W index for a multiple, complex watershed has no real physical significance. It is widely used for large watersheds, but the error of assuming that it represents an actual infiltration rate should be avoided.

For the minimum infiltration rate, reached after long continued rainfall, the W index is termed W_{min}. Used in estimating maximum flood possibilities, it is usually computed for the last rain in a long storm made up of several rain periods. For such extreme conditions, the values of ϕ and W_{min} are almost identical. Because all parts of a basin are likely to be producing runoff in extreme floods, the W_{min} index approaches the average weighted-by-area minimum infiltration rates for all complexes in the watershed.

VII.5.2.3   The f_{ave} Method (Butler, 1957, pp. 265-6)

Of the index methods under consideration, the f_{ave} method is the most detailed and rational. The factor f_{ave} is defined as "the average infiltration rate during a period of generally continuous supply for infiltration." A storm may consist of several such rain periods, in which case each period is analyzed separately for net rain. Since natural storm rain is not continuous, some limiting permissible length for a rainfall intermission within a rain period must be established.

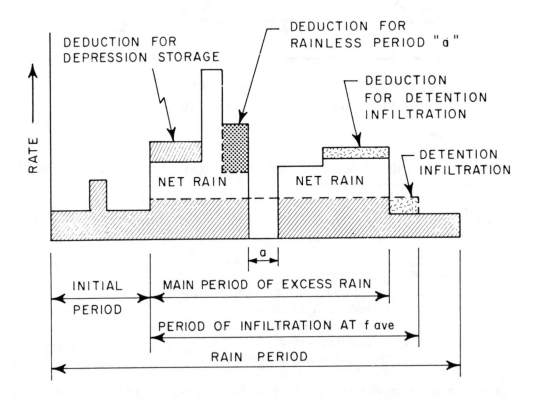

Fig. VII.12 The f_{ave} Method of Computing Net Storm Rain

In determining net storm rain by the f_{ave} method, the first step is to divide the effective rain storm hydrograph into separate rain periods. Then each rain period of the storm is analyzed individually as follows. Based on the rain data and the design value for f_{ave} which is to be used, an estimate is made of the main time period (within the rain period) during which excess rain is produced. Rain before and after this period is considered to be entirely lost. This period is assumed to be continuous for any given rain period, and includes any rainfall-intermission periods which may fall within it (see Fig. VII.12). Infiltration is assumed to occur at the f_{ave} rate for the duration of this main excess-rain period, plus a short additional period in allowance for residual infiltration. The next step is to adjust for depression storage, which is deducted from the first part of the excess rain. The final step is to deduct for infiltration during rainfall intermissions and after the end of excess rain. Intermission-period infiltration and residual infiltration are deducted, in each case, from the immediately-preceding excess rain. The remainder is net rain. It will be noted that the f_{ave} method is somewhat similar to the f-capacity method, but less precise.

VII.6   LITERATURE CITED

Barnes, B.S. 1939. The structure of discharge-recession curves. Trans. Amer. Geophys. Union 20:721-725.

Butler, S.S. 1957. Engineering hydrology. Prentice-Hall, Inc., Englewood Cliffs, N.J.

Commons, G. 1942. Flood hydrographs. Civil Engr. 12:571-572.

Dooge, J.C.I. 1956. Synthetic unit hydrographs based on triangular inflow. Unpublished M.S. thesis. State University of Iowa Library, Iowa City.

Horner, W.W. and Flynt, F.L. 1936. Relation between rainfall and runoff from urban areas. Trans. Amer. Soc. Civil Eng. 101:140-206.

Horton, R.E. 1935. Surface runoff phenomena, Part I: Analysis of the hydrograph. Horton Hydrological Laboratory, Voorheesville, N.Y.

Horton, R.E. 1938. The interpretation and application of runoff plat experiments with reference to soil erosion problems. Proc. Soil Sci. Soc. Amer. 3:340-349.

Horton, R.E. 1945. Erosional development of streams and their drainage basins: hydrophysical approach to quantitative morphology. Bul. Geol. Soc. Amer. 56:275-370.

Langbein, W.B. 1949. Storage in relation to flood waves. In: Meinzer, O., ed. Hydrology, pp. 561-571. Dover Publications, Inc., New York.

Langbein, W.B. and others, 1947. Topographic characteristics of drainage basins. U.S. Dept. Interior. Geological Survey Water-Supply Paper 968-C:125-155.

Linsley, R.K., Kohler, M.A. and Paulhus, J.L.H. 1949. Applied hydrology. McGraw-Hill Book Co., Inc., New York.

Linsley, R.K., Kohler, M.A. and Paulhus, J.L.H. 1958. Hydrology for engineers. McGraw-Hill Book Co., Inc., New York.

Ramser, C.E. 1927. Runoff from small agricultural areas. J. Agr. Res. 34:797-823

Sherman, L.K. 1932. Streamflow from rainfall by the unit-graph method. Engr. News-Record 108:501-502.

Sherman, L.K. 1940. The hydraulics of surface runoff. Civil Eng. 10:165-166.

Wisler, C.O. and Brater, E.F. 1959. Hydrology. John Wiley and Sons, Inc., New York.

Section VIII

PEAK FLOW – RAINFALL EVENTS

by

Donald M. Gray and John M. Wigham

TABLE OF CONTENTS

LIST OF TABLES

LIST OF FIGURES

Section VIII

PEAK FLOW – RAINFALL EVENTS

VIII.1   **INTRODUCTION**

The single and perhaps most important property of the hydrograph that is important to the hydraulic design of most structures or projects is the peak flow rate. Floods that have resulted in great damage, and in certain cases, loss of lives, have been recorded throughout history. Logically, this has meant that methods of flood prediction have had to be developed. The earliest method for estimating peak flow was by using empirical formulae that involved various physiographic characteristics of the basin. Today, unit hydrograph, flood routing and flood frequency analyses are commonly used to predict flood flows.

The method selected to determine peak flows and their frequencies depends on the following factors:

1. *The Desired Objective.* A distinction can be made here, in that the method used for determining the magnitude of peak flow may be different than that used to determine the maximum volume of flow during a flood period. The peak may be important in one design problem, and the volume important in another.

2. *The Available Data.* For example, long-term records of hydrologic data permit the rational application of statistical procedures, but success in the use of these techniques is inhibited by short-term records.

3. *The Area and Characteristics of the Watershed.* These factors govern the way in which runoff, and hence the peak flow, occurs.

4. *The Importance of the Project and Time Available for Analysis.*

Present practice for estimating peak discharges from catchment areas may be grouped as follows:

Catchment area in sq. mi.	Methods commonly used.
< 1	Infiltration approach, rational method.
<100	Overland flow hydrograph (up to a few sq. mi.); rational method; unit hydrographs; flood frequencies; flood peaks versus drainage area.
100–2000	Unit hydrograph; flood frequencies; flood peaks versus drainage area.
>2000	Flood routing; flood frequencies; flood peaks versus drainage area.

VIII.2   **THE RATIONAL METHOD**

The rational method of estimating peak flow on small watersheds (<5 sq. mi.) is based, in concept, on the criterion that for storms of uniform intensity, distributed evenly over the basin, the maximum rate of runoff occurs when the entire basin area is contributing at the outlet and that this rate of runoff is equal to a percentage of the rainfall intensity. In equational form

$$Q_p = ciA \dotfill \text{VIII.1}$$

where Q_p = rate of runoff (cfs),
 c = runoff coefficient,
 i = rainfall intensity (in/hr) of a storm whose duration
 is equal to the time of concentration of the basin, and
 A = area of the watershed (acres).

Although the rational method has been widely accepted (in small watershed work and urban storm sewer design), it is important for proper application of the method to understand its limitations.

The method assumes that the rainfall intensity is uniform over the entire watershed during the duration of the storm. This condition is seldom, if ever, completely fulfilled in nature. Theoretically, runoff from a watershed caused by a uniform rain of constant intensity as visualized in the rational method is shown in Fig. VIII.1.

The reasoning of the method states that if a rainfall of uniform intensity and unlimited duration falls on a basin, the runoff rate per unit area will reach a maximum

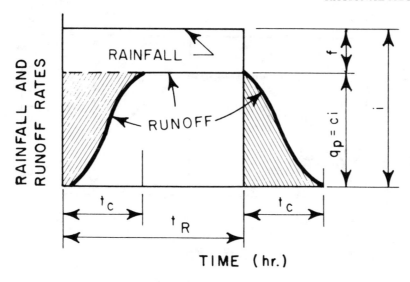

Fig. VIII.1 Runoff from Uniform Rainfall as Visualized in the Rational Method

$q_p = Q_p/A = ci$ at the time of concentration, t_c, (time required for a particle of water to move from the most hydraulically remote part of the basin to the outlet) and then will remain constant. Thus conceived, c represents the ratio q_p/i. Further, it can also be noted that the volume of water accumulated in storage must be equal to the volume enclosed by the recession limb (shaded areas). Thus, c, also represents a volumetric coefficient or the ratio of the total volume of runoff to rainfall.

Losses due to depressional storage and initial infiltration must be supplied before runoff can begin. In addition, there is retardance to flow caused by the effects of surface detention and channel storage. These factors tend to cause the hydrograph to take the shape indicated by curve 2 in Fig. VIII.2 that approaches a limit equal to i - f.

If rainfall should cease suddenly when an equilibrium condition is reached, runoff does not stop immediately, but there is a time lag, t_ℓ, caused by the momentum effects of flow which still exist between the cessation of rainfall to the time runoff begins to decrease. It follows that whereas the rational theory properly evaluates lag effects due to travel time, it does not allow for retardations by storage and momentum of flow in channels. These discussions are particularly significant when applying the method to areas such as the Prairies, which are characterized by flat topography with poorly defined drainage ways. On such areas, the rational method will only provide, at best, approximate estimates of peak flows.

In many instances, on drainage basins with long t_c values, the peak discharge rates from short-duration, high-intensity rains will exceed the discharges from lower-intensity rains of duration equal to t_c. This may be true even though, in the former case, only a portion of the drainage basin area contributes to flow. For Prairie conditions, it is extremely difficult to define the effective areas that contribute during a given storm. For

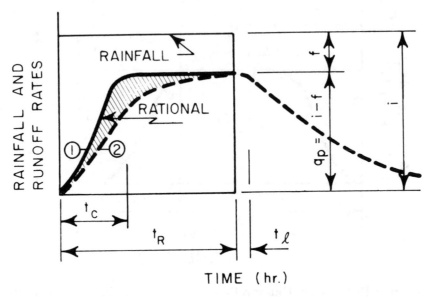

Fig. VIII.2 Runoff Hydrographs for a Block of Uniform Rainfall

further discussions of this subject, the reader is referred to the works of Stichling and Blackwell (1957) and Durrant and Blackwell (1959).

VIII.2.1   Evaluation of Components of the Rational Formula

To facilitate the use of the rational method for peak flow determinations, the following information may be used.

VIII.2.1.1 Runoff Coefficient, c

Judgment should be used in selecting the runoff coefficient, c. The magnitude of this coefficient varies with factors such as (a) the nature of the surface, (b) surface slope, (c) surface storage, (d) degree of saturation and (e) rainfall intensity. Usually, in selecting values of 'c' from prepared tables, the first four factors receive main consideration, whereas little attention is paid to the effect of rainfall intensity. It should be recognized that if 'c' is assumed constant and independent of 'i', then the infiltration rate must increase linearly with 'i'. Conversely, if the infiltration rate, 'f', is assumed constant then 'c' must vary linearly with 'i' to give a constant percentage of runoff. Horn and Schwab (1963) suggest that a more rational proposition is that 'c' increases with 'i' non-linearly. They suggest that this relationship may be of the form

$$c = a + b \ln i \quad \dots\dots\dots\dots\dots\dots\dots\dots\dots\dots\dots\dots\dots\dots\dots\dots\dots\dots\text{VIII.2}$$

Unfortunately, quantitative information of the influence of 'i' on 'c' is completely lacking. To evaluate 'c', resort must be made to use of tabulated values, such as those given in Tables VIII.1 and VIII.2.

Table VIII.1 Runoff Coefficients for Urban Areas

Description of Area	Runoff Coefficient
Flat, residential, with about 30% of area impervious	0.40
Moderately steep, residential, with about 50% of area impervious	0.65
Moderately steep, built up, with about 70% of area impervious	0.80

(Horner and Flynt 1936)

Table VIII.2 Deductions from Unity to Obtain the Runoff Coefficient for Agricultural Areas

Type of Area	Values of c'*
Topography	
Flat land, with average slopes of 1 ft. to 3 ft. per mi.	0.30
Rolling land, with average slopes of 15 ft. to 20 ft. per mi.	0.20
Hilly land, with average slopes of 150 ft. to 250 ft. per mi.	0.10
Soil	
Tight impervious clay	0.10
Medium combinations of clay and loam	0.20
Open sandy loam	0.40
Cover	
Cultivated lands	0.10
Woodland	0.20

(Bernard 1935)

* In Table VIII.2, the magnitude of c is obtained by adding values of c' for each of the three factors: topography, soil and cover; and then by subtracting the sum from unity.

VIII.2.1.2.   **Time of Concentration, t_c**

The time of concentration of a basin may be estimated from the following equations:

Kirpich (1940)

$$t_c = 0.0078L^{0.77} S^{-0.385} \quad \dots\dots\dots\dots\dots\dots\dots\dots\dots\dots\dots\dots\dots VIII.3$$

in which t_c = time of concentration (min),
 L = maximum length of travel of water, (ft), and
 S = slope, equal to H/L where H is the difference in elevation between the most remote point on the basin and the outlet (ft).

Mockus (1957)

$$t_c = t_L / 0.60 \dots\dots\dots\dots\dots\dots\dots\dots\dots\dots\dots\dots\dots\text{VIII.4}$$

> in which t_L = lag time or the time from the centre of mass of excess rainfall to the peak runoff rate,

and

$$P_R = \sqrt{t_c} + 0.60\, t_c \dots\dots\dots\dots\dots\dots\dots\dots\dots\text{VIII.6}$$

> in which P_R = time of peak or the time from the beginning of runoff to the time of peak runoff (hr).

VIII.2.1.3   Rainfall Intensity, i

Values for rainfall intensity, 'i', are obtained from nearby weather stations (see Section II). Selection of the value for rainfall intensity rests on estimates of the acceptable frequency of occurrence of the design flood and of the time of concentration of the basin.

VIII.3   EMPIRICAL METHODS AND FORMULAE FOR ESTIMATING PEAK FLOWS

Empirical formulae and methods of determining peak flows should only be used when there is insufficient available hydrologic data that may be used to perform a detailed and precise analysis. Difficulties in the application of empirical relationships arise not so much from the empiricism of equations but more so from the lack of knowledge of exact conditions under which they may be applied. More reliance can be placed, of course, on equations or relationships developed for the region of study.

VIII.3.1   Flood Flows–Area Relationships

Probably, the simplest formulae in general use for estimating flood flows are those that use the parameter: drainage area. These relationships usually take one of the following forms:

$$Q_m = CA^n \dots\dots\dots\dots\dots\dots\dots\dots\dots\dots\dots\dots\dots\text{VIII.7}$$

$$Q_m = CA^m A^{-n} \dots\dots\dots\dots\dots\dots\dots\dots\dots\dots\dots\text{VIII.8}$$

$$Q_m = \frac{CA}{(a + bA)^m} + dA \dots\dots\dots\dots\dots\dots\dots\dots\dots\text{VIII.9}$$

> in which Q_m = flood flow rate,
> A = drainage area size, and
> C, a, b, d; and n,m are coefficients and exponents respectively which must be evaluated for a given region.

A list of the more common of these relationships, which cover world conditions, and those of the Prairies, are shown in Table VIII.3, and in Figs. VIII.3 and VIII.4–from

United Nations ECAFE, Flood Control Series No.7—and Fig.VIII.5—from McKay (1962) and Durrant and Blackwell (1959).

The equations and curves presented above do not contain explicit terms representing frequency factors, and this is a major drawback in making an economic assessment of the design. In general, the relationships are used to define extreme or major flood events that have occurred during the period of record. Under such conditions it would be expected that correlation between peak discharges and drainage areas would be successful, as it is likely that during these events the gross drainage area contributes to flow. Further, as shown by Gray (1961), if watersheds are stratified to regional groupings, then basin size also reflects the influence of several other important lithological features of the basin that influence hydrograph shape.

The popularity of such expressions is enhanced because of their simplicity and because the size of a basin can usually be readily obtained from topographic maps.

VIII.3.2   Peak Flow Formulae—Area Relationships with Frequency Relation Expressed

These formulae have an explicit term for the frequency of occurrence included in the relationship. Some of the most common of these relationships are the following:

Fuller (1914)

$$Q_{t_r} = CA^{0.8} (1 + 0.8 \log t_r) \dots\dots\dots\dots\dots\text{VIII.10}$$

where Q_{t_r} = greatest 24-hr flow, cfs,
A = basin size, mi^2,
t_r = return period, yr, and
C = river coefficient.

Horton (1914) (Eastern Pennsylvania)

$$q_{t_r} = 4021.5\, A^{-0.5}\, t_r^{0.25} \dots\dots\dots\dots\dots\text{VIII.11}$$

where q_{t_r} = flood equalled or exceeded, cfs/mi^2,
A = drainage area, mi^2, and
t_r = return period, yr.

Forsaith (1949) (Prairie Provinces)

$$Q_{t_r} = C\,(32.3\, A^{0.5} t_r^{0.444}) \dots\dots\dots\dots\dots\text{VIII.12}$$

in which Q_{t_r} = peak flood in cfs equalled or exceeded on an average once in a period of t_r years,
C = runoff coefficient whose magnitude depends on watershed characteristics and the location of the area, and
A = drainage area, mi^2.

TABLE VIII.3   COMPARISON OF FORMULAE FOR ESTIMATING FLOOD DISCHARGE IN GENERAL USE

Number	Country	Equations	Units	Particulars about equations	Author	Designation of curve	Remarks
1	2	3	4	5	6	7	8
1.	The World	$Q_m = \dfrac{131,000\ A}{(107+A)^{0.78}}$	E	Max. recorded flood throughout the world.	Baird and McIllwraith	W	From paper by Baird and McIllwraith
2.	Australia	$Q_m = \dfrac{222,000\ A}{(185+A)^{0.9}}$	E		"	A	"
3.	France	$Q = (10 \text{ to } 70)\ A^{0.5}$	M	Mild rain, A between 3,000 and 160,000 sq. km.		$F_1 F_2$	From paper by A. Coutagne
		$Q_m = 150\ A^{0.5}$	M	Violent rain, A between 400 and 3,000 sq. km.		F_3	"
		$Q_m = 54.6 \cdot A^{0.4}$	M	River Garonne, A between 300 and 55,000 sq. km.		F_4	"
		$Q_m = 200\ A^{0.4}$	M	A between 30 and 10,000 sq. km.		F_5	"
		$Q_m = 10.76\ A^{0.737}$	M	Existing dams of "Massif Central"		F_6	"
4.	Germany	$Q_m = 24.12\ A^{0.516}$	M	A between 15 and 200,000 sq. km.		G	"
5.	India	$Q = \dfrac{7,000\ A}{\sqrt{A+4}}$	E	For fan-shape area	Inglis	H_1	Paper by Hunter and Wilmot
		$Q = 1,795\ A^{0.75}$	E	Rain approximately 100 in.	Dickens	H_2	"

TABLE VIII.3 (CONTINUED)

1	2	3	4	5	6	7	8
		$Q = 149\,A^{0.75}$	E	Rainfall 30 to 40 in.	"	H_3	"
		$Q = 675\,A^{0.67}$	E	Maximum flood	Ryves	H_4	"
		$Q = 560\,A^{0.67}$	E	Average floor	"	H_5	"
		$Q = 450\,A^{0.67}$	E	Minimum flood	"	H_6	"
		Curve only	E	Bombay area	Whiting	H_7	"
		$Q = 2{,}000\,A^{(0.92 - \frac{1}{15}\log A)}$	E	Tungabhadra River	Madras formula	H_8	Paper by Rao, K.L.
		$Q = 1{,}750\,A^{(0.92 - \frac{1}{14}\log A)}$	E	Tungabhadra River	Hyderabad formula	H_9	"
6.	Italy	$Q_m = \left(\dfrac{1{,}538}{A+259} + 0.054\right) A$	M	A between 1,000 and 12,000 sq. km.	Whistler	I_1	Paper by Tonini
		$Q_m = \left(\dfrac{600}{A+10} + 1\right) A$	M	A smaller than	Scimemi	I_2	"
		$Q_m = \left(\dfrac{2{,}900}{A+90}\right) A$	M	"	Pagliaro	I_3	"
		$Q_m = \left(\dfrac{280}{A} + 2\right) A$	M	Mountain basins	Baratta	I_4	"
		$Q_{m} = \left(\dfrac{532.5}{A+16.2} + 5\right) A$	M	"	Giandotti	I_5	"
		$Q_m = \left(3.25\dfrac{500}{A+125} + 1\right) A$	M	A smaller than 1,000 sq. km. Max. rainfall 400 mm. in 24 hours	Forti	I_6	"
		$Q_m = \left(2.35\dfrac{500}{A+125} + 0.5\right) A$	M	Maximum rainfall 200 mm. in 24 hours	"	I_7	"
7.	New Zealand	$Q_m = 20{,}000\,A^{0.5}$	E	A smaller than 10 sq. mi.		N	Paper by Coutagne

TABLE VIII.3 (CONTINUED)

1	2	3	4	5	6	7	8
8.	United Kingdom	$Q_m = 2,700\ A^{0.75}$	E	A smaller than 10 sq. mi.	Bransby Williams	K_1	Paper by Hunter and Wilmot
		$Q_m = 4,600\ A^{0.52}$	E	A greater than 10 sq. mi.	"	K_2	"
		Curve only	E		Institute of Civil Eng. 1933	K_3	"
9.	U.S.A.	$Q = 200\ A^{5/6}$	E		Fanning	U_1	Paper by Rao, K.L.
		$Q = \left(\dfrac{46,790}{A+320} + 15\right) A$	E	A between 5.5 and 2,000 sq. mi.	Murphy	U_2	"
		$Q = 1,400\ A^{0.476}$	E	A between 1,000 and 24,000 sq. mi.	U.S. Geological Survey for Columbia	U_3	Paper by Courtagne
		$Q = \left(\dfrac{44,000}{A+170} + 20\right) A$	E	For frequent floods	Kuichling	U_4	Paper by Rao, K.L.
		$Q = \left(\dfrac{127,000}{A+370} + 7.4\right) A$	E	For rare floods	"	U_5	"
		$Q = 4,600\ A^{-0.048}\ A^{-0.048}$	E	Upper limit	Creager	U_6	Paper by Hunter and Wilmot
		$Q = 1,380\ A^{0.894}\ A^{-0.048}$	E	Lower limit	"	U_7	"
		$Q = 10,000\ A^{0.5}$	E		Myer	U_8	"

(Compiled from *Fourth Congress on Large Dams*, 1951; Vol. 2)

Note: Q_m = maximum flood. M = Metric system. (Q in cms. A in sq. km.)
A = drainage area. F = foot-pound system (Q in cfs. A in sq. mi.)

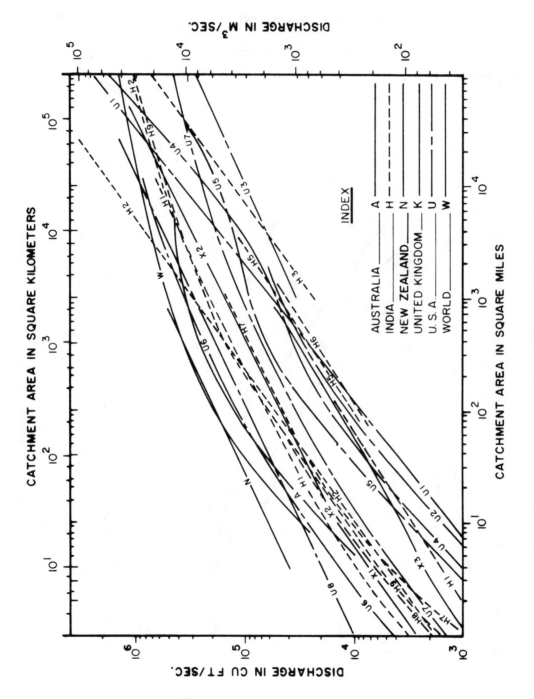

Fig. VIII.3 A Comparison of Flood Formulae in General Use (1).

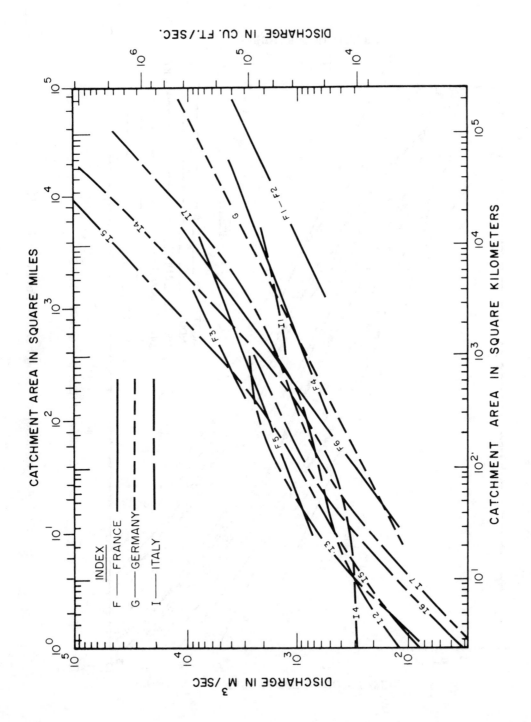

Fig. VIII.4 A Comparison of Flood Formulae in General Use (2).

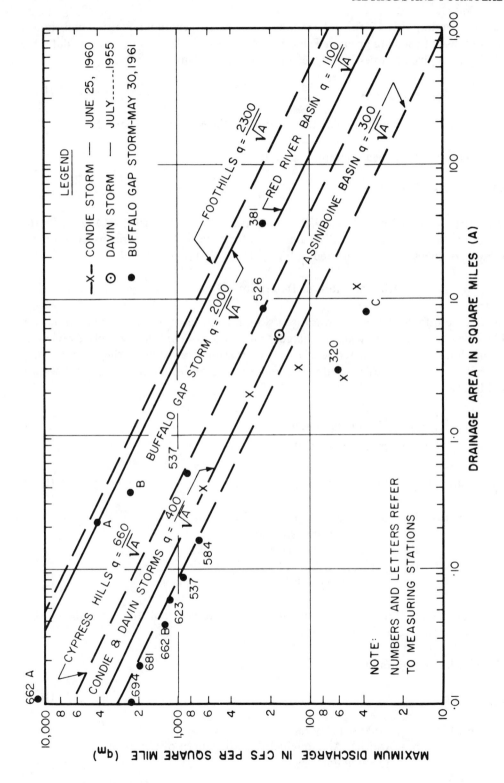

Fig. VIII.5 Envelope Curves of Extreme Floods on Prairies.

The use of Equation VIII.12 for determining peak discharges under Prairie conditions is expedited by use of the data given in Table VIII.4 and Fig. VIII.6. In using these data, the peak discharge of given frequency, Q_{t_r}, can be calculated from

$$Q_{t_r} = FQ_{10} \dots\dots\dots\dots\dots\dots\dots\dots\dots\dots\dots\dots\dots\dots\dots\dots VIII.13$$

in which F =   appropriate frequency factor for the given return period, t_r, (see Fig. VIII.6), and

Q_{10} =   10-yr peak obtained from the figure.

Durrant and Blackwell (1959)

Durrant and Blackwell (1959) reported an exhaustive study of the frequencies-of-occurrence of peak discharge rates for various Prairie streams. A summary of the results of this study are reported in Figs. VIII.7, VIII 8 and VIII.9.

VIII.3.3   Other Methods of Determining Peak Flows

As may be inferred from previous discussion, numerous other empirical methods are available for determining the peak discharges from a watershed; but a complete review of all these methods is beyond the scope of this text. Inasmuch as some of the formulae may serve a useful purpose in design, and also as guides for establishing relationships, they are reviewed briefly below.

VIII.3.3.1   Area, Rainfall and Time Parameter
(Soil Conservation Service)

$$Q_p = 484\ AP_e \Big/ P_R \dots\dots\dots\dots\dots\dots\dots\dots\dots\dots\dots\dots\dots\dots VIII.14$$

in which Q_p =   peak discharge rate (cfs),
A =   drainage area size (sq. mi.),
P_e =   rainfall excess (in), and
P_R =   period or time of rise of the hydrograph (hr).

Note: P_R can be estimated from topographic characteristics of the watershed using empirical formulae. (See Subsections VIII.2.1.2. and VIII.5.3).

VIII.3.3.2   Width and Rainfall Parameter

$$Q_{t_r} = CPb^{1.25} \dots\dots\dots\dots\dots\dots\dots\dots\dots\dots\dots\dots\dots\dots\dots\dots VIII.15$$

in which P =   100-yr, 1-day rain,
b =   width = A/L in which L is the length of the basin, and
C =   regional coefficient.

Table VIII.4 Values of Runoff Coefficient 'C'

Value of 'C'	
2.0	The extreme rate. For steep, partially wooded watersheds in the higher elevations of the Cypress Hills where conditions favor a high runoff rate.
1.5	For fairly steep, hilly topography in the Chinook Area. Use for the area south of the Cypress Hills and Wood Mountain, and for the north slope of the Cypress Hills where conditions favor a high runoff rate.
1.0	The normal rate. This value could be used with reasonable safety where there are no factors indicating an exceptionally high or low rate. — For level to moderately rolling topography south of the Cypress Hills and Wood Mountain area. — For moderately steep to hilly topography north and east of Cypress Hills. — For moderately to strongly rolling topography in Southeastern Saskatchewan and Southwestern Manitoba
0.75	For level to gently rolling topography north and east of the Cypress Hills and Wood Mountain area. For moderately rolling topography in Southeastern Sask. and Southwestern Manitoba.
0.50	The practical lower limit. For level to gently rolling topography with only fair surface drainage in Southeastern Sask. and Southwestern Manitoba.

(after Forsaith 1949)

VIII.3.3.3 Width, Rainfall Parameter and Slope

$$Q_p = C P_e^{4/3} b^{4/3} S_c^{4/9} \quad \dots\dots\dots\dots\dots\dots\dots\dots\dots \text{VIII.16}$$

in which P_e = precipitation excess for a period
equal to t_c, and
S_c = slope of the main channel.

VIII.3.3.4 Area, Slopes, and Rainfall Parameter

(Potter, U.S. Bureau of Public Roads, 1961)

Coaxial plotting of peak discharge rates of given period (usually 10-year) with a precipitation index (depth of rainfall in inches equalled or exceeded during 60-minute

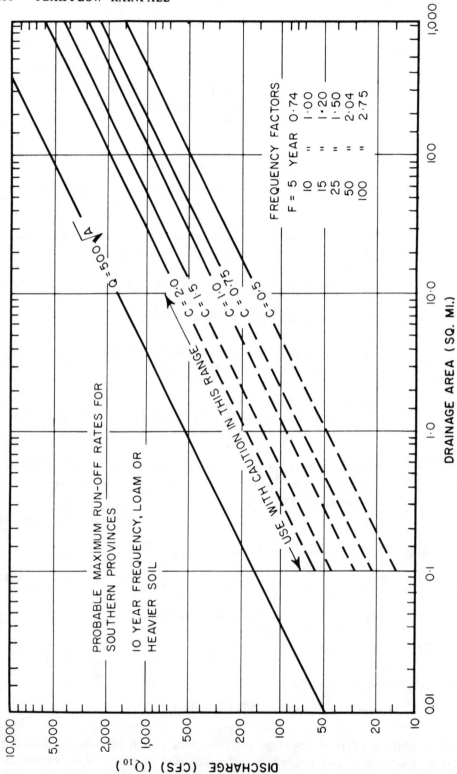

Fig. VIII.6 Forsaith Runoff Curves for a 1:10 Year Event.

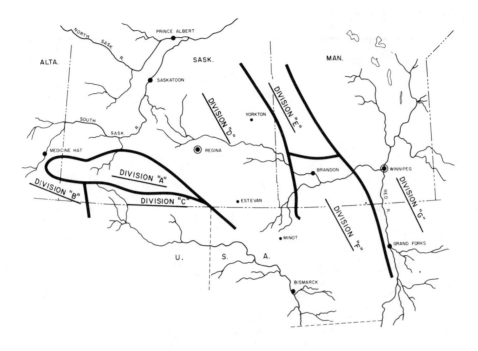

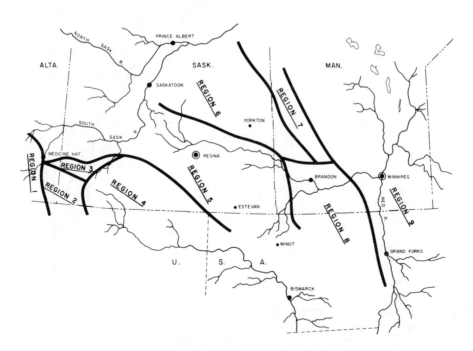

Fig. VIII.7 Flood Frequency Divisions and Mean Annual Flood Regions
on the Canadian Prairies.

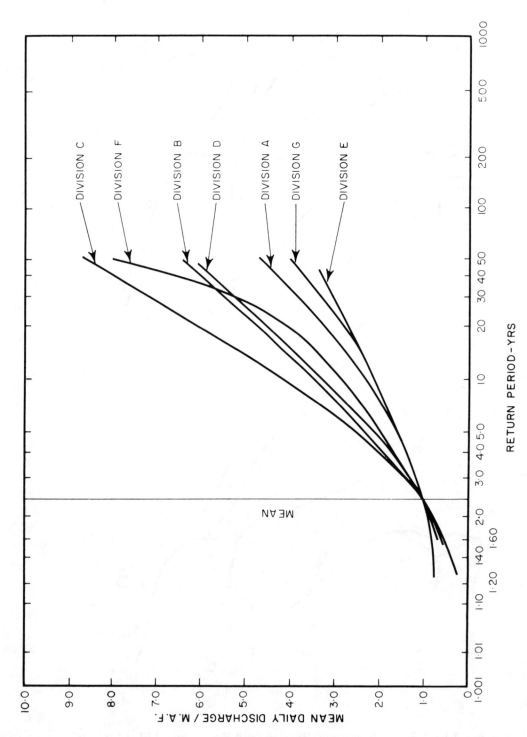

Fig. VIII.8 Ratio of Mean Daily Discharge to Mean Annual Flood with Return Period by Divisions.

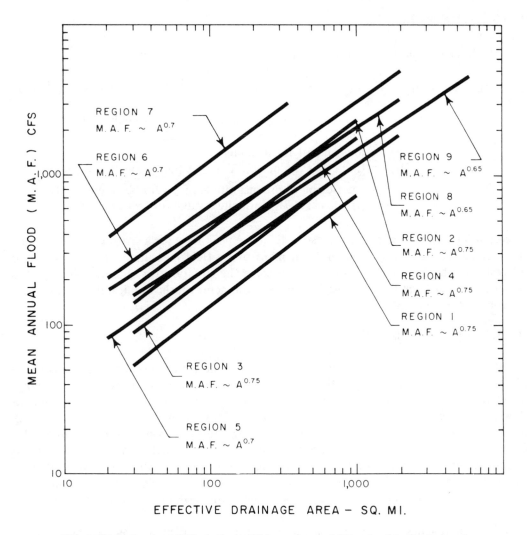

Fig. VIII.9 Regional Variation of Mean Annual Flood with Drainage Area.

period once in 10 years), area, and a topographic index $(0.3L/\sqrt{S_1} + 0.7L/\sqrt{S_2})$ in which S_1 and S_2 are slopes of the upper $0.3L$ and $0.7L$ of the main stream respectively.

VIII.3.3.5   Area, Elevation, Length and Lakes
(Kinnison 1945 – Minor Floods)

$$Q_p = (0.00036h^{2.4} + 124) A^{0.95} /a^{0.04} L^{0.7} \dots\dots\dots\dots\dots\dots\dots\dots VIII.17$$

in which h = median altitude of basin above outlet (ft),
A = drainage basin size (sq mi),
a = percentage of total area in lake, ponds or reservoirs, and
L = average distance water must travel to the outlet (mi).

VIII.3.3.6   Meander Equation

Inglis (1949) suggests a meander equation of the form

$$M_L = C \sqrt{Q_{t_r}} \qquad \dotfill \text{VIII.18}$$

in which M_L = the meander length measured along roughly a straight
line connecting the inflection points at the beginning
and end of the meander (ft),

Q_{t_r} = flood that can be considered as the 100-yr flood, and

C = coefficient whose magnitude may vary between 18–42
(Blench, 1957). Normally a value for $C = 28$ is used.

Equation VIII.18 is used to give an estimate of the peak flow for a stream by determining
the average meander length either from maps or by field measurement at the point of
interest, then by assuming a value for C to solve the equation for Q_{t_r}. Blench suggests a
slightly different form for the equation and his comments on the applicability of meander
equations are of interest.

VIII.3.3.7   Flow Records and Area

Discharge records published by agencies such as the Water Survey of Canada Division
of the Department of Energy, Mines and Resources, are usually in the form of average
daily discharges. Such records are of little value for studying the peak instantaneous
flows, particularly on small watersheds. The obvious method of obtaining the peak
instantaneous flows on a watershed is by review of the recorded stage hydrographs.
Generally, however, these records are not readily available. Further, the task of analysis is
extremely time-consuming if a long period of record is involved. An alternative method to
the above that is much more expedient, is to utilize the average daily discharges by
applying a conversion factor to these flows to derive the peak instantaneous discharges.

Two approaches are reported in the literature for resolving the relation between the
peak instantaneous discharge with another flow parameter and/or a geomorphic property
of the watershed. Fuller (1914) studied the relationship between the peak instantaneous
discharges and the average 24-hr. flows of floods on large basins located in the Eastern
United States. He developed a relationship between these variables of the form

$$\frac{Q_m - Q_{24}}{Q_{24}} = 2A^{-0.30} \qquad \dotfill \text{VIII.19}$$

where Q_m = maximum recorded instantaneous discharge (cfs),

Q_{24} = largest average rate of flow over a 24 consecutive hour
period which was recorded during the same period of
years as Q_m (the two discharges did not necessarily
occur during the same flood), and

A = drainage basin area (sq mi).

In establishing the *excess ratio*, $(Q_m-Q_{24})/Q_{24}$ of Equation VIII.19, Fuller rationalized that for all floods Q_m would be greater than Q_{24}. In addition, it was postulated that the excess ratio would decrease with an increase in drainage basin size. For large watersheds, the rate of runoff is generally high for at least 24 hrs because the runoff-producing storm is of considerable duration, whereas on small basins a cloudburst may cause a flood which will give runoff over a few hours resulting in a large Q_m and only a moderate Q_{24}.

Langbein (1944) employed a different approach to solve the problem by using only flow records. He established a relationship between the peak instantaneous discharge and maximum daily discharge in terms of the ratio of the daily discharge of the preceding day to the discharge of the maximum day and the ratio of the daily discharge of the

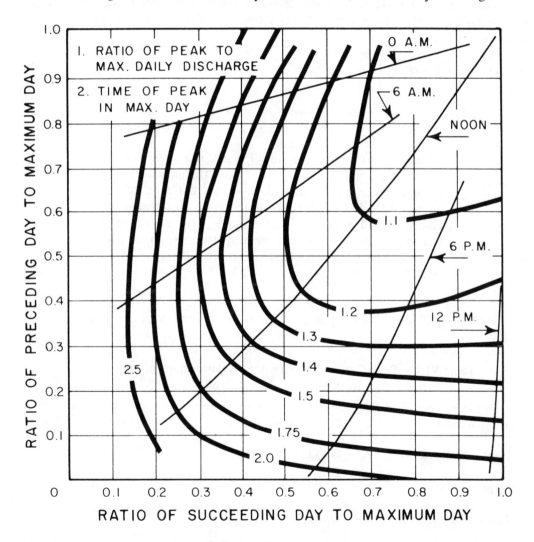

Fig. VIII.10 Discharge at, and the Time of, Crest in Relation to Daily Mean Discharges.
(Redrawn from Langbein)

succeeding day to that of maximum day. A plot of these ratios as presented by Langbein is given in Fig. VIII.10.

Ellis (1964) utilized a similar approach, as those suggested by Fuller and Langbein, to establish the interrelationship of flow characteristics of small Prairie streams. In this study, 260 hydrographs from 22 small watersheds which varied in size from 12–1000 sq mi located in the central region of Canada were studied. The results of this investigation reported by Ellis and Gray (1966) are summarized below.

Relationship Between the Peak Ratio of Summer Storm Events and Effective Area.

The average peak ratios for summer storm events from watersheds within a given physiographic region were related to the effective area of the drainage basin in equational form as

$$Q_p/Q_2 = CA_e^{-n} \quad \dots\dots\dots\dots\dots\dots\dots\dots\dots\dots\dots\dots\dots\dots\dots\dots\dots\text{VIII.20}$$

Q_p = peak instantaneous discharge rate (cfs),
Q_2 = average daily flow on the day of occurrence of the peak rate (cfs),
A_e = effective drainage area (sq mi). The effective area, as defined by Stichling and Blackwell (1957) is: "the gross area of the basin less the areas which do not contribute to the mean annual flood." That portion of the gross area which is excluded includes marshes and sloughs with no connecting channel to the stream, and areas upstream from on-stream lakes that have the capacity to detain runoff waters until the peak flow has been observed.

Values of 'C' and 'n' of Equation VIII.20 determined by Ellis are given in Table VIII.5.

Table VIII.5 Peak Ratio–Effective Area Relationships of Watersheds in Different Regional Groups.

Region	Equation	Range of Validity
Foothills (Rockies)	$Q_p/Q_2 = 3.9 \, A_e^{-0.22}$	60–300 mi²
Cypress Hills	$Q_p/Q_2 = 10 \, A_e^{-0.46}$	50–200 mi²
Central Plains	$Q_p/Q_2 = 11 \, A_e^{-0.36}$	45–225 mi²
Manitoba Escarpment	$Q_p/Q_2 = 3.7 \, A_e^{-0.38}$	15– 50 mi²

Langbein's Approach

In accordance with the approach suggested by Langbein a graph was constructed using the following flow data: (a) the average daily flow for the day preceding the peak rate, Q_1 ; (b) the average daily flow on the day following the peak, Q_3 ; and (c) values of Q_p and Q_2. These data, plotted as dimensionless ratios, are shown in Fig. VIII.11.

The most significant feature of the figure is that the discharge characteristics of the streams analyzed are shown to bear a close resemblance to those of the more amenable streams studied by Langbein (Fig. VIII.10). In general, because of scarcity and limitations of the original data, it was impossible to locate contours for values of Q_p/Q_2 greater than 2.5. Other areas of fewest points, and thus of least reliability, are delineated on the Figure. The practical use of Fig. VIII.11 is as an aid for determining Q_p when Q_1, Q_2 and Q_3 are known.

VIII.4   UNIT HYDROGRAPH

It has been pointed out that the Rational Method is limited to small watersheds on which storage effects are relatively unimportant to the surface runoff phenomena. On larger areas, on which the effect of variations in areal distribution of the storm, storage effects, and other factors on the runoff pattern cannot be neglected, resort can be made to the use of the infiltration theory in conjunction with the unit hydrograph principle and flood routing methods to determine peak flows.

VIII.4.1   Basic Principles

In 1932, L.K. Sherman advanced the theory of the unit hydrograph (or unit graph) now recognized as one of the most important contributions to hydrology related to surface-runoff phenomena. A unit graph is a discharge hydrograph resulting from *one inch of direct runoff generated uniformly over the tributary area at a uniform rate during a specified period of time.*

The theory is based in principle on the criteria (Johnstone and Cross, 1949):

1. For a given watershed, runoff-producing storms of equal duration will produce surface runoff hydrographs with approximately equivalent time bases, regardless of the intensity of the rain.
2. For a given watershed, the magnitude of the ordinates representing the instantaneous discharge from an area will be proportional to the volumes of surface runoff produced by storms of equal duration.
3. For a given watershed, the time distribution of runoff from a given storm period is independent of precipitation from antecedent or subsequent storm periods.

The first criterion cannot be exactly correct because the effect of channel storage will vary with stage. However, since the recession curve approaches zero asymptotically, a practical compromise is possible without excessive error (Linsley *et al.*, 1949) p. 445.

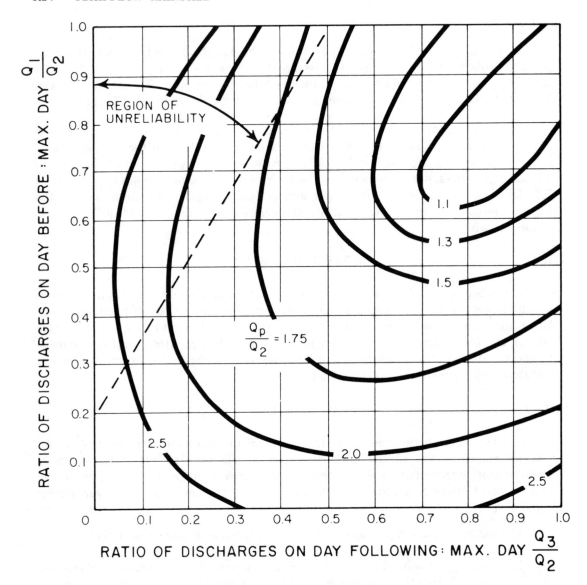

Fig. VIII.11 Peak Ratio Contours in Terms of Day Before, Day of, and Day Following Maximum Discharge.

In addition, the effective gradient and the resistance to flow change with the magnitude of the flood wave.

Sherman (1940) confirmed the hypothesis regarding the proportionality of ordinates provided that the selected time unit is less than the minimum concentration period. This was accomplished by reducing the quantitative phenomena of rainfall, loss, pondage and runoff to a problem of hydraulics that could be solved by well-known and accepted hydraulic formulae.

With respect to the third criterion, antecedent precipitation is important to the runoff phenomena primarily because of its effect on the soil infiltration rate, depressional and detention storage, and the resultant total volume of runoff occurring from a given storm.

The unit graph theory has been generally accepted by most hydrologists. Its use as a hydrologic tool is perhaps best summarized by Mitchell (1948):

> There has been developed no rigorous theory by which the unit-hydrograph relations may be proven. However, the results which have been obtained by a judicious application of the relationship have been so predominantly satisfactory that there can be no doubt that it is indeed a tool of considerable value for resolving to some extent the complex relations of rainfall and runoff and for advancing the science of hydrology.

VIII.4.2   Unit Storm and Unit Graph Duration

Theoretically, an infinite number of unit graphs are possible for a given basin because of the effects of rainfall duration and distribution. It is necessary for practical considerations, however, to know the tolerance or range of unit-storm periods within which a given unit graph is applicable. This information is required for the synthesis of a hydrograph for a storm of long duration and the development of a representative unit graph for an area.

Several investigators have expressed different opinions, based on experience, regarding the critical rainfall duration for a given basin. Wisler and Brater (1959) employ a unit storm, defined as a storm of such duration that the period of surface runoff is not appreciably less than for any storm of shorter duration. The authors found that an appropriate duration of the unit storm varies with characteristics of the basin. For small watersheds (areas less than 10 sq mi), unit hydrographs result from short, isolated storms whose durations are less than the period of rise. For large watersheds, however, the unit-storm duration may be less than the period of rise, possibly no more than half as long. Wisler and Brater recommend, in applying the distribution graph to a given storm sequence on *small watersheds,* that

> The volume of rainfall excess may be converted to runoff by means of a single application of the distribution graph, if its duration is no longer than the period of rise. The graph resulting from a longer rain must be derived by successive applications of the distribution graph to unit durations of rainfall excess.

For the *larger areas* they conclude

> The distribution graph is not a sufficiently precise tool to be sensitive to differences in duration of rainfall excess that are small compared with the period of rise.... It will require further research before enough

experimental evidence is available to establish the nature of the variation for small changes in duration.

The more common principle is to associate the unit graph with the storm from which it was produced. For example, for a given area there may be a 2-hr. unit graph or a 6-hr. unit graph, depending on whether the unit-storm duration was either 2 hrs. or 6 hrs., respectively, and provided that the time of concentration of the basin had not been exceeded. Unit graphs for various storm durations can be developed from one of known duration using the S-curve technique (Linsley *et al.*, 1949, p. 451 ff.).

The selection of a proper time period for unit graphs is important. Sherman (1949), suggests the following criteria be used in its selection:

> For areas over 1000 square miles use 12-hour units in preference to 24 hours. For areas between 100 and 1000 square miles use units of 6, 8 or 12 hours. For areas of 20 square miles use 2 hours. For smaller areas use a time unit of about one-third or one-fourth of the approximate concentration time of the basin.
>
> (p. 524)

Mitchell (1948) recommends that the storm duration or unit-graph duration which is most convenient for use on any basin is about 20 per cent of the time between the occurrence of a short storm of high intensity and the occurrence of peak discharge. He states:

> The effect upon the unit hydrograph becomes significant only when there is substantial variation between the unit-hydrograph duration and the storm duration .... It is usually permissible to allow the storm duration to vary between 50 per cent and 200 per cent of the unit-hydrograph duration before any correction factor for this effect will become necessary.
>
> (p. 30)

Linsley *et al.* (1949), p. 195, cite that in practical applications, experience has shown that the time unit employed should approximate one-fourth of the basin lag time (time from the centre of mass of effective precipitation to the peak of the unit graph). They suggest that the effect of small differences in storm duration is not large and that a tolerance of ±25 per cent from the adopted unit-hydrograph duration is acceptable.

Yet another criterion is adopted by the United States Army Corps of Engineers (1948), p. 8. They found that, for drainage areas of less than 100 sq. mi., values of the unit-storm duration equal to about half the basin lag time appear to be satisfactory.

VIII.4.3   Distribution Graph

As an outgrowth of the unit-graph principle, Bernard (1935) conceived the concept of the distribution graph. A distribution graph is a unit graph of surface runoff modified to show the proportional relation of its ordinates, expressed as percentages of the total surface-runoff volume. In accordance with the unit graph principle, if the base time of the unit graph is divided into any given number of equal time increments, the percentage of the total volume of flow that occurs during a given time interval will be approximately the same, regardless of the magnitude of total runoff.

Since the area under each distribution graph is equal to 100 per cent, differences in the runoff characteristics between watersheds are reflected in the respective shapes of their distribution graphs. The distribution graph is used in preference to the unit graph when hydrograph characteristics from areas of different size are compared.

VIII.4.4   Unit-Graph Derivation for a Basin from Streamflow Records

The data required for deriving a unit graph are simultaneous measurements of rainfall and runoff from the basin for a number of years, and also, preferably some estimate of the infiltration rate. The procedure is, basically, to choose several (preferably at least four or five) rain-storm periods for which the resulting runoff hydrographs are available. These hydrographs are subsequently reduced to unit graphs from which an average unit graph for the basin is derived.

The storm or rain periods chosen should be those of high intensity and with similar areal distributions of rainfall, preferably those which cover the whole basin. Preference should also be given to isolated storms of uniform intensity, as this simplifies the analysis.

Assuming, for simplicity, that there is one isolated storm of uniform, high intensity with a duration meeting the requirements outlined previously, the procedure to obtain the unit hydrograph for the storm is as follows:

The hydrograph associated with the storm is first separated into its component parts (as described in Section VII.4.1) and the surface runoff or direct runoff hydrograph is obtained. The volume of surface runoff, V, is obtained by planimetering the area under the hydrograph. This volume of runoff, which normally is expressed in units of cfs. days can be converted to a depth, P_n, in inches of net storm rain over the basin by the equation:

$$1.008 \, P_n A = 24V \quad \dots \dots \dots \dots \dots \dots \dots \dots \dots \dots \dots \dots \dots \dots \dots \dots \text{VIII.21}$$

or

$$P_n \cong 24 \, V/A \quad \dots \dots \dots \dots \dots \dots \dots \dots \dots \dots \dots \dots \dots \dots \dots \dots \text{VIII.22}$$

in which P_n =   net storm rain (in),
     A =   drainage basin size (acres), and
     V =   volume of surface runoff (cfs-days).

To reduce the hydrograph to a unit volume of excess rain (1-in.) each of the ordinates (discharges) is divided by P_n. The result is a unit graph for a storm of duration equal to the effective duration of the rain. The effective duration can be considered to be the duration of excess rain, or net storm rain, or the time period from the point where surface runoff becomes apparent on the graph to the end of the rainfall period. If this period is, for example, six hrs. then the unit graph is called a six-hr. unit graph.

VIII.4.4.1   Unit Graphs for Complex Storms

Often it is necessary to use complex storms for derivation of unit graphs. In such cases, if the storm pattern is comprised of several isolated periods of rainfall, then hydrograph separation methods as previously described for complex storms can be applied, and unit graphs obtained for each rain period. However, a storm will often consist of consecutive periods of varying rainfall intensity, and the effects of these variations cannot be separated in the hydrograph. In these cases the unit graph is obtained in one of three ways: (a) by successive approximations or trial and error techniques (Collins 1939); (b) by graphical procedures (Volker 1964); or (c) by solving systems of equations simultaneously (Linsley *et al.* 1958).

The Collins Method

A detailed description of this method follows (Water Conservation and Irrigation Commission—Australia, Mr. J.J. Tooney):

> This method involves the finding of a set of coefficients for a rain occurring in more than one time interval by a series of trial and error approximations.

The basic steps of the method are:

1. Assume a unit graph by considering all rains combined into one average rain and apply it to all effective rains except the largest.
2. Subtract the resulting hydrograph from the actual hydrograph of the surface runoff and reduce the residual to unit graph terms.
3. Use the computed coefficients as a revised approximation for the next trial.
4. Repeat steps 1, 2 and 3 until the residual unit graph is in agreement with the assumed unit graph.

The means of determining the portion of each rain that should be considered as runoff and the manner of plotting a distribution graph from its coefficients are now discussed.

The procedures are as follows:

1. Prepare an isohyetal map for the rainfall period which produced the hydrograph under study.
2. Prepare mass curves of rainfall from the synoptic stations' reports and pluviometer traces (if any) in/or adjacent to the catchment. These mass curves

are then converted to a percentage basis so that they may be compared, and if possible, a single mean mass curve is determined to apply to the catchment generally.

3. The mean mass curve is converted to an hyetograph with a unit time period equal to the unit time period of the unit graph. (See Table VIII.6, col. 2.)

4. To the hyetograph apply a loss rate (ϕ index) and determine the excess rainfall in each unit period. (See Table VIII.6, cols. 3 and 4.)

5. The base flow is separated from the hydrograph (usually by drawing a straight line from the point of rise to the end of surface runoff).

6. The resulting hydrograph, representing surface runoff, is converted into histogram form (distribution graph) with each time period equal to the selected unit period and tabulated. (See Table VIII.6, col. 5.)

7. The time base of the unit graph is equal to the time from the end of excess rain to the end of surface runoff plus one unit period.

8. The unit period should be equal to or less than 1/4 of the period of rise, and the total number of periods should be sufficient to define the hydrograph and unit graph clearly.

9. The distribution coefficients of the unit graph (percentage of total runoff occurring in unit time) are assumed and arranged. (See Table VIII.6, cols. 6 to 17.)

10. The first excess rainfall is multiplied by a constant which is equal to the discharge (cfs) flowing for the unit period adopted necessary to produce 1 in. of runoff from the whole catchment. In the example (Table VIII.6), the catchment area is 690 sq. mi. and the unit time period is 6 hrs. Therefore 1 in. of excess rainfall over the catchment =

$$\frac{1}{12} \times \frac{690 \times 640 \times 24}{2 \times 6} = 73,600 \text{ cfs}-1/4 \text{ days.}$$

11. The product obtained from procedure 10 is multiplied by each coefficient of the assumed unit distribution graph in turn and arranged diagonally across the page. (See Table VIII.6, cols. 6 to 17.)

12. This is repeated for all the excess rainfalls except the largest, and arranged as in procedure 11. (The first discharge produced by the first excess rainfall should be entered opposite the first excess rainfall). (See Table VIII.6, col. 6.)

13. The resulting discharges of cols. 6 to 17 are summed horizontally, and each individual sum is entered in col. 18.

14. The discharges of col. 18 are subtracted from the corresponding discharges of col. 5, and the residuals entered in col. 19.

15. The residuals are converted into unit distribution graph coefficients by dividing each residual by the largest rainfall expressed in cfs. unit days and by multiplying by 100. It should be noted that the algebraic sum of these computed coefficients should total 100 per cent.

16. The algebraic sum of the computed coefficients that do not apply to the discharges produced by the largest rainfall are redistributed over the remaining

Table VIII.6 Example Calculation of Unit-Graph Synthesis from a Complex Storm by the Collins Method

1	2	3	4	5	6	7	8	9	10	11	12	13	14	15	16	17	18	19	20
Time Period	Total Rain	Loss Rate (φ 6 hr.)	Excess Rain	Net Q			\<--------------------- Trial Coefficients % ---------------------\>										M	Residuals	Coefficients
							12.4	17.4	15.7	13.0	12.3	11.5	7.0	5.5	3.0	2.2			
1	0.14	0.30	0	0	0	0													
2	0.27	0.30	0	0	0	0													
3	0.37	0.30	0.07	0	0	0													
4	0.51	0.30	0.21	0	0	0													
5	0.88	0.30	0.58	100	0	0													
6	1.58	0.30	1.28	1,250	-	-	639											639	-539	-0.5
7	2.02	0.30	1.72	7,080	0	0	1,917	896										2,813	-1,563	-1.2
8	1.13	0.30	0.83	18,920	0	-	5,294	2,690	809									8,793	-1,713	-1.3
9	0.62	0.30	0.32	41,330	0	0	11,682	7,428	2,427	670	633							22,207	-3,287	-2.6
10	0.35	0.30	0.05	54,580	0	0	-	16,393	6,702	2,010	1,902	592						25,738	15,592	12.3
11	0.16	0.30	0	54,420	0	0	7,575	10,630	14,791	5,550	5,251	1,778	361	283				30,410	24,170	19.0
12				49,120			2,920	4,098	9,591	12,247	11,588	4,909	1,082	850	155			33,187	21,233	16.7
13				43,670			456	640	3,697	7,942	7,514	10,834	2,988	2,348	464	113		32,007	17,113	13.5
14				36,500					578	3,062	2,897	7,025	6,595	5,182	1,281	340		27,106	16,564	13.1
15				26,750						478	453	2,708	-	3,360	2,826	939		20,674	15,826	12.0
16				19,000								423	4,276	1,295	-	2,073		17,203	9,547	7.5
17				11,250									1,649	202	1,833	1,344		11,201	7,799	6.1
18				6,670									258		-	518		7,505	3,745	2.7
19				2,920											110	81		3,386	3,284	2.6
20				830														628	202	0.5
21				0														81	-81	0.2
					0	0	11.8	18.2	16.0	12.9	12.5	11.5	7.2	5.8	2.6	2.5	Adjusted Coefficients			-
					0	0	11.8	18.2	16.0	12.9	12.5	11.5	7.2	5.8	2.6	2.4				
					0	0	12.1	17.8	15.8	13.0	12.4	11.5	7.1	5.6	2.8	2.3	Accepted Coefficients			
					0	0	12.0	17.7	15.7	13.0	12.3	11.5	7.1	5.6	2.8	2.3				

coefficients, which are then adjusted until they total 100 per cent.

17.   If the differences between the assumed coefficients (Table VIII.6, cols. 6 to 17.) and the adjusted coefficients are marked, then the adjusted coefficients should be adopted for the next trial. When the adjusted and assumed coefficients vary only by a small amount and a closer agreement is desired, the new trial coefficients are a weighted average of the adjusted and assumed coefficients (see example of weighting).

If

N = the sum of residuals produced by the largest rainfall,

M = the sum of the discharges over the period when the largest rainfall would be producing runoff had it been included in the calculations,

C_1 = trial coefficients

C_2 = computed coefficients

$\overline{C}$ = coefficients to be adopted for the next trial;

then
$$\overline{C} = \frac{M\,C_1 + NC_2}{M + N}$$

When the excess rainfalls are few in number and the largest is very large compared with the rest, the coefficients should converge rapidly. This is true because an error made in assuming the trial coefficients does not cause as large a percentage error in the result, since the trial coefficients are applied to smaller runoff amounts than that from which the new coefficients are found.

In the example, the duration of excess rainfall extends over a long period compared with the duration of surface runoff, and some of the excess rainfall volumes are not small compared with the largest rainfall volume. Consequently a large number of trials was necessary before agreement between the trial and computed coefficients was reached.

In this particular example, the presence of negative coefficients prior to the coefficients effected by the largest rainfall volume (shown in column 20) is attributed to an error in the determination of the first excess rainfall periods, or in the surface runoff for the early periods.

Conclusion

Due to the variation of intensity and areal distribution of precipitation within individual periods, no storm will give distribution coefficients suitable to all other storms. It is therefore considered necessary to analyse several storms of different areal distribution and intensity. Either a mean unit graph is constructed by averaging the derived unit graphs, or the unit graph derived from the storm with similar characteristics is used.

Care should be exercised in the selection of loss rates applicable to small rainfalls. It appears that this factor has the greatest effect on the accuracy of the method when the rainfalls are small.

VIII.4.5   Derivation of Unit Graphs of Different Durations
           from One of Known Duration by the S-Curve Technique

The unit graphs obtained by analysing a number of storm events will normally have different durations. If the variation in storm duration is small (± 25%), then the unit graphs can be compared, and a single representative unit graph obtained for the basin having a duration equal to the average duration of the storms. If there are wide variations in the periods of precipitation excess from which the unit graphs have been developed, then the graphs should be reduced to a common unit duration before a representative unit graph is derived.

The basic concepts of the unit graph (discussed in Subsection VIII.4.1) indicate how unit graph durations may be changed. If, for example, a six-hr. unit graph is added to the same six-hr. unit graph beginning six hours later, the resulting hydrograph is that which would be obtained from a storm producing one in. of rainfall in each of two, six-hr. periods. The hydrograph would be a twelve-hr. hydrograph with a volume of two in. of rainfall. This can be reduced to a twelve-hr. unit hydrograph by dividing the ordinates by two—the volume of surface runoff in inches over the basin.

As can be seen it is simple to increase the duration of a unit graph by integer factors. It is more difficult however to reduce or increase the duration by fractional parts. For these latter cases, the S-curve technique is used. The S-curve is the hydrograph that would result from an infinite series of runoff increments of one inch in t_O hrs., or in other words, the hydrograph for a continuous storm with an intensity of $1/t_O$ in/hr. (See Fig. VIII.12.)

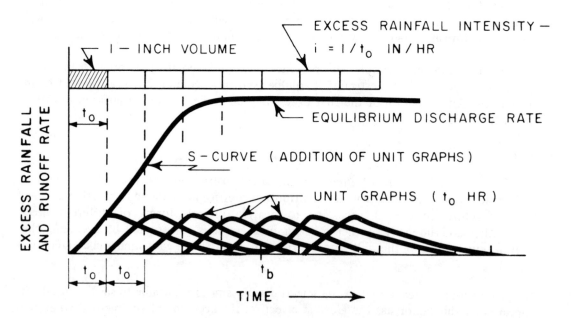

Fig. VIII.12 Development of an S-Curve Hydrograph

The S-curve is constructed by adding together a series of t_0-hr. duration unit hydrographs each lagged, t_0 hr., with respect to the preceding one. If the time base of the unit graph is t_b hr., then a continuous excess rainfall with intensity $1/t_0$ would produce a constant outflow after about t_b hrs., of A/t_0 acre-in./hr. (A is drainage area in acres, and note: 1.008 x acre-in./hr. = 1 cfs. Thus, only t_b/t_0 unit graphs need be combined to produce the S-curve up to the equilibrium flow conditions.

In actual practice one will usually find that there is some fluctuation of the S-curve about the equilibrium discharge. Such fluctuations usually occur because of the lack of precision in selecting the unit graph duration. That is, the actual duration of the unit storm may differ slightly from the duration used in the calculation. Nevertheless, an average S-curve can usually be drawn through the points without too much difficulty.

The difference between two S-curves displaced by t_0' hrs. gives a hydrograph of duration t_0'. The subtraction of the second S-curve effectively means that the excessive rainfall occurs at a rate of $1/t_0$ inches per hr. for t_0' hrs. producing a volume of t_0'/t_0 in. The unit graph for a duration of t_0 hrs. then is produced by multiplying the ordinates of the t_0'-hr. hydrograph by t_0/t_0'.

As mentioned previously, a representative unit graph for the basin may be developed from the graphs that have been reduced to a common duration by the S-curve method by comparing them on a single plot. In constructing the representative graph, averages of the unit hydrograph ordinates at a particular time should not be taken, instead the average peak value and average time of peak is obtained, and a unit graph similar in shape to those analysed is sketched through the peak, and its volume (area under the curve) adjusted to unity.

VIII.4.6   Instantaneous Unit Graph

The S-curve method can be applied to give unit graphs of any finite duration, but an extension of the method is necessary to give the instantaneous unit graph for a basin. The instantaneous unit graph can be thought of as the hydrograph of surface runoff that would result if one inch of excess rainfall was precipitated instantaneously over the whole basin. The resulting hydrograph would have a finite peak and base time, as some time would be required for the flow to reach the outlet from the furthest reaches of the basin. In fact, the base time of the instantaneous unit graph must be the time of concentration of the basin. The instantaneous unit graph is important as it is indicative of the basin storage characteristics, since rainfall duration effects are eliminated.

The instantaneous unit graph may be determined from the S-curve for the basin—see Fig. VIII.13. Assuming an S-curve has been derived for an excess rainfall intensity of 1 in./hr., or in other words, from a 1-hr. unit graph, the hydrograph for a duration of dt hrs. could be derived by subtracting the ordinates of the S hydrograph shifted dt hrs. to the right. The ordinate to this hydrograph at time t would be $S_2 - S_1$ (see Fig. VIII.13). The volume of the new hydrograph would be idt where i is the rainfall excess intensity (1 in./hr. in this example). The unit graph for a duration dt would then have an ordinate of $(S_2 - S_1)/$ idt or $(S_2 - S_1)/dt$ at time t hrs. It is evident that as dt approaches zero,

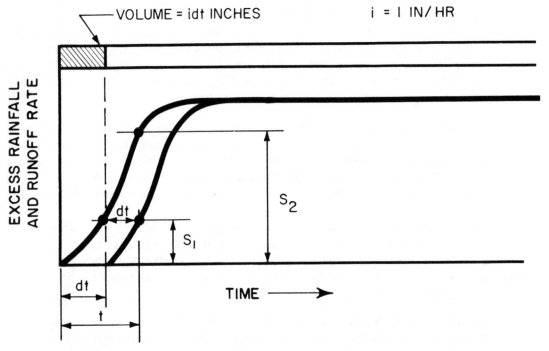

Fig. VIII.13 Definition Sketch for Determination of an Instantaneous Hydrograph

$(S_2 - S_1)$ also approaches zero, but $(S_2 - S_1)/dt$ approaches dS/dt or the slope of the S curve at time, t. Therefore, the ordinate, u_t of the instantaneous unit graph is

$$u_t = \left(\frac{dS}{dt} \right)_{dt \to 0} \quad \dots \dots \dots \dots \dots \dots \dots \dots \dots \dots \dots \dots \dots \dots \dots \dots \dots \dots \dots \text{VIII.23}$$

and

$$dS = u_t \, dt \dots \dots \dots \dots \dots \dots \dots \dots \dots \dots \dots \dots \dots \dots \dots \dots \dots \dots \text{VIII.24}$$

That is, the ordinate of the instantaneous unit graph at any time, t, can be taken as the slope of the S-curve (derived from a 1-hr. unit graph) at that time.

Further, the difference between the ordinates of two S-curves, S_2, S_1, which are displaced a time interval $t_2 - t_1$ apart, can be obtained by integrating Equation VIII.24. That is

$$S_2 - S_1 = \int_{t_1}^{t_2} u_t dt \cong \frac{1}{2} \left(u_{t_1} + u_{t_2} \right) \left(t_2 - t_1 \right)$$

or

$$\frac{S_2 - S_1}{t_2 - t_1} = 1/2 \left(u_{t_1} + u_{t_2} \right) \quad \dots \dots \dots \dots \dots \dots \dots \dots \dots \dots \dots \dots \dots \dots \dots \text{VIII.25}$$

Note: according to Equation VIII.25 the difference between two S-curves of any time (unit graph co-ordinate) is approximately equal to the average of the instantaneous unit graph co-ordinates in the selected time interval.

VIII.4.7   Application of the Unit Graph

Once a unit graph has been developed for a basin, it can be used to obtain the surface runoff hydrographs for storm events on the basin. The runoff records can be extended, then, for periods in which rainfall was measured but runoff was not. If the unit graph is applied to the maximum probable rain storm for the basin, then the maximum probable flood peak may be obtained. The procedures for obtaining hydrographs from rainfall data are as follows:

1. The net storm rain is determined using one of the methods outlined previously. If the unit graph was used to develop, for example, the ϕ index for the basin, then close reverse-similarity in storm pluviographs is necessary for accurate runoff determinations.
2. The net storm rain pluviograph is split into periods of approximately uniform excess rainfall intensity, with durations that meet the requirements of the unit-storm duration.
3. The surface runoff hydrograph is obtained for each rainfall period by applying the appropriate unit graph, and then by multiplying the ordinates by the volume of rainfall in the period in inches.
4. The surface runoff hydrographs are plotted, lagged appropriately in the correct time sequence, and their ordinates are added.
5. Base flow is added to the total or composite surface runoff hydrograph to give the total flow hydrograph. The magnitude of the base flow must be estimated considering antecedent moisture conditions.

VIII.4.8   Limitations of the Unit Graph

In manner of summarizing the applicability of the unit-graph technique for determining peak flows, it is appropriate that the hydrologist recognize there are certain limitations:

1. Unit graphs should only be applied to storm patterns similar to those from which they were developed. In this regard, the nonuniformity in the areal distribution of precipitation imposes a maximum limit on the size of the drainage area to which the method is applicable. The answer to the question: "How large is this maximum limit? "—is yet unresolved; some hydrologists indicate 4000–5000 sq. mi., others 2000 sq. mi. Nevertheless, if only daily rainfall and average daily discharge records are available, then the lower limit of applicability is to basins of about 1000 sq. mi. in area. However, in certain cases it may be permissible to use unit graphs from areas of 400–500 sq. mi. for flood routing in larger basins.
2. Unit graphs are strictly applicable when channel conditions remain unchanged, and to areas which do not have appreciable storage. In unit graph theory, the

proportionality between rates and volume of runoff necessitates that storage is a linear function of discharge. This condition is frequently violated when the drainage area contains a large number of reservoirs, or when the flood overtops the bank and overflows to the flood plain which may produce reservoir-type action.

3.   There is some question as to the applicability of the method for determining the hydrograph from snowmelt events.

4.   Under conditions where the nonuniformity of precipitation and storage effects may be important, it is advisable to employ flood-routing techniques in conjunction with the unit graph method to determine the flood hydrograph.

It may be noted that unit graph principles and flood routing procedures were used (Berry *et al.*, 1961) to determine a maximum probable flood for the South Saskatchewan River Project. Unit graph concepts were used to a degree on the Brazeau River development (Reid and Brittain, 1962) to obtain a maximum probable flood.

VIII.5   SYNTHETIC UNIT GRAPHS

VIII.5.1   Introduction

Unit graph synthesis for ungauged basins is based on empirical expressions that relate pertinent physical characteristics of the watershed to geometric aspects of the unit graph. These relationships are predicated on the basis that the unit graph of an area represents the integrated effect of all the sensibly constant basin factors and their modifying influence on the translation and storage of a runoff volume from a uniform excess rain occurring during a unit period of time.

VIII.5.2   Basin Physiography

The field of quantitative geomorphology has received considerable attention in recent years. These studies were stimulated by the work of Horton (1945) who suggested that the development of morphological characteristics depends on three main factors: surface resistivity to sheet erosion, runoff intensity, and ground slope. Since these factors vary with soil and bedrock conditions, vegetal cover, and climatic conditions, watersheds in different regions would be expected to exhibit wide differences in the degree of development of their drainage systems. For example it is pointed out (Miller 1953, p. 12) that the differences in morphological character of watersheds in areas of sandy soil underlain by porous and permeable sandstone, and in areas of dense clayey soils underlain by dense shale, may be attributed to the differences in the lithology of the two areas. That is:

...for maturely dissected topography; in the sandstone areas, streams of a given order should be longer and their drainage basins larger; drainage density should be less, slopes should be longer and possibly gentler than in the shale areas.

Furthermore, in a dolomite area the characteristics would be expected to differ from those in either of the areas described above.

VIII.5.2.1   Interrelationships of Watershed Characteristics

The interrelationships of various geometric characteristics of fluvially-eroded streambeds and watersheds are expressible in mathematical terms following the laws of stream numbers, stream lengths and stream slopes proposed by Horton (1945), and the law of stream areas suggested by Schumm (1956). According to these laws, the number, length, slope and drainage area of streams of given order are related either by a direct or inverse geometric series with stream order. Accordingly, it may be surmised that the interrelationships between these variables for watersheds of given order may be expressed by a series of simple power equations.

Gray (1961) reported the results of the interrelationships of watershed characteristics of 47 small watersheds in Illinois, Iowa, Missouri, Nebraska, North Carolina, Ohio and Wisconsin. A summary of his findings, with those of other investigators, is given in Table VIII.7.

The significance of the information provided in Table VIII.7 is that it shows many geomorphic properties to be highly related. This is especially true if these features are compared on watersheds which have been stratified according to region, or to areas, that have similar lithological and climatological characteristics. In certain respects, the data show the reason why drainage basin size has been an effective parameter for defining flow characteristics; that is, basin area reflects, to a certain degree, other parameters which are important to the time-distribution of runoff. In addition, the results suggest that attempts to establish correlations between such factors as basin shape will prove successful only if careful consideration is given in selecting watersheds to obtain a wide range of data.

The conclusions suggested above are predicated on results obtained primarily from small watersheds (<40 sq. mi.). One must recognize that in these discussions, as is the case in all other aspects of hydrologic science, certain anomalies may prevail to refute these relationships.

VIII.5.3   Procedures for Deriving Synthetic Unit Graphs

Numerous procedures have been derived whereby the unit graph for an ungauged area can be constructed. Each procedure, however, differs somewhat from another—either in the relationships established or in the methodology employed. The ensuing discussions are confined to brief summaries of several of the pertinent synthetic techniques published in the literature.

VIII.5.3.1   Snyder

Snyder (1938) was the first hydrologist to establish a set of formulae relating the physical geometry of the basin to properties of the resulting hydrograph. In a study of watersheds located mainly in the Appalachian Highlands, which varied in size from 10

Table VIII.7 Interrelationships of Watershed Characteristics

Reference	Relationship[a]	Correlation	Geographic Location
Gray (1961), Taylor and Schwarz (1952)	$L = 1.40A^{0.568}$	0.97	Ill., Iowa, Mo., Nebr., Nor.Cal., Ohio, Wisc., North and Middle U.S.
Gray (1961)	$L_{ca} = 0.54L^{0.95}$ or $L_{ca} = 0.90A^{0.56}$	0.99	Ill., Iowa, Mo., Nebr., Nor.Cal., Ohio, Wisc.
Langbein and others (1947)	$L_{ca} = 0.74A^{0.55}$		340 Watersheds from Northeastern U.S.
Gray (1961)	$S_c = 15.95L^{-0.610}$ $S_c = 2.61L^{-0.770}$ $S_c = 1.57L^{-0.662}$ $S_L = 0.86s_l^{0.67}$	0.94 0.93 0.97 0.96	North Carolina Nebraska - Western Iowa Ohio Iowa
	$A/A_c = 0.61$	$C_v = 23\%$	Ill., Iowa, Mo., Nebr., North Cal., Ohio, Wisc.
	$D/L' = 0.68$	$C_v = 20\%$	Ill., Iowa, Mo., Nebr., North Cal., Ohio, Wisc.

a/

L = Length of main stream from gauging station to outer most point (mi).

A = Plane area of the watershed enclosed within the topographic divide (mi²).

L_{ca} = Distance along the main stream from the gauging station to a point on the stream nearest the mass centre of area (mi).

S_c = Slope of the main stream i.e. slope of a line drawn along the longitudinal section of the main channel in such a manner that the area between the line and a horizontal line drawn through the channel outlet elevation is equal to the area between the channel grade line and the same horizontal line (%).

S_L = Mean land slope of a watershed as determined by the grid-intersection method (ft./ft.).

s_l = Average slope of a number of first order streams (ft./ft.).

D = Diameter of circle with equal area as the basin (mi).

A_c = Area of a circle with equal perimeter of the basin (sq. mi.).

L' = Maximum length of basin parallel to the principle drainage lines (mi).

to 10,000 sq. mi., he found that three points of the unit graph could be defined by the following expressions:

$$t_L = C_t (LL_{ca})^{0.3} \dots\dots\dots\dots\dots\dots\dots\dots\dots\dots\dots\dots\dots VIII.26$$

> where t_L = the basin lag (time difference in hours between the centroid of rainfall and the hydrograph peak),
> L = the length of the main stream in miles from the outlet to divide,
> L_{ca} = the distance in miles from the outlet to a point on the stream nearest the centre of area of the watershed, and
> C_t = coefficient whose magnitude was found to vary in the range from 1.8 to 2.2.

$$Q_p = (640\ C_p A)/t_L \dots\dots\dots\dots\dots\dots\dots\dots\dots\dots\dots\dots\dots VIII.27$$

> where Q_p = the peak discharge of the unit graph in cfs, and A is the drainage area in sq mi, and
> C_p = a coefficient whose magnitude varies from 0.56 to 0.69.

$$t_b = 3 + 3\ (t_L/24) \dots\dots\dots\dots\dots\dots\dots\dots\dots\dots\dots\dots\dots VIII.28$$

> where t_b = the length of the base of the unit graph in days.

Equations VIII.26, VIII.27 and VIII.28 define points of a unit graph produced by an excess rain of duration, $t_o = t_L/5.5$. For storms of different rainfall durations, t_R, an adjusted form of lag, t_{LR}, determined by the equation

$$t_{LR} = t_L + (t_R - t_o)/4 \dots\dots\dots\dots\dots\dots\dots\dots\dots\dots\dots\dots VIII.29$$

must be substituted in Equations VIII.27 and VIII.28.

Once the three quantities, t_L, Q_p and t_b are known, the unit graph can be sketched. It is constructed so that the area under the curve represents a 1-in. volume of direct runoff accruing from the watershed. As an aid to the sketching process, the Corps of Engineers (1948) has developed a relation between the peak discharge and the width of the unit graph at values of 50 per cent and 75 per cent of the peak ordinate.

A study similar to that of Snyder's was conducted by Taylor and Schwarz (1952) on 20 watersheds, which varied in size from 20 to 1,600 sq. mi., located in the Atlantic States. In this study, the relationships given for lag and peak discharge included a weighted slope term.

VIII.5.3.2   Commons

Commons (1942) suggested that a dimensionless hydrograph, the so-called basic hydrograph, would give an acceptable approximation of the flood hydrograph on any basin. This hydrograph was developed from flood hydrographs in Texas. It is divided so that the base time is expressed as 100 units, the peak discharge as 60 units and the area as a constant 1,196.5 units.

The absolute values for a hydrograph are established once the volume of runoff and peak discharge are known. The volume in second-foot-days is divided by 1,196.5 to establish the value of each square unit. Dividing the peak flow by 60 gives the value of one unit of flow in cfs. The magnitude of one time unit is then computed by dividing the value of the square unit by that of the flow unit. Finally, the hydrograph is synthesized by converting listed coordinates of the basic graph to absolute time and discharge readings according to the calculated conversion factors.

VIII.5.3.3   Soil Conservation Service
(USDA, SCS, 1957)

The method of hydrograph synthesis presently used by the United States Department of Agriculture, Soil Conservation Service (SCS) originated from consideration that a hydrograph could be represented in simple geometric form as a triangle (see Fig. VIII.14).

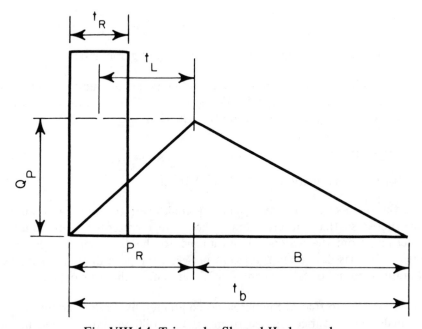

Fig. VIII.14 Triangular-Shaped Hydrograph.

From Fig. VIII.14 it is obvious that the volume of runoff, Vol.,(the area under the hydrograph) can be expressed as

$$\text{Vol.} = \frac{Q_p P_R}{2} + \frac{Q_p B}{2} \qquad\ldots\ldots\ldots\ldots\ldots\ldots\ldots\ldots\ldots\ldots\ldots\ldots\ldots\ldots\ldots\ldots\text{VIII.30}$$

$$= \frac{Q_p}{2} (P_R + B) \quad \dotfill \text{VIII.31}$$

or $\quad Q_p = \dfrac{2 \text{ Vol.}}{P_R + B} \quad \dotfill \text{VIII.32}$

From review of a large number of hydrographs, it was found that

$$B \cong 1.67 P_R \quad \dotfill \text{VIII.33}$$

Therefore Equation VIII.32 can be reduced to

$$Q_p = \frac{2 \text{ Vol.}}{2.67 P_R} = \frac{0.75 \text{ Vol.}}{P_R} \quad \dotfill \text{VIII.34}$$

or $\quad Q_p = \dfrac{0.75 \times 640 \, AP_n \times 1.008}{P_R}$

$$Q_p = \frac{484 \, AP_n}{P_R} \quad \dotfill \text{VIII.35}$$

where A = area of the basin (sq mi),
$\quad P_n$ = net storm rain (in), and
$\quad P_R$ = period of rise (hr).

Using Equations VIII.33 and VIII.35 and knowing that

$$P_R = \frac{t_R}{2} + t_L \quad \dotfill \text{VIII.36}$$

the hydrograph can be defined when t_L is known. The time lag, t_L, can be evaluated from equations VIII.4 or VIII.26 or by the expression

$$t_L = \frac{\Sigma A_x P_{nx} t_x}{\Sigma A_x P_{nx}} \quad \dotfill \text{VIII.37}$$

in which A_x and P_{nx} are respectively the area and depths of runoff from subarea 'x' of the basin, and 't_x' is the time required for the water to travel from the centroid of the subarea to the outlet of the watershed.

Currently, a more sophisticated procedure is used by the SCS that employs an average dimensionless hydrograph developed from an analysis of a large number of natural unit graphs for watersheds varying widely in size and geographical location. This dimensionless hydrograph has its ordinate values expressed as the dimensionless ratio, Q/Q_p, and its abscissa values as the dimensionless ratio, t/P_R. Q is the discharge at any time, t, and P_R is the period of rise (see Table VIII.8). For a given watershed, once the values of Q_p and P_R are defined using Equations VIII.35, etc., the unit graph can be constructed.

Table VIII.8 Ratios for the Basic Dimensionless Hydrograph of the SCS.

Time ratios t/P_R	Hydrograph discharge ratios (Q/Q_p)
0	0
0.1	0.015
0.2	.075
0.3	.16
0.4	.28
0.5	.43
0.6	.60
0.7	.77
0.8	.89
0.9	.97
1.0	1.00
1.1	.98
1.2	.92
1.3	.84
1.4	.75
1.5	.66
1.6	.56
1.8	.42
2.0	.32
2.2	.24
2.4	.18
2.6	.13
2.8	.098
3.0	.075
3.5	.036
4.0	.018
4.5	.009
5.0	.004
infinity	0

VIII.5.3.4 Hickok, Keppel and Rafferty

The approach to hydrograph synthesis given by Hickok *et al.* (1959) is very similar to that employed by the SCS. However, the investigations were confined entirely to small watershed areas. The runoff characteristics of 14 watersheds, which varied in size from 11 to 790 acres, located in semiarid regions, were investigated, and an average dimensionless graph (Q/Q_p versus t/t_L') was developed. In this study lag, t_L' was taken as the time difference between the centroid of a limited block of intense rainfall and the resultant peak discharge. The authors presented two different methods of determining lag.

For reasonably homogeneous semiarid rangelands up to about 1,000 acres in area,

$$t_L' = K_1 \, (A^{0.3}/S_L \, \sqrt{DD} \,)^{0.61} \dots\dots\dots\dots\dots\dots\dots\dots\dots\dots \text{VIII.38}$$

where S_L = the average land slope of the watershed, and
 DD = the drainage density.

With A in acres, S_L in per cent, DD in feet per acre, and K_1 equal to 106, lag is given in minutes.

For watersheds with widely different physiographic characteristics,

$$t_L' = K_2 \left[\frac{L_{csa} + W_{sa}}{S_{La}\sqrt{DD}} \right]^{0.65} \dots\dots\dots\dots\dots\dots\dots\dots\dots\dots \text{VIII.39}$$

where L_{csa} = length from the outlet of the watershed to the centre
$\phantom{where\ L_{csa} =}$ of gravity of the source area (ft),
W_{sa} = average width of the source area (ft),
S_{La} = average land slope of the source area (%).

The source area was considered to be the half of the watershed with the highest average land slope. The coefficient, K_2, is taken equal to 23.

The authors suggested that Q_p could be obtained from the relation

$$\frac{Q_p}{P_n} = \frac{K_3}{t_L'} \dots\dots\dots\dots\dots\dots\dots\dots\dots\dots\dots\dots\dots\dots\dots \text{VIII.40}$$

which gives Q_p in cfs when P_n is expressed in acre feet, t_L' in minutes, and K_3 taken equal to 545.

VIII.5.3.5   Clark

Clark (1945) suggested that the unit graph for an area could be derived by routing its time-area concentration curve through an appropriate amount of reservoir storage. In the routing procedure, an instantaneous unit graph (hydrograph resulting from an instantaneous rainfall of 1-in. depth and duration equal to zero time) is formed. The unit graph for any rainfall duration is obtained from the instantaneous graph by averaging the ordinates of the instantaneous graph (see Equation VIII.25). This procedure is outlined in detail in Subsection VIII.6.4.3.

VIII.5.3.6   Gray

Gray (1962) employed the equational form of a two-parameter gamma distribution to define a modified form of the unit graph for small watersheds. The development of this technique was predicated on the rationalization of the runoff process proposed by Edson (1951).

Considering the time-area curve of a watershed, it can be assumed that for a uniform excess precipitation rate, the area contributing to flow at the outlet, A, is proportional to some function of the travel time or storm duration, t. That is

$$A \propto t^x \dots\dots\dots\dots\dots\dots\dots\dots\dots\dots\dots\dots\dots\dots\dots \text{VIII.41}$$

in which x = exponent.

But, for a uniform excess rate, $Q \propto A$ and therefore Equation VIII.41 can be written as

$$Q \propto t^x \dots\dots\dots\dots \quad \dots\dots\dots\dots\dots\dots\dots\dots\dots \text{VIII.42}$$

Further, it is known that the shape of the recession limb follows the equation of the form

$$\ln Q = \ln Q_o - Kt$$

and thus for this limb

$$Q \propto e^{-Kt} \dots\dots\dots\dots\dots\dots\dots\dots\dots\dots\dots\dots\dots\dots\dots \text{VIII.43}$$

Combining Equations VIII.42 and VIII.43 it follows that

$$Q \propto t^x \, e^{-Kt}$$

or $Q = Bt^x \, e^{-Kt}$ $\dots\dots\dots\dots\dots\dots\dots\dots\dots\dots\dots\dots\dots \text{VIII.44}$

but the volume of runoff is

$$\text{Vol.} = \int_o^t Q dt = \int_o^t Bt^x \, e^{-Kt} \, dt \dots\dots\dots\dots\dots\dots\dots\dots\dots \text{VIII.45}$$

By substituting into Equation VIII.45 the equalities x = m-1 and z = Kt, and integrating, one obtains the expression for Q as

$$Q = \frac{(\text{Vol.}) K^m}{\Gamma(m)} \, e^{-Kt} \, t^{m-1} \dots\dots\dots\dots\dots\dots\dots\dots\dots \text{VIII.46}$$

Equation VIII.46 is identical in form to the incomplete gamma frequency distribution.

Gray (1962) reduced the distribution graphs from several small watersheds to a dimensionless form, which could be defined by the equation

$$Q_{t/P_R} = \frac{25 \, (\gamma)^q e^{-\gamma t/P_R} \, t/P_R^{\,q-1}}{\Gamma(q)} \dots\dots\dots\dots\dots\dots\dots\dots\dots \text{VIII.47}$$

in which Q_t/P_R = % Flow/$0.25P_R$ or the % of the total volume of flow which occurs during a time increment equal to $0.25P_R$ at a specific value of t/P_R. That is from Fig. VIII.15.

$$\% \text{ Flow}/0.25P_R = \frac{Q_1}{Q_1 + Q_2 + \dots \dots Q_N} \times 100$$

$$= \frac{Q_1}{\Sigma \text{ cfs}} \times 100$$

$$Q_1 = \frac{\% \text{ Flow}/0.25P_R}{100} \times \Sigma\text{cfs}$$

$$Q_1 = \frac{\% \text{ Flow}/0.25P_R}{100} \frac{A\, P_n}{0.25P_R} \quad \dots\dots\dots\dots\dots\dots\dots\dots\dots\dots\dots\dots \text{ VIII.48}$$

in which γ, q = parameters,
Γ = gamma function, and
A is in sq ft, P_n is in ft and P_R in secs.

The components of Equation VIII.47 can be evaluated from the following relationships:

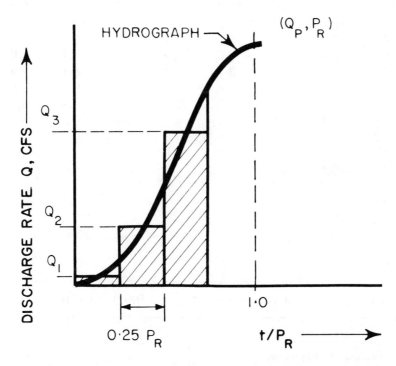

Fig. VIII.15 Interpretation of % Flow/0.25 P_R (After Gray)

$$P_R/\gamma = a\left(L/\sqrt{S_c}\right)^n \quad \dots\dots\dots\dots\dots\dots\dots\dots\dots\dots \text{VIII.49}$$

where L = length of main stream (mi), and
S_c = slope of the main channel (%).

Values of the coefficient 'a' and exponent 'n' of Equation VIII.49 obtained were: 11.4 and 0.531—Ohio watersheds; 7.40 and 0.498—Nebr. and Western Iowa watersheds; and

Table VIII.9 Dimensionless Graph Coordinates for Different Values of the Parameter, γ.

	%Flow/0.25 P_R [a]								
t/P_R	$\gamma=2.0$	$\gamma=2.5$	$\gamma=3.0$	$\gamma=3.5$	$\gamma=4.0$	$\gamma=4.5$	$\gamma=5.0$	$\gamma=5.5$	$\gamma=6.0$
0.000	0.0	0.0	0.0	0.0	0.0	0.0	0.0	0.0	0.0
0.125	1.2	0.8	0.5	0.3	0.2	0.1	0.1	0.1	0.1
0.375	6.6	6.3	5.8	5.2	4.7	4.2	3.7	3.2	2.8
0.625	11.2	12.0	12.6	13.0	13.3	13.6	13.6	13.6	13.5
0.875	13.3	14.9	16.4	17.7	18.9	20.0	21.0	22.0	22.9
1.000	13.5	15.3	16.8	18.2	19.5	20.8	21.9	23.0	24.1
1.125	13.4	15.0	16.5	17.8	19.0	20.1	21.1	22.1	23.1
1.375	12.1	13.3	14.2	14.9	15.6	16.1	16.6	16.9	17.2
1.625	10.3	10.7	11.1	11.2	11.2	11.1	10.9	10.7	10.5
1.875	8.3	8.2	8.0	7.7	7.3	6.8	6.4	6.0	5.5
2.125	6.5	6.0	5.5	5.0	4.4	3.9	3.4	3.0	2.6
2.375	4.9	4.3	3.6	3.1	2.5	2.1	1.7	1.4	1.1
2.625	3.6	2.9	2.3	1.8	1.4	1.0	0.8	0.6	0.5
2.875	2.6	2.0	1.4	1.0	0.8	0.6	0.4	0.3	0.2
3.125	1.9	1.3	0.9	0.6	0.4	0.3	0.2	0.1	
3.375	1.3	0.9	0.5	0.3	0.2	0.1	0.1		
3.625	1.0	0.6	0.3	0.2	0.1				
3.875	0.6	0.3	0.2	0.1					
4.125	0.4	0.2	0.1	0.1					
4.375	0.3	0.1	0.1						
4.625	0.2	0.1							
4.875	0.1	0.1							
5.125	0.1								
5.375	0.1								
Sum [b]	100.0	100.0	100.0	100.0	100.0	100.0	100.0	100.0	100.0

(After Gray)

[a]   Rounded to the nearest 0.10 per cent.

[b]   Sums do not include peak percentages.

9.27 and 0.562 –Ill., Mo., Wisc., and Central Iowa watersheds.

$$q = \gamma + 1 \quad \dots\dots\dots\dots\dots\dots\dots\dots\dots\dots\dots\dots\dots\dots \text{VIII.50}$$

and

$$P_R/\gamma = 1/ \left[(2.676/P_R) + 0.0139\right] \quad \dots\dots\dots\dots\dots\dots\dots\dots \text{VIII.51}$$

in which P_R/γ and P_R are in min.

Although P_R and γ can be readily solved using the above equations, the solution to Equation VIII.47 is extremely cumbersome without appropriate mathematical tables. For this reason, Table VIII.9 was prepared to expedite this computation.

Example Calculation

Calculate the peak ordinate of a unit graph on which

A = 5 sq. mi., P_R = 58 min. and γ = 3.5

From Table VIII.9 with t/P_R = 1 and γ = 3.5, the % Flow/0.25P_R = 18.2%.

Using Equation VIII.48

$$Q_p = \frac{18.2 \times 5 \times 640 \times 43,560}{100 \times 0.25 \times 58 \times 60} \times \frac{1}{12} = 2440 \text{ cfs.}$$

A comparison of the different synthetic unit graph techniques is given by Hanson and Johnson (1964).

VIII.6   FLOOD ROUTING

VIII.6.1   Introduction

Flood routing is the prediction of the hydrograph at one point, or reach of a stream, from the hydrograph observed upstream on the stream. These procedures are used to suffice several objectives:

1. Short term forecasting of floods.
2. Computation of unit hydrographs for various points.
3. Prediction of the behaviour of a river after a change in channel conditions (installation of a reservoir or a system of dykes).
4. Derivation of synthetic hydrographs.

Basically, flood routing may be divided to two basic types; 'Channel' or 'Streamflow

Routing' –the routing of a flood hydrograph through a reach of the river; and 'Reservoir Routing.'

VIII.6.2   **Streamflow Routing**

The movement of a flood wave in a stream channel is a highly complicated problem of non-steady and usually non-uniform flow. Not only flow varies with time as the wave progresses downstream, but channel properties and amounts of lateral inflow may also vary. Massau (1900) was perhaps the first investigator to solve the differential equations for steady-state flow in uniform channels by graphical integration of the equations of characteristics. This solution, although valid, is quite sophisticated and involves considerable work. For engineering purposes, the use of simpler, and more readily understandable, difference equations in place of the basic differential equations of hydraulics will provide realistic results. Use of the difference equations (which will be outlined later) is predicated on the assumption that the changes in flow occur gradually with time, thus channel storage can be expressed as a function of inflow and outflow in the channel reach.

Before proceeding, it is important to re-emphasize the influence of channel storage to the time-distribution of runoff. Channel storage dampens the flood wave as it travels down a reach, tending to attenuate the peak and extend the time base of the hydrograph. It is the purpose in routing to estimate the magnitude of this attenuation.

Let us now consider a reach of a river where no tributaries enter; where the supplies or losses by rainfall, effluent or influent seepage, or evaporation can be neglected; and for which the inflow hydrograph to the reach is known. In principle, the reach of the river may be visualized as shown in Fig. VIII.16.

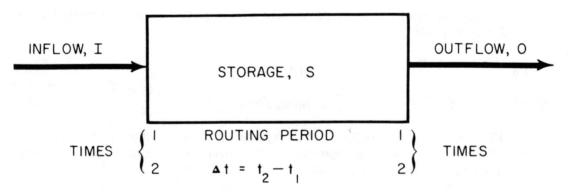

Fig. VIII.16 Visualization of River Reach Flow

As the first step in formulation of the continuity equation for the reach, the inflow hydrograph should be divided into successive periods, each of some finite duration, Δt. Δt is referred to as the routing period, and must be smaller than the time of travel through the reach so that the wave crest cannot pass completely through the reach during the routing period.

According to the equation of continuity or conservation of mass in the reach during the routing period we can write

$$I_{avg} \Delta t - O_{avg} \Delta t = \Delta S \quad \dots\dots\dots\dots\dots\dots\dots\dots\dots\dots\dots\dots\dots \text{VIII.52}$$

where

$$I_{avg} = \frac{I_1 + I_2}{2} \text{ and } O_{avg} = \frac{O_1 + O_2}{2} \quad \dots\dots\dots\dots\dots\dots\dots\dots\dots\dots \text{VIII.53}$$

in which I_1 and I_2 = inflow at the beginning and end of the routing period, and

O_1 and O_2 = outflow at the beginning and end of the routing period.

Substituting these values into Equation VIII.52 one obtains

$$\frac{(I_1 + I_2)}{2} - \frac{(O_1 + O_2)}{2} = \frac{S_2 - S_1}{\Delta t} \quad \dots\dots\dots\dots\dots\dots\dots\dots\dots\dots \text{VIII.54}$$

In Equation VIII.54, I_1 and I_2 are known from the given hydrograph to be routed, O_1 and S_1 are known from the foregoing period, and O_2 and S_2 are unknown. Thus, to solve the equation, a second relation is needed. It is in determination of this relation that the major difficulty arises.

VIII.6.2.1   Storage as a Function of Discharge

Inasmuch as it is O_2 in which we are interested, it is obvious that the second relation needed should be of the form

$$O_2 = f(S_2) \quad \dots\dots\dots\dots\dots\dots\dots\dots\dots\dots\dots\dots\dots\dots\dots\dots \text{VIII.55}$$

However, in the case of a reservoir, such a relationship may be explicitly defined. In channels, storage is usually a function of both inflow and outflow. Linsley *et al.* (1949) show the storage in a channel reach may be considered to be the sum of two portions: *prism storage* or that water below an imaginary line drawn parallel to the channel bottom; and *wedge storage* or that water between this line and the actual water surface profile (see Fig. VIII.17). Wedge storage increases the total volume during the rise, and decreases it during falling stages. Hence, if a plot is made of O_2 versus S_2 there will be a hysteresis loop between rising and falling stages.

In order to account for wedge storage, both inflow, I, and outflow, O, are considered in the expression. For steady, uniform flow in a channel, I, O, and S will be functions of the depth of flow, d. That is

$$I = ad^n \text{ and } O = ad^n \quad \dots\dots\dots\dots\dots\dots\dots\dots\dots\dots\dots\dots\dots \text{VIII.56}$$

$$S_I = bd^m \quad \text{and} \quad S_O = bd^m \quad \dots\dots\dots\dots\dots\dots\dots\dots\dots\dots \text{VIII.57}$$

where subscripts I and O refer to inflow and outflow, and a,b,n, m are coefficients and exponents. Expressing S_I and S_O in terms of I and O, one obtains

$$S_I = b\left(\frac{I}{a}\right)^{m/n} \text{and} \quad S_O = b\left(\frac{O}{a}\right)^{m/n} \quad \dots\dots\dots\dots\dots\dots\dots\dots\dots \text{VIII.58}$$

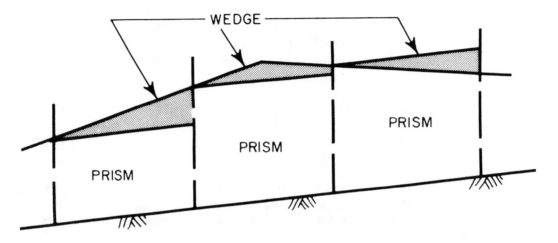

Fig. VIII.17 Profile of Flood Wave

A general expression for storage in terms of outflow can now be written as

$$S = xS_I + (1 - x) S_O \quad \dots\dots\dots\dots\dots\dots\dots\dots\dots\dots\dots \text{VIII.59}$$

or

$$S = xb\left(\frac{I}{a}\right)^{m/n} + (1 - x)b\left(\frac{O}{a}\right)^{m/n} \quad \dots\dots\dots\dots\dots\dots\dots\dots\dots \text{VIII.60}$$

in which x is a weight factor to account for the relative effect of inflow and outflow on storage. For uniform flow in prismatic channels, m = 1 and n = 5/3. Then

$$S = xb\left(\frac{I}{a}\right)^{0.6} + (1 - x)b\left(\frac{O}{a}\right)^{0.6} \quad \dots\dots\dots\dots\dots\dots\dots\dots\dots \text{VIII.61}$$

and in natural channels the ratio of exponents, m/n, can exceed unity.

If storage is only a function of outflow, as in the case of a reservoir, the value of x is zero. In cases where wedge storage is significant, then x is greater than zero—generally with a limiting value of 0.5 when inflow and outflow have equal weight. For many natural streams, x varies from 0.4–0.5, but the presence of large flood plains in a reach may cause reservoir-type action which makes S more a function of O than I, hence reducing x to values of 0.30 or lower.

VIII.6.2.2   Muskingum Flood Routing Method

Numerous routing techniques have been developed on the general basis of Equation VIII.60. In most cases, it is commonly assumed that m/n is unity, in which case the equation reduces to

$$S = \frac{b}{a} \left[xI + (1 - x)O \right] \quad \dots \dots \dots \dots \dots \dots \dots \dots \dots \dots \dots \dots \dots \dots \dots \text{VIII.62}$$

or

$$S = K \left[xI + (1 - x)O \right] \quad \dots \dots \dots \dots \dots \dots \dots \dots \dots \dots \dots \dots \dots \dots \text{VIII.63}$$

in which $K = b/a$. The factor K, normally referred to as the storage constant, represents the ratio of storage to weighted discharge in the reach, and has the dimensions of time. It is, in fact, a measure of the travel time through the reach and can be evaluated accordingly from hydrometric or topographic information.

Before proceeding, it should be mentioned that the assumed proportionality between storage and weighted discharge (that is $S = KQ_{ave}$) is consistent with unit graph theory which assumes constant time elements. For small basins and small discharges, this assumption is approximately true. However, for floods on large basins, the relationship is perhaps more nearly $S = KQ_{ave}^{1.3}$. That is, peaks of large floods occur earlier than those of small floods.

Using Equation VIII.63 we are now in a position to solve the continuity equation (Equation VIII.54). That is

$$\frac{I_1 + I_2}{2} - \frac{O_1 + O_2}{2} = \frac{K[xI_2 + (1-x)O_2] - K[xI_1 + (1-x)O_1]}{\Delta t} \quad \dots \dots \dots \dots \text{VIII.64}$$

Reducing this equation to a more simplified form

$$O_2 = c_0 I_2 + c_1 I_1 + c_2 O_1 \dots \dots \dots \dots \dots \dots \dots \dots \dots \dots \dots \dots \dots \dots \text{VIII.65}$$

where $c_0 = -\left[\dfrac{Kx - 0.5\Delta t}{K - Kx + 0.5\Delta t} \right]$

$c_1 = \dfrac{Kx + 0.5\Delta t}{K - Kx + 0.5\Delta t}$

$c_2 = \dfrac{K - Kx - 0.5\Delta t}{K - Kx + 0.5\Delta t}$

Equation VIII.65 is the so-called Muskingum method of flood routing developed by McCarthy (1938). One will realize that $c_0 + c_1 + c_2 = 1$.

VIII.6.2.3   Determination of Routing Constants and Parameters

Storage

Storage for the Muskingum method is determined from analysis of observed hydrographs of the reach, as shown in Fig. VIII.18.

Routing Constants K and x

The routing constants K and x may be determined graphically from observed hydrographs. This is done by plotting storage against the weighted discharge $[xI + (1-x)O]$

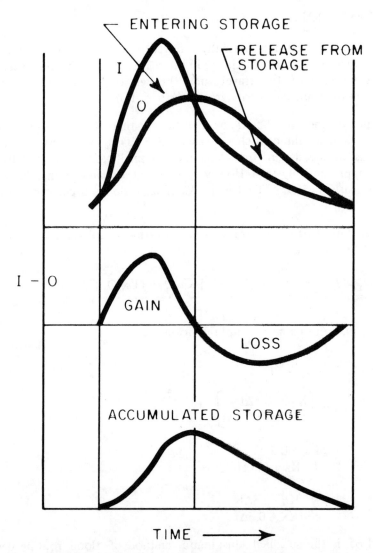

Fig. VIII.18 Computation of Storage from Observed Hydrographs

using different values of x: x_1, x_2, etc. The correct value of x for the reach to be used in the method is that value which produces a plot on which the points follow a straight line; the slope of which gives the value for the storage constant, K, (see Fig. VIII.19).

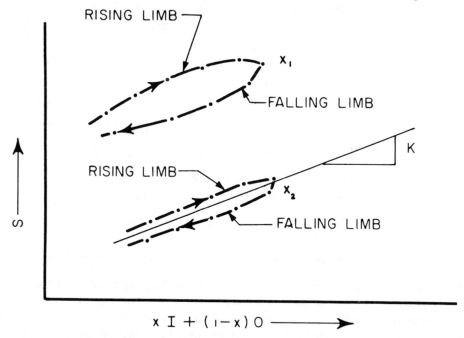

Fig. VIII.19 Determination of x and K for the Muskingum Routing Method

Applied Problem

   Given the ordinates of the inflow hydrograph, as shown in Table VIII.10, and assuming that the coefficients of the Muskingum equation as determined from characteristics of the reach are $c_0 = 0.1$, $c_1 = 0.4$ and $c_2 = 0.5$, calculate the ordinates of the outflow hydrograph.

   Using Equation VIII.65 the computation starts with period 2, with $I_2 = 300$ cfs and $I_1 = O_1 = 200$ cfs.

   The results provided in Table VIII.10 show characteristic features of flood routing through a river reach, i.e., time lag (time of maximum flow) and attenuation (from 550 to 473).

   It should be realized that there are many methods of stream flow routing. For further information the reader should consult references such as: Rutter *et al.* (1939); Linsley (1944); Johnstone and Cross (1949); and Chow (1959).

VIII.6.3   Flood Routing Through a Reservoir

   Flood routing through a large reservoir is the simplest routing problem in that the discharge from the reservoir can be expressed as a simple function of the water surface

Table VIII.10 Calculation by the Muskingum Method

Periods			Inflow	$c_0 I_2$	$c_1 I_1$	$c_2 O_1$	O_2	
1	I_1		200	—	—	—	200	O_1
2	I_1	I_2	300	30	80	100	210	O_2 O_1
3	I_2		500	50	120	105	275	O_2
4			550	55	200	138	393	
5			500	50	220	196	466	
6			400	40	200	233	<u>473</u>	
7			300	30	160	236	426	
8			200	20	120	213	352	

elevation. This is the case when the reservoir outlet or spillway is not regulated, or when the outlet gates have a fixed opening or known opening dependent on reservoir elevation. For a reservoir the storage terms of the continuity equation (Equation VIII.54) can be evaluated as single-valued functions of outflow in which the nature of the function depends on the type of spillway. In practice, storage-versus-elevation curves, and outflow-discharge-versus-elevation curves, would be obtained or would be available for the reservoir, so storage-discharge relations are explicitly known. All that remains, then, is to solve the continuity equation, and this can be done by a graphical method employing the storage-and discharge-versus-elevation curves.

The continuity equation can be put in the form:

$$\frac{I_1 + I_2}{2} \Delta t = O_1 \Delta t /2 + \Delta S + O_2 \Delta t /2 \dots\dots\dots\dots\dots\dots\dots \text{VIII.66}$$

The left-hand side of the equation is known from the inflow hydrograph and the chosen routing period, Δt. The terms on the right-hand side of the equation may be shown on the storage-elevation plot for the reservoir, as in Fig. VIII.20b.

The curves representing $S - O\Delta t/2$ and $S + O\Delta t/2$ are obtained realizing that at any particular elevation there is a particular outflow discharge, O, (Fig. 8.20a) and the storage term $O\Delta t/2$ can be added and subtracted from the storage curve S, knowing the routing period, Δt.

If the routing period begins when the water level reaches the spillway crest, O_1 is zero and $(I_1 + I_2) \Delta t/2$ is plotted along the storage axis from point A. Thus, $(I_1 + I_2) \Delta t/2 = \Delta S + O\Delta t/2$ and therefore, $(I_1 + I_2) \Delta t/2$ defines point B on the $S + O\Delta t/2$ curve.

The actual outflow at the end of the routing period can be obtained from Fig. VIII.20a from the elevation corresponding to point B. The next routing step is begun by measuring off the incremental volume $(I_1 + I_2) \Delta t/2$ for the second routing period, beginning from point C. Examination of the figure shows that this gives the outflow discharge associated with elevation D. This process is continued until the complete outflow hydrograph is obtained.

Most large reservoirs have a variety of outlets, and this necessitates that a composite outlet-discharge-versus-reservoir-elevation curve must be derived. Nevertheless, under these conditions the flood routing procedures discussed above will still apply. For gated outlets, several flood routing calculations may be necessary; one for each possible outlet gate opening. In cases where outflow from a reservoir may take place through a spillway

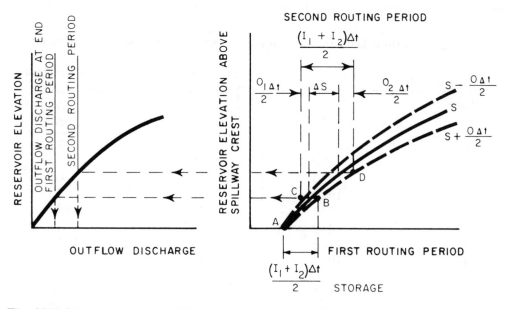

Fig. VIII.20a Reservoir Outflow Versus   Fig. VIII.20b Reservoir Storage Versus
Elevation Above Crest   Elevation Above Crest

in addition to regulated flow through some other outlet, an additional term $O_R \Delta t$, representing the volume of regulated outlet discharge, must be added to the continuity equation, and then the flood routing procedure proceeds as before.

VIII.6.3.1   Example Calculation

This example is of the routing procedure for a reservoir having the water level initially at the emergency spillway crest, with a small flow through a principal spillway.

The data sections of Tables VIII.11 and VIII.12 are obtained from reservoir surveys (storage information), knowledge of the spillway operation, and knowledge of the inflow

Table VIII.11 Storage and Outflow Data and Computations for a Reservoir

Data						Computations (Δt = 0.20 hr.)	
Elevation ft.	Storage ac. ft	Storage cfs - hr	Spillway Discharge cfs. Principal	Emergency	Total	$\frac{S+O\Delta t}{2}$ cfs - hr.	$\frac{S-O\Delta t}{2}$ cfs - hr.
77.4	308.0	3730	25	0	25	3733	3727
78	327.0	3960	26	154	180	3978	3942
79	358.0	4330	26	570	596	4390	4270
80	394.0	4760	26	1220	1246	4885	4635
81	430.0	5200	26	1960	1986	5399	5001
82	469.0	5670	27	2915	2942	5964	5376

Table VIII.12 Inflow Data and Computations and Outflow Results for a Reservoir

Data		Computations $\left[\frac{(I_1 + I_2)}{2}\right]\Delta t$ cfs - hr	Outflow Results from Fig. VIII.21. O cfs
Time hrs	I cfs		
0.0	0	0	25
0.2	120	12	30
0.4	620	74	60
0.6	1340	196	190
0.8	2140	348	490
1.0	2480	462	940
1.2	2460	494	1370
1.4	1940	440	1600
1.6	1620	356	1660
1.8	1120	274	1570
2.0	980	210	1420
2.2	620	160	1250
2.4	580	120	1060
2.6	340	92	940
2.8	340	68	740
3.0	180	52	630
etc.	etc.	etc.	etc.

hydrograph. The outflow hydrograph results in Table VIII.12 are obtained using Fig. VIII.21, which is a plot of the total spillway discharge versus the computed $\frac{S+O\Delta t}{2}$ and $\frac{S-O\Delta t}{2}$ values in Table VIII.11.

For further references on reservoir routing techniques, the reader should consult the works of Posey (1935), Goodrich (1931), Linsley et al. (1949), and Chow (1959).

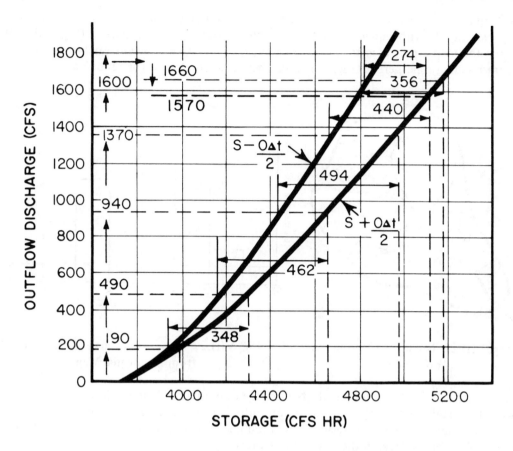

Fig. VIII.21 Graphical Solution of Reservoir Routing Problem

VIII.6.4   Synthetic Hydrographs Based on Flood Routing

VIII.6.4.1   Introduction

The theory of synthetic hydrographs based on flood routing techniques can be considered the modern version of established principles of the unit graph. Perhaps the most notable contribution to this field was the work by Clark in 1945. Other advocates of the approach are hydrologists such as Johnstone and Cross (1949), Horton (1941), Dooge (1956), and Nash (1958, 1959).

The theory is based in principle on the fact that the rainfall impulse (net storm rain) is modified by two factors: (a) the translation or time of travel of the volume in channel and overland flow, and (b) storage. Both factors occur simultaneously. However, it is the basic premise of the method to account for each factor individually. In using the technique to develop a unit graph for a basin, it is assumed that the shape of the unit graph bears a logical resemblance to the time-area concentration curve of the watershed, hence no knowledge of the runoff producing rain is necessary other than its time of cessation.

The time-area curve is constructed from a map of the watershed on which isochrones defining points with equal travel times to the outlet have been marked (Fig. VIII.22). From the map a plot is made of the area enclosed between adjacent isochrones and the corresponding time period (Fig. VIII.23). These elemental areas (or runoff) are subject to

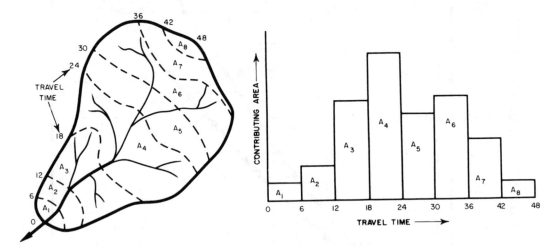

**Fig. VIII.22 Isochronal Map of
a Watershed**

**Fig. VIII.23 Time-Area Curve Constructed
from Isochronal Map**

an average lag effect that is equal to the average time values of the isochrones. Actually, it should be realized that the time-area concentration curve depicts precisely the hydrograph of runoff that would result from a rainfall excess, generated instantaneously and uniformly over the area, in the absence of storage.

VIII.6.4.2   Determination of Base Time of the Time Area Curve

The base time of the time area curve is essentially the time of concentration of the basin. Its magnitude may be evaluated from purely hydraulic considerations: by hydrograph analysis, or by empirical formulae. When rainfall and runoff data are available, the base time can be determined by measuring the time from cessation of rain to the inflection point on the hydrograph (see Fig. VIII.24).

When rainfall-runoff data are not available, the parameter is generally approximated from basin lag formulae, such as those given previously.

VIII.6.4.3   Clark's Method

Development of Routing Equations

Clark (1945) was the first to show that the routing of a flood wave in a reach could be successfully accomplished (within practical limits of accuracy) by translating the wave into a time equal to the travel time of the reach, and then by routing it through an amount of reservoir storage equivalent to that in the reach. This is referred to as the 'lag and route technique'. From this, Clark conceived that this approach could be used to derive the instantaneous unit graph for a basin by routing the time-area concentration curve of a basin (which properly evaluates travel time) through a given amount of linear reservoir storage. Hypothetically, this simply infers placing a reservoir at the outlet of a stream that has storage characteristics such that $S = KO$. In reservoir routing, the weight factor, x of Equation VIII.63, is zero; hence the continuity equation for linear storage is

$$\frac{I_1 + I_2}{2} - \frac{O_1 + O_2}{2} = \frac{K(O_2 - O_1)}{\Delta t} \quad \ldots\ldots\ldots\ldots\ldots\ldots\ldots\ldots\ldots\ldots\ldots\ldots\text{VIII.67}$$

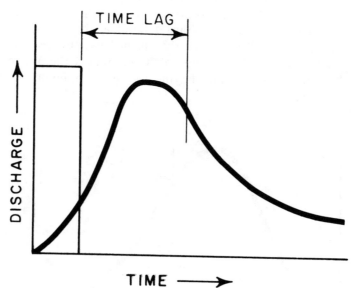

Fig. VIII.24 Determination of Base Time.

which will reduce to

$$O_2 = c_0 I_2 + c_1 I_1 + c_2 O_1 \ldots\ldots\ldots\ldots\ldots\ldots\ldots\ldots\ldots\ldots\ldots\ldots\ldots\ldots\text{VIII.68}$$

where $c_0 = \dfrac{0.5\Delta t}{K + 0.5\Delta t}$

$$c_1 = \frac{0.5\Delta t}{K + 0.5\Delta t}$$

$$c_2 = \frac{K - 0.5\Delta t}{K + 0.5\Delta t}$$

Evaluation of the Storage Constant, K

Equation VIII.67 can be written in differential form as

$$I - O = \frac{K dO}{dt} \quad \ldots\ldots\ldots\ldots\ldots\ldots\ldots\ldots\ldots\ldots\ldots\ldots\ldots\ldots\ldots\ldots\ldots\ldots . \text{VIII.69}$$

Assuming $I = o$, the equation further reduces to

$$K = -O \Big/ \frac{dO}{dt} \quad \ldots\ldots\ldots\ldots\ldots\ldots\ldots\ldots\ldots\ldots\ldots\ldots\ldots\ldots\ldots\ldots . \text{VIII.70}$$

The assumption, $I = o$, implies that inflow to the channel has ceased. This point exists on a hydrograph at the inflection point on the recession limb, since it is presumed that after this point is reached the discharge is attributed entirely to withdrawal from channel storage (see Fig. VIII.25). In effect, K thus determines the rate of withdrawal of water from channel storage after all inflow to the channel has stopped. No matter how the channel system has been filled, the value of K will reflect the physical features of the channel. It is therefore independent of the time variation in rainfall.

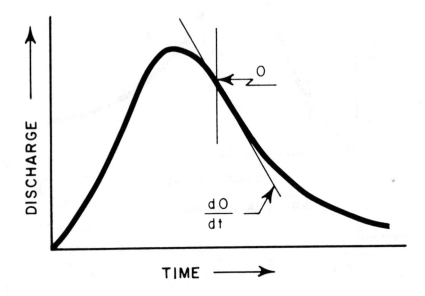

Fig. VIII.25 Determination of Point of Cessation of Inflow.

Theoretically, it should be possible to evaluate K from Equation VIII.70 at any point on the recession. However, if base flow separation of the hydrograph has not been complete, K actually increases (positively) to the right of the point of inflection, because inflow from groundwater flow does not decrease as rapidly as surface runoff. In these cases, K can be obtained from a semi-logarithmic plot of the recession limb, giving the greatest weight to points near the point of inflection.

When hydrographic data are not available, a rough estimate of the storage constant may be obtained from the empirical equations for basin lag.

Example Calculation

Problem: Determine the one-hr. unit graph from the time-area curve of a watershed by the lag and route method given:

Area of basin = 78 sq. mi.
Basin lag = 8 hr.
Storage constant = 7.7 hr.

Time Area Curve:

Hour	Subarea	Area sq. mi.
0	A_0	0
1	A_1	3.3
2	A_2	9.3
3	A_3	19.4
4	A_4	14.3
5	A_5	9.4
6	A_6	10.8
7	A_7	10.0
8	A_8	1.5
		78.0

For practical purposes, we can assume that $I_1 = I_2$ during any one routing period, $\Delta t = 1$ hr. Hence, Equation VIII.68 reduces to

$$O_2 = c_3 I + c_2 O_1 \dots\dots\dots\dots\dots\dots\dots\dots\dots\dots\dots VIII.71$$

where $c_3 = \dfrac{\Delta t}{K + 0.5\Delta t}$

Solving:

$$c_3 = \frac{\Delta t}{K + 0.5\Delta t} = \frac{1}{7.7 + 0.5} = 0.122$$

$$c_2 = \frac{K - 0.5\Delta t}{K + 0.5\Delta t} = \frac{7.7 - 0.5}{7.7 + 0.5} = 0.878$$

Hence, Equation VIII.71 becomes

$$O_2 = 0.122\, I + 0.878\, O_1$$

Solving:

Time hr.	Area sq. mi.	I cfs	$c_3 I$	$c_2 O_1$	Instantaneous Unit Graph O_2 cfs	1-hr. Unit Graph cfs
0	0	0	0	0	0	0
1	3.3	2,115	258	0	258	129
2	9.3	6,025	735	226	961	610
3	19.3	12,350	1,507	735	2,242	1,601
4	14.3	9,160	1,120	1,970	3,090	2,666
5	9.4	6,030	735	2,720	3,455	3,232
6	10.8	6,920	843	3,035	3,878	3,666
7	10.0	6,400	781	3,405	4,186	4,032
8	1.5	962	116	3,680	3,796	3,991
9				3,500	3,500	3,648
10				3,060	3,060	3,280
11				2,690	2,690	2,875
12				2,360	2,360	2,525
13				etc.		

This method of hydrograph synthesis has the advantage that temporal and areal variations in rainfall can be taken into account by modifying the time-area curve. Similarly, the curve can be adjusted to consider the effects of difference in channel slope.

VIII.6.4.4   Conceptual Model of Instantaneous Unit Graph According to Nash and Dooge.

Theory

It is obvious from previous discussions that for a 'linear' reservoir one can write the equations:

$$S = KO \quad \text{and} \quad I-O = \frac{dS}{dt}$$

Combining the equations

$$I-O = \frac{KdO}{dt} \quad \dots\dots\dots\dots\dots\dots\dots\dots\dots\dots\dots\dots\dots\dots\dots\dots \text{VIII.72}$$

If Equation VIII.72 is integrated considering the limits $O = o$, when $t = o$ and $O = I$ at $t = \infty$, then

$$O = I(1 - e^{-t/K}) \quad \dots\dots\dots\dots\dots\dots\dots\dots\dots\dots\dots\dots\dots \text{VIII.73}$$

However, if inflow terminates at some time t_0 after outflow began, then a similar derivation will give the outflow O at time t in terms of the outflow O_0 as

$$O = O_0 e^{-\Delta t/K} \quad \dots\dots\dots\dots\dots\dots\dots\dots\dots\dots\dots\dots\dots\dots\dots\dots \text{VIII.74}$$

in which $\Delta t = t - t_0$.

For an instantaneous inflow which fills a reservoir of storage S instantaneously (that is $t_0 = o$), then $O_0 = S/K$ and Equation VIII.74 becomes

$$O = \frac{S}{K} e^{-t/K} \quad \dots\dots\dots\dots\dots\dots\dots\dots\dots\dots\dots\dots\dots\dots\dots\dots\dots \text{VIII.75}$$

In 1957, Nash proposed that a drainage basin could be represented by a system of 'n' identical reservoirs placed in series (see Fig. VIII.26). Using Equation VIII.75, it is evident that the discharge rates through the reservoirs can be determined as

$$O_1 = \frac{S}{K} e^{-t/K}$$

$$O_2 = \frac{1}{K} e^{-t/K} \int_o^t O_1 \, dt$$

$$= \frac{1}{K} e^{-t/K} \int_o^t \frac{S}{K} e^{-t/K} \, dt$$

$$= \frac{S}{K^2} e^{-t/K} t$$

After successive routings through 'n' reservoirs, the discharge rate, O_n becomes

$$O_n = S e^{-t/K} t^{n-1} / K^n (n-1)! \quad \dots\dots\dots\dots\dots\dots\dots\dots\dots \text{VIII.76}$$

The components K and n of Nash's equation for the instantaneous unit graph (Equation VIII.76) can be evaluated by the method of moments (Nash, 1959).

As the model proposed by Nash does not involve the concept of translation of flow, attempts have been made to incorporate this concept into the analysis. For example, Dooge (1959) proposed using a system of linear channels (translation only with no storage) in series with linear reservoirs to represent the basin system. Further studies by Rosa et al. (1961), Singh (1961), Diskin (1964) and Kulandaiswamy (1964) have used other techniques to define the instantaneous unit graph.

Further discussions concerning the applications of conceptual models employing flood routing principles are given in Section X.

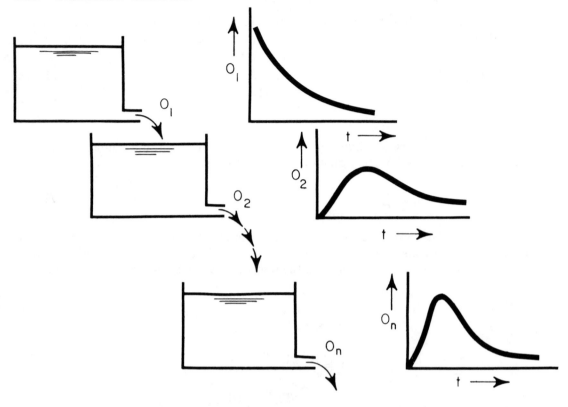

Fig. VIII.26 Systematic Routing of Instantaneous Inflow Through 'n' Linear Reservoirs.

VIII.7   RIVER HYDRAULICS AND FLOW MEASUREMENT

The measurement of water flow or discharge in a channel is normally accomplished by measurements of flow velocities at selected points in the channel. These measured velocities yield an average flow velocity which, when multiplied by the associated flow area, gives the discharge through a particular portion of the cross section. The total discharge of the stream is obtained from a summation of the discharges in each portion. Actual flow measurements are only obtained periodically, say once every two to five weeks, so the intervening flows are determined from a stage-discharge (water surface elevation-discharge) relationship developed from the actual measurements of discharge. This stage-discharge relationship may have to be adjusted for non-uniform flow conditions, and extrapolation of the curve may be checked by calculation of discharge using flow formulae.

The brief description of flow measurement methods outlined above serves to indicate the aspects of river hydraulics that are important in determining discharge. These are velocity distributions in a channel, estimation of channel roughness for use in flow formulae, and the effects of non-uniform flow on the stage-discharge relationship.

VIII.7.1   **Velocity Distributions in Channels**

VIII.7.1.1 Laminar or Turbulent Flow

Flow in open channels may be either laminar or turbulent, hence there are two possible velocity distributions. The criterion for laminar flow is that the Reynolds Number is less than 500, and for turbulent flow the Reynolds Number should be greater than about 2,000, where the Reynolds Number is defined as:

$$N_R = \frac{VR}{\nu} \quad \cdots\cdots\cdots\cdots\cdots\cdots\cdots\cdots\cdots\cdots\cdots\cdots\cdots\cdots\cdots\cdots\cdots\cdots\cdots\cdots\cdots \text{VIII.77}$$

where V = the average flow velocity (ft/sec),
       R = the hydraulic radius A/P (ft),
       A = the flow area (sq ft),
       P = the wetted perimeter (ft), and
       ν = the kinematic viscosity (ft^2/sec).

The kinematic viscosity of the water depends on the temperature, but will be of the order of 1×10^{-5} ft^2/sec. The hydraulic radius is approximately equal to the flow depth, d, for broad shallow channels, so an approximation of the Reynolds Number is

$$N_R = Vd \times 10^5$$

It can be seen from this that if the term Vd is 0.02 ft^2/sec. or larger, the flow will be turbulent, so turbulence is the normal condition in most open channels. Laminar flow conditions can occur in overland flow.

VIII.7.1.2 Velocity Distributions in Natural Channels

Field data on velocity distributions have been used to define certain rules that allow determination of discharge with a relatively small number of velocity measurements. In addition, these data provide information as to the applicability of theoretical velocity distributions. Examples of velocity distributions in vertical planes are shown in Figs. VIII.27 and VIII.28 (from *River Discharge* by Hoyt and Grover, 1924). Figure VIII.27 shows the velocity distribution in a plane perpendicular to the flow direction. The curves of equal velocity were obtained by sketching contour lines and by taking into account the positions and magnitude of measured velocities. A plot such as this could be used to calculate the discharge by determining the area between lines of equal velocity, and by multiplying by the average velocity between the lines. However, the number of velocity measurements required and the work involved in calculation precludes the use of this method, even though it would probably be accurate.

The vertical velocity curves (see Fig. VIII.28) are of greater interest. These curves show the general shape of the velocity distribution, and the fact that the point of maximum velocity often occurs below the water surface. Examination of such curves by

DISTANCE IN FEET

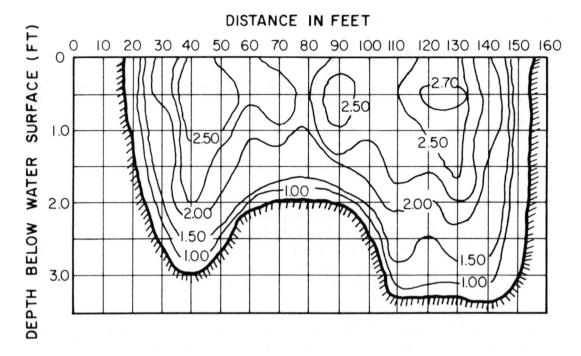

Fig. VIII.27 Contours of Equal Velocity in Natural Channels

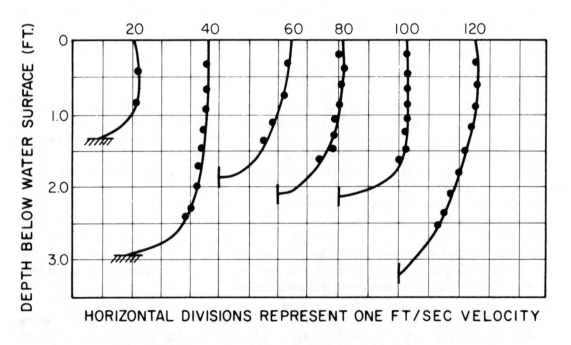

HORIZONTAL DIVISIONS REPRESENT ONE FT/SEC VELOCITY

Fig. VIII.28 Vertical Velocity Distributions in Natural Channels

many individuals and agencies for a wide variety of channel sizes, shapes and roughnesses indicated the following points:

(1) the average velocity determined by averaging the measurements taken at 0.2 and 0.8 of the total depth below the surface is a very good approximation of the average velocity in the vertical profile;

(2) the velocity at 0.6 of the total depth below the surface is a good approximation of the average velocity in the vertical; and

(3) the average velocity is between 0.85 and 0.95 of the surface velocity.

The points above are listed in order of decreasing accuracy for the determination of average velocity.

VIII.7.1.3 Theoretical Velocity Distributions

Although it is possible to use theoretical velocity distributions to calculate the discharges of natural streams, actual flow measurements or other methods of calculation are usually more practical. A theoretical velocity distribution is of value in analysis of other types of problems, and is important for an overall understanding of the hydraulics of the system.

If the flow is laminar, then an exact equation for the velocity distribution can be obtained theoretically. This is not the case for turbulent flow, as a number of assumptions are necessary to define turbulent conditions, and in addition, the conditions may change from point to point in the channel. The theoretical velocity distributions for turbulent flow often do agree fairly well, however, with observed distributions.

The commonly accepted velocity distribution for uniform (constant discharge) turbulent flow is the one developed using concepts proposed by Prandtl (1953). He indicated that the shearing stress at any point in a fluid moving over a solid boundary in turbulent flow can be evaluated by

$$\tau = \rho l^2 \left(\frac{dv}{dz}\right)^2 \quad \dots \dots \dots \dots \dots \dots \text{VIII.78}$$

where τ = the shearing stress,
ρ = the mass density $= \dfrac{\gamma}{g}$ where γ is
the specific weight of the fluid and g is the gravitational acceleration,
l = the characteristic length known as the mixing length, and
$\dfrac{dv}{dz}$ = the velocity gradient at a distance z from the solid surface.

Prandtl introduced two assumptions for the flow region near the solid surface: that the mixing length is proportional to z or $l = kz$, and that the shearing stress is constant. The shearing stress at a boundary in uniform flow is

$$\tau_0 = \gamma RS \quad \dots\dots\dots\dots\dots\dots\dots\dots\dots\dots\dots\dots\dots \text{ VIII.79}$$

where S is the channel slope (also the slope of the total energy line and the water surface), and the other terms are as previously defined.

The second assumption therefore states that $\tau = \tau_0$, therefore $\tau_0 = \rho l^2 \left(\dfrac{dv}{dz}\right)^2$ The equation can then be put in the form:

$$dv = \frac{1}{k_o} \sqrt{\frac{\tau_o}{\rho}} \; \frac{dz}{z} \quad \dots\dots\dots\dots\dots\dots\dots\dots\dots\dots\dots \text{VIII.80}$$

by rearranging and substituting $l = k_o z$. The term $\sqrt{\dfrac{\tau_o}{\rho}}$, has the dimensions of velocity, and since it varies with the boundary shearing stress, it is called the 'shear velocity' or 'friction velocity', and is normally replaced with the symbol V*. Equation VIII.80 can therefore be written

$$dv = \frac{V^*}{k_o} \; \frac{dz}{z}$$

and the equation for velocity at any point in the flow can be obtained by integration. If the limits of integration are chosen from $v = v$ to $v = V_{max}$, and the corresponding limits for depth are $z = z$ to $z = d$ (total flow depth), then the equation becomes

$$V_{max} - v = \frac{V^*}{k_o} \; (\ln d - \ln z)$$

or

$$v - V_{max} = \frac{V^*}{k_o} \ln z/d \quad \dots\dots\dots\dots\dots\dots\dots\dots\dots\dots \text{..VIII.81}$$

or in terms of logs to the base 10

$$v - V_{max} = \frac{2.3 \, V^*}{k_o} \; \log_{10} z/d$$

This equation is normally shown in the form:

$$\frac{v - V_{max}}{V^*} = \frac{2.3}{k_o} \log_{10} z/d \dots\dots\dots\dots\dots\dots\dots\dots\dots\dots \text{.VIII.82}$$

The equation indicates that the velocity in the turbulent region is a logarithmic function of the distance z from the boundary. This concept is commonly known as the Prandtl-Von Karman universal-velocity-distribution law, as Von Karman (1930) developed a similar equation using somewhat different methods and assumptions.

This law has been verified by several experiments, such as those by Vanoni (1941). Observed and computed distributions agree quite well, particularly for laboratory flume data (see Fig. VIII.29). One obvious drawback to the equation is that the calculated velocity approaches a value of minus infinity near the boundary whereas the actual velocity at the boundary is zero. The logarithmic velocity distribution can, therefore, only be considered a good approximation to the actual profile over most of the flow depth.

The average velocity, V, can be obtained from the logarithmic equation for velocity at any point by integrating over the depth of flow, and by dividing by the total depth. The lower limit of integration must be chosen with a finite, non-zero value ($z = \delta$) such that the lower limit of the integral vanishes. Thus

$$V = \frac{1}{d} \int_{z=\delta}^{z=d} \left(V_{max} + \frac{V^*}{k_o} \ln z/d \right) dz$$

$$V = \frac{1}{d} \left[V_{max} z + \frac{V^*}{k_o} (z \ln z - z) - \frac{V^* z}{k_o} \ln d \right]_{\delta}^{d}$$

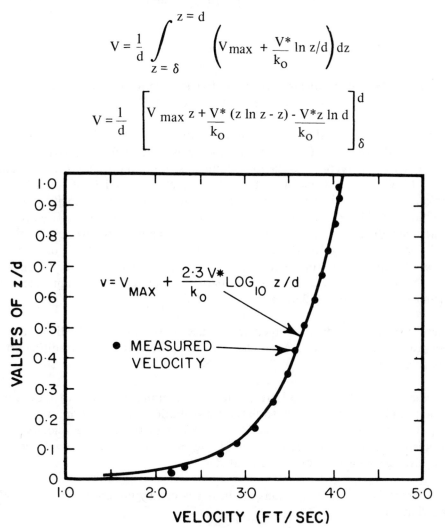

Fig. VIII.29 Velocity Profile at Centre of a Flume 2.77 ft Wide for a Flow 0.59 ft Deep
(From Vanoni 1941)

$$V = V_{max} - \frac{V^*}{k_o} \quad \dots \dots \dots \dots \dots \dots \dots \dots \dots \text{VIII.83}$$

If V_{max} is eliminated from the general equation for velocity, it can be written

$$v = V + \frac{V^*}{k_o} \; (1 + 2.3 \log_{10} z/d) \; \dots \dots \dots \dots \dots \dots \dots \text{VIII.84}$$

The location of the point at which the velocity is equal to the average velocity can be found by substituting $v = V$ into Equation VIII.84, yielding

$$\ln z/d = -1 \quad \text{or} \quad z/d = \frac{1}{e} \quad \text{and}$$

$$z = \frac{d}{e} = 0.368d \; \dots \dots \dots \dots \dots \dots \dots \dots \dots \dots \text{VIII.85}$$

This indicates that the distance from the free surface to the point of average velocity is 0.632d and is not affected by the value of k_o, or of V^*.

As mentioned previously, experience in stream gauging has shown that the velocity measured at a depth equal to 0.6 of the total depth is a good approximation to the average velocity in the vertical, so the logarithmic distribution does not agree exactly with field data. In addition, the mean of the velocities at 0.2d and 0.8d for a logarithmic distribution will give a velocity somewhat higher than the average velocity. The mean of the velocities at 0.84 and 0.16 of the depth would give the average velocity according to the logarithmic distribution. The reason for these apparent discrepancies between the points of average velocity for the logarithmic distribution and actual points of average velocity in streams is that the logarithmic velocity assumes that the maximum velocity occurs at the free water surface. In fact, in natural streams the maximum velocity often occurs some distance (up to 0.33d) below the surface.

These discrepancies do not preclude the use of the logarithmic distribution as an assumed distribution for solving open channel flow problems, as the resulting errors are generally small.

Equations VIII.81 to VIII.84 are useful for examining the form of the logarithmic distribution, and for comparing it to natural distributions, but they are not very useful for computing stream discharge. Equation VIII.83 gives the average velocity for a logarithmic distribution, but a measurement of the maximum velocity would be required before the average velocity could be calculated.

A different form of the equations can be developed by integrating Equation VIII.80, as an indefinite integral

$$v = \int \frac{V^*}{k_o} \frac{dz}{z} = \frac{V^* \ln z + C}{k_o}$$

The constant C can be assumed to be a logarithm so $C = \dfrac{-V^*}{k_0} \ln z_0$ where z_0 is a constant.

The equation for v then becomes

$$v = \frac{V^*}{k_0} \ln z/z_0 \quad \dotfill \quad \text{VIII.86}$$

The value of z_0 depends on whether the boundary surface is hydraulically smooth or hydraulically rough. The boundary between smooth and rough conditions can approximately be defined by the equation

$$k_c = \frac{100\nu}{V} \qquad (k < k_c = \text{smooth})$$

where k_c is a critical roughness height, and k is the actual effective roughness height. For smooth pipes, Nikuradse found that

$$z_0 = \frac{m\nu}{V^*} \quad \dotfill \quad \text{VIII.87}$$

where m is a constant equal to about 1/9. For rough pipes

$$z_0 = mk \quad \dotfill \quad \text{VIII.88}$$

where m is a constant equal to approximately 1/30.

If these constants are also assumed to apply for open channel flow, then the resulting equations are

$$v = \frac{2.3V^*}{k_0} \log_{10} \frac{(9\,zV^*)}{\nu} \quad \text{for smooth boundaries,}$$

$$v = \frac{2.3\,V^*}{k_0} \log_{10} \left(\frac{30z}{k} \right) \quad \text{for rough boundaries.}$$

These equations can be integrated over the depth to give the average velocity in the vertical, or they can be integrated over the whole flow area to give the average velocity at a given cross section. The latter integration requires certain assumptions regarding the position of lines of equal velocity in the cross-section. However, Keulagan (1938) has derived equations for the average velocity of turbulent flow for the complete-flow cross section. These have been modified somewhat by Chow (1959), so that the final form of the equation is

$$V = V^* \left(A_0 + \frac{2.3}{k_0} \log_{10} \frac{mR}{z_0} \right) \dotfill \quad \text{VIII.89}$$

where z_0 is as defined in either Equation VIII.87 or VIII.88, depending on the boundary conditions. A_0 is a function of the channel shape. Keulagan, from a study of Nikuradses' data, found A_O to be 3.25 for smooth boundaries; and from a study of Bazin's data (1965), he found A_O to be an average of 6.25 for rough boundaries. The resulting equations for average velocity over the whole flow cross section are then

$$V = V^* \left(3.25 + \frac{2.3}{k_O} \ \log_{10} \ \frac{RV^*}{\nu} \right) \text{ for smooth boundaries} \dots\dots\dots\dots\dots \text{VIII.90}$$

$$V = V^* \left(6.25 + \frac{2.3}{k_O} \log_{10} \frac{R}{k} \right) \quad \text{ for rough boundaries } \dots\dots\dots\dots\text{VIII.91}$$

If these velocities are multiplied by the flow area, they will give the discharge for the channel. It should be noted that the channel information includes values for the hydraulic radius, R; for the channel or water surface slope, S for the flow depth, d; and for rough channels, the effective boundary roughness height, k.

The values of A_O shown in the equations are average values. Unfortunately, A_O may vary in the order of magnitude of ±50 per cent so as to produce an error in V of ±25 per cent. Iwagaki (see Chow 1959, pp. 204-5) suggests that A_O is a function of the Froude number $N_F = V / \sqrt{dg}$ and the curves (Chow 1959, p. 205) can be used to evaluate A_O.

The effective boundary roughness height, k, may be chosen equal to the median diameter of the bed material, or better yet, the diameter of a particle such that 65 per cent of the bed particles are smaller (d_{65}). The choice of k is still somewhat arbitrary, and the values mentioned above only apply if there is no movement of the bed.

Because of the variation in A_O and the arbitrary definition of k, the logarithmic equations must be used with caution. In fact, they should be used in the same way that empirical equations are used; that is, with constants calculated from open channels of similar size and boundary conditions, or preferably, with constants determined from data on the open channel in question. The advantage of using a theoretical velocity distribution equation lies in the fact that only simple measurements (hydraulic radius, slope, flow depth) are required for its application.

The development of logarithmic equations has been covered in detail to show the method of analysis of any assumed distribution, and because such equations are commonly used and quoted. The same type of analysis could be used assuming, as is sometimes done, a parabolic velocity distribution. This type of distribution is worthy of consideration, as it fits observed distribution curves quite well over most of the flow depth. A parabolic distribution implies a constant eddy viscosity (increased apparent viscosity over that for laminar flow) due to turbulence.

The advantage of the parabolic distribution is that it allows placing the point of maximum velocity below the surface, and this agrees with actual conditions. If the axis of the parabola, which corresponds to maximum velocity, is placed 0.15 of the flow depth below the surface, then the point of average velocity will occur at exactly 0.6d below the surface. Hoyt and Grover (1924) report that the mean of the velocities at 0.21 and 0.79

of the depth is equal to the average velocity. This agrees well with the practical criterion that the average velocity of flow can be taken as the mean of the velocities at 0.2 and 0.8 of the depth.

Practically, both the parabolic and logarithmic velocity distributions describe the actual velocity distribution fairly well. The choice of distribution will, therefore, depend primarily on individual preference, and on the type of analytical problem to be solved.

VIII.7.2   Channel Roughness

VIII.7.2.1   Introduction

It is sometimes impractical or even impossible to obtain actual measurements of the discharges of floods. One obvious reason for this is that it is extremely difficult to predict the time at which the flood peak will arrive at a given station. Further, during these events it is often impossible to gain access to the measurement station; or if the measurement party is at the structure, floating ice or debris may still preclude obtaining a discharge measurement. Under such circumstances, the discharge is determined by extrapolation of the stage-discharge curve, or it may be calculated using existing flow formulae.

VIII.7.2.2   Flow Equations

The flow equation in common use for determining the velocity of flow in open channels on the North American continent is the Manning Equation, whereas in Europe the Chezy Equation is popular. The Regime Equations developed in India and North America, by Blench (1966), are also open-channel flow equations; however, they are not normally used to predict discharges at a specific stage of flow. Though definitely of value in the overall field of mobile-boundary hydraulics, they will not be considered for the purpose of predicting discharges.

The Manning Equation is

$$V = \frac{1.49}{n} R^{2/3} S^{1/2} \quad \dots\dots\dots\dots\dots\dots\dots\dots\dots\dots \text{VIII.92}$$

where V = the average velocity in the cross-section, (ft/sec),
   R = the hydraulic radius, (ft),
   S = the water surface slope, (ft/ft), and
   n = a roughness coefficient.

The Chezy Equation is

$$V = C\sqrt{RS} \quad \dots\dots\dots\dots\dots\dots\dots\dots\dots\dots\dots\dots \text{VIII.93}$$

where C = the Chezy roughness coefficient.

The two equations are similar in form, and by equating the two, a relationship can be developed between the roughness coefficients, as

$$C = \frac{1.49}{n} R^{1/6} \quad \dots\dots\dots\dots\dots\dots\dots\dots\dots\dots\dots\dots\dots\dots\dots\dots\dots\dots \text{VIII.94}$$

The equations are also used in a similar fashion: the hydraulic radius and water-surface slope are obtained from cross-section and slope data, the appropriate roughness coefficient is estimated, and the average discharge is calculated by multiplying the cross-sectional area by the average velocity determined from the equation. The only difficulty lies in determining the roughness coefficient. Since the Chezy and Manning coefficients are related by Equation VIII.94, and since R must be measured for the application of either equation, then it is only necessary to consider one of the roughness coefficients.

VIII.7.2.3   Determination of Manning's Roughness Coefficient

The factors that affect Manning's n have been listed and described (Chow 1959, pp. 101-6). These factors may be listed under three major headings: boundary conditions; conditions of alignment and cross-section; and obstructions. The boundary conditions are defined by the size of bed and bank material, by the amount and type of vegetation on the bed and banks, and on the form resistance created by bed material transport. Manning's n will increase with material size, and will also increase with the amount of vegetation, all other things being equal. The actual effect of the vegetation depends to a degree on the flow conditions and on prior flow events; for example, the vegetation may be flattened at high flows, producing a low value of n.

The form resistance is also affected by the discharge and prior flow conditions; for example, dunes may form on the bed at low discharges and may wash out at high discharges, causing a decrease in n with discharge. Alternatively, dunes may form at high discharges and remain for low discharges, thus producing a high n value for low flow conditions. The effects of transport of material in suspension are not well defined, though there is some evidence that the increased suspended load decreases the apparent roughness (Vanoni *et al.*, 1961).

Any change in flow direction and/or flow velocity will increase Manning's n due to the creation of secondary currents and/or increased turbulence. Changes in alignment (bends, meandering) and changes in cross-sectional shape or size therefore increase the n values over those for straight channels. The magnitude of this change depends on the rate and frequency of the changes in shape and size.

It is obvious that obstructions created by logs or other debris will also increase the flow turbulence, thereby increasing the n values.

Any method of determining n should take into account all of the above factors. The only way that this can practically be accomplished at the present time is by use of tables of n values for various channel conditions (see Table VIII.13). The use of such tabular data still involves judgment as a range of n values is given for each channel description. The U.S. Geological Survey uses pictures of various channels at different n values to aid in the choice of n. Chow (1959) also illustrates pictures supplied by various agencies that can be used to assist in the choice of an appropriate roughness factor.

Table VIII.13   Values of the Roughness Coefficient 'n'

Type of channel and description	Minimum	Normal	Maximum
A. Excavated or Dredged			
a. Earth, straight and uniform			
1. Clean, recently completed	0.016	0.018	0.020
2. Clean, after weathering	0.018	0.022	0.025
3. Gravel, uniform section, clean	0.022	0.025	0.030
4. With short grass, few weeds	0.022	0.027	0.033
b. Earth, winding and sluggish			
1. No vegetation	0.023	0.025	0.030
2. Grass, some weeds	0.025	0.030	0.033
3. Dense weeds or aquatic plants in deep channels	0.030	0.035	0.040
4. Earth bottom and rubble sides	0.028	0.030	0.035
5. Stony bottom and weedy banks	0.025	0.035	0.040
6. Cobble bottom and clean sides	0.030	0.040	0.050
c. Dragline-excavated or dredged			
1. No vegetation	0.025	0.028	0.033
2. Light brush on banks	0.035	0.050	0.060
d. Rock cuts			
1. Smooth and uniform	0.025	0.035	0.040
2. Jagged and irregular	0.035	0.040	0.050
e. Channels not maintained, weeds and brush uncut			
1. Dense weeds, high as flow depth	0.050	0.080	0.120
2. Clean bottom, brush on sides	0.040	0.050	0.080
3. Same, highest stage of flow	0.045	0.070	0.110
4. Dense brush, high stage	0.080	0.100	0.140
B. Natural Streams			
B - 1. Minor streams (top width at flood stage 100 ft.)			
a. Streams on plain			
1. Clean, straight, full stage, no rifts or deep pools	0.025	0.030	0.033
2. Same as above, but more stones and weeds	0.030	0.035	0.040
3. Clean, winding, some pools and shoals	0.033	0.040	0.045
4. Same as above, but some weeds and stones	0.035	0.045	0.050
5. Same as above, lower stages, more ineffective slopes and sections	0.040	0.048	0.055
6. Same as 4, but more stones	0.045	0.050	0.060
7. Sluggish reaches, weedy, deep pools	0.050	0.070	0.080

Table VIII.13   (cont'd)

Type of channel and description	Minimum	Normal	Maximum
8. Very weedy reaches, deep pools, or floodways with heavy stand of timber and underbrush	0.075	0.100	0.150
b. Mountain streams, no vegetation in channel, banks usually steep, trees and brush along banks submerged at high stages			
1. Bottom: gravels, cobbles, and few boulders	0.030	0.040	0.050
2. Bottom: Cobbles with large boulders	0.040	0.050	0.070
B - 2. Flood plains			
a. Pasture, no brush			
1. Short grass	0.025	0.030	0.035
2. High grass	0.030	0.035	0.050
b. Cultivated areas			
1. No crop	0.020	0.030	0.040
2. Mature row crops	0.025	0.035	0.045
3. Mature field crops	0.030	0.040	0.050
c. Brush			
1. Scattered brush, heavy weeds	0.035	0.050	0.070
2. Light brush and trees, in winter	0.035	0.050	0.060
3. Light brush and trees, in summer	0.040	0.060	0.080
4. Medium to dense brush, in winter	0.045	0.070	0.110
5. Medium to dense brush, in summer	0.070	0.100	0.160
d. Trees			
1. Dense willows, summer, straight	0.110	0.150	0.200
2. Cleared land with tree stumps, no sprouts	0.030	0.040	0.050
3. Same as above, but with heavy growth of sprouts	0.050	0.060	0.080
4. Heavy stand of timber, a few down trees, little undergrowth, flood stage below branches	0.080	0.100	0.120
5. Same as above, but with flood stage reaching branches	0.100	0.120	0.160
B - 3. Major streams (top width at flood stage 100 ft.). The n value is less than that for minor streams of similar description, because banks offer less effective resistance.			
a. Regular section with no boulders or brush.	0.025	. . .	0.060
b. Irregular and rough section	0.035	. . .	0.100

Cowan (1956) developed a procedure for estimating n that takes into account several of the factors which affect the values. The value of n may be computed by his procedure using the following equation

$$n = (n_0 + n_1 + n_2 + n_3 + n_4) \, m_5$$

where n   = the value to be used,
n_0 = a basic n value for a straight uniform, smooth channel in the natural materials involved,
n_1 = a value to correct for the effect of surface irregularities,
n_2 = correction factor for variations in the size and shape of the channel,
n_3 = correction factor for the effects of obstructions,
n_4 = correction factor for the vegetation and flow conditions, and
m_5 = correction factor for the meandering of the channel.

Values of n and m may be selected from Table VIII.14. The conditions applying for each value of n and m are explained in Chow (1959; pp. 106-8). It should be noted that this method does not consider the effects of sediment transport, consequently it must be limited to conditions where no transport occurs. In addition, the values were derived for small channels (R<15 ft.), so they should be used with caution for larger channels.

Manning's n may also be evaluated by analytical procedures based on a theoretical velocity distribution in the channel and on the data of either velocity or roughness measurements. Keulagan's Equations for flow velocities based on a logarithmic velocity distribution (see Equations VIII.90 and VIII.91) are

$$V = V^* \left(3.25 + \frac{2.3}{k_0} \log_{10} \frac{RV^*}{\nu} \right) \text{ for smooth channels,}$$

$$V = V^* \left(6.25 + \frac{2.3}{k_0} \log_{10} \frac{R}{k} \right) \quad \text{for rough channels.}$$

From the Manning Equation $V = \frac{1.49}{n} R^{2/3} S^{1/2}$, and from the definition of $V^* = \sqrt{gRS}$, it can be shown that

$$\frac{V}{V^*} = \frac{1.49 \, R^{1/6}}{n\sqrt{g}} \quad \ldots\ldots\ldots\ldots\ldots\ldots\ldots\ldots\ldots\ldots\ldots\ldots\text{VIII.95}$$

Substituting this equation into those of Keulagan, and assuming $k_0 = 0.4$, we get

$$n = \frac{R^{1/6}}{21.9 \log_{10} \left(\frac{14V \, n \, R^{5/6}}{\nu} \right)} \quad \text{for smooth channels} \ldots\ldots\ldots\ldots\ldots\ldots\text{VIII.96}$$

Table VIII.14 Values for the Computation of the Roughness Coefficient

Channel conditions			Values
Material involved	Earth		0.020
	Rock cut		0.025
	Fine gravel	n_0	0.024
	Coarse gravel		0.028
Degree of irregularity	Smooth		0.000
	Minor		0.005
	Moderate	n_1	0.010
	Severe		0.020
Variations of channel cross section	Gradual		0.000
	Alternating occasionally	n_2	0.005
	Alternating frequently		0.010—0.015
Relative effect of obstructions	Negligible		0.000
	Minor		0.010—0.015
	Appreciable	n_3	0.020—0.030
	Severe		0.040—0.060
Vegetation	Low		0.005—0.010
	Medium		0.010—0.025
	High	n_4	0.025—0.050
	Very high		0.050—0.100
Degree of meandering	Minor		1.000
	Appreciable	m_5	1.150
	Severe		1.300

$$n = \frac{R^{1/6}}{21.9 \log_{10} \dfrac{12.2R}{k}} \quad \text{for rough channels} \dots\dots\dots\dots\dots\dots \text{VIII.97}$$

The solution of Equation VIII.96 requires a trial and error procedure since n appears on both sides of the equality sign. In addition, the terms R, V and v have to be given or estimated. The values of v and R can be obtained easily, as v can be estimated and R is required information (obtained from cross-sectional data) for solution of the Manning Equation. Since V is eventually the quantity which one wishes to calculate, an assumed value must be used in the equation for n, and this value must be revised when V is calculated. This double trial and error procedure makes the solution of this equation difficult, if not impossible, as it is conceivable that several combinations of V and n would suffice, or alternatively that a solution could not be found.

Fortunately, the rough boundary equation is more frequently used, and simply requires an evaluation of R and k. The equation is often shown in the following form:

$$n = k^{1/6} f(R/k) \dotfill \text{VIII.98}$$

where k = roughness height and

$$f(R/k) = \frac{(R/k)^{1/6}}{21.9 \log_{10} \frac{(12.2R)}{k}} \dotfill \text{VIII.99}$$

A plot of f(R/k) versus R/k is shown in Fig. VIII.30 (from Chow 1959; p. 206)

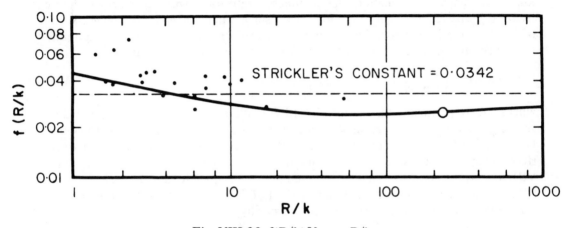

Fig. VIII.30 f(R/k) Versus R/k

This plot shows that for a wide variation in R/k, the variation in f(R/k) is small. Strickler (1923), on the basis of observations on channels in Switzerland, arrived at an equation which when compared with Equation VIII.98 gives an average value of f(R/k) = 0.0342, where k is taken as the median size of the boundary material in feet. Equation VIII.99 can then be reduced to

$$n = 0.0342 \, k^{1/6} \dotfill \text{VIII.100}$$

Several other individuals or agencies have produced equations similar in form, with the constant varying from 0.034 to 0.043. Because of this, it is recommended that the following equation be used as an approximation:

$$n = 0.04 \, k^{1/6} \dotfill \text{VIII.101}$$

where k is the median size of the bed material in feet.

The conditions to be fulfilled in proper application of these equations are that the channel be straight and free of vegetation, and that no motion of the bed material occur. In addition, the equation appears to give better results for gravel than for finer material. It should be noted that the major errors in the equation are in the value of $f(R/k)$. Variations or errors in the k value chosen will not affect n to any great degree, as the value of k is taken to the one-sixth power. The range of the equation could perhaps be extended by using figures from Table VIII.14 to correct for variations in alignment, vegetation etc.

Einstein and Barbarossa (1952) proposed a method whereby the effects of moving sediment could be considered. Equation VIII.98 is assumed to apply. And it is assumed that the hydraulic radius R for a stream consists of two parts: the hydraulic radius, R', due to surface roughness; and the hydraulic radius, $R-R'$, due to roughness caused by moving sediment beds. The roughness height is assumed to be $k = d_{65}$ for the surface roughness, where d_{65} is the particle size, such that 65 per cent of the particles of a sample are smaller in diameter. The roughness height for moving sediments is represented by the d_{35} size (35 per cent of the material is smaller in diameter). Following the concept of Einstein and Barbarossa, Doland and Chow (1952) indicated that the function $f(R/k)$ for the combined effect of the surface roughness and moving sediments is

$$f(R/k) = \frac{0.0342}{(R'/R)^{2/3}} \quad \dots\dots\dots\dots\dots\dots\dots\dots\dots\dots\dots\dots\dots\dots\dots\dots \quad \text{VIII.102}$$

where, (R'/R) depends on the measured hydraulic radius R, the slope S, and the grain sizes d_{65} and d_{35}. They prepared a plot of d_{35}/RS versus $(R/d_{65})^{1/3}$ that may be used to determine (R'/R). The plot (see Chow 1959, p. 209), is based on data from seven U.S. rivers. Once $f(R/k)$ is calculated using the (R'/R) value of the plot, then Equation VIII.98 may be used to calculate n, assuming k is equal to d_{65} in feet. The plot for (R'/R) has not been included, as this method of determining n must still be subjected to further verification before it can be used universally.

Another method of determining n, which must also be placed in the category of requiring more verification, is that used by Boyer (1954), and others. The value of n is estimated, assuming that the logarithmic velocity distribution equations apply and that the mean of the velocities at 0.2 and 0.8 of depth yields the average velocity. For a rough wide channel, these velocities may be expressed (from Equation VIII.86) as

$$v_{0.2} = 5.75 \, V^* \log_{10} \frac{24d}{k}$$

$$v_{0.8} = 5.75 \, V^* \log_{10} \frac{6d}{k}$$

where d = the flow depth, and k_0 is 0.4 (assumed).

Letting $v_{0.2}/v_{0.8} = x$ and eliminating V^* from the equations then

$$\log_{10} \frac{d}{k} = \frac{0.778x - 1.38}{1 - x} \quad \dots\dots\dots\dots\dots\dots\dots\dots\dots\dots\dots\dots\dots \quad \text{VIII.103}$$

Substituting this into Equation VIII.91, and assuming R = d (wide channel), we get

$$\frac{V}{V*} = \frac{1.78\,(x + 0.95)}{x - 1}$$

Using Equation VIII.95, with R = d, the relation can be simplified to

$$n = \frac{(x - 1)\,d^{1/6}}{6.78\,(x + 0.95)} \quad \dots\dots\dots\dots\dots\dots\dots\dots\dots\dots\dots\dots\dots\dots\dots\dots\dots \text{VIII.104}$$

The value of n can therefore be determined if the value of x is calculated from measured values of the velocities at 0.2d and 0.8d, and of course d must be measured (mean depth). The advantage of this equation is that n can be determined simply from velocity measurements, which are regularly taken in natural streams. The slope of the water surface does not have to be measured.

In summary it may be noted that the Manning n may be calculated or obtained by three methods

(1)   from tables;
(2)   from *theoretical* equations by estimating a roughness height, k, for the boundary (Equations VIII.97 and VIII.98); or
(3)   from velocity measurements at 0.2 and 0.8 of the depth from the surface (Equation VIII.104).

The first method is in general use and is recommended. The second method is limited in that the assumed velocity distribution does not always agree with field observations. In addition, these equations can only be used in the presented form for determination of n for channels free of vegetation, and obstructions having only minor changes in alignment, cross-sectional shape or size. The third method is primarily of value for computing values of n for the stream in question from actual stream data. The computed values of n could be plotted versus discharge to allow estimation of n values for unmeasured discharges. This procedure requires further verification, and it should be noted that it is difficult to extrapolate values of n because of the varying effects of sediment transport. Methods (2) and (3) are both based on the logarithmic distribution of velocity and k_0 value of 0.4. This value can change depending on the amount of suspended material transport.

VIII.7.3   **Unsteady Flow Effects**

It is recognized, in any program of flow measurement, that stream discharges cannot be measured every day, or even as often as desired because of economic considerations. The discharges, which occur between measurements, are determined from a measurement of stage (water surface elevation) and a plot of stage versus discharge drawn from data obtained from the actual flow measurements. The discharge, in a natural channel, is often changing from one time to another, and during the passage of a flood the discharges can change quite rapidly. The derived stage-discharge relationship is normally considered to apply for uniform flow conditions, so unsteady flow will cause some error (small) in the calculated discharges. The discharge can be corrected for unsteady flow conditions if the instantaneous water surface slope is measured. The equation for flow velocity for unsteady flow conditions may be obtained from the general equation for unsteady flow, which is

$$\frac{\partial V}{\partial t} + V\frac{\partial V}{\partial x} + g\frac{\partial d}{\partial x} - gS_c + gS_f = 0 \quad \ldots\ldots\ldots\ldots\ldots\ldots\ldots\ldots \text{VIII.105}$$

where $\dfrac{\partial V}{\partial t}$ = the change in average velocity, V, with time t,

$\dfrac{\partial V}{\partial x}$ = the change in average velocity with distance, x, along the channel,

g = the acceleration of gravity,

d = the flow depth,

S_c = the slope of the bed of the channel, and

S_f = the friction slope or head loss per foot of length.

The first two terms are acceleration terms, and it has been observed that accelerations are negligible in flood waves in comparison to the magnitudes of the other terms. The equation may therefore be reduced to

$$\frac{\partial d}{\partial x} - S_c + S_f = 0 \quad \ldots\ldots\ldots\ldots\ldots\ldots\ldots\ldots\ldots\ldots \text{VIII.106}$$

The term S_f is the friction term, and it may be calculated from the Manning or the Chezy equations; that is

$$V = C\sqrt{RS_f}$$

$$S_f = \frac{V^2}{C^2R} \quad \ldots\ldots\ldots\ldots\ldots\ldots\ldots\ldots\ldots\ldots\ldots\ldots \text{VIII.107}$$

Substituting this in Equation VIII.106 and solving for V, one obtains

$$V = C\sqrt{R\left(S_c - \frac{\partial d}{\partial x}\right)}$$

$$V = C\sqrt{RS_c\left(1 - \frac{1}{S_c}\frac{\partial d}{\partial x}\right)} \quad \ldots\ldots\ldots\ldots\ldots\ldots\ldots \text{VIII.108}$$

For uniform form $V = C\sqrt{RS_c}$, therefore the term $\sqrt{1 - \dfrac{1}{S_c}\dfrac{\partial d}{\partial x}}$ is the correction for non-uniform conditions. In terms of discharge, the equation may be written

$$Q = Q_0 \sqrt{1 - \frac{1}{S_c}\frac{\partial d}{\partial x}} \quad \ldots\ldots\ldots\ldots\ldots\ldots\ldots\ldots\ldots\ldots\ldots\ldots\ldots \quad VIII.109$$

where Q = the discharge at a given stage for unsteady flow conditions,

Q_0 = the discharge at the same stage for uniform steady flow conditions, and

$\dfrac{\partial d}{\partial x} =$ the change in flow depth with distance, x, at the measuring site.

The latter term may be evaluated by taking a measurement of water surface elevation upstream of the gauging site. The term $\dfrac{\partial d}{\partial x}$ is negative for the positive wave front, and positive for the recession portion of the wave. The equation indicates that the rating curve for a natural stream should be looped as shown in Fig. VIII.31.

The varying effects of vegetation, and bed roughness due to changing stage, may obscure or enhance the variations away from the steady discharge rating curve, so corrections may only have to be applied for certain streams and for periods of rapid changes in stage.

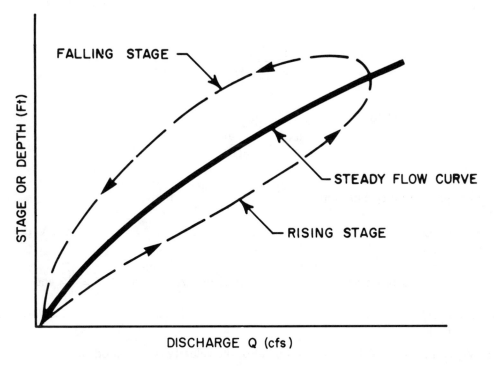

Fig. VIII.31 Stage - Discharge Relationships for a Natural Stream

Another feature of unsteady flow, which is of interest to hydrologists, is the speed of travel of the flood wave. This velocity may be determined by solving an equation of motion in conjunction with the equation of continuity. The equation of continuity for a channel of any cross-sectional shape is

$$B\frac{\partial d}{\partial t} + \frac{\partial Q}{\partial x} = 0$$

where B = the water surface width at the cross-section under study,
$\frac{\partial d}{\partial t}$ = the change in flow depth with time at the section, and
$\frac{\partial Q}{\partial x}$ = the change of discharge with distance, x, along the channel.

The first term is the rate of change in storage that must equal the second term, which is the change of discharge in the length considered.

If an observer moves with the flood wave in such a way that the discharge he observes does not change, then

$$\frac{\partial Q}{\partial x} \, dx \; + \; \frac{\partial Q}{\partial t} \, dt \, = 0 \quad \cdots\cdots\cdots\cdots\cdots\cdots\cdots\cdots\cdots\cdots\cdots \quad \text{VIII.110}$$

and

$$\frac{dx}{dt} = -\,\frac{\dfrac{\partial Q}{\partial t}}{\dfrac{\partial Q}{\partial x}}$$

This can be modified to

$$\frac{dx}{dt} = -\,\frac{\dfrac{\partial Q}{\partial d}\cdot\dfrac{\partial d}{\partial t}}{\dfrac{\partial Q}{\partial x}}$$

and from the continuity equation, since $\dfrac{\partial d}{\partial t} = -\,\dfrac{1}{B}\cdot\dfrac{\partial Q}{\partial x}$ then

$$\frac{dx}{dt} = \frac{1}{B}\cdot\frac{\partial Q}{\partial d} \quad \cdots\cdots\cdots\cdots\cdots\cdots\cdots\cdots\cdots\cdots\cdots\cdots\cdots \quad \text{VIII.111}$$

Since Q is simply a function of d and not of time or distance, then $\dfrac{\partial Q}{\partial d} = \dfrac{dQ}{dd}$ and

$$\frac{dx}{dt} = \frac{1}{B}\cdot\frac{dQ}{dd} = V_w \quad \cdots\cdots\cdots\cdots\cdots\cdots\cdots\cdots\cdots\cdots\cdots \text{VIII.112}$$

where V_w is the speed of the observer, and also, the velocity of the flood wave.

The equation shows that the velocity of the flood wave is inversely proportional to the channel width, and directly proportional to the change of discharge with depth. This latter term may be obtained from the stage-discharge curve for the channel, and it is simply the slope of the curve at any point (see Fig. VIII.32).

Equation VIII.112 strictly applies only for small wave heights. However, it will still give good results for larger waves if dQ/dd is determined for the mid-point of the flood wave rise.

For a rectangular channel, the equation can be used to show that a flood wave moves at 1.5 times the initial water velocity if the Chezy Equation is assumed to apply, or 1.67 times the initial water velocity if the Manning Equation is applied. Other velocity factors, assuming use of the Manning Equation, are 1.33 for triangular sections and 1.44 for wide parabolic sections (see Chow 1959, p. 531).

If there is overspill at high flows, then the cross-section may be divided into sections as shown in Fig. VIII.33. If the sections are essentially rectangular and the Manning Equation is used, then

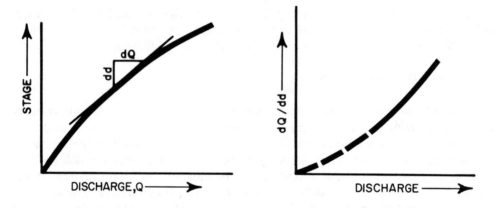

Fig. VIII.32 Rate of Change of Discharge with Stage

$$V_W = 1.67 \left[\frac{1}{B} (B_1 V_1 + B_2 V_2 + B_3 V_3) \right] \dots\dots\dots\dots\dots\dots\dots\dots \text{VIII.113}$$

where V_W = the velocity of the flood wave,
B_1, B_2, B_3 = the widths of the chosen sections, (see Fig. VIII. 33)
B = the total width, and
V_1, V_2, V_3 = the average flow velocities in the corresponding sections.

The velocities V_1 and V_3 are often nearly zero, so the equation can be reduced to

$$V_W = \frac{1.67}{B} B_2 V_2 \dots\dots\dots\dots\dots\dots\dots\dots\dots\dots\dots\dots\dots\dots\dots\dots\dots \text{VIII.114}$$

If the ratio B_2/B is less than 0.6 then the wave velocity will be equal to, or less than, the flow velocity in the main channel. This results from the storage effects on the flood plain.

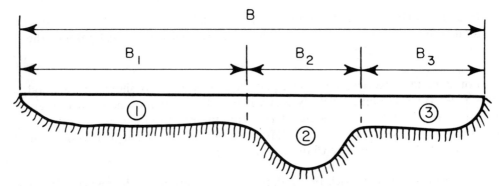

Fig. VIII.33 Illustration of 'B' Sections when Overspill Occurs

VIII.7.4   Flow Measurements

The basic methods for determining flows by measuring or calculating flow velocities have been explicitly presented in the preceding subsections, so in this context it will now suffice to list the field methods with a brief explanation of additional points.

VIII.7.4.1   Velocity Measurements

The most common method of measuring discharge in larger streams consists of measuring flow velocities at a number of points in the flow cross-section, and then of multiplying the average flow velocity, V, by the flow area, A, to obtain the discharge, Q. The measurements of velocity are normally made with a current meter, which has a set of cups or vanes that rotate about a central axis because of the action of flowing water. Thus, the rate of rotation is a definite function of the flow velocity. The meter may, therefore, be placed at a point in the flow and the flow velocity determined by measuring the number of revolutions in a known interval of time. Most of these current meters have an electrical system operated by batteries, and the rotation of the cups or vanes produces a sound in a set of earphones. The number of sounds per unit of time can be obtained, and the flow velocity determined from the calibration curve for the meter. The wide range in sizes and types of meters that are available permits the hydrologist to select one which may be used in large rivers, or one which is designed for small flumes (Corbett *et al.*, 1962).

The measurement of flow area is done simultaneously with the measurement of velocities. The flow depths at a number of points across the channel are measured by lowering the current meter on its rod or cable from the surface of the stream to the stream bed, and by recording the depth, the rod or cable having a scale or some system such that flow depths can easily be measured. Sufficient measurements of depth are taken to define the total flow area.

The following methods of velocity measurements may be used to obtain discharges:

1.   *Velocity contours.* A sufficient number of measurements of velocity are taken throughout the flow cross-section to define the velocity contours such as are shown in Figure VIII.28. The discharge is determined by methods described in Subsection VIII.7.1.2.

2. *Vertical velocity distributions.* A sufficient number of measurements of velocity are taken along a vertical line from the top to the bottom of the stream to allow plotting the distribution of velocity from the bed to the surface. The average velocity for the vertical can then be determined as the area encompassed by the velocity distribution curve divided by the flow depth. In equational form

$$V = \frac{1}{d} \int_{z=o}^{z=d} v\,dz \dots\dots\dots\dots\dots\dots\dots\dots\dots\dots\dots\dots\dots \text{VIII.115}$$

where V = the average velocity in the vertical,
       v = the velocity at a distance z from the stream bed, and
       d = the total depth of flow at the vertical section.

If the average velocities are obtained for several verticals a distance y apart across the stream, then the discharge associated with any one vertical is

$$Q = V\,dy \dots\dots\dots\dots\dots\dots\dots\dots\dots\dots\dots\dots\dots\dots \text{VIII.116}$$

where Q = the discharge associated with the vertical,
       V = the average velocity at the vertical,
       d = the flow depth at the vertical, and
       y = the horizontal distance between the verticals.

The total discharge of the stream is obtained by measuring velocities in a number of verticals, each spaced a distance y from each other across the stream; by determining the discharge associated with each vertical by Equation VIII.116; and then by summing the discharges in each section to obtain the total discharge.

3. *Measurement of the velocities at 0.2d and 0.8d.* In this case the flow depth at a vertical is first obtained. The flow velocities are then measured at 0.2 and 0.8 of the depth below the surface, and these velocities are averaged to give the average velocity in the vertical. The discharge associated with the vertical and the total stream discharge are obtained as outlined in method 2.

4. *Measurements of the velocity at 0.6 depth.* The procedure here is similar to that in method 3, except that the flow velocity is only measured at 0.6 of the depth below the surface. This is considered to be the average velocity in the vertical, and the total discharge may be obtained as in method 2.

5. *Surface velocity.* The surface velocity at a point may be obtained by measuring the time it takes for a float to travel a measured distance downstream. The average velocity may be approximated as 0.9 of the surface velocity. A number of values of

surface velocity may be taken across the stream to obtain an overall average flow velocity. The flow area may be obtained from cross-sectional data at the site for the particular water surface elevation. The discharge is then the overall average velocity times the flow area.

The methods for determining discharge outlined above are listed in order of decreasing accuracy. The first two methods are not used frequently because of the excess of time and effort involved, and because methods *3* and *4* give sufficiently accurate results. The accuracy of any method is difficult to define because it is a function of the flow conditions, the time taken to obtain the measurement, the accuracy of the current meter and the accuracy of the determinations of flow depths in the verticals. Usually the accuracy of the flow measurement obtained by either method *3* or *4* can be expected to provide an estimate within $\pm$ 5–10 per cent of the true flow.

VIII.7.4.2   Measurements of Stage

It has been mentioned previously that discharges occurring between periods of measurement are determined from stage-discharge curves, where stage is the water surface elevation above a selected arbitrary datum plane. The stage-discharge curve is obtained by measuring stage simultaneously with each measurement of discharge, and by plotting these values against the measured discharge. Thus, if the stage is known, the discharge can be obtained from the stage-discharge curve.

Depending on the flow information required for a study, the stage may be measured continuously using automatic recorders, or by only spot or periodic measurements being taken. For example, stage measurements are often taken only once a day on small streams, or at points where more sophisticated flow data are not required. In these cases the stage may be measured by a simple, graduated staff gauge, or by a wire-weight system. The wire-weight gauge consists of a weight attached to a small cable or wire. The gauge mechanism is fixed at some distance above the high water level, and readings of stage are obtained by lowering the weight down to the water surface and by recording the distance the weight moves. This distance can be related to the stage. This system may be employed by taking measurements directly from a bridge to the water surface of the stream, or by using a stilling well in which levels are taken. A stilling well is an enclosed structure sunk into the bank of the stream and connected to the stream by means of an inlet pipe. The well eliminates problems associated with measurement of stage when there is debris in the stream or where the stream is very turbulent.

Automatic stage recorders give a continuous measurement of stage. These recorders are nearly always installed in a stilling well, and the recorder is generally actuated by a float mechanism. Movements of the float with changing stage are transferred through a gear mechanism to a pen, which produces a trace on a chart moving at constant speed. For most commercially-available instruments, different combinations of chart speeds and sizes are available, so that the desired stage-time accuracy can be obtained with proper selection of equipment. Automatic recorders may also be equipped to provide a digital or other type of output, so that the data can be fed directly into a computer for analysis.

VIII.7.4.3   Calculation of Discharge

Measurements of high flows are not easily obtained (see Subsection VIII.7.2.1) and discharges must be calculated by extrapolation of the stage-discharge or similar curves, or by using a flow formula.

On some streams the stage-discharge curve is roughly parabolic, yielding an equation of the type

$$Q = c(h-a)^b \dots\dots\dots\dots\dots\dots\dots\dots\dots\dots\dots\dots\dots\dots\dots\dots \quad \text{VIII.117}$$

> where   Q = the discharge,
> a, b and c = constants, and
> h = the stage reading.

The constant, a, may either be positive or negative, and must be determined by trial and error such that the equation fits the observed data. The discharge may then be plotted versus $(h-a)$ on logarithmic paper. This will produce a straight line if the parabolic equation is correct, and thus it allows easy extrapolation to the stage reading observed for high flows.

Leopold and Maddock (1953) have shown that for some streams the relationships between width or depth and velocity or discharge may follow a straight line when plotted on logarithmic paper. Thus the plots of width vs discharge or depth vs discharge could easily be extrapolated to the observed width or depth at high flow and the discharge determined.

The curve extrapolations described above will give satisfactory results only if changes in vegetation or roughness effects, sediment transport conditions, or cross-sectional shape during the flood flow event do not differ appreciably from those conditions that prevailed for the flows from which the curve was derived.

Generally the *Slope-Area Method* is applied to determine high flows. This method makes use of Manning's Equation described in Subsection VIII.7.2.2. The equation may be put in the form

$$Q = \frac{1.49}{n} AR^{2/3} S_f^{1/2} \dots\dots\dots\dots\dots\dots\dots\dots\dots\dots\dots\dots\dots\dots \quad \text{VIII.118}$$

> where   Q = the discharge to be calculated, (cfs),
> n = the roughness coefficient,
> A = the average flow area, (ft^2),
> R = the average hydraulic radius, (ft), and
> S_f = the friction slope or slope of the
> total energy line.

The water surface slope, S_w, can be determined from levels taken along the high water mark of the channel. The flow area, A, and the hydraulic radius, R, are average values over the reach of stream for which the slope is measured.   The values of A and R are determined from cross-sectional data measured at both the beginning and end of the reach. The roughness coefficient, n, is chosen using one of the methods outlined in Subsection VIII.7.2.3. Unfortunately, the friction slope, S_f, is usually not equal to the water surface slope except when the channel is uniform in cross section throughout the reach. Infrequently does this condition ever exist in practice, thus the friction slope must be calculated from the following equation:

$$S_f = S_W + \frac{V_1{}^2 - V_2{}^2}{2gL} \quad \dots\dots\dots\dots\dots\dots\dots\dots\dots\dots\dots\dots \quad \text{VIII.119}$$

$$\text{where} \quad V_1 = \text{the velocity at the beginning of the reach,}$$
$$V_2 = \text{the velocity at the end of the reach, and}$$
$$L = \text{the length of reach under consideration.}$$

This equation must be solved by trial and error, as V_1 and V_2 are not known initially. The procedure to be used in this solution is as follows:

1.   Assume $S_f = S_W$ and solve for Q.
2.   Calculate: $V_1 = Q/A_1$ and $V_2 = Q/A_2$.
3.   Calculate a new value of S_f from Equation VIII.119.
4.   Recalculate Q from Equation VIII.118.
5.   Repeat steps 2, 3 and 4 until the Q used in step 2 is approximately equal to the value calculated in step 4.

VIII.7.4.4   Discharge Measuring Structures

Flow measurement structures are constructed in channels in an effort to ensure a constant stage-discharge relationship and to provide accurate discharge measurements without the use of velocity measurements. The discharge through such structures can be obtained from equations or tables relating discharge to the water surface elevations adjacent to or within the structure. Therefore, the only field measurements required are measurements of stage. However, initial calibration tests are normally conducted to check the discharge equation for the structure, and to determine the coefficients of discharge in the equations.

Three different types of flow measuring structures will be considered:

1.   Weirs.
2.   Contracted-section structures.
3.   Others.

Weirs

A weir may be defined in a general way as an obstruction in a channel that causes upstream storage and flow over or through the obstruction. This definition therefore includes many hydraulic structures, such as spillways and drop structures. Most, if not all, of these structures can therefore be rated and used to measure flows.

The most common types of weirs used for flow measurements are sharp-crested triangular or sharp-crested rectangular, Cipolleti and broad-crested weirs. The term *sharp-crested* means that the weir is constructed so there is only a line contact of the flow with the weir crest, whereas *broad-crested* means the flow is in contact with the crest for a finite distance parallel to the flow direction.

The equation for discharge over a weir may be determined from theoretical considerations, and modified by means of a discharge coefficient to provide a useful and practical equation. The flow equations for the weirs mentioned above are:

Triangular Weirs

$$Q = C_d(8/15) \sqrt{2g} \ \tan\frac{\theta}{2} \ H^{5/2}$$

$$= 4.25 \ C_d \tan\frac{\theta}{2} H^{5/2} \dots\dots\dots\dots\dots\dots\dots\dots\dots\dots \text{VIII.120}$$

Rectangular and Cipolleti Weirs

$$Q = C_d(2/3) \sqrt{2g} \ B \left[\left(H + \frac{V^2}{2g}\right)^{3/2} - \left(\frac{V^2}{2g}\right)^{3/2} \right]$$

$$Q = 5.35 \ C_d \ B \left[\left(H + \frac{V^2}{2g}\right)^{3/2} - \left(\frac{V^2}{2g}\right)^{3/2} \right] \dots\dots\dots\dots\dots\dots \text{VIII.121}$$

Rectangular Broad - Crested Weirs

$$Q = C_d \ (2/3)^{3/2} \ \sqrt{g} \ \left(H + \frac{V^2}{2g}\right)^{3/2}$$

$$Q = 3.09 \ C_d \left(H + \frac{V^2}{2g}\right)^{3/2} \dots\dots\dots\dots\dots\dots\dots\dots\dots \text{VIII.122}$$

where Q = the discharge over the weir, (cfs),
g = the acceleration of gravity, (32.2 ft/sec²),
H = the head or vertical distance from the weir crest to the water surface upstream of the weir, (ft),
θ = the included angle for the triangular weirs, (degrees or radians),

B = the width of weir crest, (ft),

V = the approach velocity of the flow upstream of the weir, (ft/sec), and

C_d = an appropriate discharge coefficient for the weir (dimensionless).

Equations VIII.121 and VIII.122 require trial and error solutions if the velocity of approach is appreciable. It is therefore advisable to position the structure so that the upstream flow area is at least six times the flow area over the weir crest. Under such conditions, the velocity of approach can be neglected.

The coefficient of discharge, in each equation, is a function of the friction loss due to the weir, the amount of vertical and horizontal contractions of flow, the effects of adhesion and cohesion, and the form of the weir structure. It is advisable, therefore, to rate the structure by means of a current meter. This is particularly true for large triangular, rectangular and Cipoletti weirs, as these are often constructed with reinforced concrete crests and the sharp-crested conditions may not apply. If the structures cannot be rated, then the following discharge coefficients and equations may be used:

90° Sharp-Crested Triangular Weir

$$C_d = 0.585; \quad Q = 2.5\ H^{5/2} \ldots\ldots\ldots\ldots\ldots\ldots\ldots\ldots\ldots\ldots\ldots \quad \text{VIII.123}$$

Sharp-Crested Rectangular and Cipoletti Weirs with Negligible Approach Velocities

$$C_d = 0.622; \quad Q = 3.33\ B\ H^{3/2} \ldots\ldots\ldots\ldots\ldots\ldots\ldots\ldots\ldots\ldots \quad \text{VIII.124}$$

Rating tables are available for field-size triangular weir structures of a large included angle (USDA, ARS 1962).

The coefficient of discharge for the broad-crested weir is a function of the ratio of the head to length of crest (H/L), and it may be obtained from Fig. VIII.34.

The discharge coefficients for any weir or flow-measuring structure are affected by submergence or raising of the tail-water level above the lowest point on the crest of the structure.

The sharp-crested weirs are most affected by submergence and, therefore, should be designed to have a clear drop of H or more from headwater to tailwater level. The broad-crested weir and similar critical-depth structures (H-flumes, San Dimas flumes, Parshall flumes) can be submerged up to two-thirds of the head on the structure without affecting the discharge coefficients. This is one advantage of this type of measuring structure, as they can be set lower in the channel without changes occurring in the coefficients. The effects of submergence on various structures are shown in Fig. VIII.35.

There are also equations which can be used to take care of submergence effects (see Davis, 1952).

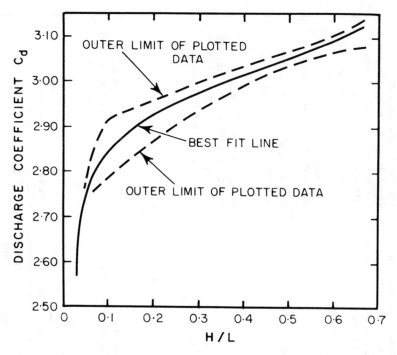

Fig. VIII.34 Discharge Coefficients for Broad-Crested Weir (From Smith 1965)

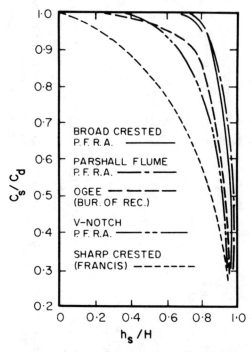

Fig. VIII.35 Effect of Submergence on Discharge Coefficient (From Smith 1965)

Table VIII.15 Artificial Hydrometric Structures

Type	Capacity	Accuracy	Siting Criteria or Advantages	Limitations
9 0° V-notch sharp-crested	39 cfs with 3 ft head Lower Limit 0.01 cfs	About 2% at low heads, and 1%at heads above 2 ft if levels are read to 0.005 ft, approach velocity is less than 0.5 ft per sec., and weir is well constructed.	Bottom of notch located the height of the weir above the stream bed. Width of approach section 6 to 8 times the height of the weir. Lends itself to winter heating.	Not recommended for streams with heavy sediment load, or where surface debris is sufficient to clog notch, and where constant attendance is not available. Required depth of approach pool may require expensive cut-off walls to limit seepage when located on fractured rock or pervious soil.
120° V-notch sharp-crested	65 cfs at 3 ft head or 133 cfs at 4 ft head.	Similar to 90° V-notch	Same as for 90°	Sediment and debris conditions still limit use.
Cipolleti - a trapezoidal sharp-crested weir with 4:1 side slopes.	323 cfs at 4 ft head for a 12 ft width	Less at low flow		Not recommended for use where extreme low flows occur. Heating for winter operation more difficult and expensive than for notches.
Rectangular Sharp-crested	Can be designed for virtually any flow.	Less at low flow	Construction less difficult than for Cipolleti.	Same as for Cipolleti.
Broad-crested triangular	1070 cfs with 5 ft head	Less accurate than any of the sharp-crested type.	More easily constructed than sharp-crested type. Less tendency for collection of debris.	Not recommended for watershed research except as an improvement to natural controls where discharges are high.
Flumes: HS, H, HL - increasing in size.	44 cfs for 3 ft head, 110 cfs for 4 ft head, for HL type. Others have less capacity.	1% through most of range if head measurement to 0.005 ft.	Does not require significant ponding as floor is laid at stream bed level. Approach section good location for calibration meter checks. Good self cleaning properties make it useful on streams carrying sediment. Construction minimizes depth required for cut-off walls and facilitates winter heating.	Not suitable for streams carrying significant surface debris.

Table VIII.15 Structures (Continued)

Type	Capacity	Accuracy	Siting Criteria or Advantages	Limitations
Parshall	139.5 cfs for 8 ft throat and 2.5 ft head	Good near capacity but drops off rapidly at very low flows.	Does not require ponding as floor laid at stream bed level. Useful where flow regime is reasonably uniform and extremely low flows absent.	Construction costly as additional precautions necessary to ensure maintenance of structure dimensions. Requires two water level recorders with separate stilling wells.
San Dimas-rectangular flume with rounded converging entrance section and sloping floor.	1,000 cfs. for a 10 ft width.	Reasonably good at flows near capacity but decreases at low flows.	Useful when extreme variations in flow are expected but where accurate low flow measurements are not important.	Large slope to floor and free fall at outlet required and therefore suitable only where steep gradients or drops in head are available.
Trapezoidal	350 cfs with 4 ft. head	Reasonably good but can be affected by shifts in the channel conditions upstream of the structure.	Good where there is a large range in stage, and stream has heavy sediment and debris load.	Requires continual calibration by current meter.

Contracted-Section Structures

A structure which causes flow constriction can be used as a flow-measuring device by relating the discharge to the change in level from the upstream water surface to the water surface elevation in the constriction. The resulting equation is of the form:

$$Q \propto \frac{A_1 A_2 \sqrt{2g (d_1 - d_2)}}{\sqrt{A_1^2 - A_2^2}} \dots\dots\dots\dots\dots\dots\dots\dots\dots\dots\dots\dots\dots \text{VIII.125}$$

where Q = the discharge,
A_1 and A_2 = the flow areas upstream and at the contracted section, respectively,
d_1 and d_2 = the flow depths upstream and at the contracted section, and
g = the acceleration of gravity.

These structures have the advantage that the floor of the structure can be placed at the level of the bed. The major disadvantage is that the term $d_1 - d_2$ is usually small and therefore difficult to measure.

Other Structures

Other flow-measuring structures in common use include the HS, H, and HL flumes, Parshall flumes, San Dimas flumes, trapezoidal flumes and broad-crested triangular weirs. These are all critical-depth structures, and therefore can be set lower in the channel than sharp-crested wiers. They all require only a single measurement of water surface level (upstream), unless the submergence becomes excessive, then essentially the structures become contracted sections and two measurements of level are required.

Each of these structures has advantages and disadvantages (see Table VIII.15 from the Secretariat, CNC / IHD, 1966). Additional advantages are that extensive rating tables are available for them, in particular for the Parshall flume. Some of the flumes are commercially available in smaller sizes, or they can be easily constructed. The choice of a flow measuring structure must, therefore, be made on the basis of the site conditions, the accuracy required, and the advantages and disadvantages of the structure.

VIII.8   **LITERATURE CITED**

Amer. Soc. Civil Engr. 1949. Hydrology handbook Manual. No. 28.

Barnes, B.S. 1940. Discussion of analysis of runoff characteristics. Trans. A.S.C.E. Vol. 105.

Barnes, B.S. 1959. Consistency in unit graphs. Proc. Amer. Soc. Civil Engr. Vol. 85, No. HY8, pp. 39-63.

Bazin, H. and Darcy, H. 1965. Recherches hydrauliques. I, Recherches expérimentales sur l'écoulement de l'eau dans les canaux découverts, Mémoires présentés par divers savants à l'Academie des Sciences, Vol. 19, No. 1, Dunod, Paris.

Bernard, M.M. 1935. An approach to determinate stream flow. Trans. Amer. Soc. Civil Engr. 100:347-395.

Bernard, M.M., Gregory, R.L. and Arnold C.E. 1932. Runoff — rational runoff formulas. Trans. ASCE 96:10-38.

Berry, W.M., Durrant, E.F. and Booy, C. 1961. Hydrologic investigations for the South Saskatchewan River Project. The Engineering Journal. Engr. Inst. Can. April.

Blench, T. 1957. Regime behaviour of canals and rivers. Butterworths Scientific Publications, London.

Blench, T. 1959. "Empirical methods". Proc. Sym. No. 1 — Spillway Design Floods. Queen's Printer, Ottawa.

Blench, T. 1966. Mobile-bed fluviology, through the Department of Technical Services, University of Alberta, Edmonton, Alta.

Chow, Ven Te. 1959. Open-channel hydraulics. McGraw-Hill Book Co., Inc., New York.

Chow, Ven Te. (ed). 1964. Handbook of applied hydrology. McGraw-Hill Book Co., Inc., Toronto, Ontario.

Clark, C.O. 1945. Storage and the unit hydrograph. Proc. Amer. Soc. Civil Engr. 69:1419-1447.

Collins, W.T. 1939. Runoff distribution graphs from precipitation occurring in more than one time unit. Civil Engr. p. 559, Sept.

Commons, G. 1942. Flood hydrograph. Civil Engr. 12:571-572.

Corbett *et al.* **1962.** Stream gauging procedure. U.S. Dept. of Interior, Geol. Survey, Water Supply Paper 888.

Cowan, W.L. 1956. Estimating hydraulic roughness coefficients. Agric. Engr., Vol. 37, No. 7, July 1956.

Davis, C.V. 1952. Handbook of applied hydraulics. 2nd. ed. McGraw-Hill Book Co. Inc., Toronto.

Diskin, M.H. 1964. A basic study of the linearity of the rainfall runoff process in watersheds. Unpublished Ph. D. Thesis. University of Illinois, Urbana.

Doland, J.J. and Chow, V.T. 1952. Discussion of river channel roughness by H.A. Einstein and H.L. Barbarossa. Trans. Amer. Soc. Civil Engr. Vol. 117.

Dooge, J.C.I. 1956. Synthetic unit hydrographs based on triangular inflow. Unpublished M.S. Thesis. State University of Iowa Library, Iowa City.

Dooge, J.C.I. 1959. A general theory of the unit hydrograph. J. Geophys. Res. 64:241-256.

Durrant, E.F. and Blackwell, S.R. 1959. The magnitude and frequency of floods on the Canadian Prairies. Proc. Sym. No.1—Spillway Design Floods. Queen's Printer, Ottawa.

Edson, C.G. 1951. Parameters for relating unit hydrographs to watershed characteristics. Trans. Amer. Geophys. Union, 32:591-596.

Einstein, H.H. and Barbarossa, H.L. 1952. River channel roughness. Trans. Amer. Soc. Civil Engr. Vol. 117.

Ellis, W.H. 1964. A study of peak instantaneous and average daily discharges from small prairie watersheds. Unpublished M.Sc. Thesis. University of Saskatchewan, Saskatoon.

Ellis, W.H. and Gray, D.M. 1966. Interrelationships between the peak instantaneous and average daily discharges of small prairie streams. Can. Agr. Engr. pp. 1-2, 38, 39, Feb.

Forsaith, T.S. 1949. Development of a formula for estimating surface runoff. Sci. Agr. 29:465-481.

Fuller, W.E. 1914. Flood flows. Trans. Amer. Soc. Civil Engr. 77:564-617.

Goodrich, R.D. 1931. Rapid calculation of reservoir discharge. Civil Engr. 417-418.

Gray, D. M. 1961. Interrelationships of watershed characteristics. J. Geophys. Res. 66:1215-1223.

Gray, D. M. 1962. Derivation of hydrographs for small watersheds from measurable physical characteristics. Iowa State University Agric. and Home Economics Expt. Sta. Res. Bull. No. 506, Ames, Iowa.

Hanson, T. and Johnson, H.P. 1964. Unit hydrograph methods compared. Trans. Amer. Soc. Agric. Engr. 4:448-451.

Hickok, R.B., Keppel, R.V. and Rafferty, B.R. 1959. Hydrograph synthesis for small watersheds for small arid-land watersheds. Agr. Engr. 40:608-611, 615.

Horn, D.L. and Schwab, G.O. 1963. Evaluation of rational coefficients for small agricultural watersheds. Trans. Amer. Soc. Agr. Engr. 6:195-198, 201.

Horner, W.W. and Flynt, F.L. 1936. Relation between rainfall and runoff from urban areas. Trans. Amer. Soc. Civil Engr. 101:140-206.

Horton, R.E. 1914. Discussion of flood flows (Fuller). Trans. Amer. Soc. Civil Engr. 77:665.

Horton, R.E. 1932. Drainage-basin characteristics. Trans. Amer. Geophys. Union 13:350-361.

Horton, R.E. 1938. The interpretation and application of runoff plat experiments with reference to soil erosion problems. Proc. Soil Sci. Soc. Amer. 3:340-349.

Horton, R.E. 1941. Virtual channel-inflow graphs. Trans. Amer. Geophys. Union 22:811-820.

Horton, R.E. 1945. Erosional development of streams and their drainage basins: hydrophysical approach to quantitative morphology. Bull. Geol. Soc. Amer. 56:275-370.

Hoyt, J.C. and Grover, N.C. 1924. River Discharge 4th ed. John Wiley and Sons Inc. New York.

Inglis, C.C. 1949. The behaviour and control of rivers and canals. Res. Publ. Cent. Bd. Irr. India. No. 13.

Johnstone, D. and Cross, W.P. 1949. Elements of applied hydrology. The Roland Press Co., New York.

Keulagan, G.H. 1938. Laws of turbulent flow in open channels. Res. Paper RP 1151, J. of Res. U.S. National Bureau of Standards, Vol. 21, December.

Kinnison, H.B. and Colby, B.R. 1945. Flood formulas based on drainage basin characteristics. Proc. Amer. Soc. Civil. Engr. 69:849-876.

Kulandaiswamy, V.C. 1964. A basic study of the rainfall excess-surface runoff in a basic system. Unpublished Ph. D. Thesis. University of Illinois, Urbana.

Langbein, W.B. 1940. Channel storage and unit hydrograph studies. Trans. Amer. Geophys. Union 21:620-627.

Langbein, W.B. 1944. Peak discharges from daily records. U.S. Geol. Survey Bull. p. 145. August.

Langbein, W.B. and others. 1947. Topographic characteristics of drainage basins. U.S. Dept. Interior. Geol. Survey Water-Supply Paper 968-C:125-155.

Langbein, W.B. 1949. Storage in relation to flood waves. In: Meinzer, O., ed. Hydrology. pp. 561-571. Dover Publications, Inc., New York.

Leopold, L.B. and Maddock, T. 1953. The hydraulic geometry of stream channels and some physiographic implications, U.S. Dept. Interior, Geol. Sur. Prof. Paper 252.

Linsley, R.K. 1944. Use of nomographs in solving streamflow problems. Civil Engr. 14:209-210.

Linsley, R.K. 1945. Discussion of storage and the unit hydrograph. Trans. Amer. Soc. Civil Engr. 110:1452-1455.

Linsley, R.K. and Ackerman, W.C. 1942. A method of predicting the runoff from rainfall. Trans. A.S.C.E. Vol. 107.

Linsley, R.K., Kohler, M.A. and Paulhus, J.L.H. 1949. Applied hydrology. McGraw-Hill Book Co., Inc., New York.

Linsley, R.K., Kohler, M.A. and Paulhus, J.L.H. 1958. Hydrology for engineers. McGraw-Hill Book Co., Inc., New York.

Massau, J. 1900. Graphical integration of partial differential equations with special emphasis to unsteady flow in open channels. Annales de l'Association des Ingénieurs sortis des Ecoles Spéciales de Gand 23:95-214. Ghent, Belgium.

McCarthy, G.T. 1938. The unit hydrograph and flood routing. Unpublished manuscript presented at a conference of the North Atlantic Division, Corps of Engineers, War Department.

McKay, G.A. and McMorine, J.G.S. 1962. Storm rainfall and runoff at Buffalo Gap, Saskatchewan. Met. Rept. No. 2, Can. Dept. of Agr. PFRA–Hydrology Division.

Miller, V.C. 1953. A quantitative geomorphic study of drainage basin characteristics in the Clinch Mountain area, Virginia and Tennessee. Tech. Rept. No. 3, Dept. of Geol. Columbia Univ. New York.

Mitchell, W.D. 1948. Unit hydrographs in Illinois. State of Illinois, Division of Waterways, Springfield, Ill.

Mockus, V. 1957. Use of storm and watershed characteristics in synthetic hydrograph analysis and application. Paper presented at Amer. Geophys. Union, Southwest Region Meeting, Sacramento, Calif. Feb. (mimeo).

Nash, J.E. 1958. The form of the instantaneous hydrograph. Int. Assoc. Hydrology, International Union of Geodesy and Geophys., Toronto, Ont.

Nash, J.E. 1959. Systematic determination of unit hydrograph parameters. J. Geophys. Res. 64:111-115.

Posey, C.J. 1935. Slide rule for routing floods through storage reservoirs or lakes. Engineering News Record 114:580-581.

Potter, W.D. 1961. Peak rates of runoff from small watersheds. Hydraulic Design Series No. 2 of the Division of Hydraulic Res., Bur. of Public Roads. U.S. Gov't. Printing Office, Washington.

Prandtl, L. 1953. The essentials of fluid dynamics. Blackie and Son Ltd. London and Glasgow.

Ramser, C.E. 1927. Runoff from small agricultural areas. J. Agr. Research 34:797-823.

Reid, J.L. and Brittain, K.G. 1962. Design concepts of the Brazeau Development including river and hydrology studies. Engineering Journal, Engr. Inst. Can. October.

Rosa, J.M. and others. 1961. Electronic analog. USDA, ARS. Unpublished Report, Moscow, Idaho.

Rutter, E.J., Graves, Q.B., and Snyder, F.F. 1939. Flood routing. Trans. Amer. Soc. Civil Engr. 104:275-294.

Schumm, S.A. 1956. Evolution of drainage systems and slopes in badlands at Perth Amboy, N.J. Bull. Geol. Soc. Amer. 67:597-646.

Secretariat, Canadian National Committee for I.H.D. 1966. Guide lines for research basins. Proc. of the National Workshop Seminar on Research Basins Studies, Canadian National Committee for the International Hydrologic Decade, Ottawa.

Sherman, L.K. 1932. The relation of runoff to size and character of drainage basins. Trans. Amer. Geophys. Union 13:332-339.

Sherman, L.K. 1932. Streamflow from rainfall by the unit-graph method. Engineering News Record 108-401-505.

Sherman, L.K. 1940. The hydraulics of surface runoff. Civil Engr. 10:165-166.

Sherman, L.K. 1949. The unit hydrograph method. In Meinzer, O., ed. Hydrology. pp. 514-526. Dover Publications Inc. New York.

Singh, K.P. 1962. A non-linear approach to the instantaneous unit hydrograph. Unpublished Ph. D. Thesis. University of Illinois, Urbana.

Smith, C.D. 1965. Prepared lecture notes "Hydraulic Structures 404 B". University of Saskatchewan, Saskatoon, Saskatchewan.

Snyder, F.F. 1938. Synthetic unit graphs. Trans. Amer. Geophys. Union 19:447-454.

Stichling, W. and Blackwell, S.R. 1957. Drainage area as a hydrologic factor on the Canadian Prairies. Proc. Inter. Union of Geodesy and Geophys. Vol. 3.

Strickler, A. 1923. Some contributions to the problem of the velocity formula and roughness factors for rivers, canals, and closed conducts, Mitteilungen des eidgenossischen Amtes für Wasserwirtschaft, Bern, Switzerland.

Taylor, A.B. and Schwarz, H.E. 1952. Unit hydrograph lag and peak flow related to basin characteristics. Trans. Amer. Geophys. Union 33:235-246.

United Nations ECAFE. Flood Control Series No. 7.

U.S. Dept. of Agric., Agric. Research Service, 1961. Field Manual for Research in Agricultural Hydrology, Agric. Handbook No. 224. U.S. Dept. Agr., Washington, D.C.

U.S. Dept. of Agric., Soil Conservation Service. 1957. Engineering Handbook Section 4, Hydrology, Supplement A. U.S. Dept. Agr., Washington, D.C.

U.S. Dept. Army, Corps of Eng. 1948. Office of the Chief of Eng. Hydrologic and Hydrograph Analyses; flood hydrograph analyses and computations. U.S. Dept. Army, Washington, D.C.

Vanoni, V.A. 1941. Velocity distribution in open channels. Civil Engr. Vol. II. No. 6.

Vanoni, ᴠ A., Brooks, N.H. and Kennedy J.F. 1961. Lecture notes on sediment transportation and channel stability. W.M. Keck Lab. of Hydraulics and Water Resources, California Institute of Technology, Pasadena, California.

Volker, A. 1964. Lecture notes in Hydrology. International Course in Hydraulic Engineering. Delft. Holland.

Von Karman, T. 1930. Mechanical similarity and turbulence, Proc. 3rd. Int. Congress of Applied Mechanics, Stockholm, Vol. 1 pp. 85-92.

Williams, G.R. 1950. Engineering Hydraulics edited by H. Rouse. Chpt. 4. Hydrology, pp. 309-318. John Wiley and Sons, Inc., New York.

Wisler, C.O. and Brater, E.F. 1959. Hydrology, John Wiley and Sons, Inc., New York.

Section IX

PEAK FLOW – SNOWMELT EVENTS

by

Kersi S. Davar

TABLE OF CONTENTS

LIST OF FIGURES

Section IX

PEAK FLOW – SNOWMELT EVENTS

IX.1 INTRODUCTION

The generation of snowmelt and sequential streamflow from the snowpack that accumulates in winter, forms one of the most important phases of the hydrologic cycle in the northern regions of this continent. Any program for the effective control, conservation, and optimum development of water resources in these northern regions must take into account the vital contribution of snowmelt to the spring season runoff, or its potential for storage and subsequent utilization.

Despite the importance of snow hydrology, it is only in the last few decades that intensive research has led to the formation of theory which yields a partly satisfactory understanding of the complex hydrothermodynamic processes producing snowmelt and sequential streamflow. The new techniques are being widely used for predicting design estimates and seasonal yield; however, for shortrange forecasting of peak flow rates the agreement between prediction and actual observations has been found to be uncertain, and leaves room for much improvement.

This Section presents briefly the techniques currently used for predicting peak flow from snowmelt events; it also points out regions of uncertainty and the needs for further research.

IX.2 BASIN SNOWCOVER

Determination of the extent of depth of snowcover forms one of the initial steps in estimation of basin snowmelt. The measurement of snowfall and snowcover has been treated in Subsections II.2.5 and II.3.3. A few brief comments concerning the use of these measurements in estimating streamflow is highly relevant.

The information on depth of snowcover and water equivalent over the entire basin at the start of the snowmelt season is useful for estimating both the seasonal yield and the potential for high flow rates. But, for estimation of peak flows, it is essential to have direct or indirect information about the contemporaneous extent and volume of snowcover in the basin as the snowmelt season progresses and the snowcover gradually shrinks. This is necessary as currently used equations for predicting snowmelt are for point locations, and to compute snowmelt contributions from a basin the contemporaneous extent of snowcover has to be known.

In Subsection II.3.3.5, reference has been made to the use of aerial photographic methods for assessing the sequential extent of snowcover; further, colour and infra-red techniques are currently being tried for distinguishing types of snowpack and the presence of melt water. The introduction of photogrammetric methods permits estimates of volume in addition to extent of cover. These techniques are only recently moving from the research to the operational phase; they would be difficult to apply in heavily forested areas.

The absence of information relating to contemporaneous extent and depth of snowcover is a common problem; methods have been suggested for indirectly taking account of these factors in estimating streamflow from snowmelt and these will be discussed in subsequent paragraphs.

IX.3   FACTORS AFFECTING SNOWMELT AT A POINT[1]

The generation of snowmelt at a point location in a snowpack is essentially a thermodynamic process, the amount of melt produced being dependent on the net heat exchange between the snowpack and its environment. The travel of the melt water to another point in the pack and its time distribution at that point depend on physiographic (gradient, depth, etc.) and hydrodynamic (porosity, structure, storage, etc.) properties of the snowpack, and will not be considered here. After the theory of snowmelt at a point has been developed and the model incorporated in prediction equations, extensions must be made for computing basin snowmelt.

The various sources and processes influencing heat transfer to or from a snowpack are shown in Fig. IX.1, and are listed below:

1.   Absorbed shortwave (solar)radiation, R_s.
2.   Net longwave (terrestrial and atmospheric) radiation, R_b.
3.   Condensation (or vaporization) from the air, R_e.
4.   Convection heat transfer (by wind), R_h.
5.   Heat content of rain water, H_r.
6.   Conduction of heat from ground, H_g.

The melt water produced by the net transfer of heat from all sources to the snowpack may be obtained from

$$M = \frac{\Sigma H}{203B} \quad \dotfill \quad \text{IX.1}$$

[1]The development of this Subsection is reasonably consistent with the theory evolved by the U.S. Corps of Engineers (1960).

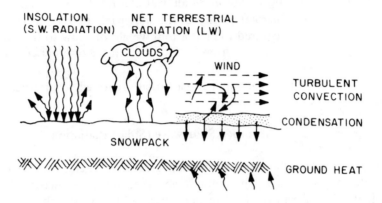

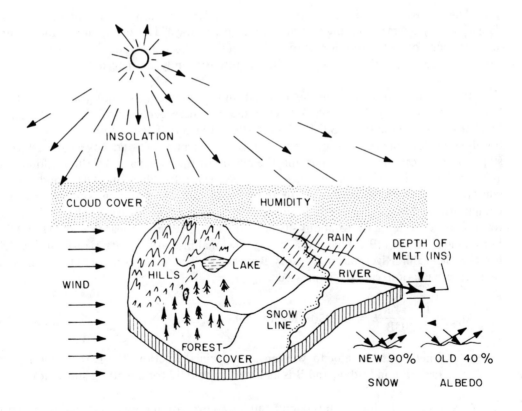

DRAINAGE BASIN

Fig. IX.1 Sources of Heat That Generate Snowmelt

where   M = water equivalent of snowmelt (in),

H = algebraic sum of all heat contributions (cal/cm²), and

B = thermal quality of the snowpack, defined as the ratio of heat required to melt a unit weight of snow to that of ice at $0°C$ (averages from 0.95 to 0.97, for 3 to 5 per cent liquid water).

The constant 203 is the heat input in cal/sq cm required to produce one inch of water from ice at $0°C$; it has the dimensions (cals/cm²/in).

IX.3.1   Absorbed Shortwave (Solar) Radiation

Incoming solar (shortwave) radiation greatly influences the environmental micro-climate above a snowpack; the amount absorbed by the snowpack forms an important component of total snowmelt. The net amount of heat absorbed by the snowpack depends on latitude, orientation and slope, season, time of day, atmospheric conditions (clouds, fog, rain), forest cover, and reflectivity of the snow (albedo).

The intensity of insolation or shortwave solar radiation is normally expressed in langleys (cal/sq cm) per unit of time (min., hr., or day). The most simple and accurate measurement of net insolation is by solarimeters or pyrheliometers[1]; these instruments are calibrated to give the shortwave radiation intensity in ly/hr or ly/day.

In those locations where net solar radiation measurements are not available, either of two approaches may be employed. One method uses generalized charts of regional insolation (as given in the U.S.C.E. Manual 1960, Fig. 5). The net insolation is estimated by allowing for transmission efficiency of the atmosphere (cloud cover), forest cover, slope, orientation and most important, the albedo of the snowpack. Of these, cloud cover and albedo are of major significance, and need careful evaluation. The cloud cover is considered to be most significant for open areas, and when known is included in melt computations. The albedo is commonly taken as 80 per cent for fresh snow and assumed to decrease exponentially to about 40 per cent for melting late-season snow (Fig. IX.2). For Canada, Mateer (1955) has given average regional insolation plots as given in Fig. IX.3 for the St. John River basin in New Brunswick. The melt component produced by shortwave radiation can be expressed as

$$M_{rs} = \frac{(1-a)\,R_{si}}{203B} = 0.00508\,R_{si}\,(1-a)\ \text{in/day} \dots\dots\dots\dots\dots\dots\dots\text{IX.2}$$

where a is the albedo written as a decimal fraction, R_{si} is the effective solar radiation in ly/day, and B is assumed to be 0.97 for a melting snowpack.

A second much simpler approach assumes a correlation between effective insolation and sunshine duration, and substitutes daily hours of sunshine with suitably modified coefficients to compute the melt. A recent study in New Brunswick by Pysklywec (1966)

[1]In Canada, Kipp and Zonen solarimeters and Eppley type pyrheliometers are currently in use.

has demonstrated this approach to be only slightly less accurate than using measured net insolation in ly /day.

IX.3.2   Net Longwave (Terrestrial and Atmospheric) Radiation

Snow is considered to be a near perfect black body with respect to longwave radiation, and longwave radiation emitted by a snow surface can be estimated from Stefan's law. However, computation of net longwave radiation involves estimation of back radiation from the atmosphere under clear or cloudy skies, and from under forest cover; these evaluations are difficult and their use is avoided in practical snow hydrology. Instead the U.S.C.E. Manual (1960) suggests the following simplified expressions using environmental temperatures:

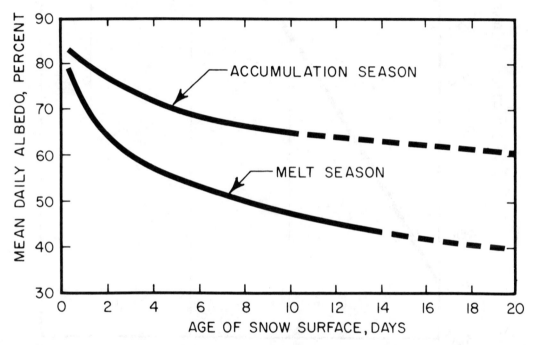

Fig. IX.2   Variation in Snow Surface Albedo with Time (U.S. Corps of Engineers, 1966)

Clear Skies in the Open: $M_{r1} = 0.0212 (T_a - 32) - 0.84$ IX.3

where M_{r1} = the daily snowmelt in inches, and
T_a = the air temperature over snow surface
at the 10-ft level in °F.

Under Forest Canopy: $M_{r1} = 0.029 (T_a - 32)$ IX.4

Complete Cloud Cover: $M_{r1} = 0.029 (T_c - 32)$ IX.5

where T_c is temperature of cloud base in °F.

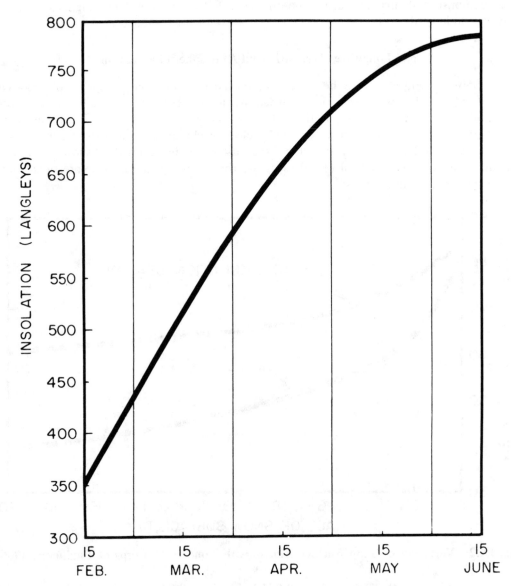

Fig. IX.3   Average Cloudless Day Insolation — St. John River Basin (Mateer, 1955)

When measured temperatures are available, the information can be used directly in these equations; otherwise suitable correlations with temperature measurements taken at a nearby meteorological station need to be established.

IX.3.3   **Condensation and Convection Melt**

When water vapour from the atmosphere condenses on a snow surface, the heat of condensation of water is absorbed by the snowpack. The vapour pressure gradient and wind speed are considered to be the principal parameters influencing this process[1]. The snowmelt produced by condensation may be estimated from the experimentally-derived relation

$$M_e = 0.054 (z_a z_b)^{-1/6} (e_a - e_o) u_b \dots\dots\dots\dots\dots\dots\dots\dots \text{IX.6}$$

> where M_e = snowmelt in ins./day,
> z_a and z_b = the heights-of-measurement (ft.) of the air vapour pressure and wind speed above the snow surface,
> e_a = the air vapour pressure in mb,
> e_o = the snow surface vapour pressure in mb (6.11 mb at a melting snow surface), and
> u_b = the wind speed in mph.

Melt produced by convection results mainly from heat transferred from warm air advected over the snow surface. The theory of turbulent transfer in the atmosphere is very complex, but experiments have indicated that simple approximations can be useful. The simplified expression considers chiefly the temperature gradient of the air above the snow surface, the wind speed, and the air density (taken as a function of air pressure). Convection melt may be estimated from the experimental equation

$$M_c = 0.00629 (p/p_o) (z_a z_b)^{-1/6} (T_a - T_o) u_b \dots\dots\dots\dots\dots\dots \text{IX.7}$$

> where M_c = the snowmelt in ins./day;
> p and p_o = the air pressure at the site and sea level, respectively;
> z_a and z_b = the heights (ft.) above the snow surface of air temperature and wind speed, respectively;
> T_a and T_o = the air and snow surface temperatures in °F, the snow surface temperature being normally taken as 32°F.

The contribution from convection is usually small relative to other factors.

The equations for condensation and convection can be conveniently combined into a single equation to give

$$M_{ce} = 0.00629 (z_a z_b)^{-1/6} \left[(T_a - 32) p/p_o + 8.59 (e_a - 6.11) \right] u_b \dots\dots\dots \text{IX.8}$$

the total condensation-convection melt in ins./day. For practical applications in hydrology this equation can be further simplified. The ratio p/p_o varies from 1.0 at sea level to 0.7 at an elevation of 10,000 ft. For areas with moderate topographic changes it may

---

[1] The theory for condensation and convection processes producing snowmelt is quite complex; for details consult the treatise *Snow Hydrology* by U.S. Corps of Engr. (1956). The expressions given here are summarized from U.S.C.E. experiments as given in their manual *Runoff from Snowmelt*.

be assumed to have a constant value of approximately 0.8. Also, the vapour pressures can be assumed to be adequately represented by dewpoint temperatures. By fixing the heights of measurement of T_a and T_d in °F at 10 ft., and that of wind speed at 50 ft., Equation IX.8 simplifies to

$$M_{ce} = 0.0084 \left[0.22 (T_a - 32) + 0.78 (T_d - 32)\right] u_b \quad \ldots\ldots\ldots\ldots\ldots\ldots \text{ IX.9}$$

where T_a and T_d = the mean air and dewpoint temperature
respectively at the 10-ft. height, and
u_b = the mean wind speed in mph. at the 50-ft. height.

In some regions a linear regression relation between T_a and T_d gives a high degree of correlation, permitting further simplification. Thus, Bruce and Sporns (1963, p. 21) obtained for Caribou, Maine, $T_d = -0.2 + 0.71 \, T_a$ during the period March to May.

IX.3.4   Heat Content of Rain

Rain falling on the snow surface at temperatures above 32°F transfers heat to the snow thus producing melt. This contribution is generally small and can be stated directly by the equation

$$M_p = 0.007 \, P(T_a - 32) \quad \ldots\ldots\ldots\ldots\ldots\ldots\ldots\ldots\ldots\ldots\ldots\ldots\ldots\ldots \text{ IX.10}$$

where M_p = the snowmelt in ins./day due to rain,
P = the mean rainfall in ins./day, and
T_a = the mean free air temperature in °F.

IX.3.5.   Heat Conduction at Ground

Melt produced by heat conduction at the ground (M_g) is generally considered insignificant, unless there is a large underground source. A nominal value of 0.02 in./day is recommended for inclusion in design estimates.

In this Subsection a semi-empirical theory, developed mainly by the U.S. Corps of Engineers, has been given in outline; it is valid for very small plots of homogeneous character. Its chief purpose has been to provide an insight into the snowmelt process as influenced by various meteorological parameters, and to quantitate the effectiveness of each of these parameters in causing snowmelt.

The total melt produced at a point location by the various factors will be

$$M = M_{rs} + M_{rl} + M_{ce} + M_p + M_g \quad \ldots\ldots\ldots\ldots\ldots\ldots\ldots\ldots\ldots \text{ IX.11}$$

IX.4   BASIN SNOWMELT

The extension of the previously developed theory to estimating basin snowmelt leads to further simplifications when assumptions are introduced compatible with practical applications in the field. That is, basin snowmelt estimation may be divided into two periods:

1. During Rain Periods.
2. During Rain-Free Periods.

This sub-division is useful in view of special simplifications that are convenient to introduce for the uniformity of meteorological conditions prevailing during rain periods.

IX.4.1   Basin Snowmelt During Rain

For these conditions, the predominant heat transfer process is that due to convection and condensation; heat input by the other processes is relatively minor.

Insolation during rain conditions is commonly assumed close to 40 ly./day for an open area, and an albedo of roughly 65 per cent. Using these values in Equation IX.2 it follows that

$$M_{rs} = 0.00508\,R_{si}\,(1 - a) = 0.07 \text{ ins./day} \dots\dots\dots\dots\dots\dots\dots\dots \text{IX.12}$$

This approximate contribution from insolation is assumed constant during rain. For densely forested areas it may be smaller.

Longwave radiation melt for rain conditions can be considered to be under a complete cloud cover; also, the cloud base temperature is nearly the same as air temperature at normal instrument height, thus $T_c \cong T_a$. From Equation IX.5,

$$M_{rl} = 0.029\,(T_a - 32) \dots\dots\dots\dots\dots\dots\dots\dots\dots\dots\dots \text{IX.13}$$

Condensation-convection melt during rain is assumed to occur under saturated air, so that $T_a = T_d$. For these conditions Equation IX.9 becomes

$$M_{ce} = 0.0084\,(T_a - 32)u_b \dots\dots\dots\dots\dots\dots\dots\dots\dots\dots \text{IX.14}$$

For basin melt estimates, this equation is modified by a basin condensation-convection coefficient, k, which varies from 1.0 for unforested plains to about 0.3 for forested regions. Rewriting Equation IX.14 to include the coefficient it follows that

$$M_{ce} = k\,(0.0084)\,(T_a - 32)u_b \dots\dots\dots\dots\dots\dots\dots\dots\dots \text{IX.15}$$

where u_b is the mean wind speed (mph.) at the 50 ft. height, and T_a is the mean air temperature in °F at the 10-ft. height. For heavily forested regions the effect of wind variability is small, and Equation IX.15 may be simplified to

$$M_{ce} = 0.045\,(T_a - 32) \dots\dots\dots\dots\dots\dots\dots\dots\dots\dots \text{IX.16}$$

Melt resulting from rain is small and represented by

$$M_p = 0.007 \; P(T_a - 32) \dots \dots \dots \dots \dots \dots \dots \dots \dots \dots \dots \dots \text{IX.17}$$

The melt produced by heat transferred from the ground is also small and taken at 0.02 in./day.

The total basin snowmelt during rain can now be written from Equation IX.11 as

$$M = M_{rs} + M_{rl} + M_{ce} + M_p + M_g$$

For Open and Partly Forested Regions (0-60% cover)
$$M = (0.029 + 0.0084 \; k u_b + 0.007 \; P)(T_a - 32) + 0.09 \dots \dots \dots \dots \dots \dots \text{IX.18}$$

For Heavily Forested Regions (60-100% cover)
$$M = (0.074 + 0.007 \; P)(T_a - 32) + 0.05 \dots \dots \dots \dots \dots \dots \dots \dots \dots \text{IX.19}$$

The latter equations are compact and convenient for use in estimating basin snowmelt during rain. However, careful attention must be given to the compatibility and correctness of the simplifying assumptions, especially where accuracy is of importance. Also, the representativeness of the meteorological parameters must be examined. These equations are valid only for that part of a basin completely covered by snow.

IX.4.2   Basin Snowmelt During Rain-Free Periods

During rain-free weather, solar and terrestrial radiation become the more important melt-producing factors; also, the degree of forest cover is of great significance in determining the extent to which simplification can be carried.

Values for insolation are best obtained by direct measurement, but such data are rarely available. Instead, indirect estimates based on generalized graphs (as given by Mateer, 1955, for Canada, and Hamon *et al.*, 1954, Fig. 5, for the U.S.A.) are used with modifications allowing for cloud cover, hours of sunshine, latitude and season.

Since values of insolation obtained indirectly are for a horizontal surface, corrections are made by introducing a basin shortwave melt coefficient k', which varies with basin slope and orientation. During spring melt season the magnitude of this factor is commonly taken to be between 0.9 to 1.1.

The amount of melt produced by longwave radiation under clear weather conditions depends on degree of forest cover (represented by the forest cover factor, F) and effective cloud canopy (represented by the cloud cover factor, N).

Melt produced by condensation and convection is relatively unimportant and adequately represented by Equation IX.9 used jointly with the convection-condensation melt coefficient k.

Total basin snowmelt is estimated from equations of varying degrees of complexity conditioned by the moderating influence of varying degrees of forest cover. Snowmelt equations based on extensive field and laboratory work conducted by the U.S.C.E. (1956, 1960) have the following form for a ripe snowpack at 32 °F :

Heavily Forested Areas (≥ 80%)

$$M = 0.074 (0.53\ T_a' + 0.47\ T_d')\ \text{in/day} \dots\dots\dots\dots\dots\dots\dots\dots\dots\dots\text{IX.20}$$

Forested Area (60-80%)

$$M = k(0.0084u_b) (0.22\ T_a' + 0.78\ T_d') + 0.029\ T_a'\ \text{in/day} \dots\dots\dots\dots\dots\text{IX.21}$$

Partly Forested Areas (10-60%)

$$M = k' (1\text{-}F) (0.004\ R_{si}) (1\text{-}a) + k(0.0084u_b) (0.22T_a' + 0.78T_d')$$
$$+ F\ (0.029\ T_a')\ \text{in/day} \dots\dots\dots\dots\dots\dots\dots\dots\dots\dots\dots\dots\text{IX.22}$$

Open Area (≤ 10%)

$$M = k' (0.00508\ R_{si}) (1\text{-}a) + (1\text{-}N) (0.0212\ T_a' \text{-} 0.84)$$
$$+ N\ (0.029T_c') + k(0.0084u_b) (0.22T_a' + 0.78T_d')\ \text{in/day} \dots\dots\dots\dots\text{IX.23}$$

where   M = snowmelt rate (in/day),
T_a' = difference between air temperature at 10-ft. height, and the snow surface temperature (°F),
T_c' = difference between cloud base temperature and snow surface temperature (°F),
T_d' = difference between dewpoint temperature at 10-ft. height, and the snow surface temperature (°F),
u_b = wind speed 50 ft. above snow surface, (mph),
R_{si} = observed or estimated insolation on a horizontal surface (ly),
a = observed or estimated average snow surface albedo,
k' = basin shortwave radiation melt factor,
k = basin condensation-convection melt factor,
F = estimated basin forest cover, and
N = estimated cloud cover.

In using these equations it must constantly be borne in mind that compactness of form has been achieved by extensive use of simplifying assumptions. The validity of these assumptions always needs to be examined before using the equations for important forecasts where precision is a prerequisite.

The sophisticated theory of snowmelt (of which we have given a résumé) was developed by the U.S. Corps of Engineers for regions in the western U.S.A., and although comprehensive and complete, it is partly based on the physics of heat transfer, and partly on experimental observations from field stations in the western U.S.A. Again, many of the coefficients have been obtained by the statistical analysis of data for the western U.S.A.— values which may be different for other regions. Inherent in their derivation is a weakness for accurately predicting snowmelt rates for a variety of geographical locations and environmental conditions. However, they do provide a fine theoretical foundation for formulating melt equations applicable to local conditions.

Even where these equations are not directly used, the formulated physical theory permits new approaches employing the same basic meteorological parameters. Anderson (1964) has incorporated this modified approach in the snowmelt segment of the Stanford Model for computer simulation of basin streamflow. Pysklywec (1966) has developed a regression type equation using essentially the same basic meteorological parameters for an experimental plot that has demonstrated satisfactory correlation with observed data.

Applications of the U.S.C.E. melt equations and other methods will be discussed later. After the melt has been estimated for the portion covered by the snowpack, losses have to be deducted, and melt excess converted into a streamflow hydrograph by application of unit-graph theory or multiple phase routing. These techniques will also be referred to later.

IX.5   PREDICTION PRACTICES

In the foregoing we have presented a comprehensive and elaborate theory that gives insight into the fundamental snowmelt processes and provides a framework of equations for estimating melt. However, methods that are currently practised for predicting snowmelt streamflow vary with (a) the purpose of the prediction, (b) the degree of accuracy required for the forecast, (c) the time available for the effective use of the forecast, (d) the variability of basin hydrological characteristics, and (e) the availability of electronic computers. The following methods are presently in use, and are listed in order of increasing complexity and refinement:

1. Temperature Index or Degree-Day Methods.
2. Degree-Days plus Recession Analysis Method (U.S. Bureau of Reclamation).
3. Generalized Snowmelt Equations (U.S. Corps of Engineers).
4. Index Plots plus Regression Analysis.
5. Hydrograph Synthesis plus Streamflow Routing.

IX.5.1   Temperature Index or Degree-Day Method

This method has certain advantages because of its simplicity and the fact that air temperature data are usually readily available for the site under study or nearby stations. In view of the theory previously developed it will be observed that air temperature is only one of several meteorological parameters influencing basin snowmelt; hence, use of temperature-index methods must be recognized as gross simplifications. Despite the approximate nature of this technique, surprisingly good results have been reported by Bruce and Sporns (1963) and by Pysklywec (1966). The normal form of a Degree-Day Equation is given in Equation IX.24 and shown in Fig. IX.4.

$$M = C\,(T_a - T_b) \text{ in./day} \dots\dots\dots\dots\dots\dots\dots\dots\dots\dots \text{IX.24}$$

where M = snowmelt in in./day,

T_a = mean daily air temperature or maximum daily air temperature, °F,

T_b = base temperature close to 32°F, to be selected by trial, and

C = coefficient determined by trial, constant or varying with season.

The values of C and T_b are obtained by trial or a regression fit to observed data.

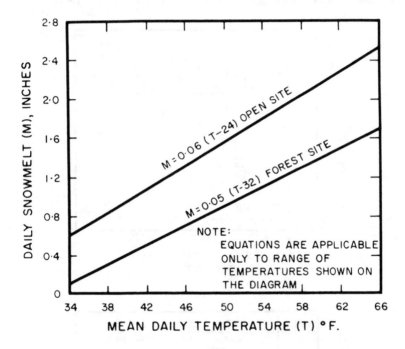

Fig. IX.4   Mean Temperature Index (U.S. Corps of Engineers, 1960)

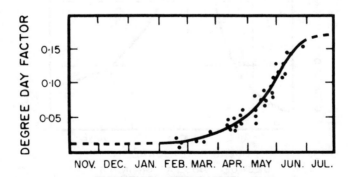

Fig. IX.5   Degree-Day Factor in the Lower San Joaquin River Basin, California (U.S. Corps of Engineers, 1956)

An alternative form uses the concept of a degree-day-factor (DDF), defined as the depth of melt-water (ins.) derived from a snow covered basin as streamflow per degree day.

$$DDF = \frac{\text{Vol. of daily snowmelt}}{\text{Number of degree-days}} \quad \text{in./deg.-day} \dots\dots\dots\dots\dots\dots\dots\dots \quad \text{IX.25}$$

The DDF normally varies with season; in the western U.S.A. it has been reported to range between 0.05–0.15 in./deg.– day, and a mean value of 0.08 is commonly used for preliminary estimates in engineering analysis (see Figs. IX.5 and IX.6).

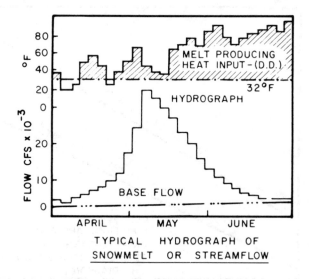

TYPICAL   HYDROGRAPH OF
SNOWMELT   OR   STREAMFLOW

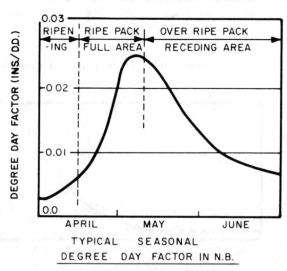

TYPICAL   SEASONAL
DEGREE   DAY   FACTOR IN N.B.

Fig. IX.6 The Degree-Day Factor Method

Bray (1965) made an extensive study of the Tobique Basin (1,670 sq. mi.) in New Brunswick, using a variety of modifications of the basic degree-day technique. His analyses included (1) the normal degree-day method, (2) the antecedent degree-day method, and (3) the accumulated degree-day method.

Attempted reconstitutions of snowmelt for the entire season, based on historically-determined, degree-day factors have proved uncertain due to seasonal bias; but projections of streamflow, based on contemporary values of the degree-day factor have appeared more feasible. Pysklywec's (1966) preliminary study conducted on a small plot in New Brunswick has shown that a degree-day equation provided accuracy of melt predictions quite comparable to estimates made by more elaborate methods.

The temperature-index methods are considered to have the best applicability to large forested basins with homogeneous hydrological characteristics. They offer the twin advantages of simplicity and speed in cases where the attainable degree of accuracy is adequate for forecast purposes. Often, even an experienced hydrologist is faced with the formidable task of predicting snowmelt streamflow with air temperatures as the only available data; for such circumstances, temperature index methods are expedient.

IX.5.2   Recession Analysis Method

This method was developed in 1953 from joint studies conducted by the U.S. Bureau of Reclamation and the U.S. Forest Service in the Fraser Experimental Forest, Colorado. A drainage area of 36 sq. mi. was instrumented with continuous recording temperature- and streamflow-measuring gauges to obtain modified correlations between degree-days and streamflow (Garstka *et al.* 1959). The fundamental feature of this method lay in developing an equation for the recession limb of the very similar diurnal hydrographs of streamflow (Fig. IX.7); once this was established, the following information was statistically derived:

(1) first day's volume of streamflow from that day's snowmelt;
(2) recession volume of streamflow from the day's snowmelt;
(3) height of hydrograph peak above previous day's trough; and
(4) height of hydrograph trough above previous day's trough.

The following correlations were determined by statistical analysis of daily observations:

$$V_1 = b_1 T_1 + b_3 T_3 - C_1 \dots\dots\dots\dots\dots\dots\dots\dots\dots\dots\dots IX.26$$

where V_1 = first day's volume in acre/ft.,
  T_1 = maximum temperature at a selected station,
  T_3 = accumulated maximum temperatures at the
    selected station, and
b_1, b_3 and C_1 = statistically derived coefficients.

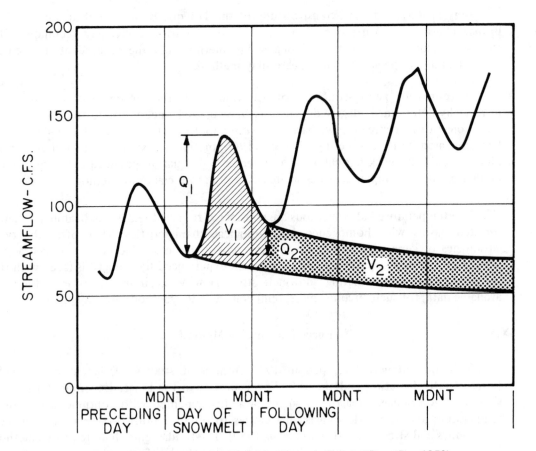

Fig. IX.7   Recession Analysis with Degree-Days (Garstka, 1959)

$$Q_1 = bV_1 + C_2 \quad \dots\dots\dots\dots\dots\dots\dots\dots\dots\dots\dots\dots\dots\dots\dots\dots \text{IX.27}$$

$$Q_2 = b_4 V_1 + C_3 \quad \dots\dots\dots\dots\dots\dots\dots\dots\dots\dots\dots\dots\dots\dots\dots \text{IX.28}$$

where Q_1 = height to peak above previous day's trough,
        Q_2 = height to trough above previous day's trough, and
   b, b_4, C_2, C_3 = statistically derived constants.

From these prediction equations, satisfactory forecasts were made for areas varying from 1 to 500 sq. mi. A comparison of predicted and observed flows will be made in the discussion. One of the principal advantages of this method is that streamflow is predicted directly from temperature correlations.

Either, due to the incompatibility of the correlations to other basin environments or climate variations, this method has not received wide acceptance. It may prove useful, however, in similar regions in the Canadian Rockies.

Both the Temperature Index Method and the Degree-Day Recession Analysis Method combine the dual advantages of simplicity and speed; they are reported to be best suited to heavily forested regions and areas where melt due to rainfall is not significant. When applicable, these methods should prove advantageous for day-to-day operational type forecasts, especially in the absence of extensive hydro-climatic data.

IX.5.3   Generalized Snowmelt Equations

IX.5.3.1   General

Recognizing the great importance of snow hydrology in the design and operation of water resources projects, the U.S. Corps of Engineers in cooperation with other agencies initiated a massive program of research in 1945 which culminated in the treatise *Snow Hydrology* (USCE, 1956). This treatise gives the most detailed and comprehensive coverage on the subject currently available. For engineering design use, a shorter manual entitled *Runoff from Snowmelt* (USCE, 1960) summarized the information generally required for estimating snowmelt runoff, with special emphasis on prediction of flood hydrographs.

Developing a general theory for the thermodynamic processes of snowmelt at a point, this method simplifies the basic theory for field applications, as already given, and provides a working framework of generalized snowmelt equations:

1.   Basin Snowmelt During Rain (Equations IX.18, IX.19).
2.   Basin Snowmelt During Rain-free Periods (Equations IX.20 − IX.23).

An alternative general theory for snowmelt-streamflow forecasting, as used in Russian practice, is given by Alekhin (1964). This material has only recently become available in translated form in North America, and its merits for practical application have not been reported in literature.

IX.5.3.2   Effective Snowcover

The generalized melt equations give estimates of snowmelt over an area or basin which is fully covered with snow. As the snowmelt season progresses and the effective snow cover shrinks, it becomes necessary for hydrograph synthesis to determine the actual amount of snowcover contributing to snowmelt generation. As already stated, effective snowcover may be determined, either directly or indirectly, through the use of different techniques.

Aerial photography permits a direct evaluation of contemporaneous basin snow cover; often, it proves to be the most reliable and economic means of obtaining this information. And aerial photogrammetry further allows volumetric determinations. However, due to lack of such services and facilities, or other difficulties in obtaining such direct information, basin snow cover is frequently estimated indirectly. Extensive measurements of snow disappearance at the Fraser Experimental Forest are reported in Garstka, 1959, Section 6, *Factors Affecting Snowmelt and Streamflow,* and suggest one method of indirectly determining basin snowcover contributing to snowmelt.

These investigations showed that for a given area there appears to be a normal pattern of snow disappearance. A most significant finding of these studies was the establishment of a close relationship between snowcover depletion and accumulated streamflow, as shown in Fig. IX.8; repeated determination of such curves can eventually lead to an average curve or set of curves which can be indirectly used for assessing effective basin cover. In Fig. IX.9 is shown an interesting comparison between a real snow cover depletion and volume depletion. Collection of such information in other project or research basins would improve confidence and accuracy in use of such indirect methods.

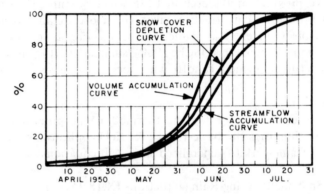

Fig. IX.8   Snowcover Depletion in Fraser Experimental Basin (Garstka, 1959)

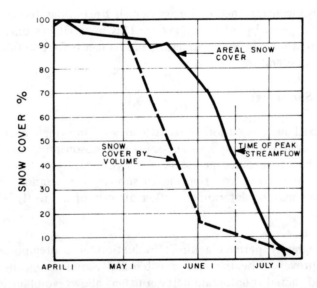

**Fig. IX.9   Relation Between Volume of Snow and Area of Snowcover in
Fraser Experimental Basin (Garstka, 1959)**

Erickson and McCorquodale (1966) have described an alternative indirect method for snowmelt runoff by computer simulation using the snowmelt runoff data for the Manicouagan River. A typical Area Contributing Curve is shown in Fig. IX.10.

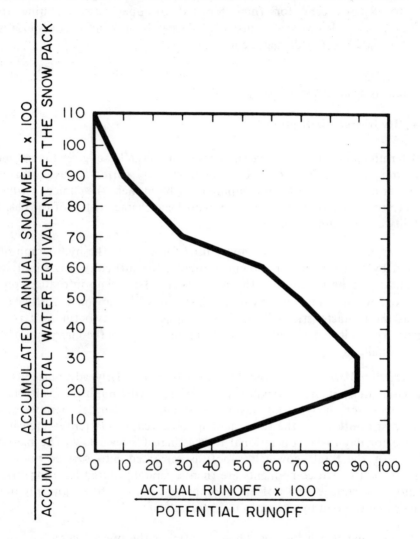

Fig. IX.10 Area Contributing Curve for Manicouagan River Basin (Erickson and McCorquodale, 1966)

IX.5.3.3   Runoff Fraction

Elaborate procedures for estimating the effect of snowpack and ground conditions on losses and runoff are given in the U.S. Corps of Engrs. Manual *Runoff from Snowmelt* (1960). Frequently, due to the inadequacy or lack of basic information needed to solve these equations, simpler techniques are employed for estimating evapotranspiration and infiltration losses.

For the Manicouagan River basin in Quebec, Erickson and McCorquodale (1966) reported that satisfactory results were obtained in their simulation program by varying the basin evapotranspiration loss from 14 per cent of the total melt for densely forested areas to 8 per cent for thinly forested or open areas. During rain periods, evapotranspiration losses were considered negligible. The infiltration loss was assumed constant at 0.057 in./day during the melt period.

Use of each assumed value needs confirmation by trial simulations, which are easily carried out using an electronic computer.

IX.5.3.4   Hydrograph Derivation

Determination of the effective snowcover and basin losses gives the net melt available for conversion to streamflow. Two methods are currently used for deriving the hydrograph of streamflow from computed net basin melt; the choice of which particular method to use depends primarily on the hydrologic characteristics of the basin and on the availability of computer facilities.

Unit Hydrograph Method–is commonly used for deriving the hydrograph of streamflow generated by rain events; for snowmelt runoff, the difficulty arises in determining the unit duration to be used. Also, the long, flat recession limb introduces problems in its application. When rainfall occurs in combination with snowmelt, further complications result as the characteristics of unit hydrographs for the two types of runoff are very different. The unit hydrograph method is preferred where the major runoff contribution is from rainfall.

Phase Routing Method–has been found to be particularly advantageous for use with digital computers. It offers flexibility in deriving hydrograph reconstitution parameters by trial-and-error analyses of historical floods. Further, it permits adjustment of computed streamflows to the flows actually observed; this latter feature can be a distinct advantage for day-to-day operational forecasting of streamflow. The method normally requires that the total snowmelt runoff is separated into surface runoff and ground contributions; the two components are then separately routed by using different storage constants. The storage times of each component need to be empirically determined by trial fits of synthesized hydrographs to historical data.

Despite obvious weaknesses, Phase Routing is the prevailing method of determining the time-distribution of snowmelt runoff when snowmelt is the major contribution. It is used principally with electronic computers. This method was initially used for the Columbia Basin Studies by Rockwood (1958), and later in reconstitutions for the St. John River by Cuthbertson and Jonker (1964), and for the Manicouagan River studies by Erickson and McCorquodale (1966).

IX.5.4   Index Plots–Regression Equation

In New Brunswick, Davar and Bray initiated a snow hydrology study on a small experimental plot — an Index Plot — located in the North Nashwaaksis watershed in an attempt to check the adequacy and accuracy of existing equations for predicting snowmelt for their geographic conditions. From the measurements of meteorological parameters and actual snowmelt at the site, Pysklywec (1966) concluded in his study that:

(1) degree-day relations were simple and satisfactory when determined from local observations;

(2) the U.S. Corps of Engrs. equations though sound in theory, could not satisfactorily predict daily snowmelt (even without the occurrence of losses or time distribution effects); and

(3) a regression equation using the same basic meteorological parameters as in the U.S. Corps of Engrs. theory, but derived from the local conditions, gave the best representation of the actual snowmelt sequence.

Pysklywec's regression equation had the following form:

$$M = 0.615 + 0.0373n + 0.00607R_b + 0.00201 (T - 36)u$$
$$+ 0.0437 (RH)u + 0.007 P(T - 32) \dots\dots\dots\dots\dots\dots IX.29$$

where M = snowmelt (in/day),
      n = sunshine (hr/day),
      R_b = longwave radiation (ly/day),
      T = temperature, (°F),
      u = wind velocity, (mph.),
      RH = relative humidity, and
      P = rainfall (in/day).

The virtue of this particular method lies in the availability of a regional experimentally derived snowmelt prediction equation of satisfactory accuracy for the specified environment. There remains, of course, the challenging problems of runoff determination and derivation of the streamflow hydrograph resulting from snowmelt.

For any major basin where long-range design programs are being instituted, a network of such simple and economically installed experimental plots should provide reliable and adequate information for the initial design phases, and for the subsequent operational phases of these projects. The alternative approach of conceptual models, the Stanford model in the U.S.A. and the Acres model in Canada, provide computer simulations of streamflow; but even these elaborate programs have their disadvantages because they employ parameters that have not necessarily been evaluated for a given region, and thus these deficiencies may lead to erroneous snowmelt segments in the main program. Many of the assumptions and coefficients are derived from historical reconstitutions, which may not be satisfactory for a particular season. Even in such analyses, data obtained from several small plots located throughout a basin would provide reliable, simple and economical sources of supplementary information that becomes increasingly important as data accumulate. Besides, these data would permit checking and adjustment of the main program segment as the current snowmelt season progresses.

IX.5.5   Hydrograph Synthesis and Streamflow Routing

Generally, a very large basin cannot be considered a unit of analysis because of the spatial variation of topographical, geological and hydrometeorological conditions. In such cases, the large basin must be subdivided into a number of tributary basins for which the working assumptions can be considered fairly uniform. The streamflow hydrographs, synthesized for the sub-basins, are all routed to the site on the main stream, and their combined effect investigated at the location. The Acres model for hydrologic simulation with the aid of computers (Cuthbertson and Jonker, 1964; and Erickson and McCorquodale, 1966) is representative of using such an approach.

Of course, this further complicates the whole procedure as hydraulic parameters of the stream channels must be known, which is only possible from extensive flow records; if these are not known, arbitrary assumptions need to be introduced.

The introduction of telemetry, radio-reporting precipitation and streamflow gauges, and sophisticated modern instrumentation that is currently being developed, further augment the potential of hydrologic forecasting for entire river basins from the system operation aspect. Where reliable and adequate basic data are available, modern analog and digital computers permit these complex programs to be handled successfully and speedily. Consequently, new vistas are revealed for optimum water use and economic analyses of vast water resource projects.

IX.6   LITERATURE CITED

Alekhin, Y.M. 1964. Short range forecasting of lowland river-runoff; Gideo-meteorolo-gicheskoe Izdatel stvo Leningrad, 1956; Translated from Russian by Israel Program for Scientific Translations and published for the U.S. Department of Commerce and the National Science Foundation, Washington, D.C., 229p.

Anderson, E.A. and Crawford, N.H. 1964. The synthesis of continuous snowmelt runoff hydrographs on a digital computer; Technical Report No. 36, Department of Civil Engineering, Stanford University, Stanford, California, June, 103 p.

Bader, H. 1954. Snow and its metamorphism, SIPRE translation 14, Snow, Ice and Perm. Res. Estab., Corps of Engineers, Wilmette, Ill.

Bray, D.I. 1965. Snowmelt-streamflow forecasting for the Tobique River, N.B. Master's Thesis, Dept. of Civil Eng., University of New Brunswick, Fredericton, N.B.

Bray, D.I. and Pysklywec, D.W. 1966. An evaluation of an Hydrological simulation for the Saint John River basin. A report to the New Brunswick Water Authority, Fredericton, N.B.

Bruce, J.P. 1962. Snowmelt contributions to maximum floods. Proceedings of the Eastern Snow Conference, 85-104, c/o G.R. Ayer, U.S. Geological Survey, Albany 1, N.Y.

Bruce, J.P. and Sporns, U. 1963. Critical Meteorological Conditions for Maximum Floods in the St. John River Basin, Canadian Meteorological Memoirs, No. 14, Dept. of Transport, Met. Service of Canada, Toronto, 42 p.

Chow, V.T. 1964. Handbook of applied hydrology. New York, McGraw-Hill Book Co., Inc.

Collins, E.H. 1934. Relationship of degree-days above freezing to runoff, Trans. Amer. Geophysical Union, 624-629.

Cuthbertson, W.B. and Dickison, R.B.B. 1962. Snowmelt and rainfall floods, St. John River basin, May 1961, Proceedings of the Eastern Snow Conference, 105-120, c/o G.R. Ayer, U.S. Geological Survey, Albany 1, N.Y.

Cuthbertson, W.B. and Jonker, F.H. 1964. Flood forecasting with the aid of a hydrological simulation program, Proceedings of the Eastern Snow Conference, 45-62, c/o G.R. Ayer, U.S. Geological Survey, Albany 1, N.Y.

Davar, K.S. and Bray, D.I. 1964. Preliminary results of snowmelt-streamflow studies in the Tobique Basin, Proceedings of the Eastern Snow Conference, 78-96, c/o G.R. Ayer, U.S. Geological Survey, Albany 1, N.Y.

Erickson, O.M. and McCorquodale, J.A. 1966. Application of computers to the determination of snowmelt runoff, presented at Symposium No. 5 - Statistical Methods in Hydrology sponsored by N.R.C. Subcommittee on Hydrology, Montreal.

Fritz, S. and MacDonald, T.H. 1949. Average Solar Radiation in the United States, Heating and Ventilating, Vol. 46, 61-64.

Garstka, W.U., Love, L.D., Goodell, B.C. and Bertle, F.A. 1959. Factors affecting snowmelt and streamflow, U.S. Bureau of Reclamation and U.S. Forest Service, 187 p., Supt. of Documents, Washington 25, D.C.

Geiger, R. 1965. The climate near the ground, Harvard University Press, Cambridge, Mass.

Glennie, J.F. 1963. A hydrological study of the 1948 flood in the Stuart River basin, B.C., University of British Columbia, Vancouver, B.C., 75 p.

Gold, L.W. and Williams, G.P. 1961. Energy balance during the snowmelt period at an Ottawa site, Research paper No. 131, National Research Council of Canada.

Hamon, R.W., Weiss, L.L. and Wilson, W.T. 1954. Insolation as an empirical function of daily sunshine duration, Monthly Weather Review, Vol. 82, No. 6, 141-46.

Hildebrand, C.E. and Pagenhart, T.H. 1955. Lysimeter studies of snowmelt, Research Note 25, U.S. Corps of Engr.

Hutchison, B.A. 1966. A comparison of evaporation from snow and soil surfaces, Bulletin of the International Association of Scientific Hydrology, Vol. II, No. 1, 34-42.

Johnson, O. and Boyer, P. 1959. Application of snow hydrology to the Columbia basin. A.S.C.E. proceedings, Vol. 85, HY1.61-81.

Light, P. 1941. Analysis of high rates of snow melting, Trans. Amer. Geophysical Union, Part I, 195-205.

Linsley, R.K. 1943. A simple procedure for day to day forecasts of runoff from snowmelt, Trans. Amer. Geophysical Union, Vol. 24, 62-67.

Linsley, R.K. Kohler, M.A. and Paulhus, J.L. 1947. Applied Hydrology, New York, McGraw-Hill Book Co. Inc., New York, 689 p.

Lull, H.W. and Rushmore, F.M. 1960. Snow accumulation and melt, Northeastern Forest Expt. Station, Station Paper No. 138, Upper Darby, Pa.

Mateer, C.L. 1955. Average insolation in Canada during cloudless days, Canadian Journal of Technology, No. 33, 12-32.

McCaig, I.W., Jonker, F.N. and Gardiner, J.M. 1963. Hydrological simulation of a River Basin — an aid in flood control plannings, The Engineering Journal (E.I.C.), Vol. 46, No. 6, 39-43.

McKay, G.A. 1964. Relationships between snow surveys and climatological measurements, symposium of surface waters, General Assembly of Berkeley, International Association of Scientific Hydrology, 214-227.

Potter, J.C. 1960. Density of freshly fallen snow, Proceedings of the Eastern Snow Conference, 41-47, c/o G.R. Ayer, U.S. Geological Survey, Albany 1, N.Y.

Pysklywec, D.W. 1966. Correlation of snowmelt with the controlling meteorological parameters; Master's Thesis, Dept. of Civil Engineering, Univ. of New Brunswick, Fredericton, N.B.

Rockwood, D.M. 1958. Columbia basin streamflow routing by computer, A.S.C.E. Proceedings, No. 84, WW5, Paper 1874, 15 p.

Secretariat of the World Met. Organ. 1965. Guide to Hydrometeorological Practices, W.M.O. – No. 168, TP. 82, Geneva, Switzerland.

Simmons, G.E. 1961. Snowmelt runoff, Fraser River basin, Proceedings of Symposium No. 1 – Spillway Design Floods, N.R.C. Subcommittee on Hydrology, the Queen's Printer, Ottawa 227-259.

U.S. Army, Corps of Engineers, 1956. Snow hydrology, Portland, Oregon; North Pacific Division, Corps of Engineers, 437 p.

U.S. Army, Corps of Engineers, 1960. Runoff from Snowmelt, EM 1110-2-1406, Washington 25, D.C., Supt. of Documents, 75 p.

West, A.J. 1959. Relation of snow evaporation to meteorological variables, Proceedings of the Western Snow Conference.

Williams, G.P. 1963. Evaporation from water, snow and ice. Technical paper, No. 164, Division of Building Research, National Research Council.

Wilson, W.T. 1941. An outline of the thermodynamics of snowmelt, Trans. Amer. Geophysical Union, part 1, 182-194.

In addition it should be pointed out that valuable information may be obtained from the following publications:

(1)   Proceedings of the Western Snow Conference
      Secretary: Mr. M.W. Nelson
                     Post Office Box 1247
                     Boise, Idaho.

(2)   Proceedings of the Eastern Snow Conference
      Secretary: Mr. G.R. Ayer
                     U.S. Geological Survey
                     Albany 1, N.Y.

Section X

BASIN YIELD

by

Hugh D. Ayers

TABLE OF CONTENTS

LIST OF TABLES

LIST OF FIGURES

Section X

BASIN YIELD

X.1   DEFINITIONS

In the generic sense *yield* refers to the quantity of any product resulting from exploitation of natural resources. There are several definitions in current use in the hydrological literature. Law (1955) quotes a definition of the Sub-group of the Hydrological Group of the Institution of Water Engineers of the United Kingdom as follows:

> The uniform rate at which water can be withdrawn from a reservoir throughout a dry period of specified severity without depleting the contents to such an extent that withdrawal at that rate is no longer feasible.

According to this definition, watershed yield is a property of the watershed and the reservoirs associated therewith. North American practice is to speak of the yield of watersheds and of reservoirs as separate characteristics. In general, basin yield refers to the quantity of water available from a stream at a given point over a specified duration of time. The duration of time would normally be for a period of a month or longer. The emphasis is on water volumes rather than instantaneous rates. However, the yield is the summation over the specified time period of the continuous hydrograph of flow at a particular point on a stream. It is therefore the consequence of all hydrologic events resulting in flow, including storms of all durations and intensities, and the climatic, geologic and land use environment. It includes streamflow from all sources. For this discussion only natural systems will be included. Direct man-made influences, such as storage, diversion or other regulation, will not be included as influences on basin yield.

X.1.1   Units of Watershed Yield

For the purpose of comparing the water-yielding properties of different watersheds, basin yield may be expressed in terms of an equivalent depth of water over the watershed area. This method of expression also proves convenient when considering the water balance of the basin, since the precipitation input is expressed in the same units, as is the other major output, term evaporation. Engineers, however, in water resources studies on particular basins often prefer units such as acre-ft. or cfs-days. For our discussions, depth units will be adopted.

X.2   WATER BALANCE

In the preceding Sections attention has been directed primarily to flow processes in the hydrologic cycle. Precipitation, infiltration, evaporation and runoff are flow processes. Storage parameters, such as interception, soil moisture, groundwater and snow, assume much more importance in the water balance approach to basin yield.

For convenience the hydrologic bookkeeping equation is recalled:

$$S_1 + Ss_1 + Sg_1 + Ssm_1 + {}_1P_2 = {}_1E_2 + \int_{t_1}^{t_2} Q \, dt + S_2 + Ss_2 + Sg_2 + Ssm_2 \ldots\ldots\ldots X.1.$$

(subscripts 1 and 2 indicate the beginning and end of the time period respectively)

where S = volume of water in storage in channels and reservoirs of the area under consideration,

Ss = volume of water or its equivalent, in storage on the surface of the ground, on leaves and pavements etc.,

Sg = volume of water in storage as groundwater.

Ssm = volume of water in storage as soil moisture.

${}_1P_2$ = total equivalent uniform depth of precipitation over the area between time t_1 and time t_2,

${}_1E_2$ = total equivelant depth of evaporation and evapotranspiration over the area between time t_1 and time t_2,

Q = instantaneous rate of discharge from the basin.

In this instance there is assumed to be no inflow or outflow of groundwater from the basin, and no inflow of surface water to the basin.

The term $\int_{t_1}^{t_2} Q \, dt$, provides the estimate of basin yield, which in many instances may be the most easily measured term of the hydrologic equation. However, where continuous streamflow measurements are not available, then yield may be determined as a residual in the water balance equation (Equation X.1). For this purpose, measurements or estimates of the two flow parameters — precipitation, and evaporation or evapotranspiration — and the several storage terms are required. The water balance equation may be reduced to

$$\int_{t_1}^{t_2} Q \, dt = {}_1P_2 - {}_1E_2 - \Delta S \ldots\ldots\ldots\ldots\ldots\ldots\ldots\ldots X.2$$

where ΔS is the change in storage.

With judicious selection of the time period $t_2 - t_1$, the magnitude of the term ΔS may be minimized. This time period, however, must normally be at least one year.

Even for a period of this duration the relative reliability of yield estimates decreases as the amount of precipitation decreases and as the range of storage potential increases.

The basis for this statement is that when precipitation is high and evenly distributed throughout the evaporation season, evapotranspiration will be nearly equal to the potential. Potential evapotranspiration is governed by meteorological conditions and will be fairly stable from year to year, and may be reliably determined. In this case, yield will be sensitive only to changes in storage throughout the period. If storage potential is low, then the magnitude of the storage change term will be small, and added reliability will be attached to the yield estimate.

Some empirical determinations of annual watershed yield will be examined to illustrate the implicit nature of the water balance approach adopted.

X.2.1 Linear Annual Precipitation–Runoff Relationships

Sutcliffe and Rangeley (1960) determined the annual yield of the Tongariro River of New Zealand from observed precipitation and runoff as follows:

$$Q = P - 20 \text{ inches} \quad \dots\dots\dots\dots\dots\dots\dots\dots\dots\dots\dots\dots\dots\dots\dots\dots\text{X.3}$$

> where Q = annual estimated runoff, and
> P = mean basin precipitation, (in.)

In the nine years of record the annual precipitation was never less than 64 in. and was fairly evenly distributed throughout the year so that actual evapotranspiration would be approximately equal to potential evapotranspiration at 20 in.

Ayers (1962) suggested that the annual precipitation should be approximately twice the potential evapotranspiration for this type of relationship to be used reliably. Sutcliffe and Rangeley (1960) found the standard error of estimate for the Tongariro to be 13.92 in. or 16.5 per cent of the mean.

A more versatile form of the linear relationship is

$$Q = aP - C \dots\dots\dots\dots\dots\dots\dots\dots\dots\dots\dots\dots\dots\dots\dots\dots\dots\text{X.4}$$

By equating $Q = P - E$ to this expression, we find that

$$E = C + P (1 - a) \dots\dots\dots\dots\dots\dots\dots\dots\dots\dots\dots\dots\dots\dots\dots\text{X.5}$$

so that watershed yield and evaporation on an annual basis are made to increase in a linear fashion with precipitation. There is a reasonable justification for this when precipitation is moderate and well distributed, such as in temperate, sub-humid regions. This relationship was used by Sutcliffe and Rangeley (1960) in expressing the annual yield for the Tana River of East Africa:

$$Q = 0.406 (P - 17) \text{ inches} \dots\dots\dots\dots\dots\dots\dots\dots\dots\dots\dots\dots\dots\text{X.6}$$

The Tana basin experiences. two distinct rainy seasons, during which rainfall amounts are normally in excess of evapotranspiration demands. However, in this case the standard error of estimate was 3.39 in., or 29 per cent of the mean runoff.

Because there may be distinct seasonal differences in the evaporation demands, thus resulting in soil moisture storage depletion, frequently there are large inconsistencies in the partitioning of precipitation between runoff, evapotranspiration and soil moisture, or groundwater recharge, from season to season. For this reason a multiple linear regression for yield based on seasonal precipitation values may be useful. Glasspoole (1960) for the Thames River of England gives the following expression:

$$Q = 0.18 \, P_{wl} + 0.51 \, P_{sl} + 0.73 \, P_{wo} + 0.13 \, P_{so} - 13.0 \, . \, . \, . \, . \, . \, . \, . \, . \, . \, . \, . \, . \, . \, . \, . \, \text{X.7}$$

where Q = annual yield in inches,
P_{wl} and P_{sl} = precipitation for winter and summer
respectively of the previous year, and
P_{wo} and P_{so} = precipitation for winter and summer
respectively of the current year.

Glasspoole (1960) claims that 92.9 per cent of the variability in flow of the Thames can be accounted for by the seasonal precipitation parameters. The use of seasonal values has reduced the standard error of estimate to 9 per cent of the mean runoff, compared to 25 per cent when using annual precipitation only.

X.2.2   Antecedent Precipitation Effects

Although annual precipitation values are normally considered to be independent random events, the effects unquestionably carry over into the succeeding year. In other words, unusually high precipitation during September of a given year will result in recharge of soil moisture and groundwater. Precipitation after October 1 (the start of the water year) will contribute in much larger measure to the streamflow for that year than if the water year had commenced dry. Thus, the storage parameters assume a large importance. To account for this in an indirect way, precipitation for a period of time preceding the water year may be included in a regression equation.

Siren (1960) has shown, for the Kymijoki River of Finland, that the correlation coefficient was improved for the precipitation-yield relationship from $r = 0.57$ to $r = 0.85$ on a January 1 to December 31 water year basis by also taking into account the precipitation for the preceding August 1 to December 31 period.

The selection of a water year in which storage conditions are reasonably consistent at the beginning of each water year would therefore aid in attempting to describe basin yield as a function of precipitation on the basin. Studies on the best water year have been carried out by Gold (1951) for the Thames River in the United Kingdom and Sharp *et al.* (1960) for the Delaware River Basin of Kansas. In the latter study it was found that a water year of May 1 to April 30 provided a correlation coefficient of $r = 0.946$ between annual yield and precipitation. On the other hand, the water year of August 1 to July 31 gave the poorest correlation of r = 0.817. These findings were generally in agreement with those of Gold (1951).

X.2.3   Temperature Modified Precipitation–Runoff Relationships

In recognition of the role of evapotranspiration in depleting soil moisture storage, several attempts have been made in the past to develop simple annual-yield equations incorporating mean temperature for the year as a factor in modifying the water balance for the basin.

Studies by Justin (1914), Grunsky (1922), Wundt (1937) and Turc (1954) are in this category. Although the studies reported provide a basis for broad regional assessments of average annual water yield, they do not in any way provide a satisfactory method for computing year-by-year yields for a particular basin.

X.3   GEOLOGY AND BASIN YIELD

Geology influences basin yield principally through the magnitude of storage potential and related water transmission properties. A distinction will be made between the surface mantle (for practical purposes the active rooting zone of vegetation) and the geologic materials below this depth. To some extent this is an artificial distinction, since the two regions form a continuum of water movement. However, the surface mantle is a region subject to storage depletion by evaporation and evapotranspiration, and its infiltration characteristics determine the extent to which rainfall or snowmelt may be received for storage and transmission. The residual of precipitation over and above infiltration rate at any given instant represents accretion to surface storage and subsequent overland flow. The surface mantle and its characteristics, including the cover, thus have an important role in the surface runoff portion of basin yield.

The deeper geologic materials, on the other hand, exert a strong influence through their storage potential and transmission properties on the groundwater or baseflow contribution to basin yield.

X.3.1   Soils

The surface mantle or soils of a watershed are frequently classified for agricultural purposes according to their naturally occurring characteristics. The parent material and arrangement within a profile of various horizons of differing color, texture, structure and thickness provide a considerable amount of qualitative information regarding the past history about the soil-hydrologic system. Likewise, with proper interpretation, soil profile characteristics may be expected to provide a basis for a quantitative description of the water-yielding properties of a basin. In general, texture, structure and density are the dominant soil properties affecting infiltration and transmission rates through a profile. Superimposed upon this must be the driving force associated with the soil-water system.

While the physical principles associated with infiltration and movement of water through the profile are well understood, as discussed in Section V, and scores of infiltration tests have been carried out in the field and laboratory, little progress has been made in translating these results into a generalized watershed model that accurately apportions precipitation of snowmelt to surface runoff, soil moisture recharge and groundwater recharge.

For our purposes then, some comparisons of the yields resulting from differing soils will be presented to indicate the relative influence.

Lvovitch (1957) in a discussion of streamflow formation factors in Russia quotes the following relative discharge coefficients (ratio of runoff to precipitation) for storm runoff for some of the Great Soil Groups:

Solonetzes and solonchaks 1.00
Degraded podsol clay and clayey soils . . . .   0.80 to 0.85
Chestnut soils . . . .   0.65 to 0.70
Clay and clayey chernozems . . . .   0.40 to 0.50
Sandy soils . . . .   0.20 to 0.35

This work is not well documented in the article quoted. It should be remembered that these coefficients are based on surface runoff. On the larger basins with sandy soils, it may well be that the baseflow would be sufficient to compensate for the reduced surface runoff, so that the total annual yields would not be so different.

There is evidence of this in Ontario. In a recent paper Ayers (1966) compares a number of basins in the Great Lakes drainage system. Two of these basins, at the same latitude but with extreme differences in surficial geology in glaciated peninsular Ontario, are selected to illustrate this effect:

	Fifteen-year mean accumulative runoff from October to:	
	April 30	September 30
Credit River (coarse-textured soils formed on sands and gravels)	6.9 in.	9.8 in.
Conestoga River (fine-textured soils formed on till)	11.2 in.	12.9 in.

It will be noted that the winter and spring runoff for the fine-textured soil exceeds by 4.3 in. that from the watershed of coarse-textured soils. However, summer flow from groundwater recharged during the winter and spring resulted in a final difference of only

3.1 in. The difference in the total mean yield can at least be partially attributed to winter precipitation differences.

X.3.2   **Subsurface Materials**

The unconsolidated deposits of sand and gravel with direct connection to streams are likely to be the most reliable for yielding a sustained streamflow. The continuity and extent of such deposits, their hydraulic conductivity and the opportunities for recharge, are of significance in the distribution of seasonal flow.

Most studies on the effects of subsurface geology have been concerned with the variability of streamflow rather than on total yields. Cross (1949) in a study of dry weather flow in Ohio showed that interlobate moraines and buried glacial valleys filled with permeable gravels, in addition to kame terraces, kames, eskers and outwash deposits, all had high rates of discharge. Glacial till and unglaciated areas experienced low rates of dry weather flow. These results are supported in studies by Ding (1965) in Ontario who showed that the mean monthly discharge rate exceeded 90 per cent of the time; the same criterion used by Cross (1949) was 0.55 and 0.73 cfs/sq mi respectively for two watersheds with sand and gravel subsurface materials, and 0.32 and 0.29 cfs/sq mi for watersheds predominantly of glacial till materials.

The consolidated subsurface materials are highly variable. However, the sedimentary limestone and sandstones are likely to be the most favorable for sustaining the flow during dry weather periods. Nevertheless, weathered and fractured igneous rocks sometimes yield unexpectedly high flows during the dry weather period. Stafford and Troxell (1944) attributed the higher sustained yield of San Antonio Creek and Lytle Creek, in the San Bernardino and San Gabriel Mountains of California, to the deeply-fractured granite in these watersheds, as compared to Strawberry Creek, Deep Creek and the Mojave River in the same locality.

Basalt is considered to provide very good conditions for sustained flow, as pointed out by McDonald and Langbein (1948). Serra (1954) compared the discharge coefficients for several basins in southern France, and found the following ratios of total yield to precipitation:

Basalt	0.81
Granite	0.63
Moraine	0.17

The tendency in assessing geological influences is to consider only the internal properties of the soil or rock. However, there are physiographic features associated with geological conditions that may be equally important in streamflow distribution. Slopes, drainage density, incisement of stream channels, and depressional or pond storage, each have some influence. The Precambrian region of Canada provides an example of how the lake storage influences seasonal distribution. Many of the drainage basins in this region have 20 per cent or more of their area in lakes, as compared to perhaps 1 or 2 per cent for glaciated basins in Southern Ontario. As a consequence, although the soil mantle is very shallow or non-existent and fracture storage and flow extremely limited, the lake storage acts in much the same way as groundwater aquifers to sustain flow through dry weather periods.

X.4   LAND USE INFLUENCES

Much has been written about the effect of land use on watershed yield, but unfortunately a great deal of it has been unsupported by scientific fact. The controversy has related largely to the influence of forests, and to the influence of intense farm cultivation. Attention in more recent years has been directed to other land uses, such as mining, urban and industrial development, that no doubt have exerted very pronounced local influences upon the hydrologic regime. However, little factual information is available on the influences of these uses on yield, and experimental studies are only just getting underway on the effects caused by these uses.

Land use influences can generally be associated with the manner in which the practices modify either the infiltration rate for the area or the moisture storage potential of the region. Dense vegetal canopies protect soils from the impact of rainfall, thus preventing compaction, in-wash of soil particles and sealing of the surface, and these processes cause a reduction of infiltration rates. Thus soils with scant vegetal protection would be vulnerable to increased runoff from storm rainfall and reduced soil moisture recharge. At the same time, a vigorous vegetal growth may result in greater depletion of soil moisture supplies, thus reducing the groundwater recharge potential and depleting the streamflow from this source.

X.4.1   Agricultural Practices

From the wide range of agricultural crops and cultural practices associated with each, only a few examples of the influences on yield can be cited.

Lvovitch (1957) of the U.S.S.R. reports that autumn plowing decreased runoff to one-fifth in the Transvolga Region. Apparently this would be in the region of winter freeze-up and major spring snowmelt. Similar findings have been reported by Hays (1955) from studies in Wisconsin. Ayers (1965), in reporting on the results from paired 20-acre watersheds in Ontario, indicated that a sod condition resulted in average increased winter water yields of 0.71 in. (2 years) for one watershed, and 0.82 in. (4 years) for another watershed, over the yields when watersheds were fall plowed or in corn stubble. In some years there was no runoff, or only very small amounts, from the plowed or corn stubble watersheds. The winter condition apparently is one in which the influence of cover is opposite to that experienced during the frost-free period.

Hays (1955) reported that the highest values of storm runoff were recorded from areas of spring planted grain, followed by first year hay land, and corn land, in that order. Older hay stands (2 to 3 years) yielded very little runoff from summer rains compared with corn land and spring grain. Young (1948) quoted tentative relative water yields in descending order for the following cultural practices in the Ozarks of Arkansas on freely permeable soils: cultivated crops; terraced meadow; strip cropping; and pasture and wooded lands. This would be south of the winter freeze-up zone. Sharp *et al.* (1960) tested by regression analysis the significance of several land use practices on the annual yield from the Delaware River Basin of Kansas, and found none of the following to be significant: percentage of row crops; miles of terraces per square mile; average percentage of normal pasture at start of the month; and pasture condition. In fact, Sharp, Gibbs and Owen (1966), as a result of the Co-operative Water Yield Procedures study conducted in the Great Plains agricultural area of the United States, state:

> No statistical approach was found that would consistently assess effects of land treatment on streamflow from river basins, or even prove conclusively that such effects do or do not exist. In a few cases, streamflow appeared to be increasing. In some, it appeared to be decreasing. In all cases, streamflow fluctuated considerably, .due to climatic or other causes.

The five-year study by Sharp, Gibbs and Owen (1966), however, did result in the development of a rational procedure for estimating the effects of land and watershed treatment, or streamflow, that is worthy of careful examination.

X.4.2   Forestry Practices

The controversy over the effect of forest cover on basin yield has traditionally been drawn along the line of forest *vs.* no forest. As a result, conclusions have often been drawn which may not have wide application in forest management. Recently, foresters and hydrologists have recognized that the type of forest and its management in association with the climatic and geologic environment require careful examination before any definite conclusions can be reached. Books by Kittredge (1948) and Colman (1953) present comprehensive reviews of forest hydrologic influences.

Hoover (1944) reported that stream yields were increased by 16.74 in. and 10.68 in. respectively in 1941-2 and 1942-3 following the complete cutting of forest vegetation from watersheds at Coweeta, North Carolina. The watersheds had been compared prior to cutting, so that the increases in yield could be based on the probable yield of the treated watershed, had it remained in forest cover. Only about 2 in. out of 70 in. of annual precipitation are in the form of snow. Lieberman and Hoover (1951), reporting on the flow distribution from treated catchments at Coweeta, found that the median value in the frequency distribution was increased by a factor of two when vegetation was cut and regrowth prevented. Apparently, the major factor contributing to higher base flows is the reduction in evapotranspiration, thus making more precipitation available for groundwater recharge. In these studies, the ground surface was protected by leaving the cutover material in place; thus infiltration rates were likely unaffected by the cutting operations. Regrowth of forest stands, however, usually results in an increase in evapotranspiration and reduction in water yields.

Goodell (1958,1964) reported in experiments at Fraser, Colorado, when a precalibrated watershed of lodgepole pine forest was cut over in a pattern of alternate clear-cut strips, that the annual yield was increased by 4 in. to an annual value of 17 in. during the first two years, and after eight years, that the average annual increase was 2.8 in. The treated basin has experienced increased spring discharge peaks in most years.

In situations where fires, disease or poor logging practices have prevailed, the tendency has been for increased flood peaks to occur at the expense of low flows, with little change in the total yield: Love (1955), Anderson (1955), Anderson and Gleason (1960), and Bailey and Copeland (1960).

X.5   BASE FLOW SYNTHESIS

During rainless periods, streamflow results from the depletion of groundwater storage. Dooge (1960) provides a refreshing approach to the problem of base flow synthesis by suggesting that groundwater systems consist of linear storage elements through which infiltrating rainfall may be routed either to the other storage elements in series, directly to the stream, or back to the atmosphere by evaporation.

X.5.1   Model of Flow System

The problem then resolves itself into one of constructing a mathematical model of the groundwater system that will serve as a basis for synthesis. A short prior record of streamflow for the watershed under consideration is required.

The assumptions in the derivation of routing coefficients are:

1. $S = KQ$ (Discharge varies linearly with storage) X.8

    where S = storage, and
    K = storage delay time (storage constant).

2. The rate of groundwater recharge is constant within any given time period.

3. The time periods during which recharge occurs are assumed to be equal.

The various types of linear storage elements are:

1. Deep Water Table Storage Element.
   This is subject to positive recharge only, when the soil moisture content of the soil above is at field capacity.

2. Shallow Water Table Element.
   This is subject to positive or negative recharge (evaporation).

3. Composite Type.
   A shallow element until storage is depleted to a certain degree, then acting as a deep storage element.

Combinations of storage elements may be assigned to the model in parallel, or in series, as required.

The discharge from a storage element is represented mathematically by the following Routing Equation of Dooge (1960):

$$Q_n = C_0 R_n + C_1 R_{n-1} + C_2 Q_{n-1} \dots\dots\dots\dots\dots\dots\dots\dots\dots \text{X.9}$$

where Q_n = discharge during any time period,
$\quad\quad R_n$ = recharge to the element during the period,
$\quad\quad R_{n-1}$ = recharge to the element during the previous time
$\quad\quad\quad\quad$ period, and
$\quad\quad Q_{n-1}$ = discharge from the element during the previous
$\quad\quad\quad\quad$ time period.

Development of the values for the routing coefficients is beyond the scope of this discussion. However, they are functionally related to the storage delay time, K, and routing time unit, t, as follows:

$$C_0 = 1 - \frac{K}{t} (1 - e^{-t/K}) \dots\dots\dots\dots\dots\dots\dots\dots\dots \text{X.10}$$

$$C_1 = \frac{K}{t} (1 - e^{-t/K}) - e^{-t/K} \dots\dots\dots\dots\dots\dots\dots\dots \text{X.11}$$

$$C_2 = e^{-t/K} \dots\dots\dots\dots\dots\dots\dots\dots\dots\dots \text{X.12}$$

X.5.2   Application to South Branch Thames River

The application of this procedure is made to the discharge for the South Branch of the Thames River above Ealing at gauging station 2GD1 near London in southwestern Ontario. The area is 519 sq. mi. The soils are predominantly of moderately fine texture, formed on fine sands and silts.

The mathematical model is represented diagramatically in Fig. X.1 and Fig. X.2. The elements are these:

1. A deep element discharging directly to the stream (K = 4.5 months) and consisting of 78 per cent of the drainage basin.

2. A parallel deep element (K = 4.5 months) discharging to a shallow element, that consists of 10 per cent of the drainage basin.

3. A shallow element in series with element 2 (K = 0.5 months), consisting of 10 per cent of the basin.

4. An impervious element (K = 0) consisting of 2 per cent of the basin.

The streamflow is calculated for the April – October period of 1954, 1955 and 1956.

X.5.2.1   The Soil Moisture Balance

The soil moisture balance is computed for the basin on the basis of an upper and lower moisture zone. The upper zone has a maximum moisture capacity (UZS) of 1.00 in. available for evaporation.

The upper zone moisture is first depleted at the potential rate (PE), as computed by Penman's (1948) method from meteorological data. When the upper zone moisture is depleted (UZM = 0), evapotranspiration from the lower zone takes place at a reduced

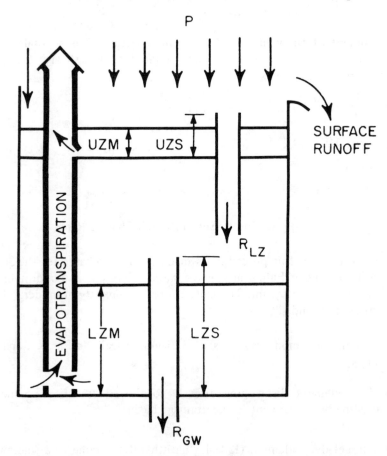

Fig. X.1   Soil Moisture Model – South Branch Thames River.

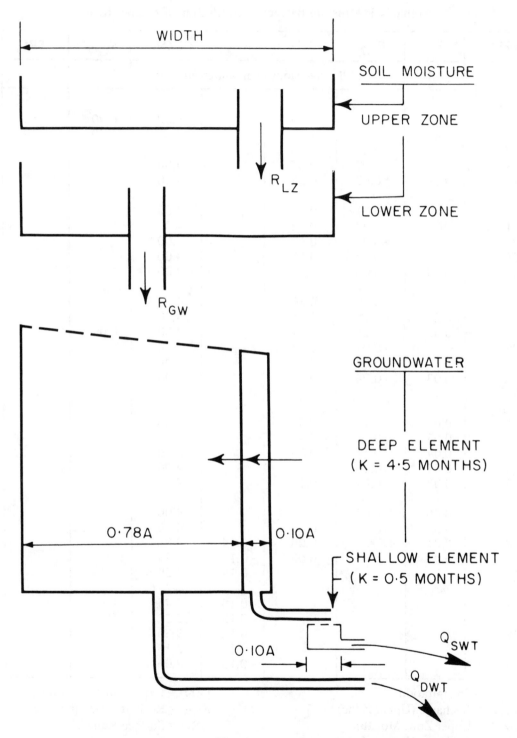

Fig X.2 Groundwater Discharge Model – South Branch Thames River.

Table X.1 Moisture Balance – South Branch Thames River

1954	P	R_{UZ}	UZM	R_{LZ}	LZM	R_{GW}	SRO
		(All dimensions in area-inches)					
April							
15	1.31	+ 0.34	1.00		4.00	0.40	0.50
16	0.66	0.00	1.00		4.00	0.40	0.19
17		- 0.07	0.93		4.00		
18		- 0.07	0.86		4.00		
19	0.09	+ 0.02	0.88		4.00		
20		- 0.07	0.81		4.00		
21		- 0.07	0.74		4.00		
22		- 0.07	0.67		4.00		
23		- 0.07	0.60		4.00		
24	0.04	- 0.03	0.57		4.00		
25		- 0.07	0.50		4.00		
26	0.22	+ 0.15	0.65		4.00		
27	0.42	+ 0.26	0.91		4.00	0.09	
28		- 0.07	0.84		4.00		
29		- 0.07	0.77		4.00		
30	0.03	- 0.04	0.73		4.00		
May							
1	0.01	- 0.14	0.59		4.00		
2	0.03	- 0.12	0.47		4.00		
3	0.04	- 0.11	0.36		4.00		
4	0.05	- 0.10	0.26		4.00		
5	0.08	- 0.07	0.19		4.00		
6	0.06	- 0.09	0.10		4.00		
7	0.05	- 0.10	0.00		4.00		
8	0.01		0.00	- 0.14	3.86		
9	0.02		0.00	- 0.13	3.73		
10				- 0.14	3.59		
11				- 0.14	3.45		
12				- 0.13	3.32		
13				- 0.12	3.30		
14				- 0.12	3.18		
15				- 0.12	3.06		

P   = Precipitation
R_{UZ} = Recharge (Upper Zone)
UZM = Upper Zone Moisture
R_{LZ} = Recharge (Lower Zone)

LZM = Lower Zone Moisture
R_{GW} = Recharge (Groundwater)
SRO = Surface Runoff

Table X.1 (continued) Moisture Balance — South Branch Thames River

Calculations based on:

UZS = Upper Zone Storage Capacity = 1.00 in.
LZS = Lower Zone Storage Capacity = 4.00 in.
AE = Evapotranspiration (Actual).
PE = Evapotranspiration (Potential — Penman's Method).

Assumptions

$AE = PE$ (when $UZM > 0$)
$AE = \dfrac{LZM}{LZS}$ x PE (when $UZM = 0$)
$R_{UZ} = P - AE$ (maximum of $1.00 - UZM$)
$R_{LZ} = P - R_{UZ} - PE$ (until $LZM = 4.00$)
$R_{GW} = P - R_{UZ} - R_{LZ} - PE$ (maximum of 0.40; when $LZM = LZS = 4.00$)
$SRO = P - R_{UZ} - R_{LZ} - PE - 0.40$ (when $LZM = 4.00$)

(All dimensions in area-inches)

rate. Evapotranspiration from the lower zone $= PE$ (LZM/LZS), where LZM is the balance of available moisture in the lower zone at any time. Table X.1 indicates the bookkeeping procedure.

A daily moisture balance is maintained throughout the season. With appropriate limitations to rates of groundwater recharge (R_{GW}), estimates of surface runoff may be made. During the typical, frequent dry summers in the east, these conditions are seldom attained. Streamflow is thus totally due to groundwater storage depletion. The moisture balance can be used to estimate the time in the autumn at which substantial increases in flow may be expected.

X.5.3   Synthetic Discharges

A sample calculation for routing recharge through the deep water table storage element is shown in Table X.2 for 1954. All dimensions are in area-inches, which are subsequently corrected for the area of the particular element as a proportion of the total watershed area. The composite discharges from impervious, shallow and deep storage elements, and from surface runoff, are shown in Table X.3.

The model is most easily established by selection of a dry year, such as 1954 or 1955, and by starting at some point where negative recharge (evaporation) will almost certainly deplete shallow water table elements so that outflow from these elements will be very small. The measured streamflow will then be from impervious surfaces, and outflow from deep storage elements. Then by adjustment, and trial and error, a suitable model is established and verified from existing flow records.

The major difficulty remains in applying the technique during years when no streamflow records are available. The apportionment of discharge from each type of

Table X.2 Routing Deep Water Table Outflow — South Branch Thames River

1954	P	AE	R_{GW}	$C_0 R_0$ ($C_0 = 0.103$)	$C_1 R_{n-1}$ ($C_1 = 0.096$)	$C_2 Q_{n-1}$ ($C_2 = 0.8$)	Q_{DWT}
			(All dimensions in area-inches)				
April	4.0	2.1	1.44				
May	0.7	3.7			0.14	0.30	0.44
June	2.6	3.0				0.35	0.35
July	1.0	2.0				0.28	0.28
Aug.	2.1	2.0				0.22	0.22
Sept.	2.4	1.7				0.18	0.18
Oct.	7.6	1.5	0.90	0.09		0.14	0.23

Table X.3 Composite Discharge — South Branch Thames River

1954	P	Q_{imp}	$0.1Q_{SWT}$	$0.78Q_{DWT}$	Q_{SRO}	Q_T(est.)	Q_T(obs.)
April	4.0	0.08	2.07	0.37	0.75	3.27	3.32
May	0.7	0.02	0.28	0.35		0.65	0.78
June	2.6	0.05	0.05	0.28		0.38	0.38
July	1.0	0.02	0.01	0.22		0.25	0.24
Aug.	2.1	0.04		0.18		0.22	0.21
Sept.	2.4	0.05		0.14		0.19	0.21
Oct.	7.6	0.15	0.28	0.18	1.55	2.16	2.39

Q_{imp} = (Discharge from Rainfall on Impervious Areas)
= $0.02P$
$0.1Q_{SWT}$ = Discharge from Shallow Water Table Elements
$0.78 Q_{DWT}$ = Discharge from Deep Water Table Elements
Q_{SRO} = Discharge from Surface Runoff
Q_T (est.) = Total Runoff (Q_{imp} + $0.1 Q_{SWT}$ + $0.78 Q_{DWT}$ + Q_{SRO})

Table X.4 A Comparison of Estimated and Observed Discharges for
the South Branch Thames River, during 1954, 1955 and 1956

Month	1954	1955	1956
	(All dimensions in area-inches)		
April	3.27 (3.32)	1.16 (1.52)	3.01 (2.67)
May	0.65 (0.78)	0.53 (0.52)	3.10 (3.42)
June	0.38 (0.38)	0.32 (0.29)	1.06 (0.50)
July	0.25 (0.24)	0.23 (0.24)	0.83 (0.82)
Aug.	0.22 (0.21)	0.24 (0.22)	1.38 (1.97)
Sept.	0.19 (0.21)	0.18 (0.23)	1.09 (1.72)
Oct.	2.16 (2.39)	0.44 (0.35)	0.92 (0.64)
TOTAL	7.12 (7.53)	3.10 (3.37)	11.39 (11.74)

Quantities in brackets are observed discharges.

storage element at the start of the dry season is particularly difficult in Canada, where winter conditions necessitate a consideration of snow storage. Discharge in March and April is believed to occur from shallow storage elements with very small storage delay time, and exists over the entire basin.

The agreement between estimated and observed discharges is quite satisfactory during the three years. Estimated and observed discharges for 1954, 1955, and 1956, are compared in Table X.4. The major discrepancies can be attributed to flood conditions, wherein a portion of the runoff is surface runoff, and another type of routing procedure is required. Temporal variation of rainfall intensity, of course, should be considered for improved estimates. A case in point is the 1956 season (a wet year). Even so, the seven-month totals agree remarkably well, and no doubt with additional refinements, the estimates could be improved still further.

The application of the model for predictive purposes is of course limited to the extent to which meteorological elements can be predicted. However, if a long record of meteorological observations is available, a streamflow record can be synthesized, and minimum flows predicted upon certain probability assumptions.

It seems that the procedure could be adopted to predict winter discharges with accuracies comparable to present gauging techniques.

Further study of the routing procedure, as applied to other streams, would probably lead to a few simple models typical of particular physiographic and geographic regions. This would then be a useful tool for synthesizing low-flow hydrographs for ungauged basins.

X.6   LITERATURE CITED

Anderson, H.W., 1955. Detecting hydrologic effects of changes in watershed conditions by double-mass analysis. Trans. Amer. Geophys. Union 36:119-125.

Anderson, H.W., and Gleason C.H., 1960. Effects of logging and brush removal on snow water runoff. Int. Assoc. of Sci. Hydrology, General Assembly of Helsinki I.U.G.G. Publ. No. 51:478-489.

Ayers, H.D. 1962. A survey of watershed yield. Report No. 63, Water Research Laboratory, University of New South Wales, Kensington, Australia, 1963.

Ayers, H.D., 1965. Effects of agricultural land management on winter runoff in the Guelph, Ontario region. Proc. of Hydrology Symposium No. 4. National Research Council Associate Committee on Geodesy and Geophysics. Subcommittee on Hydrology.

Ayers, H.D., 1966. Winter runoff in the Canadian Great Lakes Region. Paper prepared for presentation to the 21st Annual Meeting, Soil Conservation Society of America, Albuquerque, New Mexico.

Bailey, R.W. and Copeland, O.L., 1960. Low flow discharges and plant cover relations on two mountain watersheds in Utah. Int. Assoc. of Scientific Hydrology, General Assembly of Helsinki. I.U.G.G. Publ. No. 51:267-278.

Colman, E.A., 1953. Vegetation and watershed management. The Ronald Press Company.

Cross, W.P., 1949. The relation of geology to dry-weather flow in Ohio. Trans. Am. Geophys. Union 30:563-566.

Ding, J.Y.H., 1965. The flow characteristics of selected Ontario streams as related to soil and geologic conditions. Unpublished M.Sc. Thesis, University of Guelph, Guelph, Ontario, Canada.

Dooge, J.C.I., 1960. The routing of groundwater recharge through typical elements of linear storage. Int. Assoc. of Sci. Hydrology. General Assembly of Helsinki I.U.G.G. Publ. No. 52:286-300.

Glasspoole, J., 1960. Rainfall and runoff, Thames Valley, 1884-1949. Jour. Inst. of Water Engineers 14:185-186.

Gold, E., 1951. The relation between runoff and quarterly values of rainfall in the Thames Valley, Int. Assoc. of Sci. Hydrology, General Assembly of Brussels I.U.G.G. Publ. No. 34:235-236.

Goodell, B.C., 1958. Watershed studies at Fraser, Colorado. Proc. Amer. Soc. For. 42-45.

Goodell, B.C., 1964. Water management in the lodgepole pine type; Paper presented at Soc. Amer. For. Ann. Meeting, Denver, Col., September 30.

Grunsky, C.E., 1922. Rainfall and runoff studies. Trans. Amer. Soc. Civ. Engrs. 85:66-136.

Hays, O.E., 1955. Factors influencing runoff. Agr. Eng. 36:732-735.

Hoover, M.D., 1944. Effect of removal of forest vegetation upon water yield. Trans. Amer. Geophys. Union 25:Pt. 6, 969-977.

Justin, J.D., 1914. Derivation of runoff from rainfall data. Trans. Amer. Soc. Civ. Engrs. 77:346-384.

Kittredge, J., 1948. Forest influences, McGraw-Hill, New York.

Law, F., 1955. Estimation of the yield of reservoired catchments. Jour. Inst. Water Engrs. 9:467-493.

Lieberman, J.A. and Hoover M.D., 1951. Streamflow frequency changes on Coweeta Experimental Watersheds. Trans. Amer. Geophys. Union. 32:73-76.

Love, L.D., 1955. The effect on streamflow of the killing of Spruce and Pine by the Engelmann spruce beetle. Trans. Amer. Geophys. Union 36:113-118.

Lvovitch, M.I., 1957. Streamflow formation factors. Int. Assoc. of Sci. Hydrology General Assembly of Toronto I.U.G.G. Publ. No. 45:122-132.

McDonald, C.C. and Langbein, W.B., 1948. Trends in runoff in the Pacific Northwest. Trans. Am. Geophys. Union 29:387-397.

Penman, H.L., 1948. Natural evaporation from open water, bare soil and grass. Proc. Royal Soc. (London) 193:120-145.

Serra, L., 1954. Étude des facteurs géologiques conditionnant l'écoulement. Int. Assoc. of Sci. Hydrology Publ. No. 38, General Assembly of Rome I.U.G.G., 446-455

Sharp, A.I. and Gibbs, A.E., Owen, W.J., and Harris, B., 1960. Application of the multiple regression approach in evaluating parameters affecting water yield of basins. Trans. Amer. Geophys. Union 65:1273-1286.

Sharp, A.I. and Gibbs, A.E., Owen, W.J., and Harris B., 1966. Development of a procedure for estimating the effects of land and watershed treatment on streamflow. Tech. Bull. 1352, U.S. Dept. of Agric. and U.S. Dept. of the Interior.

Siren, A., 1960: Occurence of low discharge periods in rivers in Finland. Int. Assoc. of Sci. Hydrology, General Assembly of Helsinki I.U.G.G. Publ. No. 51:211-214.

Stafford, H.M. and Troxell, H.C. 1944. Differences in basin characteristics as reflected by precipitation — runoff relations in San Bernadino and Eastern San Gabriel mountain drainages. Trans. Amer. Geophys. Union 25:21-35.

Sutcliffe, J.V. and Rangeley, W.R., 1960. Variability of annual river flow related to rainfall records. Int. Assoc. Sci. Hydrology, General Assembly of Helsinki I.U.G.G. Publ. No. 51:182-192.

Turc, L., 1954. Calcul du bilan de l'eau évaluation en fonction des précipitations et des températures. Int. Assoc. of Sci. Hydrology Publ. No. 38. General Assembly of Rome I.U.G.G. 188-202.

Wundt, W., 1937. Beziehungen zwischen den mittelwerten von niederschlag, abfluss, verdunstung und lufttemperatur für die landflaechen der erde — Deutsche Wasserwirtschaft, Heft 6.

Young, V.D., 1948. Rainfall-runoff relationships for small watersheds. Agr. Eng. 29:212-214.

Section XI

SEDIMENT TRANSPORTATION

by

John M. Wigham

TABLE OF CONTENTS

LIST OF TABLES

LIST OF FIGURES

Section XI

SEDIMENT TRANSPORTATION

XI.1   INTRODUCTION

The movement and/or deposition of sediment is of interest to the hydrologist because it affects the overall management of watersheds. Fertilizer can be removed by erosion; or by excessive erosion or deposition of top soil, an area can be destroyed in an agricultural sense. Changes in sediment transport rates, or volumes, can produce changes in the flow conditions in channels. Man's activities on, or within, a watershed can accelerate or decelerate erosion, and can affect the operation of water control structures. All these items require estimates of sediment erosion, or deposition rates and volumes, and some consideration of the effects of changing flow conditions to allow for the evaluation of the effects of sediment movement.

In a broad sense, sediment may be defined as any fragmental material transported by, suspended in, or deposited by, such natural agents as water, air or ice. Water is, in general, the most widespread agent of sediment transport, so only water transport will be considered.

The process of the transportation of soil particles from a land surface begins by the action of precipitation. The impact of falling raindrops breaks down the large soil lumps into single grains, which may be thrown upward by the energy of the drops. There may be a net transport of such particles, in a particular direction, depending on the land slope and wind effects. The particles may also be carried in suspension by overland flow and the force of the water, flowing over the land surface, may loosen additional particles to be transported toward the stream channels of the watershed. The combination of these processes results in so-called *sheet erosion* from the land surface.

Generally, the overland flow concentrates in small channels or rills. This concentration of flow may increase the erosive power to more than is required to transport the incoming particles, in which case severe channel or gully erosion will occur. In any case, some of the material transported by overland flow will eventually enter the main drainage channels of the area, and be transported or deposited. The flow conditions and processes are different for channel flow than for overland flow, so this aspect of sediment transport will be considered in Subsection XI.3.

XI.2   SHEET EROSION

As previously mentioned, this process begins with the raindrops breaking down large soil lumps, and detaching and moving soil particles. Each raindrop has considerable kinetic energy, and there is a relationship between the energy available and the number of particles detached from the soil surface. The kinetic energy in any single drop is

$$KE = \frac{1}{2} m V_d^2 \quad \dots\dots\dots\dots\dots\dots\dots\dots\dots\dots\dots\dots\dots\dots\dots XI.1$$

where KE = kinetic energy,
m = mass of the drop, and
V_d = drop velocity which is normally the
terminal velocity of the drop.

The mass of a drop is proportional to the cube of the diameter. The terminal velocity is also a function of the drop diameter (see Table XI.1), so the energy of a single drop increases rapidly as the drop size increases. Some evidence obtained from experimental studies has shown that drop size increases only slightly with rainfall intensity for a given locality. However, the effect of this variation, coupled with the increased total mass of rainfall, causes the energy or erosive power to increase fairly rapidly with rainfall intensity.

Table XI.1 Terminal Velocity of Waterdrops

Drop Diameter (m.m.)	Terminal Velocities (ft./sec.)	
	(after Lenard)	(after Laws)
0.5	11.5	
1.0	14.4	
1.5	18.7	18.1
2.0	19.4	21.6
3.0	22.6	26.4
4.0	25.3	29.1
5.0	26.2	30.3
5.5	26.2	30.5
6.0	25.9	30.5
6.5	25.6	

(From Linsley *et al.*, 1949)

It is difficult to calculate the total kinetic energy available in rain striking a surface; but it is sufficient to lift the upper soil layers several feet, and is generally greater than the energy involved in overland flow. Fortunately, much of this energy is dissipated by vegetation, stones, plant residue and water covering the soil surface. In addition, some of the energy is used to compact the soil. Studies conducted by Ellison (1944, 1945) have indicated that soil particles may move as much as 5 ft. horizontally and 2 ft. vertically due to the splash effect. This movement of particles, on a ten-degree slope, resulted in about three times as much transport downhill as uphill. If the wind direction is also downslope, appreciable quantities of material could be transported downhill.

The resistance of the soil to particle detachment is important. In general, for non-cohesive materials, the resistance to movement by raindrop impact increases with particle size. For cohesive materials, which are generally fine-grained, the resistance to detachment may be very high. Vegetation may increase the apparent cohesiveness; because the root system acts as a soil binder, with fine particles of soil adhering to the system.

Soil particles detached by rainfall impact may be moved downslope by overland flow. The ability of the flow to move soil particles is related to the flow depth, velocity and turbulence characteristics; all of these are related to the rainfall intensity and volume, and the land slope and roughness. In some cases, the tractive force may be sufficient to detach and transport additional particles from the soil surface; in other cases, deposition may occur. In general, smaller particles are more easily transported than large particles. A soil may have, therefore, a high resistance to detachment (cohesive), but a low resistance to transport (small particle size) by overland flow. The net resistance to detachment and overland flow transport is a measure of the soil's resistance to erosive forces. It should be noted that, generally, only a portion of the soil eroded from an area reaches a stream channel.

XI.2.1   Equations for Predicting Rates of Sheet Erosion

Many experiments have been conducted to determine the rate of detachment of soils and their resistance to transport. In general, such tests are conducted on small plots with artificially applied or natural rainfall. Splash samplers, either horizontal or vertical, may be used to determine the weight of detached particles. The total sheet erosion from a plot may be determined by enclosing the plot in any manner that will allow entrapment of all soil particles leaving the area.

Ellison (1945) conducted a number of tests to determine the weight of soil moved by raindrop impact. The collected data allowed some definition of the relationship between weight of soil moved, the velocity and diameter of the raindrops, and the rainfall intensity. The following equation was proposed:

$$G = K \, V_d^{4.33} \, d^{1.07} \, i^{0.65} \quad\dots\dots\dots\dots\dots\dots\dots\dots\dots\dots\dots\dots\dots\text{XI.2}$$

where G = soil intercepted in splash samplers
during a 30 minute period (grams),
V_d = velocity of the drop (ft/sec.),
d = diameter of the drops (mm),
i = intensity of rainfall (in/hr), and
K = constant.

The tests were run on a soil devoid of vegetation, so the equation can be used only for the particular soil tested and for bare soil conditions. However, it shows the relative effects of rainfall intensity, drop velocity and diameter.

Musgrave (1947) determined amounts of soil loss from a number of small plots for a wide range and number of rainfall conditions. The results of the study indicated the following relationship:

$$E_S = I_S CS^{1.35} L^{0.35} P_{0.5}^{1.75} \quad \dots\dots\dots\dots\dots\dots\dots\dots\dots\dots\dots \text{XI.3}$$

where E_S = soil loss (acre-in)
    I_S = inherent erodibility of the soil (in),
    C = cover factor,
    S = degree of slope (per cent),
    L = length of slope, and
    $P_{0.5}$ = maximum 30 minute amount of rainfall with
        a two-year frequency (in).

This equation gives gross, long-term average sheet erosion volumes, and because of the conditions of derivation, should apply for a broad range of soil types. The terms I_S and C are difficult to assess for conditions not covered by the tests.

The U.S. Department of Agriculture (1961) has developed a so-called universal equation for prediction of short-term rates of soil loss. The equation is

$$E_S = P_S I_S L_S CM_S \quad \dots\dots\dots\dots\dots\dots\dots\dots\dots\dots\dots\dots \text{XI.4}$$

where E_S = average annual soil loss (tons/acre),
    P_S = rainfall factor,
    I_S = soil erodibility factor,
    L_S = slope length and steepness factor,
    C = cropping and management factor, and
    M_S = factor to evaluate supporting conservation
        practices such as terracing, strip cropping and
        contouring.

The rainfall factor, P_S, is determined by the following relationship, as reported by Wischmeier and Smith (1958):

$$P_S = \frac{\sum KEi}{100} \quad \dots\dots\dots\dots\dots\dots\dots\dots\dots\dots\dots\dots\dots \text{XI.5}$$

where KE = the storm energy (ft. tons/acre in), and
    i = maximum 30 min. intensity (in/hr).

The equation for KE, given previously (Equation XI.1), is a function of rainfall intensity.

The rainfall factor can be determined fairly simply, therefore, from existing rainfall statistics for a region.

The soil erodibility factor, I_S, is expressed as the soil loss in tons/acre for each unit of rainfall erosion index for the locality, and for continuous fallow tillage on a 9 per cent slope, 73 feet in length. This factor may be estimated from reported values for similar soil types, or may be calculated from test plot results.

The slope, length and steepness factor, L_S, corrects for slopes and lengths of plot that are different than the standard 9 per cent, 73-foot long test plots. It may be evaluated from a graph of slope length versus L_S for various slope conditions (see U.S. Department of Agriculture, 1961; or Chow, 1964).

The cropping and management factor, C, is difficult to evaluate because of the number of possible cropping sequences. There will be variations in this factor from locality to locality even for identical crop sequences, if there are differences in the rainfall-energy-time distribution curves. This term must be determined, therefore, for each specific area of interest. A base value of C = 1 is used for clean, straight row fallow, and for up and downhill cultivation. All other values of C are, therefore, less than one.

The conservation practice factor, M_S, is also equal to one for up and downhill cultivation, and will be less than one for other conditions (see Table XI.2).

Table XI.2 Some Values of Conservation Practice Factor, M_S

Condition		M_S
Contouring,	slope $< 12\%$	0.5 - 0.6
	slope 12 to 18%	0.8
	slope 18 to 24%	0.9
	slope $> 24\%$	1.0
Strip Cropping and Terracing		
	slope $< 12\%$	0.25 - 0.3
	slope 12 to 18%	0.3 - 0.4
	slope 18 to 24%	0.4 - 0.45

Equation XI.4 has found some acceptance in the United States, and apparently gives fairly good results where there are sufficient data to evaluate the constants. The constants in the equation are gradually being evaluated for different regions. However, application of the equation to Canadian conditions is difficult because of variations in soil classification systems, and because of the scarcity of data for different regional conditions.

XI.3   CHANNEL TRANSPORT

The preceding Subsection has been concerned primarily with evaluation of sheet erosion volumes. As mentioned, only a portion of this material reaches the major drainage channels because of deposition in low areas, vegetational areas, at fence lines, on the channel flood plain, and at breaks in land slope or even in the channel itself. The total amount of eroded material, which completes the journey from the source area to a downstream channel control point, is known as the sediment yield. This is normally expressed in units of weight (tons) or volume (acre-ft.) and, of course, is a function of the size of the contributing area. Comparisons of data on sediment yield are generally made on the basis of yield per unit contributing area, which is called the sediment production rate expressed in tons or acre-ft./sq. mi.

The yield and the gross erosion (sheet erosion plus erosion due to gullying, channel erosion or other causes) are, of course, related. The relationship may be expressed as the ratio of sediment yield to gross erosion; this ratio is called the sediment delivery ratio.

The yield from a particular watershed may be determined by measurement of sediment transport at a control point on the main channel, or by using empirical or semi-empirical equations. Most of the existing equations for determining rates of sediment transport in a channel have been developed either by correlating measured sediment-yield quantities with the precipitation and topographic characteristics of the watershed, or through semi-theoretical analyses relating channel flow characteristics to measured sediment yields.

The process of transportation of sediment in channels is rather complex, so measurement of transport rates still provides the best estimates of yield. However, some transport equations based on theoretical analysis are useful, and will become more so as more data become available. The accuracy of yield estimates increases, of course, with the length of period for which the yield is determined.

XI.3.1   Measurement of Sediment Transport in Channels

The sediment load transported past a given channel section can be considered to consist of wash load, suspended sediment load and bed load. The wash load consists of very fine or colloidal particles, which settle very slowly even in still water. This type of material is represented in the bed material only in very small amounts, hence the supply of wash load is limited. The normal turbulent flow in channels has a very large capacity for carrying wash load, so the amount of wash load carried is a function only of the amount supplied to the channel and available on the bed.

The suspended sediment load and the bed load are sometimes grouped together and called the bed material load, as they are made up of particles which are found in appreciable quantities in the bed material. The suspended load may be defined, however, as that sediment which does not spend any time on the bed of the channel (excluding wash load). The bed load is then defined as that part of the load which moves by rolling, sliding or saltation (hopping) along the bed. The amount of bed material load is governed by the flow conditions, as the bed material represents an adequate supply to maintain transport at the channel capacity.

The three different processes of sediment transport can affect the accuracy of transport measurements, particularly if use is made of a sediment discharge versus water discharge rating curve.

The total of the suspended load and wash load is comparatively easily measured. Since sediment particles move at essentially the same velocity as the flow; so measurements of sediment concentration, when combined with measurements of water discharge, yield the transport rate.

Special instruments and sampling methods have been developed to provide accurate estimates of suspended load. The most commonly-used instrument consists of a metal body within which there is a sample bottle connected to an intake nozzle (Subcommittee on Sedimentation, 1959). The intake nozzle (or the air exhaust port) is designed so that the flow velocities within the nozzle are the same as the water flow velocities at the sampling point.

The sampler is usually designated as either a depth-integrating or a point-integrating sampler, as there are some differences in construction depending on the type. The main difference lies, however, in the method of operation to obtain a sample. The depth-integrating sampler is lowered at a constant rate from the surface of the stream to the bed, and is raised at the same rate back to the surface. The sediment concentration in the sample is, therefore, the average concentration over the flow depth for the vertical sampled. The suspended sediment and wash load is this average concentration times the water discharge associated with the vertical. Samples are normally taken at five or more verticals across the channel (Department of Northern Affairs and National Resources, 1965), and the total suspended load plus wash load is determined by summing the loads calculated for each vertical.

The point-integrating sampler is lowered to a particular point in the flow, and the sample bottle fills at that point. The resulting sample concentration represents the time-averaged concentration at that point in the flow. Samples may be obtained for a number of points in a vertical to define the sediment concentration versus depth curve for the vertical. The concentration at any point may be multiplied by the corresponding flow velocity, and may be plotted against depth. The area beneath this curve, divided by the depth and multiplied by the flow area associated with the vertical, will give the sediment load for the vertical. This method may be used to check on the results of depth-integration sampling, and to determine particle size distribution in the vertical.

The measurement of bed load is more difficult than measurement of suspended load, because (a) the particles do not move at the same speed as the flow; (b) there can be considerable variation in the transport rate with time because of bed configuration effects; (c) any instrument placed on or near the bed can change the flow conditions locally, resulting in incorrect load measurements; and (d) if the instrument is sampled within the saltation zone, some of the sample may be suspended material. Because of these difficulties, bed-load measurements, as such, are not often obtained, though some attempts have been made to develop bed-load samplers. The quantity of bed load may be estimated, however, by comparing the total load determined from the Modified Einstein equation to the measured suspended load. Alternatively, a special structure may be constructed in the channel to cause increased flow velocities and turbulence, such that all the bed load becomes suspended. The total load may then be measured using suspended-load measurement techniques. It is often not economically feasible, however, to construct such turbulence flumes in natural channels.

The sediment transport rate, which is eventually determined, strictly applies for the time during which the measurement was taken. If the stream discharge is not changing rapidly, then one measurement of the sediment transport rate may suffice to define the average rate for a day. If the stream discharge is changing rapidly and the sediment rate is high, then several measurements of the rate should be taken to provide an accurate assessment of the average daily transport. Generally, it is too time-consuming to use either depth or point integration methods under these conditions, so only one or two samples are taken at selected points in the flow. A correlation between measured concentrations at the selected points and the overall average concentration can be developed from previous, more complete measurements. The average concentration for the whole cross section may then be obtained from the correlation. This correlation procedure is used in the Canadian sediment sampling programme to reduce the sampling time and expense.

Once daily rates of sediment transport are known, the seasonal or annual yields may be obtained for the catchment by summing the daily rates. The annual sediment yield often correlates well with the average annual water discharge. If this is the case, then variations away from the correlation may indicate and allow evaluation of the effects of changing conditions on the catchment.

Seasonal or annual sediment yields may also be determined from measurements of bed level changes in reservoirs through which the channel passes. Periodic (annual) surveys at selected cross sections of a reservoir, together with observations of the density of deposited materials, will provide estimates of the amount of sediment deposition in the reservoir. The deposited material represents only a portion of the total annual sediment transport, however, as some sediment will normally be carried through the reservoir by the outflowing water. The amount that is carried through depends on the size and extent of the reservoir, the magnitude of the outflows, the properties of the sediment and the nature of the outlet. The first two factors govern or influence the detention storage time, the time during which deposition may occur in the reservoir. The detention storage time, in conjunction with the fall velocities of the sediment particles, is the main factor influencing sediment outflow. The location of the outlet in the dam may have an effect, particularly if the outlet is at a low level so that flow can occur from regions of higher sediment concentration.

An estimate of the efficiency of storage reservoirs in trapping sediment has been made by Brune (1953), who related the percentage of sediment trapped to the ratio of reservoir capacity (acre-ft.) and annual inflow (acre-ft.)—see Fig. XI.1 The measured volume of sediment deposit in a reservoir may be divided by the estimated percentage of

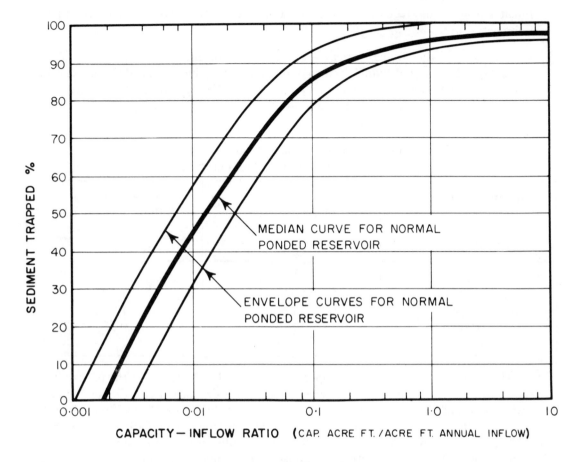

Fig. XI.I Trap Efficiency as Related to Capacity-Inflow Ratio.

sediment trapped to give the sediment yield for the catchment. The results of such calculations should be checked, if possible, by measurements of sediment transport downstream of the reservoir, as the percentage of sediment trapped may be different from that given by Brune because of differences in reservoir shape and operation.

XI.3.2   **Sediment Yield as Related to Rainfall and**
                    Topographic Characteristics

The equations that have been developed relating sheet erosion to rainfall, topographic, soil and vegetation characteristics have, logically, encouraged attempts to relate sediment yield to similar basin features. The results of such attempts have proved useful in the regions from which they have been derived, but the wide variation in conditions that affect sediment yield practically preclude using such equations on other basins. It is instructive to examine the basin characteristics, which researchers have found to be important, as this may provide a guide to the development of yield-rainfall-topographic relationships on other basins. It may also indicate the features to be considered to obtain estimates from an ungauged catchment by comparison with measured yields from nearly catchments.

The size of the drainage area is obviously important with respect to the total sediment yield from a catchment. It may or may not be important with respect to the sediment production rate, or the delivery ratio. Table XI.3 produced by Gottschalk (see Chow, 1964) would indicate that it is only the apparent relationship which may result through the action of other variables.

Table XI.3 Arithmetic Average of Sediment-Production Rates for Various Drainage Areas in the U.S.

Watershed size range (sq. mi.)	No. of Measurements	Average Annual Sediment Production Rate (acre-ft./sq. mi.)
< 10	650	3.80
10 – 100	205	1.60
100 – 1000	123	1.01
> 1000	118	0.50

The indicated relationship is supported, however, by the fact that the probability of deposition of sheet erosion particles will increase with drainage area size, partly because of wider flood-plain areas for deposition. In addition, the probability of complete storm coverage, with resulting high erosion rates, is lower for larger areas. Gottschalk and Brune (1950) found that the sediment production of the Columbia River Basin varies as the 0.8 power of the drainage area. Other researchers have found sediment production to vary from the 0.6 to 1.1 power of the drainage area.

The delivery of the sheet erosion quantities to streams defines the sediment yield of a catchment. The channel density within the catchment should, therefore, be important as a high channel density means that the distances the erosion products must move to reach streams are shortened. This characteristic is borne out through studies by Gottschalk (1946), and by Stall and Bartelli (1959). The significant factor was channel density expressed in ft./acre.

The land slope within a watershed also influences the delivery of eroded particles to stream channels; larger slopes increase the possibility of movement through the watershed. Several methods have been proposed to define slope effects, and a number of researchers have found the relief ratio, defined as the maximum relief (difference between maximum and minimum elevations in the watershed) divided by the maximum length of watershed, to correlate very highly with sediment production rates.

Obviously, the precipitation conditions with respect to amount, intensities and seasonal distribution, and the resulting runoff conditions, affect the sediment yield. These factors may be taken into account implicitly through the other characteristics mentioned, as water yield from a catchment may also be related to drainage area, channel density and watershed relief.

XI.3.3   Sediment Transport Equations

There have been many attempts made to relate sediment transport rates to the flow conditions in open channels, since it is relatively easy to obtain measurements of the flow parameters. A great deal has been accomplished, in this regard, in developing theories of suspended-load transportation through consideration of the mechanics of turbulent flow of liquids. The development of theoretical equations for bed-load transport has been less successful because of the complexity of the transport process, and because there are really no theories which adequately describe the flow conditions in the region of the bed. The wash load portion of the sediment load cannot be evaluated from consideration of flow properties, as the wash load is simply a function of the amount of fine material supplied to the channel. In many rivers, the wash load forms the largest percentage of the total load, and since the amounts cannot be calculated from equations, measurements must be made and statistical analyses used to predict future loads.

XI.3.3.1   Suspended Sediment Transport Equations

These equations were developed from turbulence theories as presented by Reynolds, Boussinesq, Von Kármán and others. Vanoni (1946) indicated the theoretical development of an equation for suspended sediment concentration at any point in the flow, and showed that experimental data agreed well with the theory. The basic concept is that sediment particles are supported and distributed in the flow through the mechanism of turbulent exchange. The vertical velocity fluctuations are equal, both up and down; but the upward velocity components originate from regions of high sediment concentration, so there can be a net upward exchange. Superimposed on this action is the condition that the sediment particles are continuously settling toward the channel bed. The net effect is a condition of stability in which the sediment concentration decreases from the bed of the channel towards the surface. The concept is similar to that of momentum exchange for two-dimensional turbulent flow in channels, so the sediment concentration equations should be analagous to those for momentum transfer.

The Von Kármán equation for the apparent turbulent shearing stress in turbulent flow is

$$\tau = \beta \rho u' l \frac{dv}{dz} \quad \dots\dots\dots\dots\dots\dots\dots\dots\dots\dots\dots\dots\dots\dots\dots\dots\dots\dots\dots \text{XI.6}$$

where τ = shearing stress on a horizontal plane
        in the fluid,
β = correlation coefficient (momentum),
u' = average of the absolute values of the velocity
        fluctuations normal to the main flow,
l = mixing length, and
$\frac{dv}{dz}$ = change in flow velocity with distance, z,
        from the channel bed.

This equation expresses the rate of transfer of momentum due to turbulent mixing, so by analogy, a similar equation may be written for the transfer of suspended particles:

$$G = -\beta \, u' \, l \frac{dC}{dz} \quad \dots \dots \dots \dots \dots \dots \dots \dots \dots \dots \dots \dots \dots \text{XI.7}$$

where G = rate of transfer of mass of suspended
 particles per unit area, and
 $\dfrac{dC}{dz}$ = change in sediment concentration with
 distance, z, from the bed.

For steady conditions, the upward transport of material due to turbulence is balanced by the settling due to the force of gravity. This is expressed by

$$-\beta u' \, l \frac{dC}{dz} = wC \quad \dots \dots \dots \dots \dots \dots \dots \dots \dots \dots \dots \dots \dots \text{XI.8}$$

where C = concentration of the fraction of the sediment
 with a fall or settling velocity, w.

The differential equation for suspended sediment becomes

$$wC + E_s \frac{dC}{dz} = 0 \dots \dots \dots \dots \dots \dots \dots \dots \dots \dots \dots \dots \dots \text{XI.9}$$

where E_s = $\beta u' \, l$ and is the sediment transfer coefficient
 which is assumed equal to the momentum
 transfer coefficient.

Integrating Equation XI.9, we get

$$\ln\left(\frac{C}{C_a}\right) = -w \int_a^z \frac{dz}{E_s} \quad \dots \dots \dots \dots \dots \dots \dots \dots \dots \dots \text{XI.10}$$

where C_a = concentration at an arbitrary reference
 level z = a.

If it is assumed that $E_s = E_m = \beta u' \, l$, then Equation XI.6 shows that

$$E_s = \frac{\tau}{\rho \dfrac{dv}{dz}} \quad \dots \dots \dots \dots \dots \dots \dots \dots \dots \dots \dots \dots \text{XI.11}$$

Introducing this into Equation XI.10 yields the relation:

$$\ln\left(\frac{C}{C_a}\right) = -w\rho \int_a^z \left(\frac{1}{\tau}\frac{dv}{dz}\right) dz \quad \dots \dots \dots \dots \dots \dots \dots \text{XI.12}$$

The shearing stress, τ, in a uniform open-channel flow with a large width to depth ratio, is given by

$$\tau = \tau_O \left(\frac{d-z}{d} \right) \dots\dots\dots\dots\dots\dots\dots\dots\dots XI.13$$

where d = total flow depth,

z = distance of a point, above the channel bed, and

$\tau_O = \Upsilon \, dS$, the shear at the bed of the channel.

An expression for the velocity gradient $\frac{dv}{dz}$ may be obtained, assuming a logarithmic velocity distribution, as

$$\frac{dv}{dz} = \frac{1}{k_O} \sqrt{\frac{\tau_O}{\rho}} \, \frac{1}{z} = \frac{V*}{k_O z} \dots\dots\dots\dots\dots\dots XI.14$$

Substituting Equations XI.13 and XI.14 in Equation XI.12 gives

$$\ln \left(\frac{C}{C_a} \right) = - \frac{w}{k_O V*} \int_a^z \frac{dz}{z \left(\frac{d-z}{d} \right)} \dots\dots\dots\dots\dots\dots XI.15$$

Integrating this equation gives

$$\frac{C}{C_a} = \left[\frac{a(d-z)}{z(d-a)} \right]^{\mathbf{z}} \dots\dots\dots\dots\dots\dots\dots XI.16$$

where $\mathbf{z} = \frac{w}{k_O V*} \dots\dots\dots\dots\dots\dots\dots\dots XI.17$

Equation XI.16 has been found to describe the sediment distributions very well, both in the laboratory and in prototype streams, even under conditions of high sediment concentrations (above 100 grams per litre) near the bed. It cannot apply for a = o as this would indicate an infinite concentration near the bed. At z = d (at the surface) the equation indicates zero concentration, whereas actual observations show a certain concentration perhaps due to secondary currents and eddies.

Equation XI.17 contains the Von Kármán coefficient, k_O, which is taken equal to 0.4 for clear water, but may be quite variable for sediment-laden streams. Thus, the accuracy of Equation XI.16 depends on the accuracy with which k_O can be evaluated.

The value of 'a' may be assumed to be 0.05d or 2D, where D is a representative bed-material grain size normally taken to be d_{65}.

The total rate of transport, q_s, over part of a vertical, or between z = a and z = d, may be obtained by integration of the product of the flow velocity, v , and the concentration, C, for a unit width of the channel; that is

$$q_s = \int_a^d vC \, dz \dots\dots\dots\dots\dots\dots\dots\dots XI.18$$

The flow velocity, v, may be considered to obey a logarithmic distribution (to be consistent with the assumed value of dv/dz). Assuming the Keulagan equations apply, then

$$v = 5.75 \, V^* \log\left(\frac{30.2z}{d_{65}}\right) \dots\dots\dots\dots\dots\dots\dots\dots\dots\dots\dots\dots\dots\dots \quad \text{XI.19}$$

Substituting this in Equation XI.18 and integrating, gives

$$q_s = 5.75 \, V^* \, d \, C_a \left(\frac{A}{1-A}\right)^z \left[\log\frac{30.2dx}{d_{65}} \int_A^1 \left(\frac{1-z}{z}\right)^z dz \right.$$
$$\left. + \int_A^1 \left(\frac{1-z}{z}\right)^z \log z \, dz \right] \quad\dots\dots\dots\dots\dots\dots\dots\dots\dots\dots\dots \quad \text{XI.20}$$

where $A = \frac{a}{d}$ and x is a correction factor for hydraulically-smooth conditions.

If the integrals are defined as

$$I_1 = 0.216 \frac{A^{z-1}}{(1-A)^z} \int_A^1 \left(\frac{1-z}{z}\right)^z dz$$

and

$$I_2 = 0.216 \frac{A^{z-1}}{(1-A)^z} \int_A^1 \left(\frac{1-z}{z}\right)^z \ln z \, dz$$

then

$$q_s = 11.6 \, V^* \, C_a \, a \left[\ln\left(\frac{30.2dx}{d_{65}}\right) I_1 + I_2\right] \quad\dots\dots\dots\dots\dots\dots\dots \quad \text{XI.21}$$

The integrals, I_1, and I_2, have been evaluated and plotted (see Chow, 1964) for z values from zero to five and for A values from 10^{-5} to 10^{-1}, with 'a' being equal to twice the diameter of a representative bed particle.

The flow and sediment data required are flow depth and water surface slope, sediment concentration at the reference level, a, a value for the fall velocity of the sediment, and the size distribution of the bed particles. The value of the fall velocity, w, (and therefore the value of z) changes with particle size, so Equation XI.21 is applied for several increments of particle size as defined by the bed material size distribution.

The fall velocity, w, for spherical grains is given by the following equations:

Laminar Flow

$$w = \frac{g \, D^2}{18\nu}\left(s - 1\right) \dots\dots\dots\dots\dots\dots\dots\dots\dots\dots\dots\dots\dots\dots \quad \text{XI.22}$$

Turbulent Flow

$$w = \sqrt{\frac{4}{3}\frac{gD}{C_r}(s\text{-}1)} \quad \dots\dots\dots\dots\dots\dots\dots\dots\dots\dots\dots\dots\dots \text{XI.23}$$

where w = settling or fall velocity (cm/sec),
          s = specific gravity of the grain with respect to the fluid,
          D = grain diameter, (cm),
          g = acceleration of gravity, (980 cm/sec²),
          ν = kinematic viscosity of the fluid, (cm²/sec), and
         C_r = coefficient of resistance which varies with the Reynolds Number.

Natural sediment grains are not spherical, but the form of Equations XI.22 and XI.23 may be used with a correction or consideration for particle shape. Rubey (1933) derived the constants, in Equations XI.22 and XI.23, for natural grains, so his equations or graphs of settling velocity versus grain size may be used to define w in Equation XI.21. (Fig. XI.2)

XI.3.3.2 Bed Load and Total Load Equations

Motion of the sediment particles in the bed layer cannot be described by the theory

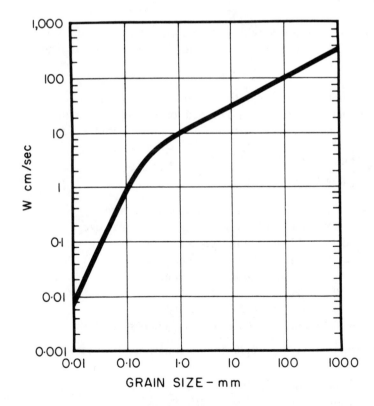

Fig. XI.2, Settling Velocity, w, for Quartz Grains of Various Sizes (Rubey, 1933)

of suspension; the reason is that these particles are not suspended in the fluid. While moving, their weight is supported by the non-moving bed.

From laboratory experiments it has been suggested that the motion of the bed particles in water is governed by statistical laws, which can be stated:

1.   The probability of a given-sediment particle being moved by the flow from the bed surface depends on the particle's size, shape and weight, and on the flow pattern near the bed.

2.   The particle moves if the instantaneous hydrodynamic lift force overcomes its weight.

3.   Once in motion, the probability of the particle being redeposited is equal in all points of the bed where the local flow would not immediately remove the particle again.

4.   The average distance travelled by a bed-load particle between consecutive points of deposition in the bed is constant and independent of the flow condition, the rate of transport, and the bed composition.

5.   The motion of bed particles by saltation may be neglected.

6.   The disturbance of the bed surface by moving sediment particles may be neglected.

Therefore, the variables, which at any spot in the bed determine the bed load, are the composition of the bed in the area, and the flow condition near the bed in this same area.

Einstein (1950) derived equations defining the bed load from consideration of the possibilities of particle movement. The bed-load transport per unit of time and channel width for a particular size fraction was given in a dimensionless function Φ_* where

$$\Phi_* = \frac{i_B}{i_b} \left[\frac{q_B}{\rho_s g} \left(\frac{\rho}{\rho_s - \rho} \right)^{1/2} \left(\frac{1}{g D^3} \right)^{1/2} \right] \quad\dots\dots\dots\dots\dots\dots\dots XI.24$$

where i_B = fraction of bed load in particular size,
$\quad\quad i_b$ = fraction of bed material in particular size,
$\quad\quad q_B$ = bed load rate in weight per unit time
$\quad\quad\quad\quad$ and per unit width of channel,
$\quad\quad \rho_s$ = mass density of the sediment,
$\quad\quad \rho$ = mass density of the fluid,
$\quad\quad g$ = acceleration due to gravity, and
$\quad\quad D$ = grain diameter of the particular size fraction.

Einstein found that Φ_* was a unique function of a hydraulic parameter ψ_* which is defined by

$$\psi_* = \xi \, Y \left[\frac{\log 10.6}{\log \left(\frac{10.6 Xx}{d_{65}} \right)} \right]^2 \frac{\rho_s - \rho}{\rho} \frac{D}{R'_b S_e} \quad\dots\dots\dots\dots\dots\dots\dots XI.25$$

where ξ = a correction of effective flow for
various grains,

Y = a correction of lift force in transition
between hydraulically rough and smooth beds,

X = a reference grain size for the particular bed,

R'_b = hydraulic radius of the bed for grain roughness
or surface drag, and

S_e = slope of the total energy line for the given
flow conditions.

Einstein also derived an equation for the reference concentration of suspended load, C_a, at a distance 2D from the bed as

$$C_a = C' \frac{i_B q_B}{2DV*} \qquad \qquad \text{XI.26}$$

where C' is a constant with an experimentally determined value of 1/11.6.

If this is substituted in Equation XI.21 with a = 2D, then the equation for suspended load becomes

$$q_s = i_B q_B \left[2.30 \log \left(\frac{30.2\ dx}{d_{65}} \right)\ I_1 + I_2 \right] \qquad \qquad \text{XI.27}$$

The total bed sediment load (suspended plus bed load) may then be computed as

$$i_t q_t = i_B q_B \left[1 + 2.30 \log \left(\frac{30.2dx}{d_{65}} \right)\ I_1 + I_2 \right] \qquad \qquad \text{XI.28}$$

The Einstein equations may be applied with minima field data as the system of equations basically implies that the discharge relationship can be obtained from equations. The calculation procedure follows:

1. Assume a value of R'_b.

2. Calculate $V'* = \sqrt{g\ R'_b\ S_e}$ using field data, assuming S_e constant for all flows (steady discharge). If the test reach is uniform in cross section, S_e will equal the water surface slope.

3. Calculate δ', the thickness of the laminar sub-layer, as

    $\delta' = 11.6\ \dfrac{\nu}{V'*}$ where ν = kinematic viscosity of the fluid, (cm^2/sec).

4. Calculate the value d_{65}/δ' using the value of d_{65} determined from a grain-size distribution analysis of the bed material.

5. Select the correction factor, x, from a plot of d_{65}/δ' versus x. [See Fig. XI.3(a)].

6. Calculate the average flow velocity, V, from the equation

    $$V = 5.75\ V'* \log \left(\frac{12.27\ R'_b\ x}{d_{65}} \right)$$

7.   Calculate the term, ψ_{35}, as

$$\psi_{35} = \left(s - 1\right) \frac{d_{35}}{R_b' S_e}$$

8.   Select the appropriate value of $\dfrac{V_*''}{V}$ from Fig. XI.3(b) and V_*'' may then be calculated using the calculated value of V from step 6.

9.   Compute the value for R_b'' as

$$R_b'' = V_*''^2 / g\, S_e.$$

10. The overall hydraulic radius for the bed is then defined as $R_b = R_b' + R_b''$. This may be assumed to be the flow depth for the particular conditions assumed.

11. Determine the total flow area and wetted perimeter from field survey data relating flow depth to these quantities. The required field data are average cross-sectional information through the test reach.

12. The discharge may then be calculated, as it is the total flow area times the average flow velocity, V.

13. The reference grain size, X, is the smallest of the grain sizes in a given bed that is fully affected by the turbulent flow. It is defined by

$$X = 0.77 \frac{d_{65}}{X} \quad \text{when} \quad \frac{d_{65}}{X} > 1.80 \ \delta'$$
$$\text{(rough bed)}$$

or

$$X = 1.39 \ \delta' \quad \text{when} \quad \frac{d_{65}}{X} < 1.80 \ \delta'$$
$$\text{(smooth bed)}$$

14. Obtain the correction factor, Y, from Fig. XI.3(c). The grain size distribution curve for the bed sediment may then be split into a number of size fractions (about eight), and the representative grain size for a particular fraction determined as the average value (if the size range of the fraction is small). The representative grain size may then be divided by X, and ξ determined from Fig. XI.3(d).

15. Using Equation XI.25, calculate ψ_* and enter Fig. XI.3(e) to find Φ_*. With Φ_* known, Equation XI.24 can be used to compute $i_B q_B$ as i_b may also be calculated for the particular size fraction considered.

16. The value of $z = \dfrac{w}{0.4\,V_*}$ may also be calculated and the integral values I_1 and I_2 determined from charts [Chow (1964) p. 17-47].

17. The value of $i_t q_t$ may then be calculated from Equation XI.28. If all the length dimensions (R_b', V_*, δ', d_{65}/x, R_b'', R_b, stage, wetted perimeter, X, and D) are in feet, area in sq. ft. and velocity in ft./sec, then $i_t q_t$ will have the units of lbs/ft-sec. The total bed material discharge in tons/day is $i_t q_t$ times the bed width times a conversion factor of 43.1.

The calculation steps 1–17 must be carried out for each size fraction of bed material. In addition, the whole sequence must then be repeated for each desired or

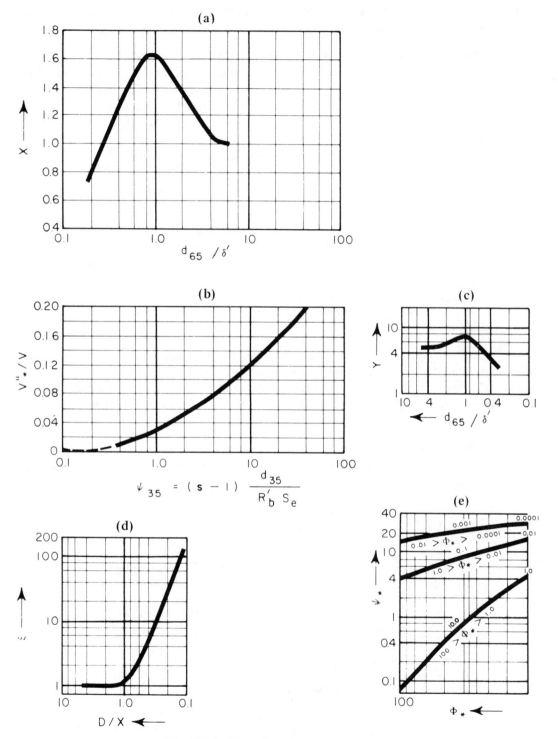

Fig. XI. 3 Einstein Transport Relations

assumed value of R'_b. If five values of R'_b are assumed, and eight size fractions considered, then 40 calculation sequences will be necessary.

By this method the amount of work necessary to determine the sediment transport is considerable, but it has given good results in certain cases. The obvious failings of the method are that it is based on assumed velocity distributions and on sediment transfer relationships, and that it does not make use of field data which are often obtained. The modified Einstein method, proposed by Colby and Hembree (1955), attempts to make more use of easily obtained field data, and it is becoming widely used. The calculation procedure is essentially the same as for the Einstein method with the following modifications:

1. The calculation is based on a measured mean velocity rather than on the slope, and the depth is observed for each velocity.
2. The friction velocity V^* and the corresponding suspended-load exponent z are determined from the observed z value for a dominant grain size. Values of z for other grain sizes are derived from that of the dominant size, and are assumed to change with the 0.7 power of the settling velocity.
3. A slightly changed ξ curve against D/X is introduced.
4. R'_b is replaced by the depth, d, in the logarithmic equation for average velocity.

The first two modifications should improve the accuracy of the equations, as less reliance is placed on assumed velocity distributions. The third modification also agrees with some observed flume data where a wide range of grain sizes was considered. Einstein (Chow, 1964) indicates that the fourth modification may result in large errors for low sediment rates.

Colby and Hembree (1955) also indicate a means whereby the total load, $i_t q_t$, may be calculated from measured values of the suspended sediment load. Einstein shows that the relationship may be given by

$$\frac{i_t q_t}{i_{sm} q_{sm}} = \left(\frac{E}{A}\right)^{z-1} \left(\frac{1-A}{1-E}\right)^z \frac{(1 + PI_1 + I_2)}{(PI_1 + I_2)_E} \quad \dots\dots\dots\dots\dots\dots\dots\dots\dots \text{XI.29}$$

where $i_t q_t$ = total load in a given size range of bed material load,
$i_{sm} q_{sm}$ = measured (depth integrated) suspended load in the same size range,
A = 2D/d (ratio of 2 grain diameters to the depth),
E = ratio of the unmeasured layer thickness (suspended load measurements can only be taken about 3 in above the bed) to the depth.
$P = 2.30 \log\left(\frac{30.2 \, dx}{d_{65}}\right)$, and
I_1 and I_2 = are the integral values previously defined.
The subscript E indicates that E should be substituted for A in evaluating the integral.

The value of z is determined from a trial and error method from the measurement and for the predominant grain size range. Any other z values derived are considered to be proportional to the 0.7 power of the particular settling velocity.

The modified Einstein method does require further confirmation of its applicability for a wide range of conditions, but it apparently has given good results in particular instances.

XI.3.3.3 Other Bed Load Equations

Most of the older equations for sediment transport were derived primarily for conditions of bed load transport. Some of them are based on the concept that transport begins when the shearing stress or tractive force at the bed reaches a critical value. The transport rate is then considered to be a function of the difference between the actual shearing stress and the critical value. Other equations have been derived by considering the forces that act on the bed particles, and the similarity or dimensional homogeneity of the system.

These equations have the advantage of simplicity and ease of calculation, but they can not be applied to all conditions. A few of these equations are presented below:

Meyer-Peter Formula (1948)

$$q_s^{2/3} = 39.25 \, q^{2/3} \, S - 9.95 \, D \dots\dots\dots\dots\dots\dots\dots\dots XI.30$$

$$D = \frac{1}{100} \sum_{i=1}^{N} p_i d_i, \dots\dots\dots\dots\dots\dots\dots\dots\dots\dots XI.31$$

where q_s = sediment discharge in lbs/sec/ft. width of channel,
q = water discharge in cfs/ft width,
S = slope of the channel in ft/ft,
D = mean size of sediment in feet as defined by Equation XI.31, and
p_i = weight percent of bed sediment with a mean size d_i, where d_i is in ft.

D is determined from a size analysis of a representative sample of bed material. The size distribution is divided into a convenient number of size fractions, and the mean size, d_i, and the weight percentage, p_i, of each fraction is determined.

The Meyer-Peter formula, as written, is only valid for the ft.-lb.-sec. system of units and for bed-load transport.

Schoklitsch Equation (Shulits 1935)

$$q_s = \sum_{i=1}^{N} \frac{p_i}{100} \frac{86.7}{\sqrt{d_i}} \, S^{3/2} \left(q - F \right) \dots\dots\dots\dots\dots\dots\dots\dots XI.32$$

$$F = 0.00532 \frac{d_i}{S^{4/3}} \quad \ldots \ldots \ldots \ldots \ldots \ldots \ldots \ldots \ldots \ldots \ldots \text{XI.33}$$

where d_i = mean size of a fraction of the bed sediment
in in., and all other terms are as defined in
Equation XI.31.

Duboys (Rouse 1950)

$$q_s = \psi \, \frac{\tau_0}{\gamma} \left(\frac{\tau_0 - \tau_c}{\gamma} \right) \quad \ldots \ldots \ldots \ldots \ldots \ldots \ldots \ldots \ldots \text{XI.34}$$

where ψ = coefficient depending on the mean size of the sediment,
τ_0 = $\gamma \, dS$ = bed shear stress (lbs/sq. ft),
τ_c = critical bed shear stress (lbs/sq. ft),
γ = specific weight of water in lbs/cu. ft,
d = water depth, ft, and
S = slope of channel.

A few values of ψ and τ_c are shown in Table XI.4.

Table XI. 4 Values of ψ and τ_c for use in Duboy's Equation

Mean diameter of sediment (m.m.)	ψ	τ_c (lbs/sq. ft)
0.125	523,000	0.0162
0.25	312,000	0.0172
0.50	187,000	0.0215
1	111,000	0.0316
2	66,200	0.0513
4	39,900	0.089

Shields Equation (Rouse 1950)

$$q_s = 10q \, S \left(\frac{\tau_0 - \tau_c}{(s_s - 1)^2 D} \right) \quad \ldots \ldots \ldots \ldots \ldots \ldots \ldots \text{XI.35}$$

where τ_c = critical bed shearing stress from Fig. XI.4,
D = mean size of the bed sediment, and
s_s = specific gravity of the sediment.

The Shields equation is dimensionally homogeneous, so any consistent set of units may be used.

Laursen Formula (1958)

$$\overline{C} = \sum_{i=1}^{N} p_i \left(\frac{d_i}{d}\right)^{7/6} \left(\frac{\tau_i}{\tau_{ci}} - 1\right) \left[f\left(\frac{V^*}{w_i}\right)\right] \qquad \ldots\ldots\ldots\ldots\ldots\ldots \text{XI.36}$$

in which $\tau_i = \dfrac{V^2 \ d_i^{1/3}}{30 d^{1/3}}$

Thus $q_s = \dfrac{\gamma \ q\overline{C}}{100}$

where $\overline{C}$ = mean sediment discharge concentration in per cent by weight,

w_i = settling velocity of particles with a size d_i (ft/sec),

d_i = mean size of the grains in a size fraction of the bed material (ft),

$\tau_{ci} = 4d_i$ = critical bed shearing stress for a particle of size d_i, (lbs/ft^2),

V^* = $\sqrt{gdS}$ = shear velocity (ft/sec),

V = mean flow velocity (ft/sec), and

$f\left(\dfrac{V^*}{w_i}\right)$ = graphical function (see the original reference).

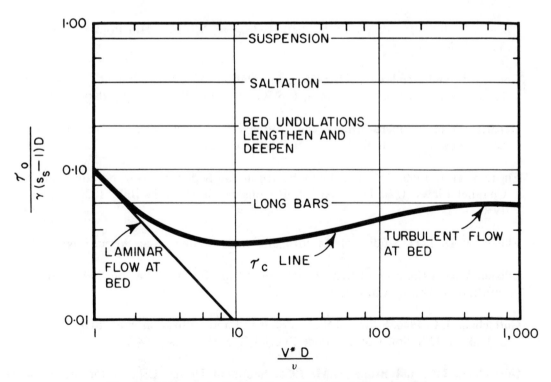

Fig. XI.4 Graph to Define the Critical Bed Shearing Stress

XI.3.3.4   General Comments

None of the sediment transport equations give accurate, short-term estimates of transport volumes and rates. Under such circumstances errors of 100 per cent or more in the calculated values may not be uncommon. The accuracy of the equations in predicting actual amounts is improved considerably, however, through consideration of transport over long-time periods (months or years) by summation of short-period volumes.

Several of the equations presented should be used when estimates of transport are required, as this will allow some estimate of the possible variation in rates and volumes. There is no real substitute for measurement of transport rates, and preferably at least, a few measurements should be obtained and compared with the results of the equations. The equation which best describes the transport may then be used with more confidence to predict occurrences for different flow conditions.

XI.4   LITERATURE CITED

Brune, G.M. 1953. Trap Efficiency of Reservoirs. Trans. Amer. Geophys. Union, Vol. 34. No. 3.

Chow, V.T. 1964. Handbook of Applied Hydrology. McGraw-Hill Book Co. Inc., New York, N.Y.

Colby, B.R. and Hembree, C.H. 1955. Computations of Total Sediment Discharge Niobrara River near Cody, Nebraska, U.S. Geol. Surv. Water Supply Paper 1357.

Department of Northern Affairs and National Resources. 1965. Sediment Data for Saskatchewan and Manitoba. Water Resources Paper No. S–2.

Einstein, H.S. 1950. The Bed-Load Function for Sediment Transportation in Open Channel Flows. U.S. Dept. Agr. Soil Conserv. Serv. Tech. Bull. 1026. September 1950.

Ellison, W.D. 1944. Studies of Raindrop Erosion. J. Amer. Soc. Agric. Engr. Vol. 25.

Ellison, W.D. 1945. Some Effects of Raindrops and Surface Flow on Soil Erosion and Infiltration. Trans. Am. Geophys. Union, Vol. 26. No. 3.

Gottschalk, L.C. 1946. Silting of Stock Ponds in Land Utilization Area. SD-LU-2 Pierre, S. Dakota, U.S. Soil Cons. Service Special Report 9. May 1946.

Gottschalk, L.C. and Brune, G.M. 1950. Sediment Design Criteria for the Missouri Basin Loess Hills, U.S. Soil Conservation Service, SCS–TP–97.

Laursen, E. 1958. The Total Sediment Load of Streams. Journal of Hyd. Division, Am. Soc. of Civil Engrs. Vol. 54. No. HY1. Feb. 1958.

Meyer, Peter E. 1948. Formulas for Bed Load Transport Report on Second Meeting, International Assoc. for Hyd. Structures Res. Stockholm.

Musgrave, G.W. 1947. The Quantitative Evaluation of Factors in Water Erosion: A First Approximation. Jour. Soil and Water Conservation Vol. 2. No. 3.

Rouse, H. 1950. Engineering Hydraulics. John Wiley and Sons, Inc. New York, N.Y.

Rubey, W.W. 1933. Settling Velocities of Gravel Sand and Silt Particles. Am. Journal of Science, Vol. 25. No. 148.

Shulits, S. 1935. The Schoklitsch Bed-Load Formula. Engineering, June 1935.

Stall, J.B. and Bartelli, L.J. 1959. Correlation of Reservoir Sedimentation and Watershed Factors. Springfield Plain, Illinois. Illinois State Water Surv. Div. Dept. Invest. 37.

Subcommittee on Sedimentation, 1959. A Study of Methods Used in Measurement and Analysis of Sediment Loads in Streams. Report AA Federal Inter-Agency Sed. Inst.

U.S. Department of Agriculture, 1961. A Universal Equation for Predicting Rainfall-Erosion Losses. U.S. Agr. Res. Service, Special Report 22-26, March 1961.

Vanoni, V.T. 1946. Transportation of Suspended Sediment by Water. Trans. Am. Soc. of Civil Engineers. Vol. 111.

Wischmeier, W.H. and Smith D.D. 1958. Rainfall Energy and Its Relationship to Soil Loss, Trans. Am. Geophys. Union, Vol. 39. No. 2.

Section XII

STATISTICAL METHODS –
FITTING FREQUENCY CURVES,
REGRESSION ANALYSES

by

Donald M. Gray

TABLE OF CONTENTS

LIST OF TABLES

LIST OF FIGURES

Section XII

STATISTICAL METHODS–FITTING FREQUENCY CURVES, REGRESSION ANALYSES

XII.1   **FREQUENCY ANALYSES**

XII.1.1   **Introduction**

Expression of hydrologic data on a probabilistic basis is a widely-used and generally-accepted procedure. The applicability of the Extreme-Value Distribution to define rainfall intensities was discussed in Section II. And in the engineering design of different hydraulic structures, it is standard practice to design on the basis of discharge rates of a given return period; that is, the period in years *on the average* during which a given discharge rate can be expected to be equalled or exceeded (equalled or less than in low flow analysis) – see Table XII.1. By definition, the return period can be calculated by

$$t_r = \frac{1}{p} \quad \dots\dots\dots\dots\dots\dots\dots\dots\dots\dots\dots\dots\dots\dots\dots\dots \text{XII.1}$$

in which t_r = return period, and

              p = probability of occurrence of the given event
                 (expressed as decimal).

The usual procedure followed in frequency analysis is to assume the specific frequency distribution the event is likely to follow, and to proceed to evaluate the parameters of the equation from experimental observations. Using the statistically-derived sample estimates, probability levels can be assigned to any specific event, and then predictions or inferences can be made for subsequent events that might occur.

It should be noted that most hydrologic populations are not of infinite extent in their values, and that the number of samples available (in time) from which the population parameters must be estimated are usually very small; thus any one of a large number of theoretical curves may often be fitted to a hydrologic event equally well. For this reason, it is important that the selection of an appropriate frequency curve to define a population should be supported by an understanding of the natural system, or by experience with other similar frequency series. Also, it should be remembered that realistic probability levels can be assigned to a given event only when a large number of accurate observations are used in the analysis, and that no amount of statistical sophistry can improve the quality of the data to produce a more fortuitous or amenable result.

XII.1.2   Linearizing Frequency Functions

Chow (1951) suggested that most frequency functions applicable for hydrologic analyses can be resolved to the generalized linear form

$$\frac{X}{\overline{X}} = 1 + KC_v \dots\dots\dots\dots\dots\dots\dots\dots\dots\dots\dots\dots\dots\dots\dots \text{XII.2}$$

in which   X = observed variate,

$\overline{X}$ = mean value of a set of X variates,
K = frequency factor which is a property of a given frequency distribution at a given probability level, and
C_v = coefficient of variation of a set of X variates.

Equation XII.2 can be written in alternate form as

$$X = \overline{X} + Ks \dots\dots\dots\dots\dots\dots\dots\dots\dots\dots\dots\dots\dots\dots\dots\dots\dots \text{XII.3}$$

in which s = standard deviation of a set of X variates or s = $\overline{X}C_v$

Simplified techniques of fitting different frequency distributions to hydrologic data employing Equations XII.2 and XII.3 are detailed in the booklet prepared by McGuinness and Brakensiek (1964).

Table XII.1 Return Periods Commonly Used in the Design of Different Hydraulic Structures.

Structure	Return Period
Bridges on important highways where back-water may cause excessive property damage or the loss of the bridge.	50 - 100 yr
Bridges on less important roads or culverts on important roads.	25 yr
Culverts on secondary roads, storm sewers, side ditches.	5 - 10 yr
Storm-water inlets, gutters.	1 - 2 yr

XII.1.3   Basic Theoretical Concepts Underlying Different
                    Frequency Distributions Used to Define
                                Hydrologic Data

XII.1.3.1   General

The following notes on distribution theory are intended primarily for hydrologists in an attempt to fill some of the gaps in their basic knowledge of the field, so that they may be more successful in applying the different frequency functions in their work. In most cases, the rigorous mathematical theory underlying each of the distributions discussed is not presented. This material may be found in most texts concerning the theory of statistics.

XII.1.3.2   Normal or Gaussian Distribution

Unquestionably, the commonest frequency distribution that is most important in statistics is the Normal or Gaussian distribution. This distribution, which has been applied successfully in studies of the errors of measurements, has a geometric shape of a symmetrical, bell-shaped curve whose limits extend from $X = -\infty$ to $X = +\infty$ (See Fig. XII.1). In the normal distribution, the mean (weighted centre $= \Sigma\, X/N$), the mode (the

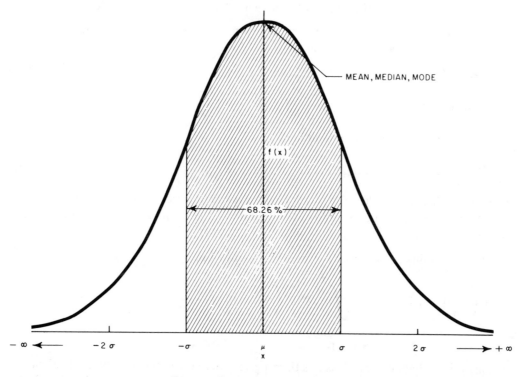

Fig. XII.1 Normal Distribution

X-value which occurs most frequently), and the median (the central value of the X variates), are coincident. The equation of the curve is given by

$$f(X) = \frac{1}{\sigma\sqrt{2\pi}} \exp\left[-\frac{(X-\mu)^2}{2\sigma^2}\right] \quad \dots\dots\dots\dots\dots\dots\dots\dots \text{XII.4}$$

in which f(X) = probability density function,
  σ = population standard deviation which is estimated by s, the standard deviation of a set of X variates, and
  μ = population mean which is estimated by the sample mean, $\overline{X}$.

The constants of the curve are chosen such that the area enclosed by the curve and the X-axis is unity. In addition, it should be noted that 68.3 per cent of the area under the curve is enclosed within $\mu \pm \sigma$, 95.4 per cent within, $\mu \pm 2\sigma$, and 99.7 per cent within $\mu \pm 3\sigma$. The probability on random drawing that the value of X will be less than a given value, X_1, is given by

$$p(X \leqslant X_1) = \int_{-\infty}^{X_1} f(X)\,dX = \int_{-\infty}^{X_1} \frac{1}{\sigma\sqrt{2\pi}} \exp\left[-\frac{(X-\mu)^2}{2\sigma^2}\right] dX \dots\dots\dots \text{XII.5}$$

or equal to the area under the cumulative distribution curve between the limits $X = -\infty$ to $X = X_1$. Further, the probability of obtaining a value of $X > X_1$ is $1-p(X \leqslant X_1)$.

Usually, it is convenient to express the variable in terms of the population mean, μ, and the standard deviation, σ; that is

$$z = \frac{X-\mu}{\sigma} \quad \dots\dots\dots\dots\dots\dots\dots\dots\dots\dots\dots\dots\dots\dots\dots\dots\dots \text{XII.6}$$

in which z = standard deviate.

Note, Equation XII.6 can be presented as

$$X = \mu + z\sigma \dots\dots\dots\dots\dots\dots\dots\dots\dots\dots\dots\dots\dots\dots\dots \text{XII.7}$$

which is similar to the form of the frequency distribution equation given by Chow (Equation XII.3). That is, for a normal distribution the frequency factor is the standard deviate.

By combining Equations XII.5 and XII.6, one obtains

$$p\left(\frac{X-\mu}{\sigma} \leqslant z_1\right) = \frac{1}{\sqrt{2\pi}} \int_{-\infty}^{z_1} \exp\left(-z^2/2\right) dz \dots\dots\dots\dots\dots \text{XII.8}$$

Note: $dX = \sigma\,dz$.

That is, the probability that X is less than a given value is the same as the probability that z will have a value less than the corresponding value given by Equation XII.8. The importance of the substitution may be summed up as follows: since the standard deviate is a dimensionless number, any variable that is normally distributed regardless of the magnitude of the variable or its standard deviation and if expressed in terms of its deviation from the mean divided by the standard deviation, it will have a distribution identical with any other variable similarly expressed. In effect, this leads to the development of one common curve for all normally distributed variables—The Standard Normal Curve. Values of the standardized normal deviates, corresponding to different probability levels, are reported in most texts on Statistics.

Estimation of Parameters

Obviously, in Equation XII.4 it can be observed that the equation of the normal curve contains two parameters, μ and σ, whose magnitude usually needs to be estimated from a random sample. It is well known that there are many ways of estimating the parameters of a statistical population; however, they may not be equally good. In essence, if a method provides an estimator T_1 which is more efficient than another estimator T_2 given by a second method, then one means that the variance of T_1 is less than that of T_2; hence method 1 provides a better estimate of T than method 2. Frequently, either the method of moments or the method of maximum likelihood is used to obtain estimators of hydrologic populations. In this regard, most scientists are familiar with the basic procedure in application of the method of moments (e.g., in calculating the moment of inertia of a body or the standard deviation), thus no further discussion of this method is presented.

Maximum Likelihood Estimates

Kendall (1967) details the principle of Maximum Likelihood introduced by Fisher (1922) as applied to a normal distribution. According to this principle, for a statistical population with a probability density function $Y = f(X, \theta)$ where θ is a parameter which must be estimated, the probability of obtaining a given value of say X_1 on random sampling is proportional to $f(X_1, \theta)$ and the joint probability, L, of obtaining a sample of N values $X_1 X_2 \ldots \ldots X_N$ is proportional to the product

$$L = f(X_1, \theta) \; f(X_2, \theta). \ldots f(X_N, \theta) \quad \ldots\ldots\ldots\ldots\ldots\ldots\ldots\ldots \text{XII.9}$$

This is called the likelihood; the principle of maximum likelihood states the estimate of θ is chosen such that it maximizes the likelihood. For a normal distribution, the likelihood function is

$$L = \frac{1}{(2\pi\sigma^2)^{N/2}} \exp\left[-\frac{1}{2\sigma^2} \sum_{i=1}^{N} (X_i - \mu)^2 \right] \quad \ldots\ldots\ldots\ldots\ldots\ldots\ldots \text{XII.10}$$

or

$$\ln L = \frac{-N}{2} \ln 2\pi - \frac{N}{2} \ln \sigma^2 - \frac{1}{2\sigma^2} \sum_{i=1}^{N} (X_i - \mu)^2 \quad \ldots\ldots\ldots\ldots\ldots \text{XII.11}$$

Differentiating Equation XII.11 with respect to μ and σ^2 and setting the resulting equations equal to zero to maximize the likelihood, one obtains the expressions:

$$-\frac{N}{2\sigma^2} + \frac{1}{2\sigma^4} \sum_{i=1}^{N} (X_i - \mu)^2 = 0 \quad\dots\dots\dots\dots\dots\dots\dots\dots\text{XII.12}$$

$$-\frac{1}{2\sigma^2} \sum_{i=1}^{N} -2(X_i - \mu) = 0 \quad\dots\dots\dots\dots\dots\dots\dots\text{XII.13}$$

Solving Equations XII.12 and XII.13 it follows that

$$\hat{\mu} = \sum_{i=1}^{N} X_i / N = \overline{X} \quad\dots\dots\dots\dots\dots\dots\dots\dots\dots\dots\dots\dots\text{XII.14}$$

and

$$\hat{\sigma}^2 = \frac{1}{N} \sum_{i=1}^{N} (X_i - \overline{X})^2 = s^2 \quad\dots\dots\dots\dots\dots\dots\dots\dots\text{XII.15}$$

in which case $\hat{\mu}$ and $\hat{\sigma}^2$ are the statistical likelihood estimators of μ and σ^2, respectively.

Equation XII.15 is considered to be a biased estimator since its mean value is $\sigma^2 (N-1)/N$ The maximum likelihood estimate is therefore

$$\hat{\sigma}^2 = \frac{1}{N-1} \sum_{i=1}^{N} (X_i - \overline{X})^2 = s^2 \quad\dots\dots\dots\dots\dots\dots\dots\text{XII.16}$$

It can be observed from Equations XII.14 and XII.16 that the maximum likelihood estimates of μ and σ^2 are the same as those obtained by the method of moments (common in the literature). However, it should be recognized that these results may not be the case for distributions other than the normal. Using the estimators of μ and σ^2 obtained from a sample these Equations can be used in XII.5 or XII.8 to evaluate the probability of a given value of X.

A somewhat different technique for fitting hydrologic data to a normal distribution is to fit the data to the linear equation proposed by Chow (see Equation XII.3). McGuinness and Brakensiek (1964) show that the least-squares estimators of the parameters are given by

$$\overline{X} = \sum_{i=1}^{N} X_i / N \quad\dots\dots\dots\dots\dots\dots\dots\dots\dots\dots\dots\text{XII.17}$$

$$ s = \sum_{i=1}^{N} (X_i - \bar{X})^2 \Big/ \sum_{i=1}^{N} X_i K_y \dots\dots\dots\dots\dots\dots\dots\dots\dots\dots \text{XII.18} $$

in which K_y = frequency factor for the normal distribution, or the standard normal deviate corresponding to the frequency plotting position determined by the equation $F_n = m/(N + 1)$ in which m is the rank of the event in decreasing order of magnitude. Values of F_n and K_y for different values of N are given in Tables XII.2a to XII.2g inclusive.

Further details of fitting the distribution are given in example calculations by the authors.

Confidence Limits of Parameter Estimates

Since the parameter estimates are in themselves random variables, it is possible to use standard statistical procedures to test the confidence with which, using the sample estimate, we can estimate the population parameter. For example, for a normal distribution, if we make the statement that the interval

$$ \bar{X} \pm 1.96\sigma/\sqrt{N} \dots\dots\dots\dots\dots\dots\dots\dots\dots\dots\dots\dots\dots\dots\text{XII.19} $$

contains the population mean, μ, we would be right for 95 per cent of the samples taken from the population; or another way of stating this result is that we are 95 per cent

Table XII 2a. Plotting Positions, F_n, and Standardized Normal Deviates, ($K_y = z$), for a Range of Sample Sizes of Number of Years of Record, N.

Rank No (m)	N = 9 F_n	K_y	N = 10 F_n	K_y	N = 11 F_n	K_y	N = 12 F_n	K_y	N = 13 F_n	K_y
	Percent		*Percent*		*Percent*		*Percent*		*Percent*	
1	10.0	1.28	9.1	1.34	8.3	1.38	7.7	1.43	7.1	1.46
2	20.0	.84	18.2	.91	16.7	.97	15.4	1.02	14.3	1.07
3	30.0	.52	27.3	.60	25.0	.68	23.1	.74	21.4	.79
4	40.0	.25	36.4	.35	33.3	.43	30.8	.50	28.6	.57
5	50.0	.00	45.5	.11	41.7	.21	38.5	.29	35.7	.37
6	60.0	−.25	54.5	−.11	50.0	.00	46.2	.10	42.9	.18
7	70.0	−.52	63.6	−.35	58.3	−.21	53.8	−.10	50.0	.00
8	80.0	−.84	72.7	−.60	66.7	−.43	61.5	−.29	57.1	−.18
9	90.0	−1.28	81.8	−.91	75.0	−.68	69.2	−.50	64.3	−.37
10			90.9	−1.34	83.3	−.97	76.9	−.74	71.4	−.57
11					91.7	−1.38	84.6	−1.02	78.6	−.79
12							92.3	−1.43	85.7	−1.07
13									92.9	−1.46

confident that the interval contains the population mean unless a one-in-twenty chance (5%) has occurred in sampling.

Usually, however, the standard deviation is unknown and must be estimated by the sample variance, s^2. Because s^2 is also a random variable, when this value is used to define the interval (Equation XII.19), the same confidence statement cannot be made and it is necessary to employ a 't' test to test the mean. That is

$$t = \frac{\overline{X} - \mu}{s/\sqrt{N}} \quad \dots\dots\dots\dots\dots\dots\dots\dots\dots\dots\dots\dots\dots \text{XII.20}$$

The variable, t, known in statistics as Student's 't', is normally distributed independent of the population parameters with $N-1$ degrees of freedom, therefore we are able to make the probability statement

$$p\left(-t_{0.025} < \frac{\overline{X} - \mu}{s/\sqrt{N}} < t_{0.025} \right) = 0.95 \quad \dots\dots\dots\dots\dots\dots\dots\text{XII.21}$$

from which we can say that we are 95 per cent confident that μ falls in the interval

$$\overline{X} - t_{0.025}\ s/\sqrt{N} \quad \text{to} \quad \overline{X} + t_{0.025}\ s/\sqrt{N}$$

Table XII 2b. Plotting Positions, F_n, and Standardized Normal Deviates, ($K_y = z$), for a Range of Sample Sizes of Number of Years of Record, N.

Rank No (m)	N = 14 F_n	N = 14 K_y	N = 15 F_n	N = 15 K_y	N = 16 F_n	N = 16 K_y	N = 17 F_n	N = 17 K_y	N = 18 F_n	N = 18 K_y
	Percent		Percent		Percent		Percent		Percent	
1	6.7	1.50	6.2	1.53	5.9	1.57	5.6	1.60	5.3	1.62
2	13.3	1.11	12.5	1.15	11.8	1.19	11.1	1.22	10.5	1.25
3	20.0	.84	18.8	.89	17.6	.93	16.7	.97	15.8	1.00
4	26.7	.62	25.0	.67	23.5	.72	22.2	.76	21.1	.80
5	33.3	.43	31.2	.49	29.4	.54	27.8	.59	26.3	.63
6	40.0	.25	37.5	.32	35.3	.38	33.3	.43	31.6	.48
7	46.7	.08	43.8	.16	41.2	.22	38.9	.29	36.8	.33
8	53.3	−.08	50.0	.00	47.1	.07	44.4	.14	42.1	.20
9	60.0	−.25	56.2	−.16	52.9	−.07	50.0	.00	47.4	.07
10	66.7	−.43	62.5	−.32	58.8	−.22	55.6	−.14	52.6	−.07
11	73.3	−.62	68.8	−.49	64.7	−.38	61.1	−.29	57.9	−.20
12	80.0	−.84	75.0	−.67	70.6	−.54	66.7	−.43	63.2	−.33
13	86.7	−1.11	81.2	−.89	76.5	−.72	72.2	−.59	68.4	−.48
14	93.3	−1.50	87.5	−1.15	82.4	−.93	77.8	−.76	73.7	−.63
15			93.8	−1.53	88.2	−1.19	83.3	−.97	79.0	−.80
16					94.1	−1.57	88.9	−1.22	84.2	−1.00
17							94.4	−1.60	89.5	−1.25
18									94.7	−1.62

In this example, the limits of the interval are referred to as the 95 per cent confidence bands. Values for 't' corresponding to different probability levels are readily obtainable in most texts on Statistics.

Slivitzky (1967) points out that a certain degree of caution must be exhibited when applying confidence tests to certain types of hydrologic data. In this regard, the presence of persistence (measured by the coefficient of autocorrelation) in records will decrease the effective number of years-of-record, resulting in an increase in the variability and a decrease in the information content of the mean. Thus, it is important that tests of persistence should be made on data prior to evaluating the confidence limits (see Matalas and Langbein 1962).

It should also be recognized that confidence limits can be applied to other statistical estimates, for example the variance, s^2, by employing other statistical tests.

Transformations

Since the properties of a normal distribution are completely defined, if the observed data do not follow a normal distribution it is common procedure to attempt to transform

Table XII 2c. Plotting Positions, F_n, and Standardized Normal Deviates, $(K_y = z)$, for a Range of Sample Sizes of Number of Years of Record, N.

Rank No (m)	N = 19		N = 20		N = 21		N = 22		N = 23	
	F_n	K_y	F_n	K_y	F_n	K_y	F_n	K_y	F_n	K_y
	Percent		*Percent*		*Percent*		*Percent*		*Percent*	
1	5.0	1.64	4.8	1.67	4.5	1.69	4.3	1.71	4.2	1.73
2	10.0	1.28	9.5	1.31	9.1	1.34	8.7	1.36	8.3	1.38
3	15.0	1.04	14.3	1.07	13.6	1.10	13.0	1.12	12.5	1.15
4	20.0	.84	19.0	.87	18.2	.91	17.4	.94	16.7	.97
5	25.0	.67	23.8	.71	22.7	.75	21.7	.78	20.8	.81
6	30.0	.52	28.6	.57	27.3	.61	26.1	.64	25.0	.67
7	35.0	.39	33.3	.43	31.8	.47	30.4	.51	29.2	.55
8	40.0	.25	38.1	.30	36.4	.35	34.8	.39	33.3	.43
9	45.0	.13	42.9	.18	40.9	.23	39.1	.27	37.5	.32
10	50.0	.00	47.6	.06	45.5	.12	43.5	.16	41.7	.21
11	55.0	−.13	52.4	−.06	50.0	.00	47.8	.05	45.8	.10
12	60.0	−.25	57.1	−.18	54.5	−.12	52.2	−.05	50.0	.00
13	65.0	−.39	61.9	−.30	59.1	−.23	56.5	−.16	54.2	−.10
14	70.0	−.52	66.7	−.43	63.6	−.35	60.9	−.27	58.3	−.21
15	75.0	−.67	71.4	−.57	68.2	−.47	65.2	−.39	62.5	−.32
16	80.0	−.84	76.2	−.71	72.7	−.61	69.6	−.51	66.7	−.43
17	85.0	−1.04	81.0	−.87	77.3	−.75	73.9	−.64	70.8	−.55
18	90.0	−1.28	85.7	−1.07	81.8	−.91	78.3	−.78	75.0	−.67
19	95.0	−1.64	90.5	−1.31	86.4	−1.10	82.6	−.94	79.2	−.81
20			95.2	−1.67	90.9	−1.34	87.0	−1.12	83.3	−.97
21					95.5	−1.69	91.3	−1.36	87.5	−1.15
22							95.7	−1.71	91.7	−1.38
23									95.8	−1.73

these data in such a way that the distribution of the transformed data can be described by a normal curve.

Cube Root Transformations

As shown by Stidd (1953), one of the most useful of the transformations, applicable to all types of precipitation series, is the cube root transformation, that is

$$Y = X^{1/3} \quad \dots\dots\dots\dots\dots\dots\dots\dots\dots\dots\dots\dots\dots\dots\dots\dots\dots\dots\dots \text{XII.22}$$

Kendall (1960) applied this transformation with success in studies of summer precipitation amounts recorded at a number of Canadian stations. To alleviate some of the work involved in the analysis, the Meteorological Service of Canada has developed a

Table XII 2d. Plotting Positions, F_n, and Standardized Normal Deviates, $(K_y = z)$, for a Range of Sample Sizes of Number of Years of Record, N.

Rank No (m)	N = 24		N = 25		N = 26		N = 27		N = 28	
	F_n	K_y	F_n	K_y	F_n	K_y	F_n	K_y	F_n	K_y
	Percent		*Percent*		*Percent*		*Percent*		*Percent*	
1	4.0	1.75	3.8	1.77	3.7	1.79	3.8	1.80	3.4	1.82
2	8.0	1.40	7.7	1.43	7.4	1.45	7.1	1.47	6.9	1.48
3	12.0	1.18	11.5	1.20	11.1	1.23	10.7	1.24	10.3	1.26
4	16.0	.99	15.4	1.02	14.8	1.04	14.3	1.07	13.8	1.09
5	20.0	.84	19.2	.87	18.5	.90	17.9	.92	17.2	.95
6	24.0	.71	23.1	.74	22.2	.77	21.4	.79	20.7	.82
7	28.0	.58	26.9	.61	25.9	.65	25.0	.67	24.1	.70
8	32.0	.47	30.8	.50	29.6	.54	28.6	.56	27.6	.59
9	36.0	.36	34.6	.40	33.3	.43	32.1	.46	31.0	.50
10	40.0	.25	38.5	.29	37.0	.33	35.7	.37	34.5	.40
11	44.0	.15	42.3	.19	40.7	.24	39.3	.27	37.9	.31
12	48.0	.05	46.2	.10	44.4	.14	42.9	.18	41.4	.22
13	52.0	−.05	50.0	.00	48.1	.05	46.4	.09	44.8	.13
14	56.0	−.15	53.8	−.10	51.9	−.05	50.0	.00	48.3	.04
15	60.0	−.25	57.7	−.19	55.6	−.14	53.6	−.09	51.7	−.04
16	64.0	−.36	61.5	−.29	59.3	−.24	57.1	−.18	55.2	−.13
17	68.0	−.47	65.4	−.40	63.0	−.33	60.7	−.27	58.6	−.22
18	72.0	−.58	69.2	−.50	66.7	−.43	64.3	−.37	62.1	−.31
19	76.0	−.71	73.1	−.61	70.4	−.54	67.9	−.46	65.5	−.40
20	80.0	−.84	76.9	−.74	74.1	−.65	71.4	−.56	69.0	−.50
21	84.0	−.99	80.8	−.87	77.8	−.77	75.0	−.67	72.4	−.59
22	88.0	−1.18	84.6	−1.02	81.5	−.90	78.6	−.79	75.9	−.70
23	92.0	−1.40	88.5	−1.20	85.1	−1.04	82.1	−.92	79.3	−.82
24	96.0	−1.75	92.3	−1.43	88.9	−1.23	85.7	−1.07	82.8	−.95
25			96.2	−1.77	92.6	−1.45	89.3	−1.24	86.2	−1.09
26					96.3	−1.79	92.9	−1.47	89.7	−1.26
27							96.4	−1.80	93.1	−1.48
28									96.6	−1.82

special cube-root normal graph paper.

Stidd (1953) also noticed that when the precipitation series for different times were plotted on the same cube-root normal graph paper, the straight lines formed were nearly parallel at a distance apart defined by the equation

$$P = at^n \quad \dots\dots\dots\dots\dots\dots\dots\dots\dots\dots\dots\dots\dots\dots \quad XII.23$$

which relates the recorded rainfall, P, to the storm duration, t. Kendall (1967) reports that this procedure appears valid in defining precipitation for Canadian stations.

Table XII 2e. Plotting Positions, F_n, and Standardized Normal Deviates, ($K_y = z$), for a Range of Sample Sizes of Number of Years of Record, N.

Rank No (m)	N = 29 F_n	K_y	N = 30 F_n	K_y	N = 31 F_n	K_y	N = 32 F_n	K_y	N = 33 F_n	K_y
	Percent		Percent		Percent		Percent		Percent	
1	3.3	1.84	3.2	1.85	3.1	1.87	3.0	1.88	2.9	1.90
2	6.7	1.50	6.5	1.51	6.2	1.54	6.1	1.55	5.9	1.56
3	10.0	1.28	9.7	1.30	9.4	1.32	9.1	1.33	8.8	1.35
4	13.3	1.11	12.9	1.13	12.5	1.15	12.1	1.17	11.8	1.18
5	16.7	.97	16.1	.99	15.6	1.01	15.2	1.03	14.7	1.05
6	20.2	.84	19.4	.86	18.8	.88	18.2	.91	17.6	.93
7	23.3	.73	22.6	.75	21.9	.78	21.2	.80	20.6	.82
8	26.7	.62	25.8	.65	25.0	.67	24.2	.70	23.5	.72
9	30.0	.52	29.0	.55	28.1	.58	27.3	.60	26.5	.63
10	33.3	.43	32.3	.46	31.2	.49	30.3	.52	29.4	.54
11	36.7	.34	35.5	.37	34.4	.40	33.3	.43	32.4	.46
12	40.0	.25	38.7	.29	37.5	.32	36.4	.35	35.3	.38
13	43.3	.17	41.9	.20	40.6	.24	39.4	.27	38.2	.30
14	46.7	.08	45.2	.12	43.8	.16	42.4	.19	41.2	.22
15	50.0	.00	48.4	.04	46.9	.08	45.5	.11	44.1	.15
16	53.3	−.08	51.6	−.04	50.0	.00	48.5	.04	47.1	.07
17	56.7	−.17	54.8	−.12	53.1	−.08	51.5	−.04	50.0	.00
18	60.0	−.25	58.1	−.20	56.2	−.16	54.5	−.11	53.0	−.07
19	63.3	−.34	61.3	−.29	59.4	−.24	57.6	−.19	55.9	−.15
20	66.7	−.43	64.5	−.37	62.5	−.32	60.6	−.27	58.8	−.22
21	70.0	−.52	67.7	−.46	65.6	−.40	63.6	−.35	61.8	−.30
22	73.3	−.62	71.0	−.55	68.8	−.49	66.7	−.43	64.7	−.38
23	76.7	−.73	74.2	−.65	71.9	−.58	69.7	−.52	67.6	−.46
24	80.0	−.84	77.4	−.75	75.0	−.67	72.7	−.60	70.6	−.54
25	83.3	−.97	80.6	−.86	78.1	−.78	75.8	−.70	73.5	−.63
26	86.7	−1.11	83.9	−.99	81.2	−.88	78.8	−.80	76.5	−.72
27	90.0	−1.28	87.1	−1.13	84.4	−1.01	81.8	−.91	79.4	−.82
28	93.3	−1.50	90.3	−1.30	87.5	−1.15	84.8	−1.03	82.4	−.93
29	96.7	−1.84	93.5	−1.51	90 6	−1.32	87.9	−1.17	85.3	−1.05
30			96.8	−1.85	93.8	−1.54	90.9	−1.33	88.2	−1.18
31					96.9	−1.87	93.9	−1.55	91.2	−1.35
32							97.0	−1.88	94.1	−1.56
33									97.1	−1.90

Logarithmic

Another most useful transformation used to normalize hydrologic data, particularly stream flow data, is the logarithmic transformation; that is

$$Y = \ln X \dots\dots\dots\dots\dots\dots\dots\dots\dots\dots\dots\dots\dots\dots \text{XII.24}$$

Inserting this equality into the normal equation, one obtains the equation for the

Table XII 2f. Plotting Positions, F_n, and Standardized Normal Deviates, ($K_y = z$), for a Range of Sample Sizes of Number of Years of Record, N.

Rank No (m)	N = 34		N = 35		N = 36		N = 37		N = 38	
	F_n	K_y	F_n	K_y	F_n	K_y	F_n	K_y	F_n	K_y
	Percent		*Percent*		*Percent*		*Percent*		*Percent*	
1	2.9	1.90	2.8	1.91	2.7	1.93	2.6	1.94	2.6	1.94
2	5.7	1.58	5.6	1.59	5.4	1.61	5.3	1.62	5.1	1.64
3	8.6	1.37	8.3	1.39	8.1	1.40	7.9	1.41	7.7	1.43
4	11.4	1.21	11.1	1.22	10.8	1.24	10.5	1.25	10.3	1.26
5	14.3	1.07	13.9	1.08	13.5	1.10	13.2	1.12	12.8	1.13
6	17.1	.95	16.7	.97	16.2	.99	15.8	1.00	15.4	1.02
7	20.0	.84	19.4	.86	18.9	.88	18.4	.90	17.9	.92
8	22.9	.74	22.2	.77	21.6	.79	21.1	.80	20.5	.82
9	25.7	.65	25.0	.67	24.3	.70	23.7	.72	23.1	.74
10	28.6	.56	27.8	.59	27.0	.61	26.3	.63	25.6	.66
11	31.4	.48	30.6	.51	29.7	.53	28.9	.56	28.2	.58
12	34.3	.40	33.3	.43	32.4	.46	31.6	.48	30.8	.50
13	37.1	.33	36.1	.36	35.1	.38	34.2	.41	33.3	.43
14	40.0	.25	38.9	.28	37.8	.31	36.8	.34	35.9	.36
15	42.9	.18	41.7	.21	40.5	.24	39.5	.27	38.5	.29
16	45.7	.11	44.4	.14	43.2	.17	42.1	.20	41.0	.23
17	48.6	.04	47.2	.07	45.9	.10	44.7	.13	43.6	.16
18	51.4	−.04	50.0	.00	48.6	.04	47.4	.07	46.2	.10
19	54.3	−.11	52.8	−.07	51.4	−.04	50.0	.00	48.7	.03
20	57.1	−.18	55.6	−.14	54.1	−.10	52.6	−.07	51.3	−.03
21	60.0	−.25	58.3	−.21	56.8	−.17	55.3	−.13	53.8	−.10
22	62.9	−.33	61.1	−.28	59.5	−.24	57.9	−.20	56.4	−.16
23	65.7	−.40	63.9	−.36	62.2	−.31	60.5	−.27	59.0	−.23
24	68.6	−.48	66.7	−.43	64.9	−.38	63.2	−.34	61.5	−.29
25	71.4	−.56	69.4	−.51	67.6	−.46	65.8	−.41	64.1	−.36
26	74.3	−.65	72.2	−.59	70.3	−.53	68.4	−.48	66.7	−.43
27	77.1	−.74	75.0	−.67	73.0	−.61	71.1	−.56	69.2	−.50
28	80.0	−.84	77.8	−.77	75.7	−.70	73.7	−.63	71.8	−.58
29	82.9	−.95	80.6	−.86	78.4	−.79	76.3	−.72	74.4	−.66
30	85.7	−1.07	83.3	−.97	81.1	−.88	78.9	−.80	76.9	−.74
31	88.6	−1.21	86.1	−1.08	83.8	−.99	81.6	−.90	79.5	−.82
32	91.4	−1.37	88.9	−1.22	86.5	−1.10	84.2	−1.00	82.1	−.92
33	94.3	−1.58	91.7	−1.39	89.2	−1.24	86.8	−1.12	84.6	−1.02
34	97.1	−1.90	94.4	−1.59	91.9	−1.40	89.5	−1.25	87.2	−1.13
35			97.2	−1.91	94.6	−1.61	92.1	−1.41	89.7	−1.26
36					97.3	−1.93	94.7	−1.62	92.3	−1.43
37							97.4	−1.94	94.9	−1.64
38									97.4	−1.94

log-normal frequency distribution as

$$f(Y) = \frac{1}{\sigma_Y \sqrt{2\pi}} \exp \left[- \frac{(Y - \bar{Y})^2}{2\sigma_Y^2} \right] \quad \dotfill \quad XII.25$$

or

$$\phi(X) = \frac{1}{\sigma_{\ln X} \sqrt{2\pi}} \exp \left[- \frac{(\ln X - \overline{\ln X})^2}{2\sigma^2_{\ln X}} \right] \quad \dotfill \quad XII.26$$

The rationality of using the logarithmic transformation stems from the fact that most hydrologic data have a lower limiting value of zero, and it is unlimited, within certain physical limits, in the values above. In taking logarithms, therefore, the lower limit $\ln(0)$ is set at $-\infty$ (comparable to the lower limit of the normal curve).

In fitting data to a log normal distribution it is necessary to determine both the coefficients of variation of the original data, C_v and of the transformed data, CV. These calculations may be carried out in a similar manner as outlined for a normal distribution, using the original data. Frequency factors corresponding to different probability levels for different values of the coefficient of variation of the transformed variate, as listed by McGuinness and Brakensiek (1964), are given in Table XII.3.

Modified Logarithmic Normal Distribution

Perhaps a transformation which has wider application in analyzing hydrologic events is of the form

$$Y = a + b\ln X \quad \dotfill \quad XII.27$$

in which a and b are constants. The distribution arising from the use of this transformation, referred to as a modified log-normal distribution, requires for its description the estimation of a transformation constant, so that the data may be described by a log-normal distribution. That is, whereas the log-normal curve plots as a straight line on log-normal paper, the modified log-normal distribution may plot as either a concave or convex curve.

Complete details of a graphical procedure for fitting observed data to a modified log-normal distribution are given in the publication by McGuinness and Brakensiek (1964). An example of the results of this technique applied to runoff volumes from two small watersheds in Saskatchewan, as reported by Nicholaichuk (1965), is shown in Fig. XII.2.

XII.1.3.3   Extreme Value Distribution

In 1941 Gumbel developed the Extreme Value Distribution. This distribution has been used with success to describe the populations of many hydrologic events. As applied to the largest extreme, the fundamental theorem can be stated:

If $X_1, X_2, X_3 \dotfill X_n$ are independent extreme values observed in 'n' samples of equal size, N, and if X is an *unlimited* exponentially-distributed variable, then as n and N approach infinity, the cumulative probability, q, that any of the extremes will be less than a given value of X; is given by

$$q = e^{-e^{-y}} \dots\dots\dots\dots\dots\dots\dots\dots\dots\dots\dots\dots\dots\dots\dots \text{XII.28}$$

Table XII. 2g. Plotting Positions, F_n, and Standardized Normal Deviates, $(K_y = z)$, for a Range of Sample Sizes of Number of Years of Record, N.

Rank No (m)	N=39 F_n	K_y	N=40 F_n	K_y	N=41 F_n	K_y	N=42 F_n	K_y	N=43 F_n	K_y
	Percent		*Percent*		*Percent*		*Percent*		*Percent*	
1	2.5	1.96	2.4	1.98	2.4	1.98	2.3	1.99	2.3	2.00
2	5.0	1.64	4.9	1.66	4.8	1.67	4.7	1.68	4.5	1.69
3	7.5	1.44	7.3	1.45	7.1	1.47	7.0	1.48	6.8	1.49
4	10.0	1.28	9.8	1.29	9.5	1.31	9.3	1.32	9.1	1.34
5	12.5	1.15	12.2	1.16	11.9	1.18	11.6	1.19	11.4	1.21
6	15.0	1.04	14.6	1.05	14.3	1.07	14.0	1.08	13.6	1.10
7	17.5	.93	17.1	.95	16.7	.97	16.3	.98	15.9	1.00
8	20.0	.84	19.5	.86	19.0	.88	18.6	.89	18.2	.91
9	22.5	.76	22.0	.77	21.4	.79	20.9	.81	20.5	.83
10	25.0	.67	24.4	.69	23.8	.71	23.3	.73	22.7	.75
11	27.5	.60	26.8	.62	26.2	.64	25.6	.66	25.0	.67
12	30.0	.52	29.3	.54	28.6	.57	27.9	.59	27.3	.60
13	32.5	.45	31.7	.48	31.0	.50	30.2	.52	29.5	.54
14	35.0	.39	34.1	.41	33.3	.43	32.6	.45	31.8	.47
15	37.5	.32	36.6	.34	35.7	.37	34.9	.39	34.1	.41
16	40.0	.25	39.0	.28	38.1	.30	37.2	.33	36.4	.35
17	42.5	.19	41.5	.21	40.5	.24	39.5	.27	38.6	.29
18	45.0	.13	43.9	.15	42.9	1.8	41.9	.21	40.9	.23
19	47.5	.06	46.3	.09	45.2	.12	44.2	.15	43.2	.17
20	50.0	.00	48.8	.03	47.6	.06	46.5	.09	45.5	.11
21	52.5	−.06	51.2	−.03	50.0	.00	48.8	.03	47.7	.06
22	55.0	−.13	53.7	−.09	52.4	−.06	51.1	−.03	50.0	.00
23	57.5	−.19	56.1	−.15	54.8	−.12	53.5	−.09	52.3	−.06
24	60.0	−.25	58.5	−.21	57.1	−.18	55.8	−.15	54.5	−.11
25	62.5	−.32	61.0	−.28	59.5	−.24	58.1	−.21	56.8	−.17
26	65.0	−.39	53.4	−.34	61.9	−.30	60.5	−.27	59.1	−.23
27	67.5	−.45	65.9	−.41	64.3	−.37	62.8	−.33	61.4	−.29
28	70.0	−.52	68.3	−.48	66.7	−.43	65.1	−.39	63.6	−.35
29	72.5	−.60	70.7	−.54	69.0	−.50	67.4	−.45	65.9	−.41
30	75.0	−.67	73.2	−.62	71.4	−.57	69.8	−.52	68.2	−.47
31	77.5	−.76	75.6	−.69	73.8	−.64	72.1	−.59	70.5	−.54
32	80.0	−.84	78.0	−.77	76.2	−.71	74.4	−.66	72.7	−.60
33	82.5	−.93	80.5	−.86	78.6	−.79	76.7	−.73	75.0	−.67
34	85.0	−1.04	82.9	−.95	81.0	−.88	79.1	−.81	77.3	−.75
35	87.5	−1.15	85.4	−1.05	83.3	−.97	81.4	−.89	79.5	−.83
36	90.0	−1.28	87.8	−1.16	85.7	−1.07	83.7	−.98	81.8	−.91
37	92.5	−1.44	90.2	−1.29	88.1	−1.18	86.0	−1.08	84.1	−1.00
38	95.0	−1.64	92.7	−1.45	90.5	−1.31	88.4	−1.19	86.4	−1.10
39	97.5	−1.96	95.1	−1.66	92.9	−1.47	90.7	−1.32	88.6	−1.21
40			97.6	−1.98	95.2	−1.67	93.0	−1.48	90.9	−1.34
41					97.6	−1.98	95.3	−1.68	93.2	−1.49
42							97.7	−1.99	95.5	−1.69
43									97.7	−2.00

Table XII.3 — Theoretical Log-Probability Frequency Factors [1]

CV	K values for probability in percentage greater than the given variate								
	99 −	95 −	80 −	50 −	20 +	5 +	1 +	0.1 +	0.01 +
0.000	2.33	1.64	0.84	0.00	0.84	1.64	2.33	3.09	3.72
.010	2.31	1.64	.85	.01	.84	1.65	2.35	3.13	3.79
.020	2.29	1.63	.85	.02	.84	1.66	2.38	3.17	3.85
.030	2.27	1.63	.85	.02	.84	1.67	2.40	3.22	3.91
.040	2.24	1.62	.85	.03	.84	1.68	2.42	3.26	3.98
.050	2.22	1.61	.85	.03	.83	1.69	2.44	3.30	4.05
.060	2.20	1.60	.85	.04	.83	1.69	2.46	3.35	4.12
.070	2.18	1.59	.85	.04	.83	1.70	2.49	3.40	4.19
.080	2.16	1.58	.85	.05	.82	1.71	2.51	3.45	4.26
.090	2.13	1.57	.85	.05	.82	1.72	2.53	3.50	4.34
.100	2.11	1.56	.85	.06	.82	1.72	2.55	3.56	4.42
.125	2.06	1.54	.85	.07	.81	1.74	2.61	3.67	4.61
.150	2.01	1.51	.86	.08	.81	1.76	2.66	3.80	4.81
.175	1.96	1.49	.85	.09	.80	1.78	2.72	3.92	5.03
.200	1.91	1.46	.85	.10	.79	1.79	2.78	4.05	5.24
.225	1.86	1.44	.85	.11	.78	1.81	2.83	4.18	5.48
.250	1.81	1.41	.84	.12	.77	1.82	2.88	4.32	5.71
.275	1.77	1.39	.84	.13	.76	1.83	2.94	4.46	5.96
.300	1.72	1.36	.84	.14	.75	1.84	2.99	4.60	6.19
.325	1.68	1.34	.83	.15	.74	1.85	3.04	4.74	6.44
.350	1.63	1.32	.83	.16	.73	1.86	3.10	4.88	6.70
.364	1.61	1.30	.82	.16	.73	1.87	3.12	4.94	6.82
.375	1.59	1.29	.82	.17	.72	1.87	3.15	5.02	6.96
.400	1.55	1.27	.82	.18	.71	1.87	3.20	5.16	7.22
.425	1.51	1.24	.81	.19	.70	1.88	3.25	5.30	7.49
.450	1.47	1.22	.81	.20	.69	1.88	3.30	5.44	7.78
.475	1.43	1.20	.80	.20	.68	1.89	3.34	5.58	8.06
.500	1.40	1.17	.80	.21	.66	1.89	3.38	5.71	8.35
.525	1.37	1.15	.79	.22	.65	1.89	3.42	5.86	8.65
.550	1.34	1.13	.78	.22	.64	1.89	3.46	6.00	8.96
.575	1.31	1.11	.78	.23	.63	1.89	3.50	6.14	9.27
.600	1.28	1.09	.77	.24	.61	1.89	3.53	6.27	9.57
.625	1.24	1.07	.76	.24	.60	1.89	3.57	6.41	9.88
.650	1.22	1.05	.76	.25	.59	1.89	3.60	6.55	10.20
.675	1.19	1.03	.75	.26	.58	1.89	3.63	6.70	10.51
.700	1.16	1.01	.74	.26	.57	1.88	3.66	6.83	10.83
.725	1.13	.99	.73	.27	.56	1.88	3.69	6.97	11.15
.750	1.11	.98	.73	.27	.54	1.87	3.71	7.10	11.48
.775	1.08	.96	.72	.27	.53	1.87	3.74	7.24	11.80
.800	1.06	.94	.71	.28	.52	1.86	3.76	7.37	12.12
.825	1.03	.92	.70	.28	.51	1.85	3.78	7.50	12.45
.850	1.01	.91	.70	.28	.50	1.85	3.80	7.62	12.79
.875	.99	.89	.69	.29	.48	1.84	3.82	7.74	13.11
.900	.97	.88	.68	.29	.47	1.83	3.84	7.86	13.44
.925	.95	.86	.67	.29	.46	1.82	3.86	7.98	13.77
.950	.93	.85	.67	.29	.45	1.81	3.88	8.00	14.08
.975	.92	.83	.66	.29	.43	1.79	3.90	8.20	14.39
1.000	.90	.82	.65	.29	.42	1.78	3.91	8.30	14.70

[1] Adapted from Chow (1954). Each column of Chow's Table 2 was plotted against CV. Values in this table were then read from these graphs. Additional values may be found by a graphical method desribed by Chow (1959).

where e = base of the natural logarithms, and
y = reduced variate.

However, the probability of the non-occurrence of a given event, is

$$q = 1 - p \dots \dots \dots \quad \dots \dots \dots \dots \dots \dots \dots \dots \quad \text{XII.29}$$

where p = the probability of occurrence that the value of X
will be equalled or exceeded.

Substituting Equation XII.29 into Equation XII.28, one obtains

$$(1 - p) = e^{-e^{-y}} \dots \dots \dots \dots \dots \dots \dots \dots \dots \dots \text{XII.30}$$

Equation XII.30 can be linearized by taking an iterated natural logarithm of both sides, that is

$$-\ln\left[-\ln(1 - p)\right] = y \dots \dots \dots \dots \dots \dots \dots \dots \dots \dots \text{XII.31}$$

in which y = reduced variate.

Table XII.4, taken from Powell (1943), lists values of the reduced variate, y, for given values of p, as calculated from Equation XII.31.

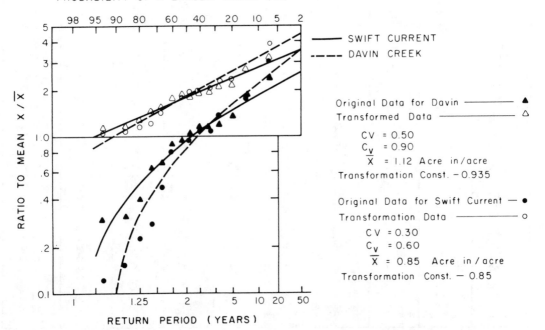

Fig. XII .2 Modified Log – Normal Probability Distribution of Runoff Volumes for Davin and Swift Current Watersheds (after Nicholaichuk 1965)

The reduced variate, y, can also be written as

$$y = a(X - X_f) \quad \ldots\ldots\ldots\ldots\ldots\ldots\ldots\ldots\ldots\ldots\ldots\ldots\ldots\ldots \text{XII.32}$$

where a   = dispersion factor, and
  X_f = mode.

For an infinitely large sample, the skewness of the extreme value distribution is constant, that is

$$\frac{\overline{X} - X_f}{s} = 0.45005 \quad \ldots\ldots\ldots\ldots\ldots\ldots\ldots\ldots\ldots\ldots\ldots\ldots \text{XII.33}$$

Hence, from Equation XII.33

$$X_f = \overline{X} - 0.45005s \ldots\ldots\ldots\ldots\ldots\ldots\ldots\ldots\ldots\ldots\ldots\ldots\ldots \text{XII.34}$$

The dispersion factor, a, is given by the relation

$$a = \frac{1.28255}{s} \quad \ldots\ldots\ldots\ldots\ldots\ldots\ldots\ldots\ldots\ldots\ldots\ldots\ldots\ldots \text{XII.35}$$

By substituting Equations XII.34 and XII.35 into Equation XII.32, one obtains the new relation

$$X = \overline{X} + s (0.7797y - 0.45005) \ldots\ldots\ldots\ldots\ldots\ldots\ldots\ldots\ldots\ldots \text{XII.36}$$

Table XII.4. Plotting Positions for the Extreme Value Distribution

P	y	P	y
(% larger)		(% larger)	
99	−1.527	20	1.500
95	−1.097	15	1.817
90	−0.834	10	2.250
80	−0.476	5	2.970
70	−0.186	4	3.197
60	0.087	3	3.491
55	0.225	2	3.902
50	0.367	1	4.600
45	0.514	0.5	5.296
40	0.672	0.4	5.520
35	0.842	0.3	5.808
30	1.031	0.2	6.214
25	1.246	0.05	7.600

Equations XII.32 and XII.36 plot as straight lines on coordinate paper. The parameters, a and X_f, of Equation XII.32 are somewhat difficult and cumbersome to derive from observed data. For this purpose, Gumbel has shown that the least-squares estimators of these values are given as

$$X_f = \bar{X} - s\,\frac{\bar{y}_N}{s_y} \quad\dots\dots\dots\dots\dots\dots\dots\dots\dots\dots\dots\dots\dots\dots \text{XII.37}$$

and

$$a = \frac{s_y}{s} \quad\dots\dots\dots\dots\dots\dots\dots\dots\dots\dots\dots\dots\dots \text{XII.38}$$

in which $\bar{y}_N$ and s_y are theoretical quantities which are a function of the sample size (see Table XII.5). Combining Equations XII.32, XII.37 and XII.38, we have

$$X = \bar{X} + \frac{s}{s_y}\left(y - \bar{y}_N\right) \dots\dots\dots\dots\dots\dots\dots\dots\dots\dots\dots\dots \text{XII.39}$$

or

$$\frac{X}{\bar{X}} = 1 + K\,C_v \dots\dots\dots\dots\dots\dots\dots\dots\dots\dots\dots\dots\dots\dots\dots\dots \text{XII.40}$$

in which the frequency factor of Equation XII.40, K, is equal to $(y - \bar{y}_N)/s_y$ whose magnitude for a particular sample size and probability level may be evaluated from information given in Tables XII.4 and XII.5. The coefficient of variation, C_v, for the extreme value distribution is calculable from the equation

$$C_v = \frac{1}{N-1}\left[\sum_{i=1}^{N}\left(\frac{X_i}{\bar{X}}\right)^2 - N\right]^{1/2} \dots\dots\dots\dots\dots\dots\dots\dots\dots\dots\dots \text{XII.41}$$

The above relationships can be used in fitting data to an extreme-value distribution for all conditions. A somewhat different procedure is outlined by McGuinness and Brakensiek

Table XII.5. Expected Means and Standard Deviation of Reduced Extremes.

N	$\bar{y}_N$	s_y	N	$\bar{y}_N$	s_y
20	0.52	1.06	80	0.56	1.19
30	0.54	1.11	90	0.56	1.20
40	0.54	1.14	100	0.56	1.21
50	0.55	1.16	150	0.56	1.23
60	0.55	1.17	200	0.57	1.24
70	0.55	1.19		0.57	1.28

(1964), utilizing the fact that the extreme value distribution is approximately the same as a log-normal distribution with CV = 0.364.

XII.1.3.4   Incomplete Gamma Distribution

Another frequency curve that has proved useful in analyses of hydrologic data, particularly drought frequencies and low flow data, is the Incomplete Gamma Distribution. This distribution, which originates out of a Pearson Type III curve, passes through the origin and is defined by the equation

$$f(X) = \frac{1}{\beta^\gamma \, \Gamma \, (\gamma)} \; e^{-X/\beta} \; X^{\gamma-1} \quad\dots\dots\dots\dots\dots\dots\dots\dots\dots\dots\text{XII.42}$$

Thom (1958) suggests that the parameters, β and γ of Equation XII.42 can be approximated by

$$\overline{X} = \gamma \beta \dots\dots\dots\dots\dots\dots\dots\dots\dots\dots\dots\dots\dots\dots\text{XII.43}$$

$$\gamma = \frac{1 + \sqrt{1 + 4A\beta}}{4A} \quad\dots\dots\dots\dots\dots\dots\dots\dots\dots\dots\text{XII.44}$$

$$\text{where } A = \ln \overline{X} - \frac{1}{N} \sum_{i=1}^{N} \ln X_i \, .$$

Integrated values of Equation XII.42 have been derived by Pearson (1951).

XII.1.4   Testing the Goodness-of-Fit

Having fitted the sample data to the assumed frequency distribution, the question remains: How good is the fit? And also: Is the agreement between sample and theoretical data such that we can accept the hypothesis that the sample is of the same population as the frequency distribution? Tests for goodness of fit are most commonly based on the discrepancy between the observed, or calculated, frequencies of values and the expected, or theoretical, frequencies of the same values for a given distribution. Obviously, different tests result from different ways of measuring the discrepancy. Two of the most popular of the 'goodness-of-fit' tests used are the classical Chi-square test and the Kolmogorov-Smirnov test.

In the Chi-square test, the statistic, χ^2 , is calculated by the equation

$$\chi^2 = \sum_{i=1}^{k} \frac{(O_i - E_i)^2}{E_i} \quad\dots\dots\dots\dots\dots\dots\dots\dots\dots\dots\dots\text{XII.45}$$

where O_i = observed value of the X variate, and
E_i = expected value of X.

The significance level of the statistic can be tested by comparing the calculated value with tabulated values (which can be found in most statistics books) with $k-b$ degrees of freedom. In this relation, k is the number of comparisons between observed and expected values, which have been made in computing x^2. The value b is the number of degrees of freedom lost in the calculation, or ways in which the observed and fitted data have been forced to agree with each other. For example, for a normal distribution this would be with respect to the mean, $\bar{X}$, the standard deviation, s, and the number of observations, N. In the test, if the calculated value of x^2 is greater than the value tabulated for a given sample size and preselected probability level, the hypothesis that the sample is of the same population as the assumed theoretical curve is rejected. The particular advantages of the test are that it is applicable for testing either discrete or continuous functions, and that it can be used to test either a completely or partially specified distribution. A particular disadvantage of the test is that it is valid only for large samples.

Another test that may be employed to test the goodness-of-fit of hydrologic data to a frequency distribution is the Kolmogorov-Smirnov statistic; this is particularly useful for small samples of data. In this test, the statistic, D_n, or the ratio of the maximum discrepancy between the theoretical distribution function and the actual distribution to the number of samples, is calculated $(E_i - O_i/N)_{max}$ and compared with tabulated values of K, as shown in Table XII.6 (taken from Lindgren and McElrath 1959). If $D_n > K$ at a given acceptance limit, then one usually rejects the hypothesis that the population is of the assumed type.

Although the Kolmogorov-Smirnov test is only applicable in a strict sense to those cases where the distribution is continuous, it is often applied to analyze discrete populations. Acceptance of this procedure is inherent, since it has been found that such practice is safe in the sense that the actual significance level of the resulting test is no larger than the one assumed in using the Tables.

XII.1.5   Application of Frequency Curves

On the basis of the preceding discussions it will suffice to say that there are numerous frequency distributions which may be used to describe populations of hydrologic data; however, it must be reiterated that care should be used in the choice of a particular frequency curve, so that whenever possible its selection is based on physical reasoning. Also, be careful to use observations which were caused by, or involve, the same physical processes, so that the population is homogeneous. For example, in analyses of runoff extremes, if all floods from snowmelt, rain, or snowmelt and rain, are included in a single series, the data will probably not follow any of the standard frequency curves. Further, it should be recognized that a frequency curve cannot be extrapolated by the fitted line to a level of zero probability. That is, there is a physical upper limit to the extrapolation, which is governed by the physical limitations and boundaries imposed on the hydrologic phenomenon by physiographic conditions.

Frequency curves give the probability of occurrence that the magnitude of the event is equalled, or exceeded (or less than), within a given year. That is, we say the return

Table XII.6. Acceptance Limits for the Kolmogorov-Smirnov Test of Goodness-of-Fit

Sample Size (N)	Significance Level				
	0.20	0.15	0.10	0.05	0.01
1	0.900	0.925	0.950	0.975	0.995
2	.684	.726	.776	.842	.929
3	.565	.597	.642	.708	.829
4	.494	.525	.564	.624	.734
5	.446	.474	.510	.563	.669
6	.410	.436	.470	.521	.618
7	.381	.405	.438	.486	.577
8	.358	.381	.411	.457	.543
9	.339	.360	.388	.432	.514
10	.322	.342	.368	.409	.486
11	.307	.326	.352	.391	.468
12	.295	.313	.338	.375	.450
13	.284	.302	.325	.361	.433
14	.274	.292	.314	.349	.418
15	.266	.283	.304	.338	.404
16	.258	.274	.295	.328	.391
17	.250	.266	.286	.318	.380
18	.244	.259	.278	.309	.370
19	.237	.252	.272	.301	.361
20	.231	.246	.264	.294	.352
25	.21	.22	.24	.264	.32
30	.19	.20	.22	.242	.29
35	.18	.19	.21	.23	.27
40				.21	.25
50				.19	.23
60				.17	.21
70				.16	.19
80				.15	.18
90				.14	
100				.14	
Asymptotic Formula:	$\dfrac{1.07}{\sqrt{N}}$	$\dfrac{1.14}{\sqrt{N}}$	$\dfrac{1.22}{\sqrt{N}}$	$\dfrac{1.36}{\sqrt{N}}$	$\dfrac{1.63}{\sqrt{N}}$

period of the event is $1/p$ years. For example, if a flood equal to or greater than a given discharge, Q, occurs in a certain stream 100 times in a 1000 years of record, we say that its average return period, t_r, is $1000/100 = 10$ years; or that there is the probability, p, of the discharge, Q occurring within a given year of $1/t_r$ or 0.10. Further considerations of probability analysis, with particular reference to flood events, are given below.

Propositions

1.   Probability of a t_r yr event is

$$p = \frac{1}{t_r} \qquad \dots\dots\dots\dots\dots\dots\dots\dots\dots\dots\text{XII.46}$$

2.   Probability of occurrence of event less than t_r yr event is

$$q = 1 - p \qquad \dots\dots\dots\dots\dots\dots\dots\dots\dots\dots \text{XII.47}$$

3.   If r is the length of record or the life of the structure (service period), the probability that Q will occur once in the next r years is

$$p_r = pq^{r-1}\, r \dots\dots\dots\dots\dots\dots\dots\dots\dots \text{XII.48}$$

4.   Probability of the event occurring n times in r years is

$$p_{nr} = p^n q^{r-n}\, \frac{r\,!}{(r-n)!\,n!} \qquad \dots\dots\dots\dots\dots\dots\dots \text{XII.49}$$

or in r years $p_{nr} = p^r$ (Note $0! = 1$)

5.   The probability that the event will occur *at least* once in r years is

$$p_r = \sum_{n=1}^{r} p_n = pq^{r-1}\, r + p^2 q^{r-2}\, \frac{r(r-1)}{2} + \dots\dots +p^r \dots\dots\dots \text{XII.50}$$

or . $p_r = (p+q)^r - q^r$

But $p + q = 1$. Therefore

$$p_r = 1 - q^r \qquad \dots\dots\dots\dots\dots\dots\dots\dots\dots \text{XII.51}$$

It follows from Proposition 5, that the probability of Q occurring at least once within its return period can be evaluated by

$$p_{t_r} = 1 - \left(1 - \frac{1}{t_r}\right)^r = 1 - \left(1 - \frac{1}{t_r}\right)^{t_r} \text{ (for } r = t_r) \dots\dots\dots\dots \text{XII.52}$$

Values of p_{t_r} for different values of $r = t_r$ are given below:

$r = t_r$	p_{t_r}
2	0.75
3	0.704
4	0.684
5	0.672
10	0.651
50	0.636
100	0.634

It can be observed in the tabulation that, as the return period increases, the probability, p_{t_r}, approaches a limit. Mathematically it has been shown that, as t_r approaches ∞, the expression $\left(1 - 1/t_r\right)^\infty$ obtains a value of e^{-1}. Therefore

$$\underset{t_r \to \infty}{p_{t_r}} = 1 - e^{-1} = 1 - \frac{1}{e} = 0.63 \dots\dots\dots\dots\dots\dots \text{XII.53}$$

For example, this suggests that if a structure is designed to withstand a long-time flood, there is a 63 per cent chance that this flood event will be exceeded at least once within its return period.

This analysis provides a basis on which a structure may be designed to accommodate the discharge from a flood event whose return period is selected to correspond to a given probability of risk, U, that a flood of equal or greater magnitude will occur at least once during the service period of the structure, r; that is

$$p_r = U = 1 - \left(1 - \frac{1}{t_r}\right)^r \dots\dots\dots\dots\dots\dots \text{XII.54}$$

When r is greater than 10, the expression can be simplified to

$$U = \frac{r}{t_r + \frac{r}{2}} \quad \text{or}$$

$$t_r = r\left(\frac{1}{U} - \frac{1}{2}\right) \dots\dots\dots\dots\dots\dots\dots\dots\dots\dots \text{XII.55}$$

Application

Select the return period of a flood event for design of a spillway where there is only a 10 per cent chance (acceptable risk) that the flood will be exceeded once within 50 years.

$$t_r = 50 \left(\frac{1}{0.10} - \frac{1}{2} \right) = 475 \text{ yrs}$$

Table XII.7. Required Return Periods, t_r in Years, Corresponding to Different Levels of Acceptable Risk, U, for Different Service Periods, r.

Risk U	Service Period, r.					
	2	5	10	20	50	100
0.75	1.7	4.0	6.7	14.9	35.6	72.7
0.50	3.4	7.7	14.9	29.4	72.6	144.8
0.40	4.0	10.3	20.1	39.7	98.4	196
0.30	5.7	14.5	28.5	56.5	141	281
0.20	9.0	22.9	45.3	90.1	225	449
0.10	19.5	48.1	95.4	190	475	950
0.05	39.5	98	195	390	976	1949
0.01	198	498	996	1992	4975	9953

XII.2   REGRESSION ANALYSES

Frequently, in hydrologic work, we are concerned with the dependence of a given variable, Y, on one or more other independent variables, $X_1, X_2, X_3 \ldots \ldots \ldots X_N$, so that we may predict the value of Y knowing the values of $X_1, X_2, X_3 \ldots \ldots X_N$. For example, we may be interested in estimating the volume of surface runoff Y from observations of factors such as depth of precipitation X_1, rainfall intensity X_2, infiltration rate X_3, and others. The analysis used to determine the association between variables may be referred to as regression analysis, correlation analysis, or simply, graphical-data analysis.

XII.2.1   Method of Least Squares

Suppose two variables Y and X (say water yield and precipitation), when plotted as a scatter diagram, indicate linear association. In equational form, therefore, Y = a + bX. The problem in analysis of these data involves: (a) determining the line of best fit to the data; and (b) determining a measure of scatter of the points from the relationship, or how well the relationship explains the variation of the independent variable.

Probably, the most common method of fitting lines to experimental data is to employ the Method of Least Squares. By this principle is meant that the line is fitted in such a manner that the sums of squares of the deviations of the individual points from the line are minimized. Hence, for a linear relation

$$\sum_{i=1}^{N}(Y_i - Y)^2 = \sum_{i=1}^{N}(Y_i - a - bX_i)^2 = \text{minimum} \dots \dots \dots \text{XII.56}$$

in which Y_i and X_i are observed values.

The sums of squares given by Equation XII.56 can be minimized by differentiating the equation with respect to a and b, and by setting the results equal to zero, the minimum. That is

$$\frac{\partial}{\partial a}\sum_{i=1}^{N}(Y_i - a - bX_i)^2 = 0$$

$$\frac{\partial}{\partial b}\sum_{i=1}^{N}(Y_i - a - bX_i)^2 = 0$$

from which we obtain the normal equations

$$Na + b\sum_{i=1}^{N}X_i = \sum_{i=1}^{N}Y_i \dots \dots \dots \text{XII.57}$$

$$a\sum_{i=1}^{N}X_i + b\sum_{i=1}^{N}X_i^2 = \sum_{i=1}^{N}X_iY_i \dots \dots \dots \text{XII.58}$$

Solving Equations XII.57 and XII.58 simultaneously, one obtains the expressions

$$b = \frac{\sum xy}{\sum x^2} \dots \dots \dots \text{XII.59}$$

$$a = \overline{Y} - b\overline{X} \dots \dots \dots \text{XII.60}$$

in which $x = (X_i - \overline{X})$

$y = (Y_i - \overline{Y})$

$$\sum xy = \sum_{i=1}^{N} X_i Y_i - \left[\sum_{i=1}^{N} X_i \sum_{i=1}^{N} Y_i \Big/ N \right]$$

$$\sum x^2 = \sum_{i=1}^{N} X_i^2 - \left[\left(\sum_{i=1}^{N} X_i \right)^2 \Big/ N \right]$$

$$\overline{Y} = \sum_{i=1}^{N} Y_i \Big/ N$$

$$\overline{X} = \sum_{i=1}^{N} X_i \Big/ N$$

XII.2.2   Interval Estimates and Test of Hypotheses

Since the 't' test and the use of confidence bands have already been discussed briefly, only the computation procedures related to these statistics as applied to linear regression will be listed below.

't' Test Applied to Slope – H_0: $b = \beta = 0$

$$t = \frac{b - \beta}{s_b} \quad \dots\dots\dots\dots\dots\dots\dots\dots\dots\dots\dots\dots \text{XII.61}$$

in which s_b is the standard deviation of the slope whose magnitude may be calculated as follows:

Variance of Slope, $\quad s_b^2 = s_{y.x}^2 \Big/ \sum x^2 \dots\dots\dots\dots\dots\dots\dots\dots\dots\dots$ XII.62

where $s_{y.x}^2$ is the variance of the individual points from the fitted line. That is

$$s^2{}_{y.x} = \sum_{i=1}^{N} d_i{}^2 \,/\, N - 2 = \left[\sum y^2 - \left(\sum (xy)^2 \Big/ \sum x^2 \right) \right] \Big/ N - 2 \dots \text{XII.63}$$

where $d_i = (Y_i - a - bX_i)$,

$$\sum y^2 = \sum_{i=1}^{N} Y_i{}^2 - \left[\left(\sum_{i=1}^{N} Y_i \right)^2 \Big/ N \right], \text{ and}$$

$y = Y_i - \overline{Y}$.

Sources of Variation and Their Standard Deviations

The Mean $\qquad = \sqrt{s^2_{y.x} / N}$ XII.64

The Fitted Line $\quad = {}^{s}y.x \sqrt{1/N + (X_i - \overline{X})^2 / \Sigma x^2}$ XII.65

Individual $\qquad = {}^{s}y.x \sqrt{1 + 1/N + (X_i - \overline{X})^2 / \Sigma x^2}$ XII.66

An illustration of confidence bands as they appear on a linear plot is shown in Fig. XII.3.

XII.2.3   **Partitioning the Sum of Squares**

Another most useful analysis of the data is to determine the relative proportions of the total variation of the individual points from their mean that can be explained or attributed to the fitted line, and the proportion which represents the deviations of the points from the line. Obviously, the smaller the variance of the points from the line, the better is the prediction equation. As shown in Fig. XII.4, each individual value of Y_i may

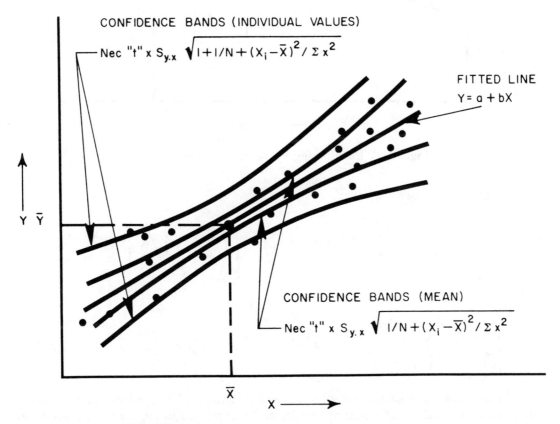

Fig. XII.3 Confidence Limits For Linear Regression

be divided to three components: (a) the mean $\overline{Y}$; (b) the difference from the mean due to the slope of the line, ΔY_i; and (c) the difference from the line, d_i. Thus

$$Y_i = \overline{Y} + \Delta Y_i + d_i \dots\dots\dots\dots\dots\dots\dots\dots\dots\dots\dots\dots\dots\dots\text{XII.67}$$

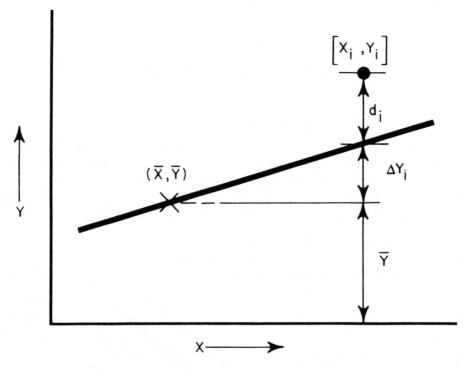

Fig. XII.4 Partitioning Variance in Linear Models

Further it can be shown that

$$\sum_{i=1}^{N} Y_i^{\,2} = \sum_{i=1}^{N} \left(\overline{Y} + \Delta Y_i + d_i\right)^2 = \left[\left(\sum_{i=1}^{N} Y_i\right)^2 \Big/ N\right] + b^2 \sum x^2 + \sum d_i^{\,2}$$

or

$$\sum_{i=1}^{N} Y_i^2 - \left(\sum_{i=1}^{N} Y_i\right)^2 \Big/ N = \sum y^2 = b^2 \sum x^2 + \sum d_i^{\,2} \dots\dots\dots\dots\dots\text{XII.68}$$

This partitioning of the sum of squares of deviations, $\sum y^2$, to its various sources is referred to as *Analysis of Variance*, and is usually expressed in tabular form as shown in Table XII.8

The significance of the fitted curve in explaining the variation of Y with X may also be tested using an 'F' test, which employs the ratio: mean square due to regression divided by the mean square of the deviations. Values for F are tabulated in most texts on Statistics (for example, Snedecor 1959).

In the preceding analysis, it has been assumed that all values of Y_i have a common variance independent of the magnitude of X_i. Most often, however, with hydrologic data, the variance of the Y-values, σ^2, increases with X_i. For the case where the increase is proportional to the value of X (that is, $\sigma^2 \propto X$, and for a linear model passing through the origin, $Y = bX$), Snedecor (1959) has shown that the least-squares estimator of the slope, b, is given by the equation

$$b = \sum_{i=1}^{N} (Y_i/X_i) \Big/ N \dots\dots\dots\dots\text{XII.69}$$

and the variance of the coefficient, b, is

$$s_b^2 = s_{y/x}^2 \Big/ N = \frac{\sum_{i=1}^{N}\left(Y_i/X_i\right)^2 - \sum_{i=1}^{N}\left(Y_i/X_i\right)^2 \Big/ N}{N} \dots\dots\dots\dots\text{XII.70}$$

The level of significance of b can be tested using Equation XII.61 with N - 1 degrees of freedom.

Table XII.8. Analysis of Variance — Linear Regression

Source of Variation	Symbol	Degrees of Freedom	Sum of Squares	Mean Squares or Variance
Slope of Line (Regression)	b	1	$b^2 \sum x^2$	$b^2 \sum x^2$
Deviation from Line	d	N - 2	$\sum d^2$	$\sum d^2 /(N-2)$
Deviations from Mean (total)	y	N - 1	$\sum y^2$	

XII.2.4   Correlation

Another common statistic that is used to indicate the degree of association between two variables is the correlation coefficient, r. The values of r may vary within the limits from - 1 to + 1, depending on the agreement of the variables (see Figs. XII.5a and XII.5b). In effect, the more elongated the elliptical shape of the observed data, the higher the degree of correlation between the variables—thus the larger the value of r (positive or negative). Negative values of, r, like those of the slope, b, simply indicate the data plot with an inclination of the ellipse sloped downward to the right.

For linear regression, the correlation coefficient can be computed from the relation

$$r = \frac{\sum xy}{\sqrt{\sum x^2 \; \sum y^2}} \quad \dots\dots\dots\dots\dots\dots\dots\dots\dots\dots\dots\dots \text{XII.71}$$

A most interesting feature of the correlation coefficient can be obtained from consideration of the sum of squares of deviation attributable to regression; that is

$$\text{Sum of Squares for Regression} = b^2 \sum x^2$$

But $b^2 = \left(\sum xy\right)^2 \bigg/ \left(\sum x^2\right)^2$ Therefore

$$\text{Sum of Squares for Regression} = \frac{\left(\sum xy\right)^2}{\left(\sum x^2\right)^2} \cdot \sum x^2$$

$$= \frac{\left(\sum xy\right)^2}{\sum x^2} \cdot \frac{\sum y^2}{\sum y^2}$$

$$= \frac{\left(\sum xy\right)^2}{\sum x^2 \sum y^2} \cdot \sum y^2$$

$$= r^2 \sum y^2 \quad \dots\dots\dots\dots\dots\dots\dots\dots \text{XII.72}$$

Note, from Equation XII.72, that r^2, which is often referred to as the coefficient of determination, R, represents the percentage of the variation of the Y_i values from the mean, which can be explained by the slope of the fitted line or the regression. This interpretation of the correlation coefficient is perhaps most useful in hydrology. It should be realized that a statistical test, which employs only the values of the coefficient, r, to establish the level of association between two variables, is theoretically valid only for bivariate normal populations.

Another property of hydrologic data that must be taken into account when employing a correlation analysis is the problem that many events may be serially correlated. That is, the magnitude of an observed value may be related to its previous value (in time). As an example, mean daily stream flows taken on successive days are frequently serially correlated, since the flow on one day is highly dependent on the flow on the preceding day. When events are serially correlated, the information content of the statistic determined from the time series is actually less than that which would result if the value were computed using independent random observations. Matalas (1963) and Yevdjevich (1964) have outlined tests which may be used to determine the correlation between different time series (autocorrelation).

Finally, a word of warning: considerable care should be taken in correlation analysis to avoid introduction of spurious (apparent) correlation to data. By spurious correlation we mean the manipulation of observations to obtain a correlation between events which need not have any relation. For example, Slivitzky (1967) points out that the Q_{50} (daily) and Q_{50} (monthly) median flows of a series of rivers may not be related. However, if each is divided by the mean annual flow of the river, Q_a, then the ratios, Q_{50} (daily)/Q_a and Q_{50} (monthly)/Q_a, may show a spurious correlation. For further discussions of spurious correlation in hydrologic data, the reader is referred to the paper by Benson (1965).

Summary

It should be recognized that statistical techniques and tests, similar to those outlined above, have been developed for fitting and testing the association between variables, in curvilinear or multiple regression. These procedures are detailed in texts such as those of Williams (1959) and Snedecor (1959). When used properly, they prove to be a powerful tool in hydrologic investigation. In most cases, these techniques have been developed on the assumptions that each variable is independent and normally distributed with homogeneous variance. Hence, special attention should be given to determine whether the

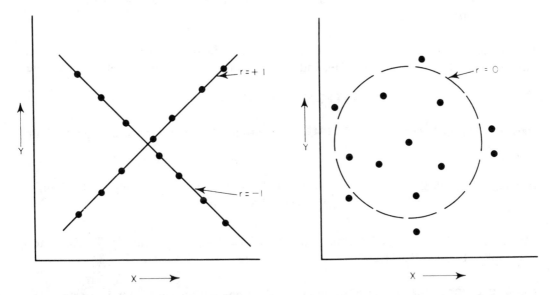

Fig. XII.5a Correlation Coefficient   Fig. XII.5b Correlation Coefficient

data fulfil these requirements. In the event that they do not, then special procedures must be used in testing the significance, and in making estimates from the correlations.

XII.2.5   Graphical Correlation

Frequently, in analyses of hydrologic data, the fitting of a curve to observations by graphical techniques will provide satisfactory results without resorting to analytical methods, such as the method of Least Squares. The particular advantages of graphical methods are that they are usually simpler and less laborious to apply than the analytical methods, and that they give proper weight to *unusual* events. Among the most common of the graphical procedures used, are the following:

1.   *Two-variable correlation*—in which a line is fitted to the data so that it passes through the mean $(\bar{X}, \bar{Y})$, and the sum of the deviations of the individual observations, d_i is approximately zero.

2.   *Multiple-variable correlation*—in which the method of deviations (Ezekiel 1941) is used. By this method a two-variable graphical correlation is first established between Y and the most important variable, X_1. Deviations from this curve, d, are then associated with the next important variable, X_2. Then the deviations from this curve d are associated with the next important independent variable, X_3, and so on until all the independent variables have been introduced into the correlation. This procedure leads to a second approximation of the relation $Y = f(X_1)$, and the procedure is repeated.

3.   *Coaxial Method* —in which a multiple-variable correlation may be plotted using common axis on the paper. This method also involves a series of successive approximations.

The details of these procedures are clearly outlined in the following papers and texts, see Linsley *et al.* (1949), Ezekiel (1941), and Solomon (1967), and will not be repeated here.

XII.3   LITERATURE CITED

Barger, G.L. and Thom, H.C.S. 1949. Evaluation of drought hazard. Agron Jour. 41:519-526.

Benson, M.A. 1965. Spurious correlation in hydraulics and hydrology. J. Hydraulics Div., Amer. Soc. Civil Engr. 91, No. HY4 Proc. Paper 4393.

Chow, V.T. 1951. A general formula for hydrologic frequency analysis. Trans. Amer. Geophys Union. 32:231-237.

Chow, V.T. 1954. The log-probability law and its engineering applications. Amer. Soc. Civil Engrs. Proc. Vol. 80, Separate 536.

Chow, V.T. 1959. Determination of hydrologic frequency factor. Amer. Soc. Civil Engrs. Proc. Vol. 85, Separate 2084.

Ezekiel, M. 1941. Methods of correlation analysis. 2nd Ed., John Wiley and Sons, Inc., New York.

Gumbel, E.J. 1941. The return period of flood flows. Ann. Math Stat. 12:163-160.

Gumbel, E.J. 1954. Statistical theory of extremes and some practical applications. U.S. Natl. Bureau of Standards Applied Math. Series 33.

Kendall, G.R. 1960. The cube-root normal distribution applied to Canadian Monthly Rainfall. Totals. Publ. 53 of the IASH Commission of Land Erosion Helsinki. p. 250-260.

Kendall, G.R. 1967. Probability distribution of a single variable. Statistical Methods in Hydrology. Proc. Hydrology Symposium No. 5. NRC Assoc. Comm. on Geodesy and Geophysics, Subcommittee on Hydrology. The Queen's Printer, Ottawa. pp. 37-51.

Lindgren, B.W. and McElrath, G.W. 1959. Introduction to probability and statistics. The Macmillan Company, Galt, Ontario.

Linsley, R.K., Kohler, M.A. and Paulhus, J.L.H. 1949. Applied Hydrology. 1st. Ed. McGraw-Hill Book Co., Inc. New York.

Matalas, N.C. 1963. Auto correlation of rainfall and streamflow minimums. U.S. Geol. Survey, Prof. Paper 434-B.

Matalas, N.C., and Langbein, W.B. 1962. Information content of the mean. J. Geophys. Res. 67, No. 9.

McGuinness, J.L. and Brakensiek, D.L. 1964. Simplified techniques for fitting frequency distributions to hydrologic data. Supt. of Documents. U.S. Gov't Printing Service, Washington, D.C.

Nicholaichuk, W. 1965. Comparative watershed studies in Southern Saskatchewan. Paper presented at Pacific Northwest Region Mtgs. Amer. Soc. Agr. Engr. Moscow, Idaho.

Pearson, K. *et al.* 1951. Tables of the Incomplete Function, Cambridge University Press.

Powell, R.W. 1943. A simple method of estimating flood frequency. Civil Engr. 13:105-107.

Slivitzky, M. 1967. Appraisal of methods of analysis. Statistical Methods in Hydrology, Proc. Hydrology Symposium No. 5. NRC Assoc. Comm. on Geodesy and Geophysics.

Subcommittee on Hydrology. The Queen's Printer, Ottawa, pp. 183-192.

Snedecor, G.W. 1959. Statistical methods. 5th ed. The Iowa State College Press, Ames, Iowa.

Solomon, S. 1967. Statistical association between hydrologic variables. Statistical Methods in Hydrology No. 5. NRC Assoc. Comm. on Geodesy and Geophysics, Subcommittee on Hydrology. The Queen's Printer, Ottawa, pp. 55-113.

Stidd, C.K. 1953. Cube root normal precipitation distribution. Trans. Amer. Geophys. Union 34:31-35.

Thom. H.C.S. 1958. A note on the gamma distribution. Monthly Weather Rev. 86:117-122.

Williams, E.J. 1959. Regression analysis. John Wiley and Sons, Inc. New York, N.Y.

Yevdjevich, V.M. 1964. Fluctuation of wet and dry years, Part II Analysis of serial correlation. Hydrology Paper No. 4. Colorado State Univ.

Section XIII

HYDROLOGY OF LAND USE

by

Walter W. Jeffrey

TABLE OF CONTENTS

LIST OF TABLES

LIST OF FIGURES

Section XIII

HYDROLOGY OF LAND USE

XIII.1   INTRODUCTION

Canada in the late 1960's still shows many characteristics of a country with pioneering conditions. The population is concentrated in the urban centres of the southern fringe, such concentrations tending to obscure the fact that elsewhere resources are being developed, and areas hitherto undisturbed are being brought into economic production.

Forestry operations now reach into watersheds previously immune from the power saw and bulldozer. Mining is being introduced to exploit deposits which either were recently discovered or were formerly uneconomic to develop. Agricultural land area is being extended. Super-highways are being built through ecosystems which are sometimes fragile and vulnerable. Increasing amounts of fertilizers and biocides are being applied in land-use practice. Existing urban areas are being extended, and new ones being developed. Urban populations in their leisure hours escape from the cities, and use the rural landscape for recreation.

All these effects are influential upon hydrology, and particularly in a small watershed context must be considered in both hydrologic research and hydrologic land management. Land use is one element of the hydrologic landscape that can deliberately be manipulated to create water resource characteristics beneficial to society. Other landscape parameters, such as physiography, geomorphology and climate, are not easily susceptible to purposive treatment.

In spite of such Canadian conditions, which make interest in land-use hydrology particularly pertinent, very few data are available that deal specifically with the Canadian environment. Few studies have been made, and those that are under way have not yet had enough time to produce definitive results. Though Canadian examples are used wherever possible, the lack of Canadian studies makes dependence upon data from other countries inevitable. There are elements of danger in this, for the environment extending over most of Canada is dissimilar to that of many areas where land-use hydrology research has been particularly active. In the extrapolation of data from these other countries, considerable caution is necessary. Land-use hydrology is still largely a regional science. It is to be hoped that, side by side with the development of more comprehensive models which will allow extrapolation of results obtained from other areas, there will be an increasing effort to study the hydrology of land use in Canada, so that information for design and development is not wanting when needed.

Study of land-use hydrology deals with virtually all elements of hydrology contained in the previous twelve Sections, for a sound knowledge of basic hydrology is needed to explore the hydrology of landscape manipulation and disturbance. Obviously then, a comprehensive treatment of the subject would require a separate text in itself. This is impossible in the present setting, though it is noteworthy that certain texts dealing with some aspects of land-use hydrology do exist, for example, Kittredge (1948), Colman (1953) and Penman (1963). To make for more concise treatment, the approach taken here is different to that of preceding Sections. A descriptive method is utilized to give the reader insight into land-use effects upon water resources, rather than to enable him actually to undertake experimentation in land-use hydrology as such. Most research into land-use hydrology utilizes techniques previously elucidated. Some techniques described were, in fact, developed for land-use hydrology research, and have entered the basic hydrology literature by that route.

Land-use hydrology deals with the effects of forestry, agriculture, mining, urbanisation, highway development, and associated activities. Most attention is given to hydrology of forestry activities, not because these are considered to be of overwhelming importance in relation to other activities but because comparatively more knowledge exists in this framework, having been developed in other countries for that purpose, and because many of the principles involved are directly transposable to other land-use activities.

To avoid an overwhelming mass of bibliography, references wherever possible are confined to major review papers rather than to original data sources. Such review papers provide sources of original research references, available for more detailed perusal.

XIII.2   PRECIPITATION

Canopy interception (as discussed in Section IV) is a function of biomass and spatial arrangement of vegetational cover, in addition to other factors. Cover modification, or removal, influences the amount of precipitation reaching the earth's surface, and accordingly the magnitude of interception loss. Interception, however, represents only one component by which land use, or vegetational cover manipulation, may influence precipitation and the other hydrologic processes. Litter removal may modify the quantity of water entering mineral soil, and may have a large effect on raindrop impact, under certain circumstances a major factor in soil compaction. Under some conditions, removal of forest cover may change the amount of occult precipitation (fog drip). In the case of snow, in addition to the canopy interception effects of land use, there are influences on snow accumulation patterns and snow evaporative losses.

XIII.2.1   **Rainfall**

Because of its greater ease of measurement, more is known of land-use effects upon rainfall input than of similar effects upon snowfall.

XIII.2.1.1   Canopy Interception (Rain)

The relationship between net precipitation P_{rn} and gross precipitation P_r is usually expressed by:

$$P_{rn} = a + bP_r. \dots\dots\dots\dots\dots\dots\text{XIII.1}$$

This equation has application for individual storm events and should not be used for periodic (weekly, monthly) summations of gross precipitation.

One may also express net precipitation as a function of certain variables:

$$P_{rn} = f\ (CC + P_r + E) \ \dots\dots\dots\dots\dots\text{XIII.2}$$

where, CC = canopy characteristics,
P_r = storm size or amount of precipitation and
E = evaporation during storm.

It is seen from this that whereas the terms P_r and E are dependent on climate, the canopy characteristic factor CC is an expression of stand biomass and spatial arrangement, and thus modifiable by land use. Management procedures, such as logging and crop harvesting, and catastrophies, such as forest fire and windthrow, have major effects upon canopy characteristics of forest stands and other types of vegetation.

Zinke (1967) has summarized the North American literature pertaining to interception losses from various forest covers. Selected values for conifers are reproduced in Table XIII.1, and examination of this table illuminates the statements made above.

Table XIII.1 Selected Interception Estimates for Coniferous Forest Stands

Abies lasiocarpa (Hook.) Nutt., single tree
Sf in cubic inches = $3583 P_r - 387$

Picea abies (L.) Karst, 20-year old plantation.
$I_v = 58\%$

Picea engelmannii Parry, single tree
Sf in cubic inches = $1035 P_r - 301$

Table XIII.1 (Cont'd)

Pinus banksiana Lamb.

P_r (storm)	0.05	0.1	0.3	>0.3
$P_{rn}\%$	60	70	80	80 + 2/3% per 0.1 increase in P_r

Pinus contorta Dougl.

$S_f < 0.01$ in. per tree for 43 summer rainstorms

Type of stand and residual volume in board feet	$P_{rn}\%$	
1. Virgin, 11,800 bd. ft./acre	69.1	$P_{rn} = 0.8046\,P_r - 0.0290$
2. Cutover, 8000 bd. ft/acre	80.1	$P_{rn} = 0.8677\,P_r - 0.0149$
3. Cutover, 4000 bd. ft/acre	85.1	$P_{rn} = 0.9055\,P_r - 0.0131$
4. Cutover, 2000 bd. ft./acre	86.5	$P_{rn} = 0.8933\,P_r - 0.0074$
5. Clear cut (all trees over 9.5 in. d.b.h.)	92.8	$P_{rn} = 0.9897\,P_r - 0.0153$

Treatment effects: in second growth

Treatment	100% canopy none	40% canopy crop tree	40% canopy single tree
Summer P_{rn}	3.68 in.	4.17 in.	4.33 in.
$P_{rn}\%$ increase	0	13.3	17.7
Winter P_{sn}	10.03 in.	11.72 in.	12.34 in.
$P_{sn}\%$ increase	0	16.8	23.0

Pinus ponderosa Laws.

Mature stand, 120 ft tall, California
    $I_{si} = 14.5\%$, 5-year average.
    $I_{si} = 0.10\,P_s + 0.09$.
Young stand, 14 ft tall, 79 percent crown coverage, California
    $I_{si} = 15.4\%$, 5-year average.
    $I_{si} = 0.11\,P_s + 0.01$
65–75-year-old stand, fully stocked, California
    $I_{si} = 12\%$ range $< 7\%$ to $> 80\%$
    $T_h = 84\%$
    $S_f = 4\%$
    $I_{si} = 0.06\,P_s + 0.09$
    $I_{ri} = 0.06\,P_r + 0.12$
    $I_v = 0.06\,P + 0.11$, all storms > 0.5 in.
Stands in Colorado
    $S_f = $ nil.
    $I_v = 16\%$, average of snow and rain storms

Table XIII.1 (Cont'd)

Pinus resinosa Ait.
Plantation, 17 years old
I_{si} = 35%
Plantation, 85 percent canopy density, twenty-one storms

P_r	I_{ri}		T_n		S_f	
in.	in.	%	in.	%	in.	%
13.4	2.5	18.7	10.7	80.1	0.1	1.2

Plantation, 31 years old, 31 ft tall, stand improvement by prunings.
Interception loss as percent of various storm sizes in pruned and unpruned.

Storm size (in.)	0.00–0.01	0.011–0.04	0.041–0.16	0.161–0.77	0.771–0.381	Average
Unpruned%	85	74	59	25	14	21.4
Pruned%	48	28	27	19	16	17.6

Pinus strobus L.
96 percent canopy I_{ri} = 43%
I_{si} = 37%, 20 year old plantation
I_{si} = 35%, 21–30 years old
I_{si} = 59%, 61–80 years old
Single tree (average of 6 storms).

Distance from trunk in feet   0   4.5   6   9
I_v 25% 38%   48%–11.2% (attributed to edge of
                                                    crown drip).

Pseudotsuga menziesii
Old growth, 199 to 248 ft tall, Vancouver Island, B.C.
   I_v = 44% annual ignoring stemflow, 57% summer with nil stemflow; average of
      5-1/2 years.
Old growth, 147 to 175 ft tall, Vancouver Island, B.C.
   I_v = 31% annual ignoring stemflow, 49% summer with nil stemflow; average of
      5-1/2 years.
Old growth, 87 to 116 ft tall
   I_v = 28% annual ignoring stemflow, 39% summer with nil stemflow; average of
      5-1/2 years.
Dense old growth, Oregon
   S_f = 0.27% of P_r considered insignificant.
   I_{ri} = 14% winter, 24% summer.

Storm size in inches	0–0.5	0.05–0.5	0.5–1.0	1.1–1.3	1.5–2.0
I_{ri} %	100	32	23	21	19

Pseudotsuga menziesii (Mirb.) Franco—*Tsuga heterophylla* (Raf.) Sarg.
Old growth, 144 to 209 ft tall
   I_v = 34% annual, 51% summer.
Old growth, 77 to 110 ft tall
   I_v = 20% annual, 30% summer.

Table XIII.1 (Cont'd)

Thuja plicata Donn.
Old growth, 125 to 196 ft tall
  I_v = 33% annual, 40% summer
  Average of 5-1/2 years with stemflow ignored.

P = gross storm precipitation	P_r = depth of storm precipitation (rain)
P_{rn} = net precipitation (rainfall)	P_s = depth of storm precipitation (snow)
P_{sn} = net precipitation (snow)	I_v = total interception loss
I_{si} = snow interception loss	I_{ri} = rain interception loss
S_f = stemflow	T_h = throughfall

Values for Douglas fir *(Pseudotsuga menziesii* Mirb., Franco) illustrate effects of both storm size and canopy characteristics; results for red pine *(Pinus resinosa* Ait.) show effects of pruning; and values for lodgepole pine *(Pinus contorta* Dougl.) demonstrate the effects of forestry treatments, showing a sizeable increase in net precipitation to result from removal of all or part of the stand biomass. Allied with similar results from other species, this has encouraged the view that a relationship exists between net precipitation and the basal area of forest stands, particularly to residual basal area after thinning. Further examination of lodgepole pine results cited in Table XIII.1 is merited.

The regression equation derived by Niederhof and Wilm (1943) for a mature lodgepole pine stand of 159 sq. ft. basal area per acre was

$$P_{rn} = -0.0290 + 0.8046P_r \dots\dots\dots\dots\dots\dots\dots\dots\dots XIII.3$$

For a stand of 96 sq. ft. basal area per acre, the regression was:

$$P_{rn} = -0.0149 + 0.8677P_r \dots\dots\dots\dots\dots\dots\dots\dots\dots XIII.4$$

Two stands, each with 65 sq. ft. basal area per acre, yielded regressions of:

$$P_{rn} = -0.0131 + 0.9055P_r \dots\dots\dots\dots\dots\dots\dots\dots\dots XIII.5$$

$$P_{rn} = -0.0074 + 0.8933P_r \dots\dots\dots\dots\dots\dots\dots\dots\dots XIII.6$$

Equations XIII.5 and XIII.6 , though similar, are not identical, and suggest that basal area is not an ideal measure of stand variability. These results were taken immediately following application of a thinning procedure.

Jeffrey (1968a) carried out studies in lodgepole pine stands which had been thinned 25 years previous to measurement. His results are shown in Table XIII.2.

The only significant difference in the regression equations of Table XIII.2 is between regressions XIII.8 and XIII.9, from plots having virtually identical basal areas, whereas no significant differences exist between XIII.7 and XIII.8, or XIII.7 and XIII.9, wherein the basal areas are quite different. Plot 1, moreover, represents a condition created by thinning the analog of Plot 2, 25 years previously. Furthermore, it can be observed that Equation XIII.9 is almost identical to the regression equation obtained for the uncut stand in the study carried out by Niederhof and Wilm (1943) (Equation XIII.3); yet these two stands are completely different.

Table XIII.2   Net Precipitation in Lodgepole Pine Stands

Plot	Stems/acre	Mean height (ft.)	Mean DBH (in.)	Basal area (sq. ft.)	Regression equation	Equation Number
1	627	53	6.7	155	$P_{rn} = -0.0352 + 0.8689\,P_r$	XIII.7
2	1729	40	3.8	140	$P_{rn} = -0.0328 + 0.8965\,P_r$	XIII.8
3	4556	27	2.5	142	$P_{rn} = -0.0245 + 0.8101\,P_r$	XIII.9

It is concluded that thinning and other forms of forest harvesting have an effect upon net precipitation. However, it is clear that the effects of thinning are not long-lasting, that the canopy recovers following thinning, and in a relatively short period (less than 25 years) no differences in net precipitation remain. It may, furthermore, be concluded that basal area is not a good descriptor of canopy characteristics for hydrologic purposes, and that where the literature uses basal area as such a descriptor it should be treated with caution.

Interception loss is one of the most thoroughly researched topics in the hydrologic literature. Recent studies have shown, however, (Goodell 1963), that compensating reductions in evapotranspiration may take place when the foliage is wet, so that the water stored on tree canopies during and following rainfall events, may not be truly regarded in total as a hydrologic loss. The final answer to this question is still awaited. The fact that wetted foliage has a lower albedo than dry foliage would indicate that, whatever the amount, evapotranspiration is reduced when the foliage is wet, and it is unlikely to be in complete compensation for the moisture intercepted by canopy, and evaporated therefrom.

XIII.2.1.2   Rainfall Interception by Surface Litter

The litter layer on the forest floor intercepts rainfall in much the same manner as the vegetal canopy. During small rainfall events very little precipitation may actually reach the mineral soil. The forest floor, however, is not a sponge; it does not have an unlimited capacity to absorb rainfall.

A forest floor weighing 5.5 tons per acre (oven dry) and with a field capacity of 200 per cent has a storage capacity of 0.1 in. rainfall. A forest floor weighing 45 tons per acre and having a 500 per cent field capacity has a mean rainfall storage capacity of 2.0 in. rainfall. Both examples represent extreme ends of the forest-floor range. The mean condition for Canadian forests probably lies in the lower half of the range described. Tree roots may occupy the forest floor, so that rainfall stored there is, in this case, available for evapotranspiration. Where logging results in litter reduction or removal, more water is made available to soil-moisture storage, since more can enter the mineral soil.

XIII.2.1.3 Raindrop Impact

The kinetic energy of falling rain has a well-known capacity to compact exposed mineral soil. Where soils have previously been protected by surface litter cover and when they are laid bare through land use, soil compaction and associated effects may result. This effect is, in balance, hydrologically much more important than litter interception.

The amounts of kinetic energy developed on a surface by storms of different intensities, as estimated by Lull (1959), are listed in Table XIII.3. This energy may be of sufficient magnitude to cause soil splash and plugging of the surface soil pores, thereby reducing infiltration and increasing surface runoff. Results of such operations as logging, slashburning and tillage of arable land, are readily envisaged.

Table XIII.3 Kinetic Energy and Number of Drops for Rainfall of Various Intensities

	Intensity (In. per hr.)	Median diameter (Mm.)	Velocity of fall (Ft. per sec.)	Drops per square foot (No. per sec.)	Kinetic energy (Ft. - lbs. per sq. ft. per hr.)
Fog	0.005	0.01	0.01	6,264,000	4.043×10^{-8}
Mist	.002	.1	.7	2,510	7.937×10^{-5}
Drizzle	.01	.96	13.5	14	.148
Light rain	.04	1.24	15.7	26	.797
Moderate rain	.15	1.60	18.7	46	4.241
Heavy rain	.60	2.05	22.0	46	23.47
Excessive rain	1.60	2.40	24.0	76	74.48
Cloudburst	4.00	2.85	25.9	113	216.9
Do.	4.00	4.00	29.2	41	275.8
Do·	4.00	6.00	30.5	12	300.7

(Lull, 1959)

Though forests intercept rainfall it does not follow that the energy of net rainfall is less than that in the open, since drops from canopy drip are larger and reach their terminal fall velocities in relatively short distances. The effects of litter removal are, therefore, more important than the effects of canopy removal. Where stands are open and lack undergrowth, the energy impact of rainfall beneath the forest is equal to that in the open. However, a heavy undergrowth may reduce the impact on the surface.

XIII.2.2   Occult Precipitation

Hydrologically, occult precipitation is rather a minor phenomenon. Occult precipitation (fog drip) is produced when airborne water vapour condenses on vegetal surfaces, and falls as liquid water to the ground. It is, therefore, an edge effect, occuring primarily where cloud and fog vapour comes in contact with high vegetational cover.

At such edges, precipitation at ground level may be 300 per cent of precipitation in the open, over a one month period. It is most common near sea coasts and at timberline. Measurements in Oregon showed that, two miles from the ocean, 25 per cent more precipitation was found beneath forests than in adjacent openings. It was believed that superior growth of forest types located close to sea coasts was dependent upon increased fog-drip precipitation; however, it is probably more correct to consider enhanced growth under these circumstances to be attributable to lower transpiration resulting from reduced solar radiation input under fog-belt conditions.

Where occult precipitation is an appreciable hydrologic factor, forest removal results in reduced precipitation at the soil surface. It is stressed, however, that such conditions are quite localized. For a more comprehensive discussion of occult precipitation, the reader is referred to the work of Penman (1963).

XIII.2.3   Snow

Snow hydrology has been treated in detail in previous Sections. The purpose here is briefly to examine the effects of land use upon snow interception, evaporation and accumulation.

XIII.2.3.1   Canopy Interception (Snow)

There is no doubt that large masses of snow are intercepted by, and adhere to, forest canopies (Jeffrey 1964, 1968b). Beyond that simple statement, however, much less certainty is possible. Volumes of 100 cu. ft. lodged in a single tree in the Australian Alps, and weights of 230,000 lbs./acre in Finland are cited in the literature. These are extreme values. Pruitt (1958) found snow to stay in the canopy of the subarctic forest for months on end.

Factors influencing the amount of snow lodging in a forest canopy include snow characteristics, wind movement, and the geometry of the forest biomass. Hoover (1962) noted

> trees with dense stiff foliage, horizontal or upturned stiff branchlets, considerable vertical spacing between branches, and closely crowded together, hold a maximum of snow in their crowns.

Leafless deciduous trees may also accumulate surprising volumes of snow.

More important than the fact of snow lodging in the canopy, are two factors: Where does it go? What influence does it have while in the canopy?

The main parameter involved in the second of these questions is the effect snow lodging exerts on the albedo of the canopy. Freshly fallen snow reflects up to 85 per cent of the incident solar radiation, whereas the albedo of the canopy without its snow load is approximately 10 per cent. Longwave absorption is probably not significantly changed.

In a series of papers, Miller (1964, 1965, 1966, 1967) has given particularly detailed attention to the first of these questions. He has identified five 'routes' by which snow may leave the canopy. Of these, only two truly represent evaporative loss transfers. These are shown in Table XIII.4.

Table XIII.4 Processes of Transport from Intercepted Snow During Storms

Transport by:					
Weather element	Falling or blowing of dry snow	Sliding or falling of partly-melted bodies of snow	Dripping or flowing of melt water	Vapour flux from melt-water film.	Vapour flux from snow
Windspeed	++a	+b	+b, c	+d	+d
Air temperature		++c	+c	+d	+d
Vapour pressure		+	++	−	− −
Insolation		++	+	+	+d

(Miller, 1966)

+   Indicates an element of storm weather that favours a mode of transport from the crowns.

++   Indicates strongly favours.

−   Indicates an element of storm weather that discourages a mode of transport from the crowns.

− − Indicates strongly discourages

a = Effect of wind is conditioned by the rate at which masses of intercepted snow are streamlined and wind packed, or develop internal cohesion.

b = Conditional on air temperature being about $0°C$.

c = Conditional on air being near saturation.

d = Conditional on low vapour pressure in air.

To melt one gram of snow at 32° F requires 80 calories, while to evaporate one gram requires 675 calories. Considering the likelihood of low radiation surpluses in winter, it does appear reasonable to consider mass movement processes as probably more important than vapour transfer processes. However, Goodell (1959), found by weighing a freshly cut lodgepole pine branch covered with freshly fallen snow that a 60 per cent snow loss occurred in less than 3 hrs; the whole weight loss was due to vapour transport. He also noted that many facets of tree crowns are perpendicular to the sun's rays, and therefore accumulate more energy than horizontal snow surfaces.

In addition to the radiant heat load, snow lodged in tree crowns may receive advected heat, either regional (from the gradient wind), or locally engendered within the ecosystem itself.

Miller (1965) concluded of snow interception:

> Recent work on snowfall interception has identified a grouping of storage and transport phenomena, none of which has had much study although notions about the total result, often inconsistent with the heat balance, are plentiful. Among the few quantitative data are some from an interesting experiment in weighing a tree during snowstorms, from which the rates of snow movements and required heat have been calculated. Removal by sliding of partly melted snow masses proceeded at about 2 mm water equivalent per hour and required 1 or 2 ly./hr.; removal by evaporation seldom proceeded faster than 1 mm/day, that is, at a rate of 60 ly./day. Larger amounts of heat are not often available in any snowy region where radiation surpluses are small and advected heat is usually accompanied by vapour pressure so high as to suppress evaporation. Reports of great amounts of evaporative 'loss' from intercepted snow may not be verified when the heat balance is cast. Forests with radiation deficit, little chance of receiving advected heat, and little post-storm wind movement are snow-covered for months at a time, and are generalized into regional patterns that take in the boreal forest, north slopes of midlatitude mountains and humid coastal mountains.

It is concluded that snow interception losses from forest canopy under most conditions, particularly Canadian, are probably rather small. Removal of forest cover, however, does result in elimination of snow interception losses, and therefore in somewhat increased water input being available for other hydrologic processes.

XIII.2.3.2   Snow Evaporation

Evaporative losses may best be understood in relation to energy input and output within the forest ecosystem. Evaporation requires energy. Energy surplus during winter in cold climates may be small. In humid marine climates vapour pressure gradients favour melting over evaporation. Diamond (1953) and Bergen (1963) present physical-mathematical analyses of snow evaporation processes. Diamond (1953) in his analysis of snowpack-evaporation opportunity concluded that ". . . very little evaporation of snow may occur without simultaneous melting, and usually much more snow will be melted than evaporated."

Hutchison (1966) measured evaporative losses from snowpacks in forest openings in the Colorado Rockies. He concluded that snow evaporation was small, and greatly exceeded by evaporation from wet, bare ground patches. In late April and early May the mean loss per 24-hr. period attributable to evaporation was 0.036 in. of water. Bergen (1963) in the same region found daily evaporation losses in a large forest opening ranging from 0.1 to 0.2 gm./sq. cm. in February and early March, with an extreme value of 0.29 gm./sq. cm. in late March.

West (1959) concluded that, in Central Sierra Nevada, snow evaporative losses from openings were higher than under forest, though night-time evaporation was greater under forest. This he attributed to a lesser 'loss of heat', pointing to 'crust' formation in openings as corroboration.

In summary, snowpack evaporation loss seems to be generally accepted as small. Even in the very favourable (to evaporation) climatic conditions of the California Sierra Nevada it is assessed as less than 3 per cent of the total snowfall (West, 1962).

The removal of forest cover will apparently result in somewhat greater evaporative losses from a snowpack. However, since such losses appear relatively low, the net hydrologic effect must likewise be considered insignificant.

XIII.2.3.3   Snow Accumulation

This topic received detailed treatment by Meiman (1968).

Less snow accumulates beneath forest canopies than in small openings in the stand, and the snow accumulation rate is inversely related to canopy density. Within the stand itself, snow accumulation may not be uniform. Beneath hardwood forests, snow cover is more uniform than in coniferous forests, where the effect of dense crowns is to create a 'ridge and hollow' effect, with hollows occurring beneath the tree crowns. Though more snow is consistently found in forest openings than within the stand itself, no conclusions about snow interception can safely be deduced, because the effect may largely be due to redistribution of snow attendant upon the reduction in wind velocity in the opening. As Hoover (1962) points out, the questions raised and not yet fully answered are:

1. Is the excess in the opening a result of evaporation of snow from tree crowns?
2. Was intercepted snow merely blown, or shaken off, into the opening?

3. Did the wind eddies due to the surrounding tree crowns cause excess snow deposition in the opening?

The difference in snow accumulation between forest and small opening has been termed ΔA by Miller (1966) who surveyed the available literature. He concluded that it was difficult to generalize concerning the magnitude of ΔA. Hoover and Leaf (1967) in Colorado lodgepole pine forests determined that cut strips accumulated 4 inches more water than uncut forest over the whole winter, an increase of about one third.

Anderson (1963) found in the California Sierra Nevada that cut strips accumulated 26-35 per cent more snow than the uncut forest. Stanton (1966) in a study conducted in the subalpine forest of the Alberta Rockies concluded that wide strips cut in the forest increased the maximum snow accumulation by 10-46 per cent of the maximum accumulation in the uncut forest. These studies dealt with heavy snowfall regions. However, the effect of openings is present in areas of ephemeral snowpack. Jeffrey and Stanton (1968) in studies on low elevation Alberta lodgepole pine forests found measurable snowpack in forest openings on 18 measurement dates spread over two winters, whereas in the adjacent forest snow could be measured on only 3 of the 18 occasions.

The porosity of stand borders has been noted to be a major variable influencing differential accumulation.

To repeat, the difference ΔA cannot be interpreted as being equal to snow interception, in spite of the occurrence of this interpretation in the older, and even in some recent, literature.

Maximum values of ΔA have been found in openings of width or diameter of 1-tree-height. It has been suggested that the finding of prevalent eddy size, during snowstorms, of less than 300 feet suggests a critical dimension for openings receiving excess snowfall.

Deep drifts generally form at the edges of forests, adjacent to open areas. This is the major source of snow-depth variation in hardwood forests. Around timberline, the zone of maximum snow accumulation is often at the edge of the forest, because of snow blowing in from adjacent alpine areas. Considerable variation in the mean snow depth exists between different types and densities of forest cover (Meiman 1968).

Meiman (1968) summarized the effects of forest canopy upon snow accumulation. A summary of his data is given in Table XIII.5, which shows that little real quantitative generalization is possible concerning forest canopy influences, and therefore of forest treatment effects upon snow accumulation, beyond that already stated. Meiman (1968) in his summary stated:

> It is obvious that those of us accustomed to working on the land surface need a broader vision both horizontally, over the land surface complex of factors, and vertically, into the atmospheric conditions.

Table XIII.5 Selected Studies on the Effect of Forest Canopy on Snow Accumulation. 1/

(1) Source and Location	(2) Forest Type	(3) Procedure	(4) General and Descriptive	Results		(7) Reviewer's Comments
				(5) Difference (inches)	(6) Diff. As % of Lesser	
Mattoon (1909) Arizona	Ponderosa pine (mature)	Compares forest with adjacent open park at 7500' elev. for max. accumulation; 10 points each type.	Forest Park Mar 17 Accum. 5.2 9.4 Concluded that forest caught more snow but it melted faster.	4.2	81	Author believed forest trapped snow similar to silt deposition in streams.
Jaenicke and Foerster (1915) Arizona	Ponderosa pine (open stand)	Sampled from Oct. 12 to April 2, 1912–1913 in a forested and non-forested area.	"No appreciable difference". Distribution differs greatly-park snow is even, forest snow is shallow under trees and in deep drifts in openings.	0	0	Measured water equivalent with overflow can of standard rain gage.
Kittredge (1953) Calif.	Mixed Conifer (MC) Ponderosa pine (PP) Red fir (RF) White fir (WF)	Related crown cover for 40' diameter circles to max. water equivalents.	Years & Type / Max H_2O Equiv. and (%) crown cover 1935&1936(MC) 15.7(0) 2.9(80) 1934 (RF) 14.5(0) 5.7(80) 1937 (WF) 15.8(30) 10.2(70) 1935&1936(PP) 7.2 3.2(40)	12.8 8.8 5.6 4.0	441 154 55 125	Values calculated from significant regression equations.
Anderson et al (1958a) Calif.	Not given	Used 57 snow courses from 6100' to 7800' in Sierra; sampled in various parts of openings and adjacent forests.	April 22, 1958: Max range in various parts of openings and adjacent forest 46—71" Mean values for density groups: 80–100% 48.1 50–80% 50.3 20–50% 57.0	25 2.2 8.9	54 5 18	Approximately 6" of ablation had already occurred by April 22.

Table XIII.5 (Cont'd)

(1)	(2)	(3)	(4)	(5)	(6)	(7)
Anderson & Gleason (1959) Calif.	True fir (100,000 ft. b.m./ac.)	Swain Mtn. Compared snow in natural opening (1/2H)[2/] and in dense forest (100,000 f.b.m) and open forest (50,000 f.b.m.)	Old dense Fir 34.6; Old open Fir 45.0; Opening 47.4	10.4, 12.8	30, 37	
	Mixed conifer	Onion Creek Compared commercial clear cut with unlogged; 25 samples in each at time of max. accumulation.	Cut / Uncut; 1958 58.3 / 51.2; 1959 26.4 / 20.1	7.1, 6.3	14, 31	Removed 35,000 bd. ft. per ac. leaving 2000.
Stanton (1966) Alberta, Canada	Spruce-fir (80')	Compared March 25, 1965 readings (20 points each) for paired courses; cut areas varied from 8H to 16H[2/], #5 was shelterwood cut removing 30% of 350 M ft. b.m./ac.	Aspect Cut Uncut; 1 S 65.5 54.0; 2 N 71.0 64.5; 3 N 58.0 50.5; 4 E 71.0 48.5; 5 E 52.5 51.0	11.5, 6.5, 7.5, 22.5, 1.5	21.5, 10.0, 14.6, 46.4, 3.0	
Wilm & Collet (1940) Colorado	Lodgepole pine (mature)	Twenty 5-ac. plots with 25 points each measured in spring 1938 and 1939.	Highly significant linear relation between H_2O equivalent and crown proximity.	4	80	Columns (5) and (6) estimated from text Fig. 2 for 30' under canopy and 30' into opening.

Table XIII.5 (Cont'd)

(1)	(2)	(3)	(4)	(5)	(6)	(7)
Niederhop & Dunford (1942) Colorado	Lodgepole pine (young)	Compared maximum accumulation in openings from 4 to 56' diameter in two stands; Utah snow tube measurements made March 13 and 14, 1941.	Comparing 4' vs 20' opening: Stand 1 4.25 vs. 6.85 Stand 2 5.75 vs. 8.35 No additional effect beyond 20'. Correlation between accum. and opening diameter highly significant.	2.6 2.6	61 45	Values for columns (5) and (6) taken from text Fig. 3.
Niederhop & Dunford (1944) Colorado	Aspen Lodgepole pine	Maximum accumulation for 1942 & 1943 measured with Utah tube at 50 points on dense young pine, in 4.3 ac. opening, and in Aspen	Aspen 10.9 Open 9.6 Pine 8.4 Diff. consistent for both years. Std. error of the av. for each type ± 0.145"	2.5 1.2	30 14	
Wilm & Dunford (1948) Colorado	Lodgepole pine (mature)	Compared precutting (1938–1939) with post cutting (1941–1943); Sampled with Utah tube between March 15 and April 1. Uncut has 11,900 ft. b.m. (see Wilm and Collet)	Avg. after cutting 6000 ft. b.m. 8.41 4000 8.61 2000 9.09 0 9.59 Uncut 7.60 Timber stand improvement: Improved 9.38 Unimproved 8.92	0.8 1.0 1.5 2.0 0.46	11 13 20 26 5	
Goodell (1952) Colorado	Lodgepole pine (second growth)	Measurements on 6 blocks of 2 1/4 ac. each; untreated-89.4 sq.ft./ac.basal area, single tree thinning – 8-1/2" apart, crop tree – 8' radius around each.	Untreated 10.03 Crop tree 11.72 Single tree 12.34	1.7 2.3	17 23	Total winter and spring snow; snow tube measurement plus late spring snow by gages.

Table XIII.5 (Cont'd)

(1)	(2)	(3)	(4)	(5)	(6)	(7)
Hoover & Leaf (1967) Colorado	Spruce-fir Lodgepole	Compared cut and uncut strips on Fraser Exp. Forest at end of March 1965; 1224 sampling points.	Cut 19.0 Uncut 15.6	3.4	22	Values in column (4) obtained by personal communication with authors.
Connaughton (1935) Idaho	Ponderosa pine	One plot with 15 to 25 samples for each condition; 1931-1933.	Reprod. (20-25') 15.4 Virgin No adv. reprod. 12.4 adv. reprod. 11.8	4.5	38	Used only 1932 and 1933 because of missing data.
Packer (1962) Idaho	Western white pine	Canopy density measured with spherical densiometer.	Average differences for 4 years between 0 and 100% canopy density; relation occurred regardless of differences in year, elevation, or aspect.	4.2	32	Author attributes canopy effect to interception rather than melt.
Gary & Coltharp (1967) New Mexico	Spruce-fir Aspen Douglas-fir		Douglas-fir vs. Aspen South slope 7.6 - 7.7 North slope 9.4 - 13.1 Spruce-fir vs. grass South slope 13.6 - 14.5 North slope 13.9 - 14	.1 3.7 .9 .1	1 39 7 1	
Miner & Trappe (1957) Oregon	Lodgepole pine (70-year old)	Five plots with 20 samples each in forest and adjacent meadow; total basal area of all trees avg. 140 sq. ft. per acre.	Comparison on 3-18-57 gave 2.1 in forest vs. 5.1 in meadow	3	143	

Table XIII.5 (Cont'd)

(1)	(2)	(3)	(4)	(5)	(6)	(7)
Rothacker (1965) Oregon	Mountain Hemlock-true fir (105')	Compared cut strips (132') with uncut at max. accumulation in March, 1964.	Uncut 24.7 Cut 33.3	8.6	35	
Maule (1934) Connecticut	Northern-Hardwood Hemlock White pine Red pine Norway-spruce	Measurements after each of 6 snowfalls in 1932–'33; hardwoods included 4 age classes and silvicultural modifications.	Accumulated depth: Norway spruce 7.3 Red pine 10.3 Hemlock 10.8 White pine 11.0 Hardwood 16.2 Open 17.5	3.0 3.5 3.7 8.9 10.2	41 48 51 122 140	Used a 'Kadel' snow tube with 2.655 I.D. cutter. Values in column (4) from test Fig. 1.
Morey (1942) Vermont	Northern-hardwood (60 year) Red spruce (30 year)	Compared one storm-April 11, 1942.	Spruce 0.46 Hardwood 0.74 Open 0.82			
Sartz & Trimble (1956) New Hampshire	Mixed hardwood	Compared openings of about 2H²/ with forest by measuring March 4, 1955 H_2O equivalent.	Open 9.8 Forest 9.1 Difference indicated as highly significant.	0.7	8	
Baldwin (1957) New Hampshire	White pine (41-50 year)	Compared snow depths under forest and in open (more than 1H²/ from forest)	Readings of 3–6–40 Forest 23.7 Open 26.7 Differences varied with years and storm conditions.	3.0	13	Readings are snow depths; no densities given.

Table XIII.5 (Cont'd)

(1)	(2)	(3)	(4)	(5)	(6)	(7)
Lull & Rushmore (1960) New York	Mixed-Hardwood Hemlock Red spruce Balsam fir	Measured 32 snow Courses Mar. 16-18, 1959; 10 points per course; control was uncut forest.	<u>Area 1</u>-Hardwood sapling vs. conifer sapling: 8.35 vs. 6.05	2.3	38	LSD given for Area: 1 = 2.05 2 = 1.25 3 = 1.73
			<u>Area 2</u>-Hardwood sapling vs. conifer sapling: 9.85 vs. 7.90	1.9	24	
			<u>Area 3</u>-Selective cutting vs. control: 10.12 vs. 8.25	1.9	23	
			r = 0.79 for accumulation vs. canopy density.			
Weitzman & Bay (1959) Minnesota	Black spruce (132 sq. ft. basal area)	Compared March 7 H_2O equivalent in clear-cut strip (66'), single tree selection (75 sq. ft. basal area), shelterwood (51 sq. ft.), and 1/2 ac. clear-cut.	Uncut 2.5	0.1	4	
			Single tree 2.6	0.7	28	
			Shelterwood 3.2	1.5	60	
			Strip 4.0	2.0	80	
			Clear-cut 4.5			
Berndt (1965) Wyoming	Lodgepole pine (12,000 bd ft/ac)	Compared 5, 10, and 20 ac. clear-cut blocks with uncut forest; mean elevation 8500'; March and April, 1961; logged in 1958-59.	Avg. max. snowpack water equivalent for all aspects, cut = 8.7" vs. uncut = 6.2"	2.5	40	

1/ Values are maximum snowpack water equivalents by snow tube measurement unless otherwise indicated.
2/ H is tree height.

Also, we need a more penetrating vision, looking at the land surface complex as the snow crystal sees it.

Gray (1968) has examined snow hydrology in the Prairie environment, while Jeffrey (1968b) has provided a summary of snow hydrology in forests.

The phenomenon of greater snow accumulation in forest openings, whether natural or artificial, has important implications for snowmelt and spring runoff, as demonstrated later.

XIII.3   INFILTRATION

The crucial role of infiltration in the hydrologic cycle has previously been elucidated. Its influence upon soil-moisture recharge and groundwater, and its interrelationships with overland flow, have already been shown along with its interaction with rainfall intensity.

Land use may affect infiltration of water in various ways by affecting soil porosity, soil structure and texture, vegetation, organic matter, soil microorganisms, soil moisture, soil frost, microtopography, and raindrop impact. Soil compaction and the removal of surface litter are two important ways in which infiltration may be modified through land use.

XIII.3.1   Soil Compaction

A prime agency of soil-compaction damage in land use is the use of machinery. Track-type tractors, such as used in logging, exert a pressure of 3-9 psi (mean, 7 psi) depending upon vehicle weight and track width. Loading from wheeled vehicles is greater, trucks exerting a pressure from 50-100 psi, and passenger cars 30 psi. By way of comparison, horses exert a mean pressure of 40 psi (Lull, 1959).

Another variable in machinery use is the number of passes made with the machinery. Wheeled tractors passing 10-20 times over an area have been found to create measurable compaction to a depth of 9 in. Another factor is the soil moisture content. For instance, track-type tractors operating on wet soil produce vibration effects so that a wider area is affected than the actual track width. Machinery compaction, like all compaction, results in a reduction of pore space (particularly macropore space) in increased bulk density, and in reduced permeability. Soils of equivalent texture compact to a greater extent when in a wet condition than when they are dry.

Logging by track-type tractors has been found to produce 35 per cent permeability loss, 2.4 per cent bulk density increase, and 11 per cent decrease in macroporosity. Lull (1959) found from measurements taken on skid roads within logging areas, a decrease in permeability of 92 per cent, a bulk density increase of 35 per cent, and a reduction of 53 per cent in macropore space.

Different logging systems exert different effects upon soil disturbance. Track-type tractors, for instance, have been found to disturb 15 per cent of the logging area deeply and 6 per cent shallowly. By comparison, cable logging disturbed soils on 2 per cent of the area deeply, and 13 per cent shallowly; and horse logging, 2 per cent of the area deeply, and only 9 per cent shallowly. The major difference between systems is, therefore, the amount of deep disturbance from logging with use of track-type tractors.

The reduction in infiltration caused by compaction is graphically shown by results obtained in the agricultural field. On irrigated land a stream of water flowed 3 times the distance in a compacted furrow than it moved in a dry uncompacted soil. If the soil was compacted when wet, the water stream flowed 15 times further than in neighbouring dry, uncompacted soil.

Trampling by animals can also create compaction. In comparison to the 40 psi pressure exerted by horses, cows exert 24 psi, and sheep 9 psi over their bearing surface. Under heavy grazing, an increase in bulk density from 1.54 to 1.91 has been documented, with accompanying reduction in macroporosity. Macroporosity reduction has been found to depths of 9 in., though the literature seems to agree that appreciable reduction occurs only to a depth of 1-4 in. The resultant reduced infiltration rates have been measured after animal compaction.

On camp grounds and other recreational areas, human compaction may result. Men exert pressures of about 6 psi, and women's shoe stylings result in pressures of 8-13 psi. On camp grounds, infiltration has been found to be reduced to one fifth to one sixth of that of undisturbed, contiguous areas.

The effects of raindrop impact were previously discussed. A 2-in. rainfall falling at a velocity of 20 mph may produce a kinetic energy of 6×10^6 ft.-lbs./acre, or 138-ft.-lbs./sq. ft. Macroporosity may be reduced by two-thirds through raindrop impact on soils unprotected by surface litter.

XIII.3.2   Litter Removal

The most important influence of surface litter (unincorporated organic matter) in forests is in its maintenance of 'good' soil conditions; that is, soil conditions favourable to infiltration. Litter on the forest floor reduces overland flow; conversely, litter removal reduces infiltration rates. The primary mechanism for these effects seems to be a phenomenon of surface sealing, resulting from a plugging of surface pores by soil particles exposed to the elements through removal of the protective litter cover. Incorporated organic matter also affects soil aggregate formation, an important variable affecting the infiltration rate.

In ponderosa pine forests, where infiltration rates as high as 1.52 in./hr. were measured in undisturbed forest, a significant difference was found after litter removal, in that the average infiltration rate on the area after disturbance was measured as being only 0.92 in./hr.

It should not be believed, however, that the undisturbed forest invariably is free of overland flow. Imbrication of leaf litter in broad-leaved forests has been found to contribute to overland flow in forested watersheds (Pierce, 1967), and a water repellency phenomenon in some forest surfaces has been shown to have similar effects (Krammes & DeBano, 1965).

In the undisturbed forest, plant-root channels do generally favour infiltration, however, and the organic matter of forested areas contributes to soil-aggregate formation.

XIII.3.3   A Case History: Slashburning

In the forests of coastal British Columbia, and to an increasing extent in other areas of Canada, it is common practice to burn the detritus left on the forest floor after logging, as a means of fire hazard abatement and of preparing the ground for tree planting. This involves a very radical disturbance of the forest soil, all the disturbance effects of timber extraction being followed by the additional disturbance of burning large quantities of debris and slash.

Slashburning is used here only as an example. The effects of slashburning on soil characteristics are present, to a greater or lesser extent, in all disturbances of land placed under human management. Slashburning is singled out merely because it illustrates most of the relevant effects (Dyrness, 1967; Willington, 1968).

It is common in slashburning studies to classify the ground surface into three categories:

1. Unburned—untouched by fire.

2. Lightly burned—surface litter charred only.

3. Severely burned—litter removed and soil surface baked.

Three studies in the U.S. Pacific northwest have shown that on slashburned areas there, 49 per cent of the area was unburned, 45 per cent lightly burned, and 6 per cent severely burned.

The logging-slashburning sequence affects soil porosity, both capillary and noncapillary. Capillary porosity is significantly increased and noncapillary porosity is significantly decreased, whereas the total porosity is not significantly changed. The mechanism causing the increase in capillary porosity is not well understood, but may result from soil aggregate breakdown. The decrease in noncapillary porosity significantly affects the infiltration process.

Soil texture and structure affect soil porosity and infiltration. Severe burning has the greatest influence upon these parameters. Under high temperatures, fusion of clay particles may occur, transforming these into stable, sand-sized, secondary particles. Colloidal organic matter is reduced by fire. Loss of this cementing agent may lead to aggregate disintegration. Of these two conflicting effects, the second is probably more common, and results in reduced infiltration rates.

Logging and slashburning removes vegetation on the soil surface. This effect on infiltration is variable. Severe burning can remove all organic matter, including shallow roots.

Logging plus slashburning affects surface organic matter. Burning on seven areas in the interior of British Columbia was found to reduce the depth of surface litter by 54 per cent, and the weight of surface litter by 61 per cent.

Incorporated organic matter is significantly reduced by severe burning. In Oregon slashburns, soils which initially contained 12 per cent incorporated organic matter, after being burnt were found to contain only 3 per cent organic matter. Light burning has little effect.

Microorganisms in the soil influence its structure. While severe burning probably has the effect of 'sterilizing' the upper soil layers, bacterial populations have actually been found to increase after burning. Fungal organisms are reduced for about two years after burning, but this effect is only associated with severe burning.

The effect of logging and slashburning, through removal of vegetal cover, reduces evapotranspiration, at least temporarily. This means, conversely, that soil moisture on logged areas is increased, so that the final infiltration (f_c) is reached more quickly than in comparable uncut areas. In addition, burning reduces the moisture-holding capacity of the soil, though this is not well understood, for—as noted previously—capillary porosity is increased by burning and this should theoretically result in a greater moisture-holding capacity. However, the removal of organic matter, having high water-holding properties, may more than compensate for the increase in capillary porosity.

Slashburning following logging may increase soil freezing. Concrete soil frost is usually found in large open areas and is considered relatively rare in undisturbed forest. Little is known concerning the effects of slashburning on soil frost.

Logging and slashburning alter microtopographic effects that are important to hydrology. For example, they reduce surface depression storage and therefore contribute to a more rapid runoff. The effects of raindrop impact have already been discussed.

The total integrated result has been established by infiltration measurements in slashburned areas in British Columbia and contiguous regions in the United States. In the interior of British Columbia, infiltration was reduced by 64 per cent over the whole of a slashburned area.

In Oregon, sandy loam soils in an unburned condition had a maximum infiltration rate of 24 in./hr. When these same soils were severely burned their infiltration rates were reduced to 7.8 in./hr. In Montana, for whole slashburned areas, infiltration was reduced by 37 per cent. In the Ozarks area of the U.S.A., seven soils showed infiltration reduction after burning, ranging from 20-60 per cent.

This reduction in infiltration, of course, disappears with time. Present evidence would indicate that soils may recover completely, insofar as infiltration is concerned, in the space of 3-5 years, though this is not yet fully documented.

It is generally agreed that logging without slashburning also reduces infiltration rates, though to a much lesser extent, determined by compaction effects.

XIII.4   EVAPOTRANSPIRATION

The general water balance equation shows the importance of evapotranspirative losses to water resources. In effect, as evapotranspiration varies so runoff varies, the sole qualification being widescale weather modification to increase precipitation input. Previous Sections have shown the restrictions of energy and soil moisture upon evapotranspiration, one or other being limiting factors in the magnitude of the process.

We have considered the precipitation and infiltration processes, so that water may now be regarded as being in the soil, having arrived at the soil surface as precipitation to move across the soil/air interface through infiltration. It is timely at this stage therefore to consider evapotranspiration. Such consideration implies examination of energy, soil moisture and groundwater, and the influence of land use upon these components.

XIII.4.1   Energy

There are basically two energy sources for evapotranspiration, radiant and advective energy. Of these two, radiant energy is usually much more important.

XIII.4.1.1   Effect of Vegetal Cover

In respect of radiant energy, forest vegetation is characterized by high absorptivity, low albedo and low transmissivity (Reifsnyder and Lull, 1965). Absorptivity of coniferous cover > hardwood cover > field crops. Conversely, conifer albedo < hardwood albedo < field crop albedo. The absorptivity of conifers to solar radiation is around 0.90, that of hardwoods 0.80-0.85 and that of grass <0.80. These are approximate values. All vegetation has high absorptivity to longwave radiation.

For radiation of wavelengths within most of the solar spectrum, the approximate albedos of different types of vegetation are:conifers 0.10-0.15; hardwoods 0.15-0.20 and field crops >0.20.

In the closed forest, little solar energy reaches the soil surface, that is its transmissivity is very low. The magnitude of transmitted shortwave radiation depends upon forest structure, composition and density, and may be as low as 1 per cent in the visible spectrum. A wide range exists. In the sense of repeated absorption and re-emission, forests deplete longwave radiation very little.

These factors mean that cover changes have large effects on energy budgets, which limit evaporation, transpiration and snowmelt.

In respect of advected energy the height of vegetal cover is a major factor. If one considers a forest, in a square block of 10 acres size having a height of 100 feet, the plane area is 435,600 sq. ft., but added to this is an edge area of 264,000 sq. ft. making a total of 699,600 sq. ft. In comparison, a grass field of the same plane dimensions, but having a height of 12 in., has the same plane area, but an edge area of only 2,640 sq. ft., making a total area of 438,240 sq. ft.

These edge areas affect radiation absorption and also influence the amount of turbulent sensible heat exchange (advected heat). Surface roughness affects air mixing, and therefore the exchange of advected heat. Forest surfaces are very rough. A change in cover, accordingly, produces a change in advected energy, and therefore modification of energy-dependent processes.

XIII.4.2   Soil Moisture

In conditions where energy supply is not the limiting factor in evapotranspiration, soil moisture becomes limiting. Soil moisture is the water source for transpiration, in vegetated conditions a much greater evaporative stream than direct soil evaporation, this resulting from the vegetational canopy being the primary energy exchange surface. It has already been seen how land use can affect infiltration, which in turn affects soil moisture recharge. It has been demonstrated how the presence, or absence, of litter (unincorporated organic matter) influences soil moisture storage.

XIII.4.2.1   Effect of Vegetal Cover

The type of vegetal cover affects soil-moisture depletion. Thus it also influences soil-moisture storage opportunity. Hence it is a major determinant of total water yield and stream regime.

In addition to the absorptivity and albedo differences previously considered, the major variables influencing differential evapotranspiration losses, according to vegetational cover, are the rooting depth of the cover and the depth of the soil mantle. Shallow soils are generally completely occupied by plant roots, regardless of the type of cover. Deeper soils may not be totally occupied by roots of one cover type, but be much more fully occupied by the roots of another cover type. One may, therefore, logically expect large differences in evapotranspiration to occur on deep soils carrying different covers, but only small differences from shallow soils with varying cover.

Much research has been done into comparisons of soil moisture depletion by forest and grass covers (Douglass,1967). The rooting depth of pine forest in Ohio was measured as 72 in. in comparison to broomsedge grass on the same site with a rooting depth of 30 in. Another study in Utah showed aspen to have a rooting depth of 6 ft., compared to herbs on the same soil, with a rooting depth of 4 ft.

Seasonal differences also occur. Grass aestivates in hot, dry climates. Water use is thereby confined to spring and early summer, whereas forests transpire for a much longer period. However, in wet climates water vapour may diffuse upwards in the soil along potential gradients. This may reduce the effects of differential rooting depths between covers.

Differences in soil-moisture depletion (evapotranspiration) between tree species are probably rather small. Differences in rooting depths and in albedo do nevertheless occur, and one study has shown that hardwood forests used 6 in. more water over a season than a spruce forest. This is thought to be an unusually large difference.

Manipulation of forest cover has large effects upon soil moisture. Logging, by removing the transpiring tree cover, reduces soil-moisture depletion particularly in deep soils, thereby increasing soil-moisture content. Partial cuttings or thinnings also increase soil-moisture content through reduction in stand density. However, this latter effect is generally not long-lasting. Soil-moisture content differences are generally proportional to the severity of treatment. In low intensity treatments, differences in stored soil moisture may be difficult to detect.

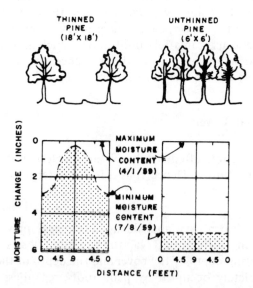

Fig. XIII.1 Effects of Cutting Patterns on Soil Moisture

Generally speaking, if a given proportion of the forest cover is removed, differences are greater if the cover is removed in patches, rather than uniformly over the whole area

(Goodell, 1967); Douglass (1967). Figure XIII.1 demonstrates differences in soil-moisture content following cutting. A major variable between patch, and uniform partial cuttings, is the factor of speed of root reoccupation of soil.

XIII.4.3   Groundwater

Where roots reach the capillary fringe of the water table, evapotranspiration takes place at the potential rate, so long as the capillary fringe remains within the rooting zone. In certain cases, phreatophytic activity may use significant quantities of water. For example, in the flood plains of the southwestern United States riparian stands of tamarisk and cottonwood produce large losses of scarce and valuable water.

Removal of vegetation in such circumstances causes the water table to rise. In certain conditions, this may lead to paludification, a condition whereby the water table rises to the soil surface itself. Poor soil-aeration conditions thus created preclude re-establishment of vegetation, so that a previously-forested area may become a swamp. This phenomenon is likely to be important in the management of some of Canada's boreal forests growing on organic soils.

Studies carried out in Finland on drained peatland (Heikurainen, 1967) in which the water table under forested conditions was originally 30-70 cm. below the soil surface (depending upon the season of year) have shown clearcutting to result in the water table rising by as much as 40 cm. The effect presumably would have been even greater if the site under study had not been drained prior to clearcutting. The same study showed the effect of partial cutting to be less than that of clearcutting, and to depend upon the severity of partial cutting.

Planting of spruce in Denmark (Holstener-Jorgensen, 1967), conversely, has been found to lower the water table on clay soils. Also in Denmark it was found that clear-cutting of a beech stand in which the water table attained a minimum depth of 70 cm. below the soil surface, dropping in late summer to 230 cm. below the surface, showed that after clearcutting the water table did not drop during the summer, in fact it rose to a minimum depth of 40 cm.

In such shallow water-table sites, species have been found to vary in their effect. In Denmark, for instance, spruce (which transpired all year round) had a greater effect upon water-table depth than deciduous beech forests, which transpired only during the summer. It has also been found that on soils with a shallow water table planted arti-ficially with trees, the water table was progressively lowered as the stands grew older, at least in the case of forests which had not yet reached maturity. This presumably is a reflection of progressively greater soil occupancy by tree roots, as the trees matured. Partial cutting has been found to have intermediate effects.

It should be borne in mind that any cutting also influences snow cover character-istics. This, in turn, affects soil moisture and ground water recharge. An integrated effect of cutting is found, combining the individual effects mentioned; that is snow, soil moisture and evapotranspiration.

XIII.4.4   Evapotranspiration Losses

Relationships between evapotranspiration losses, energy (radiant and advective) and soil moisture have already been shown. Soil-moisture depletion results from evapotranspiration drain. Accordingly, it is difficult to separate considerations of soil moisture and evapotranspiration, since evapotranspiration determines soil-moisture storage opportunity.

In green vegetal covers, transpiration is much greater than soil evaporation, and operates to greater soil depth. Therefore, if the cover is removed, thinned, or otherwise modified, soil evaporation becomes relatively more important. Direct soil evaporation is effective only from the uppermost 12-24 in. of the soil mantle, in comparison to uncut forests rooting to 6 or 8 ft. Thus after clearcutting, although evaporation becomes greater than transpiration, the total evapotranspiration decreases. This, it is restated, does not apply to very shallow soils. Disruption of rooting pattern, by partial cutting or thinning, leaves 'pockets' of soil moisture lacking root occupancy. However, in a short time, roots of remaining plants reextend, and root reinvasion takes place into such temporarily unoccupied 'pockets'. Evapotranspiration can therefore be expected to increase again, following either clearcutting or thinning, but—generally speaking—more quickly in the case of thinning.

When soil moisture is abundant (nonlimiting), the evapotranspiration rate is determined by the amount of energy available. Wind movement and vapour pressure also affect evapotranspiration by influencing water-vapour concentration gradients. Evapotranspiration may occur even when the atmosphere is saturated with water vapour, if heat is available and has to be dissipated. This is evidenced by the 'smoking' effect seen over wet vegetation immediately following rainfall, which represents the immediate condensation of evaporated water.

Where soil moisture is limiting, more energy is exchanged as sensible heat, and the air temperature rises. This accounts for the 'air conditioning' effect of forests and parks in urban situations.

Under conditions of very rapid transpiration, internal water deficits occur within the plant. These may limit transpiration, even when soil moisture is abundant.

In open forests (forests without complete canopy cover) openings between the trees receive energy. Soil surfaces in such openings are heated. As a result, soil moisture is conserved, since energy is dissipated primarily through sensible heat transfer. In such conditions, though total evapotranspiration per unit area of land surface may be lower than in forests with complete canopy, evapotranspiration rate per unit of leaf surface may be greater, because of radiation reflected from ground in the openings to foliage of surrounding trees.

As stem density increases, evapotranspiration per unit area of land surface increases, until all energy available is being used. Thereafter, increased stem density has no effect upon evapotranspiration.

In forests, the predominant variables influencing evapotranspiration may be considered as albedo, height of tree cover (through its effect on ventilation), crown depth, crown geometry (through its effect on radiation capture) and surface roughness (as it effects turbulence and wind velocity). However, actual differences may not be great between full stands of different composition.

Evapotranspiration in the natural landscape is difficult to measure. Reliable comparisons between forests of different composition, or between forests and other vegetational covers, are rather infrequent. Penman (1963) has reviewed much of the relevant literature. Only one example is given here. This is that of the Castricum lysimeters in Holland (Deij, 1954), which is considered fully by Penman (1963, 1967).

Table XIII.6 Castricum Lysimeter Data: Evapotranspiration from Different Covers

Year	Precipitation mm	Evapotranspiration			
		Bare mm	Low Vegetation mm	Deciduous mm	Conifer mm
1948	637	158	342	304	376
1950	1042	252	530	504	631
1951	995	217	494	473	579
1953	639	155	413	385	484
Mean (6 years)	832	195	448	417	538

(Deij, 1954; as given by Penman, 1963)

The postulate in this study was that afforestation of sand dunes on the Dutch coast would affect water supply. The site was sandy, with water table at 2.25 m. below the land surface. Four lysimeters were constructed, having covers of bare soil, low vegetation (natural cover of dunes), deciduous forest, and conifer forest (Austrian pine). Selected data from these lysimeters are presented in Table XIII.6. The mean free-water evaporation from the area was 530 mm/yr.

The conditions under which the Castricum lysimeter study was carried out deserve examination. The climate is oceanic: a relatively warm sea mass adjacent to the study area provides a source of advected energy, and mild winter conditions allow winter evapotranspiration. This may account for relatively high evapotranspiration from the low shrub and grass vegetation which is partially evergreen and therefore capable of transpiring during winter. The deciduous cover, consisting of oak, alder and birch, is free of leaves in the winter. The evergreen conifer forest is, under these climatic conditions, capable of overwinter transpiration.

A radical difference in vegetal cover influences evapotranspiration, whereas the effect of small differences is not detectable. The same conclusion applies to modification of vegetal cover through land use. Variables meriting consideration in evapotranspiration effects of land use include rooting depth, crop (cover) density, cover radiation characteristics, soil depth, and degree of cover modification undertaken.

XIII.5   SNOWMELT

This Subsection draws heavily upon the material presented in the paper "Snow Hydrology in the Forest Environment" (Jeffrey, 1968b). To consider snowmelt in the forests it is necessary to examine the heat exchange between atmosphere, forest, soil and snowpack. This exchange, though complex, is subject in some degree to rational analysis. The total heat exchange may be considered among the following components: (a) shortwave (solar) radiation, (b) longwave radiation, (c) convected and advected energy (sensible heat), (d) condensation, (e) conduction from underlying soil and (f) warm rain. Of these, items (e) and (f) are minor effects and relatively unchanged by forest cover. Condensation melt can be important in fog climates. Consideration of forest-snowmelt effects will largely involve the components listed as items (a), (b) and (c).

Shortwave radiation incident on forest ecosystems may be either direct-beam irradiation or diffuse-sky radiation. In either case, a large portion (greater than 80 per cent) is absorbed by the forest canopy, and a relatively small portion reflected. This is particularly true of coniferous forests. Leaf-tree forest, bare of foliage in winter, absorbs less shortwave energy and allows more to be transmitted to the snowpack. The density and structure of the forest canopy influences the transmission of solar energy to the soil or snow surface. This effect has been studied by Miller (1959) in the western United States, and by Vezina and Pech (1964) in Quebec. The curves derived by both studies are given in Figure XIII.2, and show Miller's relationship for lodgepole-pine stands, and Vezina and Pech's results for balsam fir and jack pine.

Cloudiness affects the proportion of shortwave radiation reaching the soil or snow surface; the transmissivity of canopy being greater on overcast days. Vezina and Pech's data for balsam fir and red pine are quoted in Table XIII.7, to illustrate this point.

These data show that relatively small amounts of shortwave radiation reach the snowpack, even in relatively open stands in eastern Canada. The relationship of solar energy transmitted with stand density is not linear, this resulting from repeated (multiple) reflection. Reifsnyder and Lull (1965) suggest that removal of 75 per cent of the stand basal area would still leave an absorptivity of about 0.50.

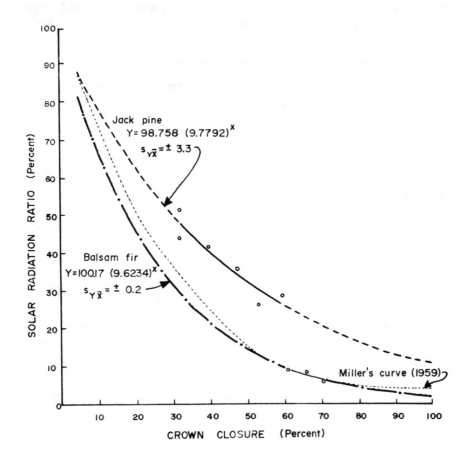

Fig. XIII.2   Crown Closure and Solar Radiation Transmitted as Percentage of Solar Radiation in Open, in Balsam Fir and Jack Pine Stands of Quebec, and Lodgepole Pine Stands of Western U.S.A.

A high proportion of the solar radiation reaching the snowpack will be reflected and only small amounts absorbed or transmitted. The albedo of freshly fallen snow is in the range from 0.85–0.95. During the melt season, however, the overwinter accumulation of needles, bark, other organic matter and deposited aerosols results in darkening and pitting of the snowpack surface, so that albedo decreases markedly, down to a value of 0.40 (Garstka, 1964).

The solar radiation reflected from the snowpack is susceptible to absorption by the superimposed biomass, so that a large proportion of this energy probably remains within the ecosystem and is not lost to the atmosphere. On the other hand, where snow is lodged in tree crowns, it increases the albedo of the canopy so that net shortwave radiation exchange of the ecosystem may be decreased by one half.

Table XIII.7   Relationship Between Daily Solar Radiation Ratios, and Degree of Cloudiness for Various Balsam Fir and Red Pine Stands Thinned to Different Grades

Tree Species	Plot and Treatment	Percent crown closure	Percentage mean solar radiation ratio*		
			Sunny days (0-2/10)[†]	Partly cloudy (3-7/10)	Overcast (8-10/10)
Balsam fir	A Unthinned	71	5.9±1.2	6.4±1.7	7.6±2.9
	B Lightly thinned	66	8.0±1.6	8.3±1.3	10.3±3.6
	C Heavily thinned	61	8.7±1.2	9.8±1.9	10.7±2.6
Red pine	D Moderate thinning	66	6.6±0.3	7.7±0.4	10.2±1.5

Source: Vezina and Pech (1964)

*   Twelve sunny, 22 partly cloudy, and 23 overcast observation days for the balsam fir stands; 37 sunny, 44 partly cloudy, and 23 overcast observation days for the redpine stand.

†   Degrees of cloudiness

In contrast to the forested situation, level open snowfields in large openings receive the direct solar beam unimpeded by canopy effects. How much of this direct beam irradiation is absorbed by the snowpack is dependent upon its albedo, which in turn is a function of the age of the uppermost snow layer. The difference between forest and open might be that of 5 per cent of the transmission of clear-day solar radiation in forest, and 100 per cent incidence in large openings. Assuming all transmitted energy in the forest to remain within the ecosystem, with a solar radiation input of 500 ly./day the difference between forest and open would be only 55 ly./day for new snow (albedo of 0.84) though this represents 220 per cent of the total forest solar radiation input, and 275 ly./day in old snow (albedo of 0.40) or a difference equivalent to 1100 per cent of the net solar radiation of forest snowpack. However, the forest canopy greatly influences longwave heat exchange and sensible heat transfer, and these partially compensate.

Longwave radiation exchange takes place between the forest canopy and the snowpack. The melting snow surface may be assumed to have a temperature of 0° C, while the canopy warmed by incoming shortwave and longwave radiation will have surface temperatures usually equal to, or greater than, the ambient air temperature.

According to Stefan-Boltzmann's law:

$$R_L = e\sigma T^4 \dots\dots\dots\dots\dots\dots\dots\dots\dots\dots\dots\dots\dots\dots\dots\dots\dots\dots\text{XIII.10}$$

where, R_L = emitted radiation,

e = emissive power,

σ = Stefan-Boltzmann's constant,

T = absolute temperature of the body.

Therefore $R_{snow} = e_s\sigma T_s{}^4$,

where $e_s \geqslant 0.82$

$T_s = 273° K$

and $R_{forest} = e_f\sigma T_f{}^4 = variable$,

the net longwave exchange between forest and snowpack will be:

$$R_b = e_f\sigma T_f{}^4 - e_s\sigma T_s{}^4 \dots\dots\dots\dots\dots\dots\dots\dots\dots\dots\dots\dots \text{XIII.11}$$

$$R_b = \sigma(e_f T_f{}^4 - e_s T_s{}^4) \dots\dots\dots\dots\dots\dots\dots\dots\dots\dots\dots \text{XIII.12}$$

$$R_b = \sigma\left[\geqslant 0.90\, T_f{}^4 - (\geqslant 0.82)\, 273^4\right] \text{ for snowmelt conditions.}$$

This is essentially the same as the equation used by the U.S. Army Corps of Engineers (1956) given below:

$$R_b = \sigma(T_a{}^4 - T_s{}^4) \dots\dots\dots\dots\dots\dots\dots\dots\dots\dots\dots\dots\dots \text{XIII.13}$$

where, T_a = ambient air temperature, and

T_s = snow temperature.

Foliar temperatures were assumed to be equal to air temperatures, and both snow and forest were considered to radiate as perfect black bodies $(e=1)$.

These equations show the dependence of longwave flux from forest to snowpack upon temperature of the biomass. Canopy temperatures vary a great deal. In forced convection situations they will be close to air temperature. In the absence of wind, canopy temperatures have been found to be 19° C higher than air temperatures on some days during the growing season and 2.5° C lower at night than air temperatures. Leaf temperatures vary also according to their position in the canopy.

Consideration has been given to instantaneous radiation only. However, it is clear that the upward and downward longwave fluxes undergo considerable diurnal change as canopy and snow surfaces are either warmed by the sun or cooled by outward re-radiation.

The fact that a snowpack in a forest ecosystem acts as an efficient sink for longwave radiation from the forest canopy considerably reduces the effect of downward shortwave flux depletion by the forest canopy.

The compensating interaction of the net shortwave and longwave streams is influenced by the degree of canopy coverage, as is shown in Figure XIII.3. Net shortwave gain decreases progressively as canopy cover increases. Net longwave gain from the forest correspondingly increases progressively as canopy cover increases. Net longwave loss from the snowpack to the atmosphere also decreases progressively (becomes less negative) with increasing canopy cover. The algebraic sum of these net streams shows a progressive decrease in the total radiation gain to a canopy closure of about 20 per cent due to the reduction of the incoming shortwave stream by canopy absorption of solar radiation without sufficient longwave reradiation to, or longwave absorption by, the snowpack to compensate for the reduction in solar income. Thereafter, at increasing canopy densities the net radiation gain increases progressively up to total canopy cover, though at this point the net all-wave gain is only a little more than half that of an open condition (canopy density = 0).

It is also noteworthy that net shortwave gain and net longwave gain from the forest are equal at a canopy density of about 45 per cent beyond which net longwave gain from the forest canopy is progressively greater than net solar income.

In comparison to the longwave budget of the forested site, the longwave budget of large open snowfields will show no gain from the forest and a large loss to the sky. This budget may be inferred from the calculated values for canopy density = 0 in Figure XIII.3. This shows a longwave budget of -0.08 ly/min. compared to +0.06 ly/min. for 100 per cent forest canopy. (Solar income compensates, being 0.20 ly/min. at zero cover, and 0.01 ly /min. under 100 per cent forest cover.)

Table XIII.8 Comparable Heat Budgets for Lodgepole Pine Stands and Open Snowfields in the California Sierra Nevada

	Snow in pine stands (ly./day)	Open snow-fields (ly./day)
Insolation absorbed by snow	160	350
Net loss of heat by longwave radiation from snow to sky and foliage	-40	-155
Convection of heat from air to snow	65	75
Heat loss by evaporation	-20	-50
Total heat surplus	165	220

(Miller, 1955, 1959)

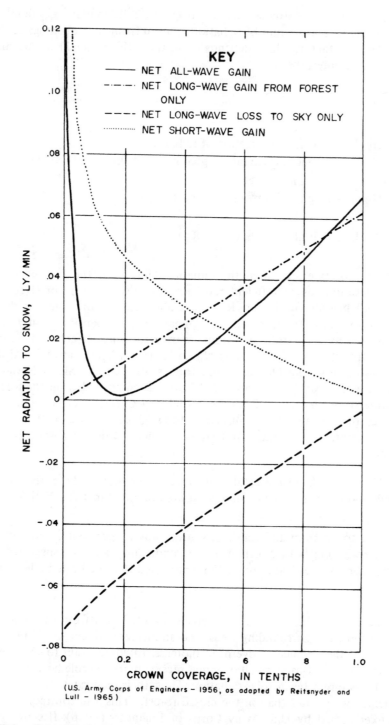

KEY
- —— NET ALL-WAVE GAIN
- —·—·— NET LONG-WAVE GAIN FROM FOREST ONLY
- — — — NET LONG-WAVE LOSS TO SKY ONLY
- ·········· NET SHORT-WAVE GAIN

(U.S. Army Corps of Engineers – 1956, as adapted by Reitsnyder and Lull – 1965)

Fig. XIII.3   Calculated Net Gains and Losses of Radiation to Snowpack in Relation to Crown Cover: Spruce-Fir Forest in Oregon

Small openings in the forest benefit from forest surroundings, gaining longwave radiation from them. The amount of radiation emitted from such openings to the sky is a function of the view factor, which depends upon the ratio of height of the surrounding trees to opening diameter, that is

$$F = \sin^2 \left(\arctan \frac{d}{2h} \right) \quad \dots\dots\dots\dots\dots\dots\dots\dots\dots\dots\dots \quad \text{XIII.14}$$

where F = view factor (fraction of radiation
leaving surface intercepted by open sky),

d = opening diameter, and

h = height of surrounding trees.

Thus, at d = 2h, F = 0.5; at d = 4h, F = 0.8

In addition to its effect upon the radiation climate of the snowpack, the forest through its functioning as an efficient radiation trap also has a major effect upon convective (sensible) heating of the air. Radiation absorbed by the forest is dissipated as reradiation, sensible heat transfer and latent heat transfer. Sensible heat transfer from forest biomass to the air results in regional warming, which affects not only the local ecosystem but the climate of the whole region. This has been shown by Miller (1955) who demonstrated how sensible heat from forests warmed the Sierra Nevada. He calculated that on a clear winter day sufficient heat was convected from the forest to raise the temperature of an air layer 300 feet in depth by 14°C. Such large scale heating effects extend to large open snowfields in primarily forested landscapes, so that melt rates in such situations contain an additional heat component which is absent in steppe and tundra climates.

Miller (1955) calculated the total heat budget over snowpacks in the Sierra Nevada for pine stands and open snowfields. His results are presented in Table XIII.8.

Melt in the forest from all local heat sources was 75 per cent of melt in the open. Miller (1959), reviewing both North American and European data, suggested this as an approximate general value. Anderson (1956) concluded that melt rate in dense forest was 64 per cent of that in the open.

Small openings in the forest also receive convected heat from the adjacent forest, longwave radiation from surrounding trees, and an increased direct solar irradiation component because of the absence of overhead biomass. In comparison to large, open snowfields in the forested landscape, small openings receive more sensible heat, more longwave radiation, and less direct beam solar radiation. Thus, their melt rates are less than in large open snowfields, but greater than in the closed forest. (This is in contradiction to earlier opinion, as exemplified by U.S. Army Corps of Engineers, (1956). If snow patches are found to persist longest in small forest 'glades', this is a function not of lower melt rates but of greater overwinter accumulation.

In a mosaic landscape, stand borders under conditions of direct beam irradiation receive increased solar radiation. A south-facing border receives 40 to 60 hrs. of sunshine per month in January, while a north-facing border receives none. Border trees on the north side around clearings radiate to the snowpack in the opening, while trees to the south shade the snowpack in the opening. Differential melt rates within openings result from such effects, in addition to variations in direct beam irradiation on different parts of openings.

These energy fluxes demonstrate that removal of forest cover results in more rapid snowmelt. In addition, where forest cover results in lowered evapotranspiration, and therefore in an increase in soil moisture content at the end of summer, runoff will be initiated sooner, and will be greater in volume from snow-zone forests following logging. The effects are less pronounced under partial cutting and thinning than for clear cutting.

XIII.6   RUNOFF

XIII.6.1   Yield

Consideration of the water-balance equation shows runoff as a residual closely dependent upon the magnitude of losses. Since it has been shown that land use can considerably affect evaporative losses, it follows that, as land use affects evaporation, so will it affect water yield. Thus such can already be inferred from previous consideration. Land use may affect runoff in its three components: total basin yield, stream regime and water quality. Insofar as total basin yield is concerned, removal of vegetation may be said to increase runoff to some degree, this being particularly true of deep soils. The persistence of water yield increase resulting from vegetation removal or other manipulation varies greatly from one ecosystem to another, and is in large measure dependent upon the speed and completeness of vegetation reinvasion. Table XIII.9, taken from Hibbert (1967), summarizes the results of experiments carried out in treated catchments designed to study the effect of vegetational manipulation upon total basin yield.

The technology of experimental watersheds has become somewhat standardized. In its simplest form the technique consists of taking a measurement of runoff in each of two watersheds over a number of years. After a considerable time, generally seven years or more, a regression is made of runoff from one watershed expressed in terms of runoff from the second. During this period when runoff data are being collected, both watersheds are maintained in undisturbed condition. If the regression is considered to be sufficiently accurate, land-use treatment is then carried out in one catchment (treatment watershed). Discharge from the treatment watershed is measured, along with flow from the untreated catchment (control watershed). Comparison of post-treatment flow from the treated watershed, with the regression line established during the calibration period, allows evaluation to be made statistically of the effect of the treatment upon stream discharge. Persistence of the treatment effect may be evaluated through continuing evaluation of actual and predicted flows on an annual basis.

Table XIII.9   Location, Description, and Results of Water-Yield Experiments (Hibbert 1967)

Catchment	Area (ha)	Mid-area elev. (m)	Slope x 100 (Elev. Diff./Length) (%)	Aspect	Vegetation and soils	Mean annual precip. (mm)	Mean annual stream flow (mm)	Description of treatment (percentage refers to portion of area treated unless otherwise stated)	1st	2nd	3rd	4th	5th
											mm		
Coweeta, N.C.													
13	16.1	810	26	NE	Mixed hardwoods. Basal area about 24 m²/ha. Granitic origin, deeply weathered sandy clay loam, up to 6 m deep, base rock tight	1829	792	1940, 100% clearcut, no removal, regrowth. 1962, experiment repeated.	370 / 371	283	279	247	203
17	13.5	885	44	NW		1895	775	1941, 100% clearcut, no removal, regrowth cut annually except years 3, 4 and 5.	408	361	256	167	245
22	34.4	1035	35	N		2068	1275	1955, 50% poisoned in alternate 10 m strips, no removal, regrowth restricted 4 years.	198	155	130	112	100
19	28.2	960	32	NW		2001	1222	1949, 22% basal area cut (understory only), regrowth.	71	64	55	47	39
1	16.1	840	34	S		1725	739	1954, 25% poisoned (cove hardwoods), regrowth restricted 3 years, 1956–7, 100% clearcut, partly burned, pine planted, regrowth restricted.	46	24	36		
3	9.2	825	32	SE		1814	607	1940, 100% area cleared for agriculture.	152	48	46	50	38
10	85.8	975	24	SE		1854	1072	1942, to 1956, 30% basal area cut by uncontrolled logging, regrowth.	127	95	59	113	80
41	28.7	1065	46	SE		2029	1285	1955, 35% basal area cut by selective logging, regrowth.	averaged 25 mm per year				
40	20.3	1035	42	SE		1946	1052	1955, 27% basal area cut by selective logging, regrowth.	averaged 55 mm per year				
6	8.8	790	35	NW		1821	831	1942, (July), 12% clearcut (streambank vegetation), regrowth.	nonsignificant				
37	43.7	1280	47	NE		2244	1583	1963, 100% clearcut, no removal, regrowth.	immediate small increases, nonsignificant on annual basis / 286				
28	144.2	1200	31	NE		2270	1532	1962–4, 51% clearcut, timber removal, 26% thinned, regrowth.	200 (approximate)				
Fernow, W. Va.													
1	29.9	755	23	NE	Mixed hardwoods. Basal area about 24 m²/ha. Sandstone and shale, stony silt loam, 1 to 1.5 m deep.	1524	584	1957–8, 85% basal area removed by commercial clearcut, regrowth.	130	86	89		
2	15.4	780	15	S		1500	660	1957–8, 36% basal area removed by diameter-limit cut, regrowth.	64	36			
5	36.4	780	14	NE		1473	762	1957–8, 22% basal area removed by extensive-selection cut, regrowth.	36				
3	34.4	805	13	S		1500	635	1957–8, 14% basal area removed by intensive-selection cut, regrowth.	8 (nonsignificant)				
7	24.2	800	13	NE		1469 (some snow)	788	1964, 50% (upper half) area cut, timber removed, regrowth not permitted.	92 (growing season only)				
H.J. Andrews, Oreg.													
1	95.9	700	28	NW	Coniferous. Volcanic tuffs and breccias, clay loams, shallow to deep.	2388	1372	1962–3, 40% commercial clearcut. 1963–4, 40% additional commercial clearcut. 1959, 8% area cleared for road construction.	small increase in low flow / small increase in low flow / small increase in low flow				
3	101.2	760	32	NW		2388	1346	1962–3, 25% clearcut and burned.	small increase in low flow				
San Dimas, Calif. Monroe Canyon	354.1	840	17	S	Chaparral with woodland riparian vegetation along streams. Granitic, rocky sandy loam, generally shallow.	648	64	1958, 1.7% cut (riparian vegetation only), sprouts controlled, grasses encouraged. 1959, additional 2.6% cut (canyon bottom vegetation), sprouts controlled, grasses encouraged.	May–December 6 mm, January–April 4 mm.				
Bell 2	40.5	885	32	S				1959, 40% poisoned (chaparral on moist sites), repeated application of herbicide.	May–December 5 mm				
Sierra Ancha, Ariz. North Fork, Workman Creek	100.4	2225	17	SW	Coniferous (ponderosa pine). Quartzite, clay loam up to 5 m deep.	813 (some snow)	86	1953, 1% cut (riparian vegetation only), sprouts controlled. 1958, 32% cleared (moist site), grass seeded. 1953–5, 30% basal area cut by selective logging.	June–September 17 mm / nonsignificant / 13	51	15	48	30
South Fork, Workman Creek	128.7	2165	8	NW		813	87	1956, 6% basal area cut by thinning. 1957, 9% basal area reduced by burning.	nonsignificant / nonsignificant				

Table XIII.9 (cont'd.)

Location	Area	Elev.	Slope	Aspect	Vegetation and soil	Precipitation	Runoff	Treatment	Yield
Fraser, Colo. Fool Creek	289.0	3200	18	N	Coniferous (lodgepole pine, spruce-fir). Granitic, sandy loam 2.5 m deep.	762 (75% snow)	283	1954–7, 40% commercially clearcut in strips, regrowth.	86 53 79 97 53
Wagon Wheel Gap, Colo. B	81.1	3110	37	NE	84% forested (aspen and conifers). Augite, quartzite, rocky clay loam.	536 (50% snow)	157	1919, 100% clearcut, some removal, slash burned, regrowth.	34 47 25 22 13
Meeker, Colo. White River	197,400				Conifers (spruce).	265		1941–6, insects killed up to 80% of timber on 30% of area.	58 (average for 5 years)
Kamabuti, Japan II	2.5	200	40	E	Conifer 60%, broadleaf 40%, Tuff, shale.	2616 (40% snow)	2075	1948, 100% cut, annual recut of sprouts.	110 (average for 3 years)
Kenya, East Africa Kericho Sambret	688.0	2200	5	NW	High montane and bamboo, Phenolite lava, deep friable clay.	1905	416	1959–60, 34% cleared for tea plantation, clear-weeded.	103
Kimaki A	35.2	2440		S		2014	568	1956, 100% cleared, pine planted, cultivation of vegetables for 3 years.	457 229 178 Reduced water yield for given year
Jonkershoek, South Africa Bosboukloof	208.0	520	30	SW	Sclerophyll scrub (chaparral type).		475	1940, 53% afforested with pine.	104 (4-yr. mean) at 16–20 yrs.
Biesievlei	32.0	365	30	SW			490	1948, 98% afforested with pine.	142 (4-yr. mean) at 8–12 yrs.
Coshocton, Ohio 172	17.6	350	14	SW	30% hardwoods in 1938. Sedimentary, silt loams.	970 (little snow)	300	1938–9, 70% reforested, mostly pine.	135 (after 19 years)
Western Tennessee Pine Tree Branch	35.7	160	5	E	23% mixed hardwoods in 1941. Sandy silt loams.	1230	255	1946, 75% reforested, mostly pine.	76 to 152 (after 16 years)
Eastern Tennessee White Hollow	694	410	6	SE	65% mixed hardwoods and pine in 1934. Limestone, cherty silt loams.	1184	460	1934–42, 34% reforested, mostly pine.	no detectable change
Central New York Sage Brook	181	525	15	SE	Mixed hardwoods and conifers.	974	535	1932, 47% reforested, conifers.	106 (after 26 years)
Cold Spring Brook	391	565	35	S	Shales and sandstones overlain by glacial till, silt loams up to 3 m deep.	1030	616	1934, 35% reforested, conifers.	172 (after 24 years)
Shackham Brook	808	520	5	S		1030	627	1931–9, 58% reforested, conifers.	130 (after 24 years)
Adirondacks, New York Sacandaga River	127,200	575	1		Northern hardwoods with conifers. Glacial till, sandy loam 1 m deep.	1143 (some snow)	770	1912 to 1950, basal area increased from 17 to 28 m²/ha.	196 (after 38 years)
Southwestern Washington Naselle River	14,245	275			Douglas-fir, western hemlock. Silty-clay loam and stony loam, 2 m deep.	3300	2690	1916 to 1954, 64% area logged at rate of 2% per year, regrowth	no detectable change

The sequence thus expressed shows experimental watershed techniques in their simplest form. Many adaptations are possible, going far beyond statistical analysis of total annual flows from treatment and control watersheds. Reinhart (1967) has summarized present knowledge dealing with watershed calibration techniques. He shows that considerable refinement can be used, using more complex regressions, and that evaluation of flow regime effects are possible.

Where only one watershed exists, that is where no watershed sufficiently similar to the proposed treatment watershed is available to serve as a control basin, it is possible to calibrate flow before treatment through regression methods using climatic variables. This procedure is more complex and contains disadvantages. Double mass analysis may sometimes be used, though the subjectivity of this simple technique presents disadvantages. However, this may be avoided somewhat through use of a covariance analysis.

A means of estimating the effects of land use on water yield from large basins, with particular reference to agricultural land use, has been presented by Sharp, Gibbs and Owen (1966). Recent developments of mathematical models simulating runoff, using climatic and other variables, offer considerable hope that new methods may be derived to allow reliable estimates of land-use effects upon hydrology in large drainage basins, both as they exist presently and as they might be modified by development in the future.

XIII.6.1.1 A Case Study: Fool Creek Watershed

The Fool Creek watershed experiment (Goodell, 1958) dealt with the effects of logging in a small watershed (714 acres) in the Rocky Mountains of Colorado, 65 miles northwest of Denver. The watershed ranges in elevation from 9,000 to almost 13,000 ft. It has a northerly aspect and is characterized by steep slopes. The geology of the watershed is metamorphic, consisting of schist and gneiss derived from granite, subjected in the past to extensive glaciation. Soils are of course, gravelly texture and low in fertility, excepting the deep alluvial soils adjacent to stream courses. Vegetation on the watershed is a dense mature stand of lodgepole pine, Engelmann spruce and subalpine fir, with timber volumes averaging 18,000 board feet/acre.

Precipitation varies with altitude. The average annual precipitation is 30-40 in., decreasing rapidly with elevation; and 60-80 per cent of the annual precipitation occurs as snow. Snow accumulation begins in October, and melt begins in April, continuing until June.

The uppermost 25 per cent of the watershed is not timbered, consisting of an alpine zone lying above timber line.

Calibration of Fool Creek watershed began in 1943 and ended in 1954. The contiguous, comparable East St. Louis Creek was used as a control. Logging of the area began in 1954 and was completed in 1956. The cutting pattern consisted of alternate clear cut strips of different widths, 1, 2, 3 and 6 chains, running normal to the contours; 40 per cent of the area was cleared in these strips.

The difference between annual predicted yield and the actual annual yield, measured in inches, is given in Table XIII.10.

Table XIII. 10   Results of Fool Creek Treatment Experiment

Years	Predicted Yield (in.)	Actual Yield (in.)	Actual Yield Minus Predicted Yield (in.)
1956	11.4	15.6	4.2
1957	19.6	23.0	3.4
1958	11.4	13.5	2.1
1959	10.5	13.6	3.1
1960	11.1	14.9	3.8
1961	8.8	10.9	2.1
1962	17.3	19.1	1.9
1963	5.4	6.9	1.5

Grateful thanks are extended to the Rocky Mountain Forest and Range Experiment Station, U.S. Forest Service, for making unpublished data available.

XIII.6.1.2   A Case Study: Coshocton Experimental Watersheds

The Coshocton experimental watersheds were set up to study the effects of agricultural practices upon water yield (Harrold et al., 1962). They are located about 80 miles northeast of Columbus, Ohio. The elevation of the watersheds is between 900 and 1,300 ft. MSL. The watersheds have a southerly aspect with slopes ranging from 14-24 per cent. Geology consists of coal, limestone, sandstone, clay and shale. The area is not glaciated. Perched water tables are common, giving rise to springs and seeps. The soils that predominate are well-drained, residual silt loams of shale or sandstone origin. On gently sloping lands, the soils, again silt loam, are more permeable and of shale origin. The area is typical of much agricultural land in the Allegheny Plateau.

Precipitation in the area is approximately 38 in. of which approximately 95 per cent occurs as rain. July is the wettest month.

Five watersheds, having areas ranging from 29-76 acres, were studied; four were treated and the fifth served as a control. The study began in 1938. Details of treatments practised are given in Table XIII.11.

The effects of the treatments after ten years of treatment on water yield, are shown in Table XIII.12. This table shows that the following treatments were carried out: reforestation in watershed 172; elimination of cross-slope tillage in 177, and its replacement by contour-strip tillage along with an increase in permanent pasture; replacement of across-slope tillage by contour-strip tillage in watershed 169; and

13.42   HYDROLOGY OF LAND USE

reduction in the amount of permanent pasture in watershed 183 (Table XIII.11). All resulted in a decrease in water yield ranging from 15 per cent to 44 per cent, the greatest decrease in water yield being experienced in the watershed converted to forest cover.

Table XIII.11 Coshocton Experimental Watersheds. Description of 1938 and 1957 Land Use in Percentage of Total Area for Five Study Watersheds

Watershed Item	172		177		169		183		196	
	1938	1957	1938	1957	1938	1957	1938	1957	1938	1957
Woodland	29.4	100.0	9.1	9.8	6.2	11.7	13.1	13.8	25.2	27.7
Rotation across slopes	0	0	47.4	0	69.3	0	42.4	50.8	45.7	38.1
contour strips	0	0	0	34.5	0	69.0	0	8.6	0	0.5
Permanent pasture	50.5	0	31.1	49.5	17.9	13.4	44.2	26.8	19.3	30.5
Idle land	20.1	0	1.8	0	0	0	0	0	2.6	0
Miscellaneous	0	0	10.6	6.2	6.6	5.9	0.5	0	7.2	3.2
Area (acres)	43.6		75.6		29.0		74.2		30.3	

Table XIII.12   Effect of Treatments on Water Yield: Coshocton Experimental Watersheds

Watersheds	172	177	169	183	196
Item					I
Average runoff (in.)					N
(1938)					D
Year	12.02	7.89	6.81	10.46	E
Growing season	3.44	2.10	2.05	7.57	X
Dormant season	8.57	5.79	4.76	2.89	
Decrease in runoff (in.)					W
(1957)					A
Year	5.32	1.62	1.08	1.60	T
Growing season	1.71	0.36	0.72	1.00	E
Dormant season	3.80	1.26	0.72	0.60	R
					S
Percent reduction					H
(1938–1957)					E
Year	44.0%	20.6%	15.9%	15.3%	D

XIII.6.1.3   A Case Study: Guelph Agricultural Watersheds

The Guelph watersheds incorporated a study of winter runoff as affected by agricultural practice. The watersheds, located at Guelph, Ontario are under the direction of the University of Guelph. The elevation of the watersheds falls between 1,100-1,400 ft. MSL. Drumlin topography predominates, and the major soil type is loam, well drained, of the Grey-Brown Podzolic Great Soil Group.

Mean annual precipitation is approximately 33 in. The mean monthly precipitation during winter is slightly lower than that which occurs in the summer months. November-March precipitation is almost equally divided between rainfall and snow.

Each of the watersheds is approximately 20 acres in size, with slopes of about 5 per cent. The study was instituted in 1952, and is described by Ayers (1964). Watersheds with sod cover were compared to basins with ploughed or corn-stubble conditions.

The results showed that runoff was decreased from the ploughed fields, and that more runoff was obtained from the sod areas. These results are at variance with the commonly-held opinion that the runoff volume is greater from agricultural watersheds having sparse vegetation. Ayers (1964) attributed the decreased runoff from ploughed fields to the rough condition of the surface, which resulted in black soil being exposed to shortwave radiation, so that melting of the relatively-shallow snow cover took place, along with a large potential for depression storage. In sod areas, the insulating layer of snow and vegetation prevented thawing of the soil surface. Concrete frost, made persistent by this insulation, prevented infiltration and resulted in greater overland flow. Over one additional inch of water per winter was obtained from sod watersheds than from ploughed watersheds, and almost one additional half-inch per winter from sod watersheds in comparison to corn-stubble watersheds.

This study is particularly valuable in that it points out one of the major difficulties in pursuit of land-use hydrology research. Results from previous work comparing agricultural watershed covers had been carried out in areas with a more temperate climate than at Guelph. The more severe winter climatic conditions at Guelph, and the soil-frost phenomenon mentioned, were sufficient to reverse completely the trend of runoff from the two cover conditions. This demonstrates one of the biggest problems in land-use hydrology, namely the dangers inherent in extrapolating research results from one geographical or climatic region to another. Because of this, land-use hydrology still remains, to a very large extent, primarily regional science.

XIII.6.1.4   First Year Water-Yield Increases from Cover Removal

The results summarized in Table XIII.8 show the first-year increase in water yield resulting from logging.

Hibbert (1967) summarized these results. He considered that the first year water-yield increase depended upon the proportion of the cutover area. Detectable water-yield increases were obtained whenever 20–100 per cent of the timber was removed on a forested watershed. Results from complete clear cutting showed that most first year water-yield increases were 300 mm. (approximately 12 in.) or less. Some results showed 300–400 mm. water-yield increase in the first year. In exceptional cases, 450 mm. were obtained. This neglects riparian cuttings which, over the area treated, can yield as much as 600 mm. Large water-yield increases were generally obtained in areas of high precipitation. However, some work carried out in Japan and Oregon shows, rather parodoxically, that low water-yield increases may be obtained in high-precipitation areas. These particular results still await explanation.

XIII.6.1.5   Yield Increase Decline

Generally speaking, increases in water yield are greatest in the first year following cutting. Thereafter, the general pattern is for the yield to decline. This is generally attributed to revegetation of the cutover area. The Coweeta, North Carolina, studies showed that all water-yield increase would have disappeared 35 years after clear cutting (Kovner, 1956; Hibbert, 1967). This is an area of relatively rapid regrowth.

At Fool Creek, Colorado, however, after 12 years, no decline in water-yield increase seems to be apparent. Regrowth in that area is slow. Initial calculations based on some years of post-treatment data estimated that the yield increase would disappear after 50 years. More recent calculations tend to show that the period during which some yield increase occurs may be even longer (Hoover, 1968).

XIII.6.1.6   Hydrologic Effects of Afforestation

Hibbert (1967) summarized the results of treatment experiments in which afforestation of previously non-forested watersheds was carried out. This is the converse of the more usual situation, in which previously-forested watersheds are denuded of their cover. Following afforestation and the establishment of a closed-forest canopy, flow decreases are general. On the rather limited range of conditions presently sampled (Table XIII.8), 220 mm. water-yield decrease seems to be usual. The anomaly between magnitude of yield increases following logging, and yield decreases following afforestation, probably results from insufficient sampling of the data.

XIII.6.2   Regime

Land use may considerably influence stream regime. Land-use effects on infiltration have already been considered. Reduced infiltration rates commonly accompanying landscape disturbance may generate increased overland flow, and be highly influential upon the pattern of stream discharge. Discharge peaks are considered to depend upon the following variables: precipitation, climate, geology, topography-physiography, and land use. Of these factors, only land use can be manipulated. It therefore represents the only — and certainly the only practical — means of controlling peak flow.

Peak flows and floods from large rivers are generally created by rainfall events, snowmelt conditions, or a combination of both, of such magnitude as to obliterate largely the effects of land use. Stream regime, therefore, in its extreme manifestations on large watersheds is predominantly climatically dependent, rather than dependent upon land use. The following treatment primarily refers to stream regime in small watersheds.

Very high peaks have been measured in small watersheds. In the U.S.A., some have been greater than 2,000 csm, and the largest documented is 12,300 csm. Flows of this magnitude are often accompanied by debris-mud flows, and by landslides. In Europe, particularly in the Alps, the effects of land use have created, or aggravated, torrent formation; thus, in the interest of protecting downstream populations, very large investments have been needed to modify and control stream regime whose extreme expression has resulted from inappropriate land use.

Stream regime may be considered to fall into two interconnected topics: Peak (normal peak) Flows, and Floods.

XIII.6.2.1   Peak Flows

Normal peak flows are best subdivided into those resulting from snowmelt and those without snowmelt.

Considering firstly the case of peak flows occurring in the absence of snowmelt, small cuttings in West Virginia showed no effect upon peak flows, while in Mississippi the peak flows from abandoned fields were greater than peak flows coming from watersheds having a cover of depleted hardwoods, which in turn were greater than peak flows originating from basins carrying 20 year-old pine. In the Little Hurricane watershed at the Coweeta Hydrologic Laboratory, North Carolina, logging and application of mountain-farming practices had an appreciable effect on peak flows, in that the magnitude of the peak flow and flood frequency were both increased. Elapsed time between beginning of a storm event and occurrence of peak flow was cut by half, and the amount of storm flow generated during the peak flow period was increased by approximately 20 per cent.

In the Pine Tree Branch watershed (T.V.A., 1962) in Tennessee (part of the Tennessee Valley Authority scheme) the watershed was planted to pine, and certain engineering control works carried out on severely eroded areas. Fifteen years after planting, dramatic reductions in peak flows were found. Summer peak flows were reduced by 80 per cent. Average maximum annual discharge in summer was 103 cfs before treatment, and 19 cfs after treatment. A reduction in winter peak flows of 70 per cent also occurred. Average maximum annual winter discharge was 111 cfs before treatment and 31 cfs following treatment. Peak discharge frequency was also reduced by about 95 per cent in summer, and by 70–90 per cent in winter. The recession curve was flattened after treatment. Time of concentration was increased by 50 per cent in the first five years after treatment, and by almost 400 per cent during the 15 years following treatment. The forested watershed (172) in the Coshocton study (Tables XIII.11, XIII.12) also showed a significant reduction in peak flows.

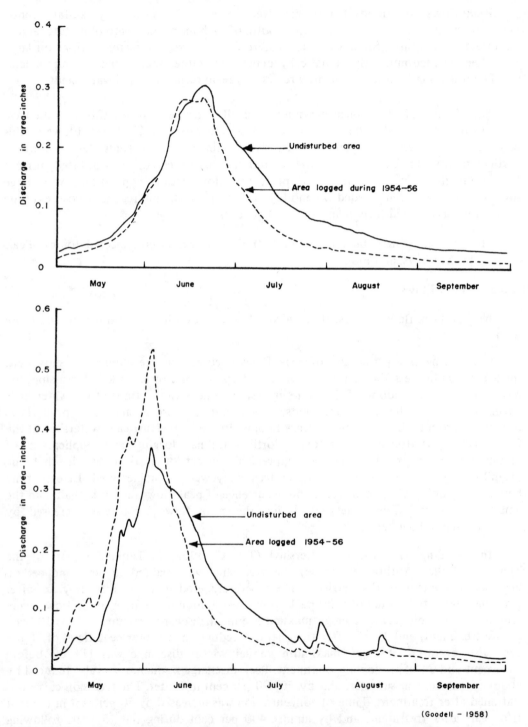

(Goodell – 1958)

Fig. XIII.4   Hydrographs on Fool Creek and Control Watershed before Cutting
(Upper), and in the First Year after Cutting (Lower)

Peak flows resulting from snowmelt are also affected by land use, as may be inferred from consideration of snowmelt, outlined previously. The Fool Creek watershed, previously summarized, showed increased peak flow in the years after treatment. Figure XIII.4 shows peak flows during the year immediately preceding treatment and the first year following treatment. One study dealing with the effect of logging on peak flows of particular interest is that conducted by Anderson and Hobba (1959). The area under investigation was the coastal forest region of Oregon and Washington states. Two rather large watersheds were compared. The Molalla watershed was 320 sq. mi. in area; in 1928–1943, 35 sq. mi. of the Molalla watershed were denuded by logging. The Clackamas watershed nearby, having an area of 665 sq. mi. remained under constant land-use conditions. Double mass analysis techniques were used to compare floods in an annual series extending over 21 years. The results showed peak flows to become progressively higher in the Molalla watershed as the series extended. Compared to the first seven years of the series, floods were 31 per cent higher in the second seven years, while peak flows in the last seven years were 56 per cent higher than in the first seven years of record. Other analyses on a number of other watersheds were also carried out.

The authors concluded of this study: (a) forests and forest fires increased peak flows for both rain-snowmelt, and snowmelt, floods; (b) where forest stocking recovered with time, peak discharge volumes decreased; (c) in two geological types, marine sediments and volcanic rocks, no variation attributable to geology was discernible; and (d) effects were similar for large storms and small storms.

This study is particularly interesting in relation to the size of watersheds. Most effects of land use on discharge pertain to relatively small watersheds. Data for larger basins are rare, and the occurrence of pronounced effects, attributable to land use, in watersheds of the Molalla size tend to indicate that some caution is desirable in scrutinizing the conventional statement that land-use effects on peak discharge are confined to small watersheds.

In passing, it is noted that the effect of land use on low flows is unpredictable (Johnson,1967), so that little generalization is possible. In the case of logging, the number of instances in which logging has increased low flows exceeds the documented cases in which low flows have decreased. Reforestation has resulted in permanent streams becoming ephemeral in some instances.

XIII.6.2.2   Floods

Floods are merely an extreme manifestation of normal peak flows. It is helpful to categorize floods according to their genesis, as an aid to understanding. Zinke (1965) has provided such a classification, considering floods to fall into four types:

1.   Floods with all storage elements full, but without snow storage.
2.   Floods with all storage elements full, and with snowmelt.
3.   Floods with storage elements not full.
4.   Floods with no precipitation, but sudden release of storage elements.

Type 1 floods emerge from a completely-saturated landscape, subjected to high precipitation for considerable periods. Land use has relatively little effect on such floods.

Type 2 floods result from snowmelt, or from warm rain falling upon a pre-existent snowpack. Land use may have important effects on snowmelt, since forest cover affects the amount of radiation and advected heat reaching snowpack. Clearcutting over large areas may result in accelerated runoff. Soil-freezing differences may also affect the rate of runoff.

Type 3 floods result from high-intensity rainfall whose rate of delivery exceeds the soil-infiltration rate. Land use has the largest influence upon this type, by adversely affecting infiltration capacity through soil compaction and litter removal.

Type 4 floods may result from extremely-rapid snowmelt occurring without coincident precipitation. Presence or absence of forest cover may be influential in this flood type, in the same way as Type 2 floods are affected. This flood type may also result from a sudden release of water impounded by temporary barriers. Where land use creates landslides which temporarily block stream channels and subsequently fail, it is influential upon Type 4 floods also, but this is relatively uncommon.

Most of the extremely damaging floods attributable to land use are Type 3 floods. In one study in California (Hoyt and Troxell, 1934), for instance, a watershed of 6½ sq.mi. was burned over while an adjacent watershed of 10½ sq. mi. was unburned. Peak flows prior to burning in Fish Creek (the basin that was burned over) ranged from 3–78 cfs. After burning, the annual series was as follows: 335, 292, 145, 15, 11, 11 cfs. The effect of forest fire in this case extended for only about three years. Flows in nearby Santa Anita Creek, in the years following the fire on Fish Creek, showed no variation from pre-fire years. In the case of forest fires, Krammes and DeBano (1965) have shown that a water-repellency condition in the upper soil mantle may result from fire, with drastic effects upon stream regime. Fire effects were extremely well demonstrated (Krammes and Rice, 1963) on the San Dimas Experimental Forest in southern California. Following wildfire, peak-storm flow from one watershed was measured as ranging from 200-800 times greater than peak flows produced by the watershed prior to burning.

Another excellent example of floods having a land-use genesis is exemplified in the data obtained in the Davis County experimental watershed (Bailey and Copeland, 1961) in Utah, in the Wasatch Mountains, about 10 miles north of Salt Lake City. The annual precipitation on the basin is about 43 in. Summer (convectional) storms are intense, the maximum rate being measured as approximately 8½ in. per hour, over a 5-min. time period.

The area was settled in the mid-1800's, and a pattern of heavy grazing and watershed denudation was established by pioneer activity. In 1923, and again in 1930, though no damaging floods had ever previously issued from the Wasatch canyons, devastating floods emerged, carrying very large mud-rock flows. As a consequence, six lives were lost, and property damage of $1,000,000 accrued. Subsequent study showed no evidence of sudden change in precipitation or in diastrophism. It was postulated that floods had their genesis in the upper reaches of the six flooding canyons. Such flood source areas totalled less than 10 per cent of the total watershed area. Their deterioration was attributed to long-continued overgrazing of livestock, to forest fire, and to depletion of plant cover.

To control floods, and to test the hypothesis of land-use genesis, control measures, consisting of control of grazing and forest fire, and more importantly of contour trenching and reseeding of postulated flood-source areas, were carried out.

In the period following flood-control works, though extremely high-intensity summer precipitation has been experienced, no floods have developed in any of the treated watersheds, thereby confirming the postulate of a land-use genesis for the catastrophic floods and debris flows from the Davis County canyons.

XIII.6.3   Water Quality

A most important effect of human activity and land use upon water resources is the influence upon quality of runoff waters. Attention here is given solely to changes in water quality resulting from management or other disturbance of the landscape. The effect of urbanization upon water quality is not considered. The whole question of pollution by urbanization and industrialization is a reflection of man's activity, but it lies outside the scope of this treatment. The background papers of the conference "Pollution and Our Environment" offer a survey of the general pollution problem in Canada (Canadian Council of Resource Ministers, 1967).

In considering the effects of landscape disturbance through human activity upon the quality of water resources, two separate though interconnected topics are Sediment and Chemical Water Quality.

XIII.6.3.1   Sediment

Sediment transported in runoff reflects both normal (geologic) and accelerated erosion. Sediment from land use falls in the latter category. It is stressed that much sediment is wholly natural, and does not result from deleterious human activity. This is particularly true in western Canada, where many streams are glacier fed. Furthermore, in the lower reaches of streams with extensive flood plains, the continual deposition and degradation of such flood plains through the natural geologic process contributes largely to sediment load. In the relatively undisturbed landscape of northern Canada, such major drainages as the Athabasca, Slave, Mackenzie and Liard Rivers can be cited as examples of streams carrying naturally-high suspended sediment loads.

Agricultural land use is one of the major sources of sediment pollution. Bullard (1966) cited the case of the wheat belt in northwestern United States as probably having the highest sustained sediment production within the United States. Some fields in this area have lost as much as 300 tons of soil/acre in one year. However 10 tons/acre/year is probably an average soil loss over an area of 2,000,000 acres. Of this, probably 3 tons/acre/year go into major streams to form sediment pollution.

Correlations of suspended sediment production with cover condition have been made, for instance in Mississippi. The sediment contribution of various cover types has been rated as follows: corn > pasture > abandoned fields; depleted hardwoods > pine plantations. The range of sediment yield in this sequence was from 43 tons/acre/year in cornfields, to negligible production in pine plantations (Ursic and Dendy, 1965).

It is clear that the regime changes previously discussed must be associated in many instances with changes in sediment production from the disturbed landscape. The debris and mud-flow production from the Davis County canyons has been mentioned. The Fish Creek study conducted by Hoyt and Troxell (1934) showed that great volumes of sediment accompanied high-peak flows generated by forest burning. In addition to agriculture and forest fire, the cases already cited, forestry operations may create increased suspended-sediment loads; road construction may have damaging effects on water quality; mining activity may also create both increased suspended sediment and severe damage to chemical water quality (U.S. Dept. Interior, 1967).

In the following discussions, forest harvesting and associated road development are used as examples of what may occur in land use, insofar as sediment production is concerned.

Undisturbed forested watersheds have been found to have relatively low levels of suspended sediment. At the Coweeta Hydrologic Laboratory, nonstorm flows were found to carry < 2 ppm turbidity, storm flows < 11 ppm turbidity in most cases, and very heavy storm flows < 80 ppm turbidity, in the latter case being mostly organic matter. At the Fernow Experimental Forest, West Virginia, natural turbidities from undisturbed watersheds were found to be < 5 ppm. Similarly, at the Hubbard Brook Experimental Forest, New Hampshire, heavy storms carried < 11 ppm turbidity. In coastal Oregon, natural turbidities are somewhat higher; 22 ppm were recorded at 5 csm., 52 ppm at 58 csm, and 94 ppm at 72 csm. The maximum turbidity allowable in drinking water is generally placed at a level of 11 ppm. The values cited are from Packer (1967).

In Canada, as mentioned earlier, the presence of glaciers and developing flood plains may lead to high natural suspended-sediment concentrations. This must be borne in mind in considering the data above. Where sediment concentrations in major streams are naturally of a high order, one may consider the additional sediment produced by land-use activity to be unimportant. However, where tributary streams are of high quality, one may alternatively consider the sustaining of such naturally high quality in these streams to become the most important property. Where high sediment levels exist in major streams, tributaries with clear water represent the more valuable resource.

Some studies have been carried out in which the forest cover was felled, but not extracted. They show that the effect of timber cutting alone did not result in any appreciable increase in sediment, despite fairly large increases in water yield. The studies in Wagon Wheel Gap (Bates and Henry, 1928) and Coweeta Watershed No. 17 (Hoover, 1944) exemplify this case.

Where timber is not only cut, but extracted as a merchantable product, the situation changes rapidly. Data presented earlier show part of the reason. Logging machinery disturbs and compacts soil. Large reductions in infiltration occur, surface litter is lost, and roads and skid trails become a major sediment source. The quality of such roads and skid trails, that is, care and planning and the standards of construction invested in them, is a major determinant of the amount of sediment produced by forest harvesting. One study carried out in the Fernow Experimental Watersheds is highly illustrative (Reinhart, Eschner and Trimble, 1963). Four harvesting experiments and a control watershed were involved. Maximum suspended sediment in ppm turbidity was as follows:

Control watershed ....   15 ppm
Intensive selection — forester planned ....   25 ppm
Extensive selection — forester planned ....   210 ppm
Diameter limit — logger's choice ....   5,200 ppm
Clearcut — logger's choice ....56,000 ppm

These results show the potential increase in sediment which may result from poor logging, and secondly, demonstrate how management quality may strongly affect sediment production. Packer (1967) and Dyrness (1967) summarize other studies of timber harvesting and associated-activity effects on water quality.

The effects of road construction on sediment production have been studied. On the H.J. Andrews Experimental Forest, western Oregon, studies carried out in an undisturbed watershed showed that maximum turbidity never exceeded 200 ppm in six years prior to road construction. Thereafter, 1.7 miles of road were constructed in a 250 acre watershed. In the first storm following construction, 1,780 ppm turbidity were measured, while in a comparable control basin the same storm produced only 22 ppm. However, three years later in a very large storm, only 439 ppm turbidity were measured in the treated basin, equal to approximately 4.7 times the turbidity measured in the control watershed. This reflects the speed of revegetation under the climatic and geologic conditions of the H.J. Andrews Experimental Forest.

The Zena Creek watersheds, Idaho, are characterized by soils derived from granite, which are highly erodible and lack internal cohesion. In these watersheds, following road construction, sediment yields as high as 12,400 tons/sq.mi. were measured in the season immediately following construction of logging roads. Watersheds having no roads yielded no sediment in the same season. Three years after logging, the rate of sediment production had undergone no measurable decrease. This again reflects revegetation conditions on the experimental area. It is stressed that sediment damage can be greatly reduced by good planning and proper forest management.

The total effect of forest operations on water quality, insofar as suspended sediment is concerned, is shown by a study carried out in Castle Creek watershed, northern California. This watershed of 2,500 acres was 25 per cent cutover; and 6 per cent of the watershed was in skid trails, while it also contained 3 miles of truck-haul road. Suspended sediment before logging was in the order of 900 tons/year for the whole watershed (normal geological sediment). In the year immediately following logging, 4,600 tons of suspended sediment were produced, while in the second year 1,800 tons of sediment were measured. It was also found that only 12 per cent of flow carried 60 per cent of total sediment.

In western Oregon, a regression analysis was carried out of sediment records in 29 streams, to establish the relationship between logging activity and sediment yield. It was predicted from this study that, if the area logged increased from 0.6 per cent/year to 1.5 per cent/year, sediment would increase by 18 per cent. The analysis also showed that, if total road area went from 0.1 per cent of watershed surface to 0.5 per cent, the total sediment would increase by 260 per cent. It was concluded that future forestry development had the potentiality of tripling present sediment production.

XIII.6.3.2   Chemical Water Quality

Bullard (1966) reviewed land-management effects upon chemical water quality. He pointed out that in 1962 more than 125,000,000 pounds of halogenated hydrocarbons (principally DDT, aldrin, toxophene and BHC) 20,000,000 pounds of organophosphate, and 14,000,000 pounds of arsenites, were used in the United States in management of agricultural and forest lands. Obviously, with such pesticide loading, the possibilities for chemical water-quality deterioration are very high, as is confirmed by recurrent fish kills, notably in the lower Mississippi River in 1963—64. The problem of long-lasting pesticide residues is well known, as is the catastrophic effect of progressive pesticide concentration at higher levels of food chains, in both aquatic and terrestrial ecosystems. DDT is a particularly noteworthy example.

Fertilizers used in agriculture, and to a lesser extend in forestry, may lead to a nutrient enrichment of streams with accompanying water-quality deterioration. About 30,000,000 tons of fertilizers are used each year in the United States. Irrigation return flows are a good example of water-quality deterioration associated with fertilizer application.

Mining is a large source of water-quality damage. Coal mining in the eastern United States has severely damaged many thousands of miles of streams, through acid-mine drainage. Oil field salt brines in the southern plain states of the United States are another example. In addition to mine-waste water, there is also a problem of leachates from mine tailing dumps.

The cases above are used strictly as examples; the list of instances wherein land use has caused damage to water quality could be extended virtually indefinitely.

One interesting example of a rather obscure forest harvesting effect on water quality is that shown by Bormann et al. (1968). This study showed that, in a small watershed in New Hampshire, clearcutting resulted in greatly increased dissolved-nitrate concentrations in stream water, to the extent that the small stream in the treatment watershed underwent eutrophication and algal blooms occurred.

XIII.7   **LITERATURE CITED**

Anderson, H.W. 1956. Forest-cover effects on snowpack accumulation and melt, Central Sierra Snow Laboratory. Trans. Amer. Geophys. Union 37:307-312.

Anderson, H.W. 1963. Managing California's snow zone lands for water. U.S. Dept. Agric., For. Serv., Pac. Southwest For. and Range Expt. Stat., Res. Pap. PSW-6.

Anderson, H.W. and Hobba, R.L. 1959. Forests and floods in the northwestern United States. Symposium, Hannoversch-Munden, Internatl. Assoc. Sci. Hydrol., Pub. 48:30-39.

Ayers, H.D. 1964. Effects of agricultural land management on winter runoff in the Guelph, Ontario region. Proc. 4th Hydrol. Sympos., Can. Nat. Res. Counc., Assoc. Comm. on Geodesy and Geophys., Subcomm. on Hydrol.: 167-182.

Bailey, R.W. and Copeland, O.L. 1961. Vegetation and engineering structures in flood and erosion control. Intermount. For. and Range Expt. Stat., IUFRO meeting, Vienna, Austria, Sept. 1961.

Bates, C.G. and Henry, A.J. 1928. Forest and streamflow experiment at Wagon Wheel Gap, Colorado. Mon. Weath. Rev. Suppl., Wash., 30, W.B. 946.

Bergen, J.D. 1963. Vapor transport as estimated from heat flow in a Rocky Mountain snowpack. Internat. Assoc. Sci. Hydrol., Publ. 66:62-74.

Bormann, F.H. *et al.* 1968. Nutrient loss accelerated by clearcutting of a forest ecosystem. Science 159:882-884.

Bullard, W.E. 1966. Effect of land use on water resources. Jour. Water Pollut. Control, Fed., April 1966: 645-659.

Canadian Council of Resource Ministers, 1967. Background papers prepared for the national conference on "Pollution and our Environment" held in Montreal from Oct. 31 to Nov. 4, 1966. Queen's Printer, Ottawa.

Colman, E.A. 1953. Vegetation and watershed management. Renalf, New York.

Deij, L.J.L. 1954. The lysimeter station at Castricum (Holland). Int. Assoc. Sci. Hydrol., Rome 3: 203-204.

Diamond, M. 1953. Evaporation or melt of a snow cover. U.S. Army, Corps of Engin., Snow, Ice and Permafrost Estab., Res. Pap. 6.

Douglass, J.E. 1967. Effect of species and arrangement of forests on evapotranspiration. Proc. Internat. Sympos. For. Hydrol., Pergamon Press: 451-461.

Dyrness, C.T. 1967. Erodibility and erosion potential of forest watersheds. Proc. Internat. Sympos. on For. Hydrol.: Pergamon Press: 599-611.

Garstka, W.U. 1964. Snow and snow survey. Section 10 *in* Handbook of Applied Hydrology, McGraw-Hill, New York.

Goodell, B.C. 1958. A preliminary report on the first year's effects of timber harvesting on water yield from a Colorado watershed. U.S. Dept. Agric., For. Serv., Rocky Mtn. For. and Range Expt. Stat., Stat. Pap. No. 36.

Goodell, B.C. 1959. Management of forest stands in western United States to influence the flow of snow-fed streams. Int. Assoc. Sci. Hydrol., Sympos. Hannoversch-Munden, IASH Public. No. 48: 49-58.

Goodell, B.C. 1963. A reappraisal of precipitation interception by plants and attendant water loss. Journ. Soil Water Cons. 18:231-234.

Goodell, B.C. 1967. Watershed treatment effects on evapotranspiration. Proc. Internat. Sympos. on For. Hydrol.: Pergamon Press: 477-482.

Gray, D.M. 1968. Snow hydrology of the prairie environment. *In* Snow Hydrology, Proc. Workshop Seminar, CNC/IHD.

Harrold, L.L. *et al.* **1962.** Influence of land use and treatment on the hydrology of small watersheds at Coshocton, Ohio. U.S.D.A. Tech. Bull. No. 1256. 194 pp.

Heikurainen, L. 1967. Effect of cutting on the ground water level on drained peatlands. Proc. Internat. Sympos. for For. Hydrol.: 345-354.

Hibbert, A.R. 1967. Forest treatment effects on water yield. Proc. Internat. Sympos. on For. Hydrol.: Pergamon Press: 527-544.

Holstener-Jorgensen, H. 1967. Influences of forest management and drainage on ground water fluctuations. Proc. Internat. Sympos. on For. Hydrol.: 325-333.

Hoover, M.D. 1944. Effects of removal of forest vegetation upon water yield. Trans. Amer. Geophys. Union, 25 (6) : 969-977.

Hoover, M.D. 1962. Water action and water movement in the forest. *In* Forest Influences, Food and Agr. Org., United Nations, Forest and Forest Prod. Stud. 15: 33-80.

Hoover, M.D. 1968. Personal communication.

Hoover, M.D. and Leaf, C.F. 1967. Process and significance of interception in Colorado subalpine forest. Proc. Internat. Sympos. For. Hydrol.: Pergamon Press: 213-222.

Hoyt, W.G. and Troxell, H.O. 1934. Forests and streamflow. Trans. Amer. Soc. Civ. Engrs. 99:1-30.

Hutchison, B.A. 1966. A comparison of evaporation from snow and soil surfaces. Internat. Assoc. Sci. Hydrol. Bull. XI (1): 34-42.

Jeffrey, W.W. 1964. Vegetation, water and climate: needs and problems in wildland hydrology and watershed research. Proc. Second Water Studies Institute Sympos., Saskatoon, Sask., Nov. 1964: 121-150.

Jeffrey, W.W. 1968a. The hydrologic significance of stand density variations in Alberta lodgepole pine forests. Ph.D. thesis, Colorado State Univ. 412 pp.

Jeffrey, W.W. 1968b. Snow hydrology in the forest environment. *In* Snow Hydrology, Proc. Workshop Seminar, CNC/IHD.

Jeffrey, W.W. and Stanton, C.R. 1968. Snowpack measurements in lodgepole pine stands at low elevations in the Kananaskis River Valley, Alberta. Can. Dept. Forestry, Unpublished report.

Johnson, E.A. 1967. Effects of multiple use on peak and low flows. Proc. Internat. Sympos. on For. Hydrol.: 545-550. Pergamon Press.

Kittredge, J. 1948. Forest influences. McGraw-Hill, New York.

Kovner, J.L. 1956. Evapotranspiration and water yields following forest cutting and natural regrowth. Soc. Am. Foresters Proc.: 106-110.

Krammes, J.S. and DeBano, L.F. 1965. Soil non-wettability: a neglected factor in watershed management. Water Resources Res. 1:283-286.

Krammes, J.S. and Rice, R.M. 1963. Effect of fire on the San Dimas Experimental Forest. Arizona Watershed Sympos., Seventh Annual Proc: 31-35.

Lull, H.W. 1959. Soil compaction on forest and range lands. U.S.D.A., For. Serv., Misc. Publ. No. 768.

Meiman, J.R. 1968. Snow accumulation related to elevation, aspect and forest canopy. In Snow Hydrology, Proc. Workshop Seminar, CNC/IHD.

Miller, D.H. 1955. Snow cover and climate in the Sierra Nevada. Pub. in Geog. vol. 11, 218 pp., Berkeley, Calif.: Univ. Calif. Press.

Miller, D.H. 1959. Transmission of insolation through pine forest canopy as it affects the melting of snow. Schweiz. Anst. f. Forstl. Versuchsw. Mitt. 35:37-79.

Miller, D.H. 1964. Interception processes during snowstorms. U.S. Dept. Agric., For. Serv., Res. Pap. PSW-18.

Miller, D.H. 1965. The heat and water budget of the earth's surface. Adv. *in* Geophysics 11:175-302.

Miller, D.H. 1966. Transport of intercepted snow from trees during snow storms. U.S. Dept. Agric., For. Serv., Res. Pap. PSW-33. Pacific Southwest Forest and Range Expt. Stat., Berkeley, Calif.

Miller, D.H. 1967. Sources of energy for thermodynamically-caused transport of intercepted snow from tree crowns. Proc. Internat. Sympos. For. Hydrol.: 201-212.

Niederhof, C.H. and Wilm, H.G. 1953. Effect of cutting mature lodgepole pine stands on rainfall interception. Jour. For. 41:57-61.

Packer, P.E. 1967. Forest treatment effects on water quality. Proc. Internat. Sympos. on For. Hydrol.: 687-699. Pergamon Press.

Penman, H.L. 1963. Vegetation and hydrology. Tech. Comm. No. 53. Comm. Bur. of Soils, Harpenden, Engl. pp. 30-76.

Penman, H.L. 1967. Evaporation from forests: a comparison of theory and observation. Proc. Internat. Sympos. For. Hydrol. Pergamon Press: 373-380.

Pierce, R.S. 1967. Evidence of overland flow on forest watersheds. Proc. Internat. Sympos. For. Hydrol.: Pergamon Press: 247-252.

Pruitt, W.O. Jr. 1958. Qali, a taiga snow formation of ecological importance. Ecol. 35: 169-172.

Reifsnyder, W.E. and Lull, H.W. 1965. Radiant energy in relation to forests. U.S.D.A. Technical Bulletin. No. 1344.

Reinhart, K.G. 1967. Watershed calibration methods. Proc. Internat. Sympos. For. Hydrol.: 715-723. Pergamon Press.

Reinhart, K.G., Eschner, A.R. and Trimble, G.R. Jr. 1963. Effect on streamflow of four forest practices in the mountains of West Virginia. U.S.D.A. For. Serv., Res. Pap. NE-1.

Sharp, A.L., Gibbs, A.E. and Owen, W.J. 1966. Development of a procedure of estimating the effects of land and watershed treatment on streamflow. U.S.D.A., Tech. Bull. No. 1352.

Stanton, C.R. 1966. Preliminary investigation of snow accumulation and melting in forested and cut-over areas of the Crowsnest Forest. Proc. 34th Ann. Meet., West Snow Conf.: 7-12.

Tennessee Valley Authority. 1962. Reforestation and erosion control. Influences upon the hydrology of the Pine Tree Branch watershed, 1941-1960. Knoxville, Tennessee. 98 pp.

Ursic, S.J. and Dendy, F.E. 1965. Sediment yields from small watersheds under various land uses and forest covers. Proc. Fed. Inter-Agency Sediment Conf. 1963. U.S.D.A., Misc. Publ. 970:47-52.

U.S. Army. 1956. Snow hydrology. U.S. Army Corps of Engineers. North Pacific Div. 437 pp., Portland, Ore.

U.S. Dept. Interior. 1967. Surface mining and our environment. U.S. Govt. Printing Office, Washington, D.C.

Vezina, P.E. and Pech, G. 1964. Solar radiation beneath conifer canopies in relation to crown closure. For. Sci. 10:443-451.

West, A.J. 1959. Snow evaporation and condensation. 27th Ann. West. Snow Conf. Proc. 1959:66-74.

West, A.J. 1962. Snow evaporation from a forested watershed in the central Sierra Nevada. Jour. For. 60:481-484.

Willington, R.P. 1968. Some effects of clearcutting, slashburning and skid roads on the physical-hydrologic properties of coarse glacial soils in coastal British Columbia. Univ. of B.C., Faculty of Forestry, M.Sc. thesis.

Zinke, P.J. 1965. Influence of land use on floods in relation to runoff and peak flows. Proc. State Board of For. Meeting, Feb. 26, 1965, State of Calif., Resources Agency, Dept. of Conservation, Div. of For.: 47-52.

Zinke, P.J. 1967. Forest interception studies in the United States. Proc. Internat. Sympos. on For. Hydrol.: 137-160. Pergamon Press.

INDEX

INDEX

INDEX

INDEX